SECOND EDITION

DESCRIPTIVE AND INFERENTIAL STATISTICS

An Introduction

HERMAN J. LOETHER
California State University, Dominguez Hills

DONALD G. McTAVISH
University of Minnesota

ALLYN AND BACON, INC.
Boston, London, Sydney, Toronto

Library of Congress Cataloging in Publication Data

Loether, Herman J
  Descriptive and inferential statistics.

  Bibliography: p.
  Includes indexes.
  1. Sociology—Statistical methods. 2. Sampling (Statistics) 3. Statistical hypothesis
testing. I. McTavish, Donald G., joint author. II. Title.
HN25.L62 1980   301'.028   80-147
ISBN 0-205-06905-3

Printed in the United States of America

10 9 8 7 6 5   85

To our wives, Louise Loether and Janet McTavish, and our children,
Chris Loether, Kathleen, Karen, and Steve McTavish,
with love and appreciation.

Contents

Preface

Sociology is coming of age. Armchair philosophizing in this field has given way to theory building and theory-testing research. Increasing use is being made of a range of scientific methods and techniques. Anyone with a serious interest in doing sociology or in understanding what sociologists do must become familiar with these tools of the sociologist's trade. An important ingredient of the sociologist's tool box is that set of techniques collectively known as statistics. Our purpose in writing this book is to acquaint the reader with the statistical techniques that sociologists commonly use and, more especially, to illustrate *how* sociologists make use of them.

Rather than present applied statistics as a field that stands by itself, divorced from any particular subject matter, we have focused our attention upon sociology and have portrayed statistics as a set of tools that sociologists use in the process of making decisions about their data. We have tapped the literature of sociology, government sources, and our own research for studies and data to serve as points of departure for the discussion of particular statistical techniques.

We believe that the common practice of teaching statistics in isolation from any particular discipline has the effect of making it seem unnecessarily abstract, complex, and purely mathematical. Too often students are turned off by the subject because they cannot see its relevance for their major field of interest or they come to see statistics in a ritualistic fashion, as if use of these techniques by themselves would certify the truth of research conclusions. Sociology students are generally interested in inquiry in sociology, but not in statistics *per se*. We have found in our years of teaching statistics that the subject can be made relevant and even exciting by presenting it as one among a number of sets of tools that sociologists use. When the subject matter of statistics is integrated into inquiry in the subject matter of sociology, students realize its value for sociology and for them.

For the most part, statistical theory and the techniques of statistics have been developed by mathematical statisticians who were not unduly concerned with applications. Consequently, the derivations of formulae tend to be based upon sets of ideal conditions such as scores free of measurement problems and error or "a normally distributed random variable." The sociologist whose primary concern is with the real world usually must work with data that do not satisfy these ideal conditions. When statistics is taught in isolation from subject matter there is a real danger students will get the impression that the ideal conditions are usually realized when research is done. By focusing on the uses of statistics in sociology it is possible to show how statistical techniques are actually applied, how these applications relate the ideal forms of many techniques to the actual data available, and what the consequences of this adjustment may be. In this volume we have made a point of calling to the reader's attention the meanings and quality of data to be analyzed. Hopefully, this will lead the reader to become a more informed user of statistics and to be more cautious about the unqualified acceptance of findings of others.

Whenever the occasion has arisen we have raised methodological and research design issues of concern to sociologists. We do not believe that students of elementary statistics need to be shielded from the problems of doing sociology. Rather, we believe that a preliminary consideration of methodological issues at an early stage gives the student a better appreciation of the interrelationship of theory, methods, and statistics. These issues are not considered in great detail here, but their introduction at this point should highlight the broader range of analytic problems facing the sociologist and raise the student's level of interest in pursuing them further.

Although statistics is based upon mathematical considerations we do not feel that it is necessary at this level to present rigorous, mathematical definitions of concepts or formal mathematical proofs of theorems. The underlying logic of the field of statistics is relatively simple and can be understood, to a large extent, from a verbal presentation. It is true, of course, that mathematical sophistication is both desirable and necessary for more extended treatments. Students who are curious about mathematical proofs and derivations of formulae should consult more mathematically oriented texts (some of these are included in the references at the end of chapters).

This volume begins by explaining the use of statistics to describe and contrast the data at hand. Often these data are sampled from a broader population and the sociologist is interested in generalizing conclusions based on descriptions of the sample at hand to conclusions about descriptive features of the population from which the sample was drawn. In the later sections of the book we focus our attention on those statistical techniques which help make such inferences possible.

Although the chapters included in this second edition carry the same titles as those in the first edition, substantial changes have been made in the contents of several of them. While minor changes have been made in all of the chapters in order to improve readability, to remove ambiguities, and to tighten up technical language, substantial changes have been made in chapters 2, 5, 7, 8, 10, and 12.

Generally, these changes have been made in order to explain more clearly the consequences of using statistical techniques to analyze data which fall short of the level of quality to be desired or to introduce new techniques or new materials which did not appear in the first edition. More specifically, in chapter 2 new material has been added to enhance the explanations of properties of ordinal and interval variables and a new section has been added dealing with the process involved in the construction of variables. The discussions of the measures of skewness and kurtosis in chapter 5 have been revised and the section on the normal curve has been expanded.

Throughout chapter 7 new material has been added to explain how the characteristics of the data being analyzed influence both the magnitude of and the interpretation of the various measures of association used. There is, for example, a new section comparing Gamma, Somers' d_{yx}, and Tau-b and a new section dealing with the assumptions of and possible interpretations of Product-moment r.

The major change in chapter 8 is the addition of a section which explains Goodman's Odds Ratios and Hierarchical Analysis, a very useful analytical technique which has become increasingly important in sociological research in the past few years. A new introduction has been added to chapter 10 to make more smooth the transition from the discussion of descriptive statistics to the consideration of inferential techniques. In addition, significant revisions have been made in the explanation of the notion of *a priori* probability.

In an effort to make more understandable the concept of a sampling distribution a new set of data has been introduced in chapter 12 to illustrate the generation of a binomial sampling distribution and, also, to illustrate the use of the normal approximation to the binomial probability distribution. Another change in chapter 12 appears in the form of a revised discussion of the concept of an unbiased estimator.

In response to an often repeated request from users of the first edition we have added computational problems to the discussion questions which appear at the end of each chapter. We have retained all of the teaching aids which appeared in the first edition and, where necessary, have revised or expanded them. These aids are the following:

1. The "diagrams" which follow immediately after the dicsussion of each technique and which summarize the essential steps in its use. These diagrams may serve as reviews of the techniques and may also be useful as "recipes" for their use.
2. The "loop-back boxes" inserted at strategic points throughout the text to flag important points in preceding discussions and to refer the reader to other sections of the text that may help clarify the subjects covered.
3. The lists of key concepts which are included at the end of each chapter. Any concept with which the reader is unfamiliar may be looked up and the explanation of it may be reviewed until the concept is understood.
4. The discussion questions at the end of each chapter which require the reader to integrate materials from the chapter or to apply what has been learned. These, as was mentioned above, have now been supplemented by computational problems.
5. An annotated list of references at the end of each chapter for those readers who wish to explore some of the topics covered in more depth.

The conscientious student who takes advantage of these learning aids will undoubtedly benefit from the effort he (she) devotes and gain a more thorough understanding of the subject matter of the book.

All cross-reference citations within the text are made with reference to the various numbered sections within each chapter. To spot a section number location in cross-reference, use the first digit of the numbered section to identify the chapter; two-digit numbers indicate major section divisions; three-digit numbers indicate minor section divisions; numbering sequence runs from major to minor divisions (if any) and back to the next major division (thus, in Chapter 3: Sections 3.1; 3.1.1; 3.1.2; 3.1.3; 3.2; 3.3; 3.3.1; etc.). In addition, major or minor numbered sections may have subsections designated by an added lower case letter (thus: Sections 3.1a; 3.1b; 3.1.1a; 3.1.1b; etc.), usually indicating a paragraph heading for a particular topic.

A student manual designed to accompany this book is available through the publisher. Among other things, the student manual includes a wealth of sociological data which can be analyzed using the statistical techniques covered here. We have found that the best way for the student to master statistical techniques is to use them to analyze data. Readers should try their hands at applying what they have learned to some of these data sets or other comparable sets of data at their disposal.

For material in the Appendix we are indebted to the literary executor of the late Sir Ronald A. Fisher, F.R.S., and to Oliver and Boyd, Edinburgh, for permission to reprint Tables III, IV, and V from their book, *Statistical Tables for Biological, Agricultural, and Medical Research.*

We are indebted to Richard J. Hill of the University of Oregon and Edmund D. Meyers, Jr., of Dartmouth College, who read the entire original manuscript, and to Marian Deininger of St. Cloud State College, who read the descriptive sections. Kirby McMaster and Charles Y. Glock gave us detailed critiques of the first edition of the book. Their suggestions for improvements in the manuscript had a major influence on the changes made in the second edition. Other colleagues, too numerous to name, also contributed valuable suggestions and pointed out errors in the first edition. The suggestions of all of these have added significantly to the quality of the finished product.

We wish to thank Beverly Farrell and James C. Smith for supplying the answers to the problems at the end of many of the chapters. We are sure that their diligence will be especially appreciated by the students who use the book.

Finally we are indebted to our students at California State University, Los Angeles, California State University Dominguez Hills, and the University of Minnesota who contributed so much over the years to the development of this book.

This book has been a joint product of the authors. We have both been so deeply involved in the writing that we are no longer able to sort out individual contributions of one or the other. The order in which our names appear on the title page should not be taken to imply that one of us is the senior author. The names could easily have been reversed. They are merely listed in alphabetical order.

Some Basic Considerations

1 $\quad$ *An Introduction to*
$\quad\quad$ *Statistics in Sociology*

The Civil Rights Act of 1964 included a section requiring the U.S. Commissioner of Education to conduct a study, ". . . concerning the lack of availability of equal educational opportunities for individuals by reason of race, color, religion or national origin in public educational institutions. . . ." Early in the following fall a team of social scientists headed by James Coleman, a sociologist at Johns Hopkins University, administered questionnaires to nearly 650,000 grade school and high school pupils, teachers, and administrators across the United States. People questioned in the study were carefully selected, using a "two-stage, stratified, probability sampling plan" (Coleman, 1966: I, 22).

Among the wealth of findings in the voluminous report is the conclusion that the achievement of minority pupils relates more to the schools they attend than does the achievement of majority-group pupils. For example, 20 percent of the variation in achievement of blacks in the South is associated with differences in the schools they attend, whereas only 10 percent of the variation in achievement of whites in the South is associated with differences in the schools they attend. If true, this finding has implications for improving educational opportunity of all the nation's youth. Majority pupils presumably would not be greatly harmed by attending poorer minority-area schools, and minority students would be aided by transfer to white schools. This suggests busing is a possible solution to educational inequality.

This report is one of many available to any citizen, and it has been quoted several times in the mass media and in other specialized reports to legislators. The report includes sections on such topics as sample design, sampling variability, correlation, percentage distributions and many others. Conclusions of such studies often find their way into texts and technical reports, as well as to the mass media, and increasingly form the basis for policy decisions.

But is the conclusion true, and does it characterize the nation as a whole? Can we afford to reject out of hand such extensive research results or, for that matter, can we afford to accept them without question? Furthermore, if we question these findings, what information do we want or need to make a reasoned judgment of the study's worth? Are the revealed differences sufficiently large so that we should say they are meaningful and should compel action? What sorts of bias and error, if any, might creep in, and were these possibilities taken into account in the study? This study is an excellent example of full and careful research reporting, but how should a concerned citizen or legislator use this information? Is any of this helpful to the school superintendent planning his (her) program for the next few years?

Clearly we would want the legislator to have sufficient understanding of statistics to make a responsible judgment.* Furthermore, we would want sufficient insight into these matters to assure ourselves both that good information is properly used by those in a position to influence important decisions, and also that unworthy information is recognized and set aside.

One of the most impressive developments in contemporary society has been the vast growth of research into all aspects of our physical environment, our society, and ourselves. Increasingly, statistically analyzed research results and statistical models are used as a basis for gaining insight and making decisions. To understand contemporary society, to understand the basis of many decisions, to follow intelligently and evaluate the output of others' study, or to handle important lines of thought and reasoning, the knowledge of elementary statistics is essential. The demand for research is growing, and research is being conducted by neighborhoods, political parties, and public and private agencies, as well as university investigators. The chances are rather high that you will become involved in the conduct and evaluation of some kind of research yourself. Certainly we cannot read critically and interpret most of the literature in sociology, much less make original contributions to knowledge in this area, without a relatively detailed knowledge of how to interpret statistical arguments and descriptions and apply them to our own ideas and research. For leader or legislator or citizen, for poet or social scientist, the need for statistical insight seems to be a fact of contemporary life and a powerful tool for understanding and decision-making.

1.1 SOME BASIC CONCEPTS IN STATISTICS

The Coleman study involved questioning a rather large number of students in grades 1, 3, 6, 9 and 12 across the nation at the end of September, 1965. Each student in the study was asked questions about such things as ethnic status, age, sex, background, and batteries of items to measure academic ability. Each of these questions was designed to measure a property or characteristic of the student. Properties such as age, ethnic status, or academic ability are called **variables,** ways individuals could differ from each other which are of interest to

*The Coleman study used a statistical technique called *multiple regression*. This technique is discussed in Chapter 9 in this book.

the investigators. In other words, a **variable** is a measurable characteristic of a unit of analysis (such as a student) that can differ in value from one unit to another. Sometimes it is helpful to refer to the scores, or measured values, on variables (such as those Coleman collected) as **statistical observations** or simply **observations.** Typically, statistical observations are expressed as numbers (*e.g.* age in years, academic ability scores) although numbers are not essential to the use of statistics.

It is important to know what one's statistical observations are, of course. The number 11 hardly has any meaning until one learns that it is the age in years of a student who was in school in the United States in September, 1965. Notice that the statistical observation was described in terms of three things: (a) what the variable was that was measured (*e.g.* age in years), (b) what the variable was measured on (*e.g.* a person who was a student in one of the five target grades in the study), and (c) where and when this student was located. The second point deserves a further comment. Part of Coleman's study did not deal with individual students, but dealt with individual teachers, and other parts of the study dealt with entire schools. Any of these parts could involve measurements in terms of age, of course, but what one concludes from a study would be vastly different if "age-of-student" were measured rather than "age-of-school." The kind of "object" being measured is called the **unit of analysis,** and in the part of Coleman's study cited earlier, the unit of analysis was the individual student, not the individual school. Sociology often deals with different units of analysis, and this makes clarity on this point very important.

The scores on all of the variables included in Coleman's study,—for all of the students in grades 1, 3, 6, 9, and 12 in the United States in September, 1965,—describe a statistical population, the population that Coleman was ultimately interested in describing. A **statistical population,** sometimes called a **universe,** is composed of scores on all the variables the investigator wants to measure on all of the "objects" at the time and place indicated; in short, a statistical population is *all relevant statistical observations.* In this case, the statistical population was **finite,** since, in principle, the statistical observations in such a statistical population could be counted. Often this is not the case, particularly in scientific pursuits which hope to develop principles applying to certain kinds of phenomena wherever and whenever they may occur. An **infinite** statistical population, sometimes called a **conceptual** or **general universe,** would be one in

---

BOX 1.1 For Those Who Like Overviews

There are certain basic *themes* and ideas which unite the various specific techniques in different chapters of this book. These themes are reviewed in Chapter 17, and it may be helpful to you to read through that chapter now. Chapter 17 should alert you to possible ways of organizing your thinking about statistics.*

---

*"Loop-back" Boxes such as this will be used here and there in the text to flag important ideas you should be sure to get at that point, or to suggest points where you should pull yourself back and reflect on your own grasp of the argument being made. We will refer and cross-refer you to points where necessary ideas can be found and reviewed.

which relevant statistical observations are not limited to any particular time and/ or location. All of the cities of the past, present, and future; all of the students any time or place they may exist; or all of the families which ever have or ever will exist would be examples of infinite statistical populations or conceptual universes.

Whether the universe being studied is infinite or finite, investigators rarely enumerate an entire statistical population (*i.e.* they seldom examine *all* statistical observations), but rather, they select a subset of the units of analysis called a **statistical sample.** Coleman's study was based on a sample from a defined statistical population. Although any subset of a statistical population would constitute a sample, in practice some kinds of samples may not represent very well the population from which they were drawn. Usually the investigator's interest is in selecting an appropriate sample which will give him (her) accurate information about the population from which it was drawn. He (She) wants to make inferences about the population, but he (she) wants to do this in a way that is sound, practical and economical. Samples are, of course, always finite, although they may also be large as in Coleman's case. The field of inferential statistics is concerned with the ways in which reliable samples may be drawn and how inferences can be based on them.*

1.2 THE FIELD OF STATISTICS

Like most words, the word *statistics* is used in several ways in everyday speech. Some use it to refer to virtually any number which describes something, such as the height of Mont Blanc, the number of babies born in Vermont, or the turning radius of a Ford. *Statistics* is also used in two more technical ways: first, to refer to a certain subject matter or field of intellectual pursuit, and, secondly, in a specific and narrow sense (which will be defined later) related to the description of statistical samples.

The field of statistics, broadly viewed, encompasses a range of techniques and human activities that border on mathematics on one side, and on the other, on the problems that investigators and people in everyday life want solved. Basically, statistics involves logical thinking about statistical samples or populations, either for the purpose of arriving at a compact and meaningful description, or for the purpose of assessing some of the risks involved in extending conclusions based on samples to conclusions about populations. Logical thinking about such collections of scores involves knowing which comparisons to make as well as how to make the comparisons in an informative way. The field of statistics includes a growing "kit" of techniques particularly useful for making these descriptions and inferences.

The subject-matter of statistics may be divided up into two parts. One part, called **descriptive statistics,** consists of tools and issues involved in describing collections of statistical observations, whether they are samples or total populations. The other part, called **inferential** or **inductive statistics,** deals with the logic and procedures for evaluating risks of generalizing from descriptions of samples to descriptions of populations.

*See Chapter 11 for a thorough survey of sampling procedures in sociology.

Roughly the first half of this book will focus primarily on descriptive statistics. An analysis of how statisticians make inferences through drawing reliable samples is pursued in the latter half of the book. One last terminological distinction may be worth making at this point. A second technical meaning of the word *statistics* refers to any summary measure of a characteristic or property of a *sample* of statistical observations. If, however, the collection of statistical observations which is described is a statistical *population,* then a summary measure of a characteristic or property of these observations would be called a **parameter.** For instance, the average achievement level of all twelfth graders in the United States in 1965 could be treated as a population *parameter,* and the average achievement level of twelfth graders obtained from Coleman's sample could be treated as a *statistic.* A statistic and a parameter may be the same kind of summary measure (*i.e.* a percentage or average or ratio). It is *what* is described — population or sample — that makes the difference between these terms. The focal problem of inferential statistics is the relationship between a statistic and a parameter. The problem of descriptive statistics is the development of ways to analyze samples *or* populations, that is, to develop different kinds of statistics and parameters which could be used on any given collection of scores both to describe and to analyze them.

The field of statistics has grown enormously in the past fifty years or so. Doctorates are offered in this specialty, and there are professional associations and professional journals which report new techniques and raise and resolve new issues. Student memberships are offered for these, and it would be worthwhile for you as a student to look at some of the ideas and publications in this field. Its history makes fascinating reading.*

*The development of the field of statistics is one of the more interesting segments of our intellectual history. One theme in statistics (a word coined in a 1770 British publication) has been the interest of politicians in characteristics of their citizenry, usually for taxation purposes. Among early "political arithmetic" was the work of Englishman John Graunt (1620–1674) who was one of the first to describe gross population data in his book *Natural and Political Observations Made Upon the Bills of Mortality,* which described 1604–1661 London death records. This was shortly followed by an analysis by English astronomer Edmund Halley to "ascertain the prices of annuities upon lives," the first life insurance tables. Quetelet (1796–1874), Belgian supervisor of statistics and "father of the quantitative method in sociology," and Galton (1822–1911) extended work in statistical description and analysis, the latter developing percentiles, the median, correlation, and regression, — all ideas we will discuss later in this book. Karl Pearson (1857–1936), English biologist, extended statistics in the areas of regression, correlation, and chi-square. A second theme is that of probability. In China as early as 300 B.C. there is reported discussion of the probability of an unborn child being a boy or girl. Early pondering on outcomes of games of chance (*e.g.* dice rolling, card dealing) led to the development of formal rules of probability and their application to the fledgling field of astronomy by such men as Galileo (1564–1642), Pascal (1623–1662), and Fermat (1601–1665). The French astronomer Demoivre (1667–1754) was first to describe the *normal curve* as a description of errors in his observations. James Bernoulli (1654–1705) is credited with what was later called the "Law of Large Numbers," a principle central to statistical inference which states what the expected sample value of proportions will be, and how the probability of a given difference between sample and true proportions decreases as the size of a sample increases. Gauss (1777–1855) extended thinking on probability and worked with the concepts of standard deviation and standard error and, under a pseudonym, "Student," William S. Gossett extended further our knowledge of sampling distributions. In the early 20th century, Ronald A. Fisher developed the analysis of variance and other statistical tests, and contributed in important ways to knowledge about the design of experiments. You will encounter the names of other pioneers in statistics as you read through this book — names such as Cochran, Yule, Yates, Spearman, Markoff, Poisson, and others. An interesting presentation of the work of some of these pioneers appears in James R. Newman's *The World of Mathematics,* Simon and Schuster, New York, 1956. The development of statistics is far from finished, and one of the exciting current topics in sociology is the tailoring and creation of new and more useful statistical techniques.

> **BOX 1.2** Terms to Consider Further
>
> Some of the terms defined here will be used in later chapters. To familiar-
> ize yourself with these, it might be helpful at this point to flip through a
> sociology book or journal to see if you know the meanings of these terms
> and can find examples of each of them. Especially important ideas are the
> following in Sections 1.1 and 1.2:
>
> > *Statistical observation*
> > *Variable*
> > *Unit of analysis*
> > *Statistical population*
> > *Statistical sample*

1.3 THE RELATIONSHIP BETWEEN STATISTICS AND SOCIOLOGY

Sociologists are interested in studying individual social actors, groups, or organizational processes, interaction patterns, norms, and social behavior of various
types. Typically, a sociologist concerned with, say, the age structure of societies
and its consequences for kinship structures might begin by simply gathering
information on age structure from each of a series of societies with existing records. In order to highlight the age structure rather than simply the size of the
society, he (she) will use statistical procedures for creating, say, averages or proportions of those over 65, or perhaps "dependency ratios" (*i.e.* the ratio of the number of young and old people to the number of middle aged). Since the data are
quite likely based on a careful sample of existing societies, he (she) may well
want to know how much confidence he (she) should place in these figures, and
again he (she) will turn to statistics, especially inferential statistics, to help
reason out the risks involved in using a sample. He (She) may also examine the
record-creating procedures in different societies for evidence of bias, and perhaps
compute an index number of some sort to show the way two or more variables are
related – a correlation coefficient, for example.

Simple statistical description frequently leads to more detailed theoretical
work and, perhaps, the creation of some type of model or theory about the way
age structure would be expected to influence kinship. Usually this process will
lead the sociologist to take account of other variables, such as the level of technology in a society, or sex ratios, and generally he (she) will then have to set out to
measure these variables on a sample of societies. If he (she) is fortunate, information may already exist in some archive, but usually he (she) will have to organize
records for his (her) own purposes or gather new data. These processes pose issues
of *validity* (*i.e.*, questions of whether you are measuring what you want to measure) and of *reliability* (*i.e.*, questions of whether the measurement process itself
is stable), and these again may be examined with statistical tools. In fact, he (she)
will probably express his (her) hypotheses or theories in terms of statistical descriptions and will compare his (her) revised theories with new data in terms of
statistical tools.

A brief look at some of the main journals in the field, the *American Socio-logical Review*, the *American Journal of Sociology*, or *Social Forces*, for example, should convince you of the centrality of statistical expression and analysis in our field. An increasingly interesting area of sociology, in fact, is in the development of more useful statistical tools for handling some interesting new questions about social behavior. One of the purposes of this book is to help you move to the forefront of sociology where interesting phenomena are described in new and different ways.

1.4 WHAT YOU CAN EXPECT IN STUDYING STATISTICS

You will do much better, we feel, if you have some knowledge of what to expect in general from your encounter with statistics. Most students seem to find the area somewhat different from others they know, requiring something of a new approach.

First and most basic, the field of statistics is oriented toward drawing logical conclusions. This implies that there is some type of *question* or *problem* which commands an interest in finding an answer. Coleman's study addressed several different questions. What is the extent of ethnic segregation? What differences are there between schools and in achievement? What is the relationship between students' achievements and the kinds of schools they attend? Procedures for supporting the claim that an answer has indeed been found are basically those of logic. Statisticians develop various measures, examine their logical implications, and then critique and develop statistical models which may be of use in finding answers in certain restricted situations. While this characterizes many of the interests scholars and citizens have about the world, it by no means exhausts the things they are interested in doing to discover or experience what their world is like. It is, nevertheless, a surprisingly useful approach.

Secondly, the field of statistics is quite similar to a language. There are several key concepts and technical terms, some of which we have already introduced, and there is a kind of grammar involved in using them. The language is relatively carefully worked out and consistent, although sometimes the same idea may be symbolized by somewhat different symbols in different contexts. Once you learn how to express yourself in the language, you will find that you have a very powerful way to make important distinctions and express logical relationships. In fact, it is often true that difficult problems disappear once the problem is expressed in a clear and consistent fashion. The early part of the book will emphasize the important "language of statistics."

Thirdly, while it is true that many of the summary measures and lines of reasoning are expressed in numerical form, there is a basic underlying logic of statistical reasoning which can be discussed and handled apart from much of mathematics. Mathematical-logical procedures, however, are such powerful and logically clear tools for handling statistical reasoning that they are nearly always of interest to the professional statistician. This means, for example, that it is often helpful to look for ways to express ideas and relationships in numeric form, although this is not always true nor formally necessary. It also means that one could neatly express the logical relationships between different ideas by mathematical notation and derivation. In this beginning treatment of statistics we will highlight the underlying logic without resorting to more than elementary algebra, but it is clear that we are imposing a limitation on expression which could not be conveniently imposed on general treatments of statistical reasoning. Other books use more mathematical expression to explain the reasoning involved, and you may want to consult one of these references from time to time.

Finally, the field of statistics focuses upon situations where there are repeated or repeatable operations—such as in taking a measurement of the organizational involvement of each of 137 people who are members of the local YMCA, or as in tossing an imaginary coin an infinite number of times and recording whether the outcome each time is "heads" or "tails," or as in drawing samples of organizations in order to measure the relationship between the height of their stratification system and productivity. In principle, these repeated or repeatable operations (*i.e.* samplings, tossings, measurings, etc.) must be such that there could possibly be a range of different outcomes. For example, in principle, people might be of different degrees of organizational involvement, coins might turn up "heads" or "tails," and an organization might have a "tall" or "flat" stratification hierarchy, or have high or low productivity.

If one's approach does not involve *logical* progress toward a solution of an interesting problem, if there is no interest in *supportable* findings, if *repeated or repeatable* operations and the possibility of *differences* in outcomes are not involved, then the language of statistics probably does not have much to offer. The surprising thing is that statistical reasoning *is* so useful in handling many of the truly important problems confronting scholars in sociology and many of the issues confronting those who govern and are governed. Basically we are interested in drawing logically supportable conclusions in situations where we have carried out some kind of measurement procedure on a number of subjects (often on samples of subjects). We need to summarize our findings clearly, assess risks of inference, and draw valid conclusions. Most of all we need to get to the point where we can use useful tools to promote important inquiry in our field without losing sight of the original purpose of our inquiry. Hopefully the organization of this volume will help.

1.5 PURPOSES AND OBJECTIVES OF THIS BOOK'S APPROACH

The aim of this beginning course in sociological statistics is to introduce you to those aspects of statistical description and inference which will be most useful to you as a consumer and potential creator of statistical information. The emphasis will be on applications of statistical reasoning, description, and inference and on some of the theory behind it, as well as on interpretation of these procedures in the context of sociological research. At the end of the course you should be able to read intelligently and evaluate statistical arguments made in the current sociological journals and monographs.

To accomplish this we are presenting an elementary text with a coverage of current techniques somewhat broader than usual; we incorporate in a subordinate role the methodological issues that you will encounter in more depth in other social science research methods courses; we attempt to emphasize a user's approach and understanding in the context of social science problems and current usage; and we minimize the more formal mathematical mode of expression as well as minimizing most aspects of clerical mathematics. We have assumed that the more clerical or computational aspect of statistics is less important at an elementary level, and in any event will most likely be handled by electronic calculator or computer when the need to compute arises. It is our view that it is much more important to gain a comprehension of statistical measures through several encounters with summary measures in real research contexts, rather than to emphasize computational routine or shortcuts on a few fictitious examples of "error-free" data.

Ideas are fun, particularly when one has the tools to rub them together in interesting ways. It is hoped that the pursuit of statistics in sociology will be interesting, challenging, and fun too. Some of the ideas behind measurement in sociology are introduced in the next chapter.

CONCEPTS TO KNOW AND UNDERSTAND

statistical observation   unit of analysis
statistical population   statistic
statistical sample   parameter
descriptive statistics   variable
inferential statistics

QUESTIONS AND PROBLEMS

1. Some sociologists study the phenomenon of power balance in families by having a family (parents and children) play a game of some type in a laboratory setting. As the game progresses the investigators record verbal and non-verbal interaction (*e.g.* who asks questions of whom, who provides information or gives direction or commands). What might be the unit of analysis, the main variable of interest to the investigator, and an example of a statistical observation he might make? Define a statistical population such an investigator might become interested in studying.

2. Select any sociological journal or research monograph which reports the results of research and then select an interesting article. Identify the statistical observation, unit(s) of analysis, and variables the investigator uses. Then describe the sample and population he (she) is dealing with. Point out situations where descriptive and inferential statistics are used, and give an example of a statistic and a parameter.

3. Usually, at this point, it is helpful to gather some actual statistical observations yourself. To do so you might do one of the following:
   a. Go to the cafeteria and observe people sitting at tables. How many people are sitting together at each table (Note that you must define what you mean

by "sitting together" in order to count numbers of people)? Also observe the gender (sex) of those sitting together. Record numbers of males and females in each "sitting" group.

b. Observe shoppers in a shopping mall or district. Observe the sizes of groups of shoppers and the distribution of shoppers by gender in each group observed. Classify the shoppers as children, young adults, middle-aged adults, and older adults. Observe the distribution of these age categories in each group of shoppers.

c. Check the vital statistics section of the newspaper for seven consecutive days. Note the age difference of each couple applying for a marriage license. Make note of whether the male is older or younger than the female and the number of years difference.

d. Take some hints on making statistical observations from a research article or from a workbook that includes some problems to suggest how you might proceed.*

GENERAL REFERENCES

Amos, Jimmy R., Foster Lloyd Brown, and Oscar G. Mink, *Statistical Concepts: A Basic Program* (New York, Harper and Row), 1965.
This is a brief programmed book covering the central argument of descriptive and inferential statistics. It is useful as a brief overview of statistics. You might read through it quickly, not worrying much about trouble in understanding the details, at this early stage.

Blalock, Hubert M., Jr., *Social Statistics,* Revised Edition (New York, McGraw-Hill), 1979.
Often it helps to read the discussion of some statistical procedure in more than one text, and the intermediate level text *Social Statistics* would be especially recommended for this purpose. It is detailed, relevant to our field and covers most of the topics included in this book.

Hagood, Margaret Jarman, and Daniel O. Price, *Statistics for Sociologists,* revised edition (New York, Holt), 1952.
Part one describes quantitative methods in sociology and some of the kinds of sources and ways to handle data.

McCollough, Celeste, and Loche Van Atta, *Introduction to Descriptive Statistics and Correlation* (New York, McGraw-Hill), 1965.
For those interested in a programmed approach to some of the earlier material in descriptive statistics.

Tufte, Edward R. (ed.), *The Quantitative Analysis of Social Problems* (Reading, Mass., Addison-Wesley), 1970.
This book is a collection of articles on social problems and studies about them which use statistics or a quantitative approach. Although some of the issues are beyond the scope of this text, the articles are all interesting and help point out some of the difficulties and successes in studies for which statistical analysis may be relevant.

Wallis, W. Allen, and Harry V. Roberts, *Statistics: A New Approach* (New York, The Free Press of Glencoe), 1956.
Chapters one and two present an interesting introduction to statistics and some of the uses to which it is put. A large number of brief illustrations are presented.

*See Loether and McTavish, *Statistical Analysis for Sociologists: A Student Manual.*

OTHER REFERENCE

Coleman, James S., *et al., Equality of Educational Opportunity* (Washington, D.C.: U.S. Govt. Printing Office), 1966.

2 Measuring Variables

A number of sociologists have argued that a change from a community-based society toward a mass society means that individuals become separated from binding social norms, and this then increases an individual's feeling of isolation and powerlessness. In a mass society, it is argued, membership in organizations which are intermediate between the family and mass society (*e.g.* unions, church groups, and various voluntary associations) are essential if a person's sense of mastery is to be maintained. Those who maintain membership of this sort will feel more in control of their lives, greater mastery, and less isolation than those who do not. This is the "mediation hypothesis" of organizations and powerlessness.

At this point, the ideas being discussed — about why individuals feel the degree of powerlessness they do — are very general and abstract. What is needed is some specific evidence and useful indicators showing that the real world behaves in this way. We could start by selecting a specific example of an intermediate organization, such as a union. If ideas about powerlessness are useful, then we might expect that workers who are members of a union will have a lower degree of powerlessness than workers who are not union members. This relationship was examined by Neal and Seeman (1964) in a study of adult males who were residents of Columbus, Ohio.

In addition to the research problems of locating and selecting research subjects, securing their cooperation, and recording their responses, there are two basic problems to be solved which lead directly to a consideration of what statistical reasoning is all about. These problems are (a) classifying subjects on their degree of powerlessness and group membership, and (b) comparing powerlessness scores for the group of subjects who are members of a task group and for those who are not members. The first problem will be discussed in the remainder of this chapter. The second problem will be dealt with in the next chapter where some of the difficulties and possibilities for making valid comparisons will be described.

2.1   MEANINGS AND MEASUREMENT

The first problem, noted above, is measuring powerlessness and group membership. **Measurement,** most simply stated, is a procedure for carefully classifying "cases" (called research subjects, respondents, or, more generally, "objects," such as a person, an interacting pair of people (a dyad), a company, a whole society, etc.) and putting them into previously defined categories of some variable. A **variable** is simply any characteristic or property of a case which has a series of two or more possible categories into which a "case" could potentially be classified. Group membership and powerlessness are two variable characteristics of individuals. Classifying a person as a "group member" or "not a group member" is an example of a simple measurement process. Alternatively, one might investigate the number of groups to which an individual belongs and use a series of categories such as 0, 1, 2, 3, or more, and "no response." The result of this process is a **statistical observation** or, simply, an **observation** which is usually labelled with an appropriate name or number.

There are two general types of problems that must be faced in deciding upon a measurement procedure. One problem concerns the *adequacy* of the actual measurement process to carry out a defined classification procedure required to classify cases on some variable characteristic. For example, it is probably useless to measure a person's blood type by asking him what it is. This procedure depends upon his knowing what it is, and he may not know. Other procedures, such as taking a blood sample for laboratory testing, are likely to lead to a better classification or measurement of blood type. Adequacy of a measurement procedure is usually discussed in terms of **validity** (does the procedure measure what it is intended to measure?) and **reliability** (would the same classification result if the same procedure were tried over again under the same conditions where there was no real change in the variable being measured?). These are difficult questions to answer, and some approaches to checking validity and reliability will be discussed in Section 2.2a of this chapter.

The second general problem posed by the measurement process concerns the *meanings* that are to be conveyed by measurement categories. We must ask, what does "group membership" or "powerlessness" refer to? What sorts of information can we glean from knowledge that a subject is classified into the "high powerlessness" category, for example, rather than another category? In our present example, this second problem will be solved as soon as we *define* what is meant by a variable like powerlessness.

2.1a   Concepts and Indicators.   The **concept,** or idea, of powerlessness refers to the degree that individuals expect not to have some control over events around them which affect their lives. It is usually considered a property of individual social actors, and in this sense it is used to describe an actor's own feelings about his (her) control of events. Individuals may differ in the feelings of powerlessness; some feel a greater sense of mastery than others. The feeling of powerlessness is one of the ways by which individuals can be said to be alienated from their surrounding social structure. The formal definition of the concept of "powerlessness" could be stated more elaborately or with examples, of course, and some investigators devote considerable effort to careful definitions of this sort. Clarity here is essential if an investigator wants to examine the validity of some measurement procedure.*

When it comes to measuring the concept of "powerlessness," one might think of a number of **indicators,** or items that will serve to measure this concept. One might, for example, ask a person if he agrees with the following statement: "We are just so many cogs in the machinery of life." If a person agrees, he would be classified as having a feeling of "more" rather than "less" powerlessness than would someone who disagreed. Sociologists often use several similar questions to form an attitude scale where the answers to all of the questions are combined in some way into an overall "powerlessness" score. Techniques for creating these scales are discussed at some length in most research methods texts. Usually the result of measurements like this is a number, datum or **score** (*i.e.* a statistical observation). A collection or group of several of these scores is called **data.**† Descriptive statistics are aimed at describing and analyzing data.

The meaning of a score resulting from a measurement process depends upon the way variables are defined (as discussed in Chapter 1) and the adequacy of the measurement process itself. The way these scores are statistically described depends on the information they contain, and it also depends upon the specific use of the variable in an overall research project.

Three differences between variables will be pointed out in the following discussion. These are (a) level of measurement (Section 2.1.1), (b) scale continuity (Section 2.1.2), and (c) the role the variable plays in a research project (Section 2.1.4).

2.1.1   Level of Measurement

To illustrate some of the different types of information a defined variable is intended to convey, it is convenient to distinguish among four **levels of measurement.** These four types of measurement are traditionally listed in statistics texts primarily as a way to reinforce the importance of knowing what information is contained in scores which are assigned.

This difference is one of several which have an important influence on the kinds of statistical summary one chooses, and it certainly influences the kinds of interpretations that are warranted. It is worth noting that some scholars distin-

---

*It is interesting to note that the problem of definition is sufficiently serious for whole books to be written about it. One interesting and useful book is by Robinson (1954).
†Note that the word *data* is plural. One way an investigator's data *are* described is by using the statistical techniques discussed in this book.

guish more levels of measurement than these four, and they may include different mixtures of these which feature some of the properties of the levels mentioned below (Stevens, 1946).

   2.1.1a Establishing Categories of Measurement. Any adequately specified measurement procedure must have a category into which each and every case can be classified; it must be **inclusive.** If only the categories 0, 1, and 2 were available to classify individuals on the number of groups to which they belong, there would be no category for those belonging to more than two groups. Such a situation could be corrected by making the last category "2 or more" or by providing a category labeled "other," if additional categories could not be defined in a more positive way. The "other" category is sometimes used simply to create a logically complete or inclusive classification system. In most actual research it turns out that categories for "don't know" (often symbolized in printed tables as D/K) and "no answer" (often symbolized as NA) are often needed as well. These categories pose problems for the thinking statistician.

   Another property of a good classification system is that its categories are **mutually exclusive** as well as exhaustive; that is, it is possible to classify an individual case into one and *only one* category of the classification system. If a variable has two categories "married" and "has children," for example, it would not be adequately defined in this sense, because married people with children could be classified into both categories. The categories are not mutually exclusive. "Ever married" and "never married," on the other hand, are mutually exclusive categories. It is true, of course, that individuals change over time, but at any given point when the measurement is made it should be possible for a case to fall in only one category.

   It is preferable, generally, to have a measurement procedure which is as **precise** as possible; that is, one which makes more rather than fewer distinctions. The measurement scheme: "never married," "currently married," and "other" would be less adequate in this regard than "never married," "currently married," "divorced," "widowed," and "separated," simply because there are more categories in the second scheme. Knowledge that a subject is "divorced" is less ambiguous and conveys more information than knowledge that he (she) is an "other." Adequately defined variables are generally classified into the following four levels of measurement.

   2.1.1b Nominal Variables. Nominal variables are those which are properly defined with logically exhaustive and mutually exclusive categories, so that equivalences or differences are clearly established. Good nominal-variable measures are *precise* and have *positively defined* categories. There is no implication, however, of any ordering of categories; there is no implication that distance exists between one category and the next; the categories are simply logically *different* or distinct from each other. Many scientific variables are of this kind. In sociology, nominal variables include marital status, gender, group membership, religious affiliation, kind of role relationship, etc.

   The idea of level of measurement refers to the relationship between *categories* of a variable. For nominal variables the categories are simply different, whether or not they are labeled with numbers. It would not be appropriate to add together, for example, the category labels of a nominal variable. Sometimes students wonder why formulas are used to compute statistics on nominal variables if this is so. The reason is that the formulas make use of category frequencies, the number of cases in each category, and not the numeric label that may be used to identify any particular category of the variable. It *does* make logical sense to say that there are twice as many "ever-married" people as "never-married" people, but it would not make sense to add up numbers which may be used to label marital status categories (although it would make logical sense to do this for a variable such as age).

2.1.1c Ordinal Variables. Ordinal variables are those which are defined to include the features of nominal variables, plus the feature that categories are *ordered* or *ranked* in a described way. For example, "friendliness," as we usually think of it, is defined as an ordinal variable. Categories generally distinguished (low, medium, and high, for example) not only imply categorical differences but also imply an order from low to high on that variable. We are not able to define *amount* of difference between these categories however; it is only sequence that we can define in an ordinal variable. Traditional class grading schemes are often thought of as ordinal properties of students. Social class is ordinal (*e.g.* lower-lower, upper-lower, lower-middle, upper-middle, lower-upper, upper-upper class), no matter how precisely it may be measured. In principle, it is not defined in such a way that a "degree" or "unit" of social class is meaningful or distinguishable, although we feel we can distinguish rankings by social class. Rank order is defined in variables such as powerlessness, desirability, segregation, and many other attitudinal variables used in sociology.

The values assigned to the categories of an ordinal scale may be expressed either in words or in numbers. Friendliness, for example, as defined above could take the values low, medium, and high. On the other hand, the prestige rankings of universities might be expresses as numbers (e.g., 1st, 2nd, 3rd, etc.). Note, however, that these numbers are ordinal rather than cardinal. The usual arithmetic operations of addition, subtraction, multiplication, and division are not permissible with ordinal numbers. It makes no sense whatsoever to add two numbers such as 1st and 3rd. Consequently, ordinal level measurement does not lend itself to computations which require the use of arithmetic. It is permissible, however, to count the numbers of observations which fall into each category of the scale.

2.1.1d Interval Variables. Interval-level variables, a third level of measurement which is usually distinguished, includes the logical and ranking features of ordinal and nominal variables, but in addition, its categories are defined in terms of a standard *unit of measurement*, such as the individual "social actor," the year of age, or the inch, ounce, liter, or degree. Definition of a unit which can be unambiguously detected and counted is a difficult undertaking. You may recall accounts of the very elaborate definitional procedures to establish the foot or meter, the degree Fahrenheit, the ton, or the U.S. Dollar. The U.S. Bureau of Standards, a special governmental agency, is concerned with the definition of certain standard measurement units such as these. Given a recognizable unit such as "one social actor," however, one is able to compare the differences between pairs of values on an interval scale. Therefore, although we cannot say that an interval value of 40 is twice an interval value of 20, we can say that a difference in values of 40 points (e.g., the difference between scores of 20 and 60) is twice as large as a difference of 20 points (e.g., the difference between scores of 20 and 40). An interval scale, then, allows us to compare the difference between families with one or two members and those with three or four members. The *differences* are the *same*. We can distinguish this sameness by virtue of having defined "one social actor." The scale of measurement of an interval variable is defined in terms of a standard unit size.

Determining the interval for a variable is an unusually important step up from nominal and ordinal variables, in that it greatly increases the amount of information which is contained in numeric scores. Given interval-level variables, a large amount of mathematical manipulation is possible, and conclusions can be drawn that are often not at all obvious from the raw scores themselves. Much of statistics implies the use of interval-level scores, and because of the added power of these procedures, there is a strong motivation to use these techniques whenever they can be justified.

2.1.1e Ratio Variables. Finally *ratio* variables include the foregoing characteristics and, in addition, the scale of measurement has a defined origin, or zero point. A zero point is a theoretically defined point on a scale representing the absence of any of the property being measured. Not all properties can easily be measured from zero. Zero social class, for example, is not a meaningful concept thus far in sociology. On the other hand zero inches of length, or families of size zero is meaningful in this sense. Prior to the development of theoretical ideas of an absolute zero point in measurement of temperature, other temperature scales used the concept of zero degrees (as in Centigrade or Fahrenheit) as a relative measurement that did not correspond to a theoretical absence of any positive features of the property "temperature." In the Fahrenheit and Centigrade scales, since degree-units were defined, but zero did not represent any absolute in temperature, their degree-unit would be classified as an interval, not a ratio variable. With the development of the Kelvin scale, temperature became a ratio variable.

If it is possible to define a variable as a ratio variable, then it is also possible to talk about ratios, (*e.g.* two years of education is half of four years and this ratio is the same as the ratio of 12 years to 24 years). These statements would not be possible if one could not count units (*e.g.* year-units) from a meaningful zero starting point. However, if one is interested in the variable "amount of learned knowledge," the number of years of education might be used as an indicator of this concept. In this case, however, because of the way "amount of learned knowledge" is usually defined, we may not have available a specified unit of "learned knowledge" nor a meaningful zero point on the "learned knowledge" scale. Notice that the way a variable is defined governs its level of measurement. We probably would not want to say, for example, that a person with four years of education has exactly twice the amount of learned knowledge that a person with only two years of education has.

In most cases, once one has defined a standard unit of measurement in an interval scale, a zero point is also established. For this reason most variables in sociology which we are able to measure with this degree of sophistication qualify as ratio scales rather than interval scales. In our work we will not make this possible distinction between interval and ratio levels of measurement, but will refer to either type as "interval/ratio" or simply "interval" variables. Sociological examples of such variables include counts like "family size," "number of levels in an organizational hierarchy," "group size," "number of years of formal education," and perhaps "social status."*

Figure 2.1 illustrates the cumulative character of levels of measurement. The important point is that the way one defines a variable has consequences for the categories developed, the relationships between them, and the meaning of an assigned score.†

2.1.1f Using Level of Measurement Information. Labels for categories can be of a wide variety of different types. Names ("married," "high-power-lessness," etc.) are often convenient to use, although sometimes letters or, more frequently, numbers are used. The utility of numbers lies in the possibility of an unlimited set of symbols and in the fact that there is a set of powerful logical rules (mathematics) for handling numbers. The symbols one uses and the available means for operating on the symbols has far greater consequences than is immediately apparent. It leads most scientists to the use of numbers and to various branches of mathematics for handling these number codes, although in principle the logical distinctions could be handled (usually inefficiently and with great effort and risk of confusion) by other code schemes such as words.

Defined Characteristics
of Category Systems

Level of Measurement	Exhaustive, Mutually Exclusive Categories	Categories Defined to Have an Ordering	Defined, Standard Unit of Measurement	Meaningful Zero Point on Scale
Nominal	X			
Ordinal	X	X		
Interval	X	X	X	
Ratio	X	X	X	X

FIGURE 2.1 PROPERTIES OF DIFFERENT LEVELS OF MEASUREMENT

*An example of work toward establishing a ratio scale of measurement for social status is Jones and Shorter, (1972). See also Hamblin (1971).

†A very helpful chapter on measurement in sociology, and on some of the meanings which may be available in scores, is Clyde Coombs' Chapter 11 in Festinger and Katz (1953).

A word of caution is necessary at this point. It is clear that number-symbols could be used to label the categories of a variable like marital status, as well as the categories of the variable age. Marital status, however, is a nominal variable, since its categories have no defined ordering and age is usually defined as an interval/ratio variable. For marital status of a group of individuals, one could not use the mathematical operation of adding up the number-labels, for example, because such a manipulation of the scores would rely upon information about ordering and relative spacing of categories, and these are not part of the (usual) definition of marital status. The adding operation is, of course, appropriate when the variable is years of age, an interval/ratio variable. To take another example, if one were to score ten individuals on sex, assigning the number code 0 for males and 1 for females, a collection of ten scores such as the following might be available.

0 0 1 0 0 1 1 1 0 1

If these ten numbers were added and then divided by ten to create a simple average score for the group, the average-sex would be 0.5. Adding and dividing scores in this manner clearly involves the assumption that females are higher — one unit higher — than males. A statement that the average sex of this group of scores is 0.5 is nonsense, but on the other hand if this were reinterpreted as a proportion or percentage (*e.g.* 50 percent of the group are female) the results can be meaningfully interpreted.

There are statistical procedures which are created in such a way that only nominal or only ordinal or only interval/ratio meanings in scores are relied upon. It is most appropriate logically to use the statistical procedure which features the kind of information contained in the scores. As indicated in Figure 2.2, if statistics appropriate for lower levels of measurement were to be used on scores defined at a higher level of measurement, no *technical* error would be made, since the properties of levels of measurement are cumulative. An interval-level measure contains all of the properties of a nominal variable

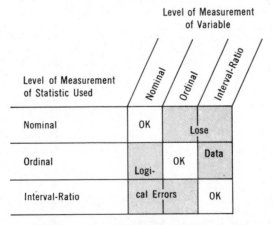

FIGURE 2.2 Correspondence of Statistical Procedures and Level of Measurement

plus added properties – added information. However, though no technical error was made, a serious research problem would arise in that this procedure did not use all of the information that was readily available in the higher-level score.

A technical "error" *is* made, however, if we use statistical procedures appropriate to higher levels of measurement when the data are defined at a lower level of measurement. In this case, we would be acting by our statistical operations as if the scores contained more information than they were defined to have – *i.e.* we would have assumed a property of order where there was none, or of defined distances where there were none. Consequences for the interpretation of research findings when this technical error is made have been the subject of detailed discussion in sociology; (see Burke, 1971; Labovitz, 1970; replies by Vargo *et al.*, 1971). Thus the strong preference of most careful investigators is to select appropriate statistics for the data at hand.

Statistical procedures appropriate to higher levels of measurement generally have the advantage of permitting more compact descriptions of data, as we shall see. In the case of our dichotomous (two-category) nominal variable of sex in the example above, the "average-sex" could be readily re-interpreted as a percentage: a meaningful interpretation. On the other hand, if we discovered that the average "marital status" scores were 2.563, we would be essentially at a loss to provide a meaningful interpretation of this figure. Some research on the issue indicates that for many purposes, a procedure that moves from ordinal to interval-level statistics while using ordinal data does not result in great errors in interpretation of the statistical results. The important point is to know what differences between scores constitute meaningful variation, and then to use statistical procedures which highlight meaningful variation in scores, ignoring undefined or meaningless differences.

An associated problem is that of not collecting sufficient information in the process of measurement to classify a case adequately. For example, if one

---

BOX 2.1 LEVELS OF MEASUREMENT – SOME REVIEW EXAMPLES

It would be well to pause at this point and think of some examples of variables which fit each *level of measurement* we have been discussing. If you are clear about this first classification of variables, then you are ready to go on to the other two ways to classify variables – by scale continuity, and by role in a research project.

As a review-starter, try defining "age" first as a nominal variable, then as an ordinal variable, and finally as an interval or ratio variable.

---

wanted to measure age (an interval/ratio characteristic) but found that employment forms only listed the person as "old," "middle-aged" or "young," one would have to treat the variable as if it were ordinal because information on number of units (years) is not available in this classification for use in computation of interval/ratio-level statistics. If the information is not contained in the data one gathers, no amount of statistical manipulation will create it.

2.1.2 Scale Continuity

Another distinction between variables in addition to level of measurement involves whether or not variables are defined to have a continuous or discrete scale of measurement. A **continuous** variable, as Figure 2.3 suggests, is defined in such a way that in principle individuals could have infinitely varied fractional scores—*i.e.*, scores located at any point on an unbroken scale. The same concept can be stated in at least three other ways. First, given any two scores, however close together, in principle a third score could always fall between them; or second, the probability that any two scores will be identical is zero; or third, an infinite number of different scores is possible between any two points on the scale. In any case, this concept implies an unbroken scale of possible score values. In sociology such variables as age, alienation, segregation and social class are defined (usually) as continuous variables. There is a well-known debate in the social stratification field over whether the phenomenon of social class is most usefully defined as discrete or continuous (Duncan and Artis, 1951).

A contrasting type of variable is a **discrete variable.** This type is defined in such a manner that, again in principle, only a *limited set* of distinct scores can be achieved. Stated differently: (a) the scale of measurement is broken by "gaps" in the numbering scale which, in principle, can not contain measured cases; only integers are needed to label scale categories appropriately, not decimal fractions; or, (b) within a finite range there are only a finite number of possible score categories. Thus a variable such as family size (as usually defined), in principle, is a discrete variable, because families can only be of size one or two or three . . . etc.; there can be no example of a fractional size such as 2.5. This is true of most "count" type variables; nominal variables are always discrete. It should be noted that "average family size" is a different variable. It is measured on "cases" which are *groups*, such as states or nations or cities; it is possible for that kind of variable to take on fractional values, since it is usually defined as a continuous variable characteristic of *groups* of observations.

2.1.3 Grouping Scores Into Classes

From the start, most measurement processes in sociology have tended to group scores into defined classes. In the case of nominal and ordinal variables, it is usually quite clear that categories of social class, anomie, marital status, or

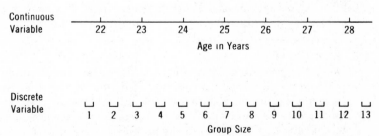

FIGURE 2.3 CLASSIFICATION OF VARIABLES BY SCALE CONTINUITY

productivity have to be defined before one can begin to measure or ask which of these categories the individual falls in. Actually this is the case with interval- and ratio-level variables as well.

2.1.3a   Units of Measurement and Nearest Whole Units. Typically an investigator, confronted with the problem of measuring an interval variable, decides upon a **unit of measurement** such as the inch, foot, year, or 1000's of people. He then counts up the number of these "units" which can be said to characterize the object being measured rather than attempt to continuously measure the exact amount of these variables.* The interpretation of continuous variable scores is different from the interpretation of discrete scores, and this difference is shown in the way continuous variables are measured.

For continuous variables, the problem is one of deciding where any given case should fall in a classification scheme, and then, secondly, characterizing the category so that its meaning can be interpreted. Take the age category, for example, that we usually refer to as "65." Assuming that age has been measured by asking individuals what age they were, at their *nearest* birthday, the category could be said to include those who are past their 64th birthday by just over a half a year (half a unit of measurement) and that category would still include a person who is almost half a year over his 65th birthday. To be precise, the category would extend from 64.5 to 65.5 years as illustrated in the following diagram.

Measuring Age To The Nearest Whole Year:

62.5	63.5	64.5	65.5	66.5	"real limits" or class boundaries
	63.0	64.0	65.0	66.0	class midpoints

A category of an interval variable is generally referred to by its **midpoint.** In this case, using one year as a *unit* of measurement, the category midpoints would be 63.0, 64.0, 65.0, 66.0, etc. Scores, or measured values, would be classified into the nearest category, which means that scores in the category 65.0 may be as small as 64.5 and as large as 65.5. These limits, extending one half unit on either side of the midpoint, are called **real limits** or **class boundaries.**†

2.1.3b   Significant Digits and Rounding. How far one should round off an answer, to how many digit places, is determined basically by the degree of precision in the original measurement process. If a measurement is exact (*e.g.* counting the number of people in a group, a discrete variable), then it is appropriate to retain all significant digits in a score and to carry out com-

---

*A thermometer measures temperature in a continuous fashion, since the physical height of a column of fluid varies with temperature. We usually record temperature, however, as if the variable were discrete, in terms of degrees or tenths of degrees. An analog computer works on the principle of continuous measurement by varying electrical current. A digital computer, on the other hand, operates in terms of discrete states where a switch is either on or off, or a magnet is polarized in one direction or not, etc. Most measurement of continuous variables involves counting units of measurement to yield discrete-looking scores. However, the differences, built in by definition, between discrete and continuous variables mean that scores can be manipulated and interpreted differently.

†Students are usually concerned about classification of scores near class boundaries — a relatively rare and minor problem in most research. What if a "super-precise" subject reported his age as 65.51 years or 65.49 or 65.50; how shall one classify these age figures? The rules of *rounding* solve the problem. These rules are given in Figure 2.4, and they would result in classifying the first age above as 66, the next as 65, and the third as 66.

putations involving these scores to as many decimal places as desired, treating the results as meaningful and significant (although this could easily be carried to the point of absurdity, *e.g.* average family size $= 2.5632567$). If, however, measurement is inexact, as it is in the case of continuous variables, then the answer should not be expressed to more digits than are significant in the original scores; (sometimes the rule is to admit one more digit than this). Thus, if one were measuring age in years, then answers should be rounded to whole years (or, perhaps, tenths of years) even though the computational process (as in computing an average) may result in a fractional number which could be carried out to many decimal places. If births are recorded to the nearest 100, then the answers from computations using these scores should be expressed to the nearest 100 births. To express the result to more digits gives readers a distorted impression of accuracy. It is an appropriate practice, however, to retain as many digits as possible throughout a computational procedure rather than rounding at each step, since this practice avoids the accumulation of small rounding errors at each computational step. Only the final answer should be rounded. On the other hand, at any point where digits are to be dropped (*i.e.*, the number is too big for your calculator) they should be rounded off rather than simply dropped so that systematic distortions are not built into the result.

These principles can be quickly illustrated if we think of an example where two ages are to be added together, say, age 65 and the age 69. A score called 65 (the midpoint of its category) could actually be anything between 64.5 and 65.5 in actual value. Likewise a score called 69 could be between 68.5 and 69.5. Adding the class midpoints of these two scores gives a total of 134 ($65 + 69 = 134$), but the actual sum of the two scores could be as small as 133 or as large as 135; ($64.5 + 68.5 = 133.0$ and $65.5 + 69.5 = 135$). Perhaps the sum should be expressed with only two significant digits as about 130. or as 134. but certainly not 134.0 which gives the impression that measurement was to the

---

If a number, such as 34.6 were to be rounded to a whole number, the digit to the right of the place to which it is to be rounded should be examined. In this case, that digit is "6." Then:

   (a) If that digit is greater than 5, round up by increasing the digit to the left by one. In the example above, this would result in a rounded number of 35.
   (b) If that digit is less than 5, round down by simply dropping digits to the right of the last place in the desired answer.
   (c) If the digit is exactly 5 (followed by zeros), then the rule is to round up if the last digit in the desired answer is odd, and round down if the last digit in the desired answer is even. This is the so-called "engineers rule" and its logic is simply that odd and even numbers are likely to occur equally frequently so that no bias in rounding up or down will occur if this is always followed.

Some examples should help.

65.60	could be rounded to	66
64.50	could be rounded to	64
65.50	could be rounded to	66
64.51	could be rounded to	65
65.49	could be rounded to	65

FIGURE 2.4   RULES FOR 'ROUNDING OFF'

nearest tenth of a year. With more scores, of course, the problem becomes potentially more serious.

When a classification scheme is published in a table or chart, it is common practice to express the category boundaries in a fashion which clearly indicates that the categories do not, in fact, overlap, but are mutually exclusive. Thus the series of age categories,

Years Age
31–35
36–40
41–45

would be preferred over an expressed system such as:

Years Age
30–35
35–40
40– 45

However expressed, the meaning of a score on a continuous variable extends from real limit to real limit.

2.1.3c Measurement to the Last Whole Unit. Measurement to the **last whole unit** is sometimes used rather than measurement to the *nearest* whole unit. This introduces a complication which is worth a word of caution. A common example of this is in the measurement of age where an individual typically is asked for or gives information on his age at his *last* birthday even though his next birthday may be nearer. The meaning, then, of the age expressed as "65" would change, and the data would have to be treated slightly differently as a result. The real limits of the age 65 measured to the *last full year* would be 65.0 and 66.0, as shown in the diagram below.

Age Measurement To The Last Whole Year:

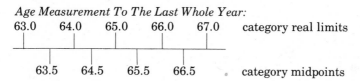

63.0   64.0   65.0   66.0   67.0   category real limits

63.5   64.5   65.5   66.5   category midpoints

Although the difference between measurement to the "nearest" or "last" whole unit may seem to be slight and may not lead to gross errors in interpretation if caution and consistency are adhered to, failure to use the appropriate midpoint will have the effect of adding (or subtracting) a constant to each score. It is important that information measured by a measurement procedure be correctly represented.

The relationship between scale continuity and levels of measurement is shown in Figure 2.5 below and it is usually helpful at this point if you read a sociology text or journal article and attempt to classify the variables used in terms of *both* scale continuity and level of measurement. An abstract from an article by Marsh is reproduced in Figure 2.5в. Can you identify the different variables, units of analysis, and types of information probably contained in scores on the different variables?

For most statistical purposes, the distinction between continuous and discrete variables, where ordinal and interval/ratio variables are concerned, is of less importance.* In some treatments of statistics and in other parts of statistics than the descriptive, the distinction would be more consistently carried through. Here we will make the distinction at points (such as in graphic presentation of data) where it is necessary to describe data accurately. Our main point here is that the scores one chooses to summarize contain certain kinds of information, and the manipulations and conclusions one is able to draw from them depend upon a clear understanding of their meaning. The results of statistical description do not contain any more information or meaning than the scores which were originally produced by measurement.

2.1.4 Roles Variables Play in Research

A third and crucial way in which variables differ from each other is in the use made of them in research. Usually two main categories of variables are distinguished: **independent** and **dependent** variables.

Frequently an investigator begins his research with a curiosity about the ways in which a certain variable fluctuates from case to case in a population. It is noted, for example, that some marriages last a long time and others last but a brief time; some persons have more power than others; some have more income than others; some are more respected; some organizations are more effective

Level of Measurement	Scale Continuity	
	Discrete	Continuous
Nominal	sex a	Logically Impossible
Ordinal	b	d
Interval-Ratio	family size c	e

To illustrate the use of the figure above, sex (a nominal, discrete variable the way it is usually defined) is an example in cell *a*. Meanwhile, family size (a ratio variable, and a discrete variable, if it is defined in terms of number of related people) is an example of a variable appropriately classified in cell *c* above.

FIGURE 2.5A CLASSIFICATION OF VARIABLES

*Note that we have used the term "variable" in a broad sense in this chapter to refer to properties we may be interested in measuring, regardless of their level of measurement or scale continuity. In many statistics texts the term "variable" is reserved for only those characteristics which are defined as interval or ratio in level of measurement and, perhaps, are also continuous. Other characteristics, in this usage, are then referred to as "attributes" or "categorical" data. Whichever usage you and your instructor prefer, the important point is that you should carefully inquire into the meaning of the scores that are assigned.

The following is an abstract from an article by Robert M. Marsh (of Brown University) *Social Forces*, 50, (December, 1971) p 214–222. Most journals include these brief abstracts or "road-maps" to a research report, and in most cases they indicate the main variables, the ideas to be examined, and some of the main conclusions. Can you classify the variables Marsh talks about in terms of Figure 2.5A above?

THE EXPLANATION OF OCCUPATIONAL
PRESTIGE HIERARCHIES

Abstract
The ranking of occupational prestige in Taiwan is highly similar to that in the United States and the numerous other societies—both modernized and relatively non-modernized—in which such studies have been conducted. Previous explanations of this important cross-societal invariant, occupational prestige evaluation, have emphasized *common structural features* of any complex society, whether industrialized or not. It is not clear how this proposition can be empirically falsified. A more satisfactory explanation is sought in terms of *specific properties of occupational roles* in any society, viz., education, responsibility (or authority), and income. The relationship of each of these variables to prestige is analyzed with data from Taiwan, Denmark, and the United States.

FIGURE 2.5B CLASSIFICATION OF VARIABLES IN RESEARCH

than others, and so on. The next question is usually: Why? Why does the variable vary as it seems to do? This then leads to explanations and predictions of the variation in one variable based on variations in other variables, and this eventually becomes the general substance of theory or scientific knowledge.

2.1.4a Dependent Variables. The variable of prime interest, the one whose variation is to be explained in the research, is called a **dependent** variable. Differences between scores on this variable are thought to depend upon certain other variables. Dependent variables are usually defined as those which are influenced by some other variable, but sometimes they are defined merely as the crucial variable whose variation an investigator is interested in examining.

2.1.4b Independent Variables. Independent variables are those which are thought to influence dependent variables; that is, they are explanatory variables which may help account for why a dependent variable fluctuates as it does in some population. If one thinks in cause/effect or influencer/influenced terms, as most people do who are interested in theory development, the relationships between independent and dependent variables are frequently expressed in terms of a directional arrow or path pointing from the independent variable at the left to the dependent variable at the right, as shown below.

An investigation of how to look at data to see whether this type of argument is reasonable will be one of the central purposes of this book. Some statistical procedures are meaningful only if we can clearly settle the role the variable plays in the research.

2.1.4c   Intervening Variables.   Referring back to the example of membership in organizations and powerlessness, cited at the beginning of this chapter, since powerlessness was the variable of central concern, and was thought to be influenced by other variables, it would clearly be the dependent variable. Whether or not an individual is a member of organized groups is then the independent variable, because it is seen as having an influence on the dependent variable. In fact, we could introduce a third variable from that study, namely, whether or not a society is a mass society. It is thought to have an effect on powerlessness which may be controlled or modified in important ways by organizational membership. This three-variable relationship could be shown as follows with organizational membership **intervening** between the other two variables:

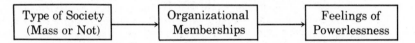

The role a variable plays in one piece of research may not be the same as the role it plays in another. For example, it may be reasonable to think about the influence of an individual's level of powerlessness at one point in time on the likelihood that he would join any group at a later point in time. In that research, the role of the variables would be just the opposite from the role they played in the Neal and Seeman study.

Research can be more complex in the sense of involving relationships between more than two or three variables. Often there are several independent (or dependent) variables, and not infrequently some variables intervene (and are called intervening variables for that reason) as illustrated in the three-variable example above and the five-variable example below.

Elder, in a study of marriage mobility of women from middle and working class families, hypothesized that the five variables shown below would all be related (Elder, 1969). Women in his sample were first rated on "attractiveness" and some other personal variables; "husband's occupational status" was measured in a re-interview several years later. In this study, "husband's occupational status" is the primary dependent variable, and the other four variables are inde-

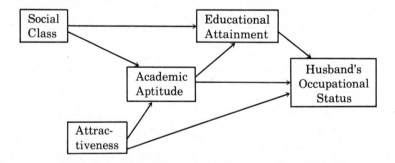

pendent variables which are thought to help explain or account for score differences on the dependent variable. One could focus on other pairs of variables, however. For example, "social class" is an independent variable with respect to the dependent variable, "educational attainment"; one could go further and speak of "educational attainment" as intervening between "social class" and "husband's occupational status." The diagram shown here in effect shows the theoretical relationships between variables which Elder believes exist, and his statistical analysis will be focused on the question of whether the data he examines behave as his model would predict they would.

Usually a great deal of clarity can be gained in reading a research report if one first draws out the relatonships hypothesized in the article in a fashion similar to that shown by Elder. The abstract presented at the beginning of most journal articles usually presents the model the investigator is considering. The important point here is that variables do play different roles in research; their role will generally determine the kinds of questions which are raised and, in turn, the kinds of statistical operations which are most appropriate in answering the questions.

2.2 MEASUREMENT IN SOCIOLOGY

The crucial role of measurement in sociology is becoming recognized, and there are a number of sociologists working on problems of how to measure significant variables adequately, how to move toward interval and ratio scales without leaving all sense behind, and how to actually carry out measurement of rapid-paced interaction on subjects who often are influenced by the measurement process itself. Bonjean, Hill, and McLemore in a recent assessment of measurement used in articles in main sociological journals between 1959 and 1963, found over 2,000 different scales were used in addition to the simple background information such as sex and age (Bonjean et al., 1965). Most of the measures were defined as ordinal, and most also as continuous variables. No defined, standard unit of measurement was used, no zero points were defined for scales in most cases, and most of the measures were not extensively examined for validity and reliability. Measurement is clearly a first priority problem.

Statistical procedures, as we will show later on, aid a great deal both in focusing attention on the need for better measurement and in providing ways to assess the quality of measurement scales. With increased sophistication in statistics and with more investigators aware and concerned about the problem, it appears to us that some exciting and significant strides will be made in sociological measurement. We already have a number of measurement devices of impressive and demonstrated quality, and many, such as social class measures, have received repeated examination and improvement.* In a few sentences we would like to point out some of the issues and developments.

Because much of what sociology is interested in predicting and explaining turns out to be variable characteristics of people, interacting pairs, and the various groups people create and engage in, it is natural that individual people are among the most helpful sources of information. Much measurement in the past

---

*For an inventory of sociological measures, see Bonjean et al., (1967); for more specialized collections of measurement scales, see Shaw and Wright, (1967), and Price, (1972).

and present has involved **questioning** people. Particularly in the area of attitude research, this questioning has taken the form of batteries of questions which are then examined for consistency and combined into attitude scales. One of the problems with this approach is that people sometimes become aware of the purpose of the questions and specifically are influenced by this knowledge. In other cases, respondents are not particularly knowledgeable and simply do not know the answer to the questions that are asked. Where this is the case, alternative measurement procedures are pursued.

With the advent of the field of electronics, a number of devices have been invented which greatly aid the measurement process. One device, for example, involves a simple ten-key keyboard much like that on an adding machine. This is connected to a portable tape recorder and, by setting up numeric codes for categories of interaction, an investigator can make records of the type of interaction in group settings. The tape, of course, moves at a certain rate, and so the record permits an examination of the shuttle of interaction through time, in terms of meaningful variables. This permits a kind of observation of interaction that pencil and paper methods cannot achieve, of course, and with the improvement in equipment comes improvement in the quality of measurements.* A very interesting book about possible ways to solve the problem of "reactive" measurements, those which also influence the measurements taken, is *Unobtrusive Measures: Nonreactive Research In The Social Sciences*, by Webb and others (1966). Such measurement problems, of course, are evident in every area of inquiry.

Personal **observation** by a trained social scientist is another approach to measurement of sociological variables. In laboratory settings, of course, more control and more options for measurement are available, but many situations of interest either cannot be readily captured in a laboratory or cannot be artifically created. War might be an extreme example. On the other hand, some aspects of competition and conflict can be simulated, and sometimes this takes the form of a respondent playing a carefully devised game where measurements have been built into the process of playing or scoring outcomes.

Records created in the normal process of agency operation, or especially created records from the U.S. Census or from vital statistics, for example, form the basis for much other measurement and research work. Unfortunately, many people think only of questionnaire research as the basic source of information for sociology, and one of the reasons this brief section is included now is to get you to think seriously about measurement possibilities and be sensitive to the issues and possibilities implied by different types of data.

2.2a   Variable Construction.   An integral part of the conceptualization process which is so vital for conducting sound research is the process of constructing variables to represent the concepts to be studied. Given a clearcut definition of a concept it should be possible to identify an empirical indicator or indicators of that concept. Although a single indicator may be employed, it is

---

*Early instruments to do this in laboratory settings were developed by Bales at Harvard. The device described here is called "MIDCARS," and it was developed and refined by Richard Sykes and other sociologists at the University of Minnesota. The cassette tapes are converted to computer tape, and the computer is able to carry out the needed statistical analysis using programs written by the researchers. This is only one example of a number of technological developments which have impressive consequences for the quality of sociological measurement.

generally the case that multiple indicators will give better results. Accordingly, many of the variables utilized in sociological research are composite variables consisting of two or more indicators of the concept of interest (e.g., socioeconomic status is a composite index which is usually constructed from the three indicators income, occupation, and education).

In a study of the "academic mind" Lazarsfeld and Thielens (1958) looked at the eminence of faculty members. In the course of their study they developed two indices of eminence, one based upon the honors a faculty member had received and the other upon the faculty member's productivity. The honors index consisted of four indicators—possession of a Ph.D., publication of three or more papers, holding office in a professional association, and experience as a professional consultant. The productivity index also consisted of four related, but somewhat different, indicators—having written a dissertation, having published at least one paper, having read three or more papers at meetings, and having published at least one book (Lazarsfeld and Thielens, 1958). The researchers found that, although their honors and productivity indices were not the same, they were closely interrelated and thus could be used interchangeably as a measure of eminence in relating that concept to variables such as age and rate of promotion in academic rank.

According to Lazarsfeld (1972), the process of constructing a variable involves the following four steps:

1. The concept of interest is put into words and is communicated by examples. The concept is defined.

2. As the concept is being analyzed and defined a number of potential indicators are identified. With expanded discussion of the concept the number of potential indicators is increased. The total number of potential indicators is known as the *universe of indicators*.

3. The universe of indicators is likely to be large; therefore, it is necessary, from a practical standpoint, to select a subset of the indicators to form the composite variable to be used for the empirical work at hand.

4. The subset of indicators is combined in some fashion to form the composite variable.*

This process is the one used by Lazarsfeld and Thielens in constructing their honors and productivity indices. As a matter of fact, they extracted two subsets of indicators and developed two separate indices of eminence.

Ideally the process of variable construction should be carried out in the conceptualization stage of a research project—before any data are collected. When this is, in fact, done, the resulting variables are more likely to be satisfactory measures of the concepts of interest than if it is not done prior to the collection of data. In point of fact, in many studies variable construction takes place after the data have been collected. When this is the case, it is often difficult to identify available indicators which adequately measure that which the researcher wishes to measure. Thus, indicators such as assessed value of a house may have to

*Formation of the composite variable can be done systematically through the use of such techniques as factor analysis. This procedure is not discussed in detail in this book. See Harman (1967) for a comprehensive treatment of factor analysis and Heise (1974), Armor (1974) and Allen (1974) for a discussion of the use of factor analytic techniques for the construction of composite variables.

be substituted for the more relevant indicator income in the construction of socioeconomic status as a composite variable. Even in this kind of an *ex post facto* situation, however, the clever researcher is often able to construct quite satisfactory composite variables.

2.2b   Validity and Reliability.   Regardless of the procedure by which a measurement is taken, the issue of quality of the procedure has to be resolved before statistical manipulation is useful or interpretable. The quality of measures is usually phrased in terms of two questions: (a) Is the procedure valid? and (b) Is the procedure reliable? **Validity** is usually taken to mean the extent to which a measurement process is able to make distinctions based (only) on the variable one intends to measure. Measurement procedures, for example, which measured family size by recording the number of people in a dwelling unit would be less valid to the extent that boarders, visitors, and domestic help are not identified and eliminated from the count.

There are several ways by which validity of a procedure can be checked. Generally these involve the use of some criterion measure and comparison of scores on this and the new measurement procedure. Statistical tools help in making these contrasts and comparisons.

The problem of **reliability** refers to the stability of the measurement process itself when applied under standard conditions. If there is a good deal of sloppiness in the measurement, then to that extent the measurement is unreliable. A classic example of an unreliable measurement instrument would be an elastic measuring tape which would be likely not to give the same reading upon repeated measurement of the same object (where the object itself is known not to have changed). One would have to separate simple change in the phenomenon under study from unreliability of a measurement instrument, of course, and again statistical procedures (often correlation coefficients, discussed in Chapter 7, or path coefficients, discussed in Chapter 9) are available to help in the evaluation of reliability.

Validity and reliability are linked. To the extent that a measurement is unreliable, it can not be said to be a valid measure of an intended variable. Again this relationship can be expressed in terms of statistical descriptions that will be developed in this volume. We will not develop the topic of validity and reliability at any great length here (that will be a main topic in a research methods course), but we will develop some of the elementary tools of thought which are helpful in handling these and other questions of inquiry.

2.3   SUMMARY

This chapter has dealt with the central problem of measurement. It is central because the scores we will be describing statistically are the result of a measurement process, and their meaning and quality depend upon what happens there. It is central too, because the information contained in scores helps determine what sorts of statistical summary are appropriate and how they can be interpreted. Finally, as sociologists, we are interested in making significant statements about the phenomena we find interesting, and that means we have a central concern for the ability of scores to capture the variations we intend to examine and feel are important.

The next chapter begins the more organized pursuit of statistics itself by showing the importance of comparison and the various ways comparisons may be made statistically. The next chapter will also establish some of the basic themes which serve to organize the whole field of sociological statistics.

CONCEPTS TO KNOW AND UNDERSTAND

measurement
level of measurement
  nominal
  ordinal
  interval
  ratio
scale continuity
  continuous variables
  discrete variables
role of variables in research
  dependent
  independent
  intervening
reliability

validity
inclusive
mutually exclusive
precision
grouping into classes
  class midpoint
  real limits
  measurement
  to nearest whole unit
  to last whole unit
significant digits
rounding
concept
indicator
variable construction

QUESTIONS AND PROBLEMS

1. Pick an interesting variable and define it (you may want to look up other definitions and critique them). Is it adequately defined and, if not, what difference does poor definition make, specifically? How can it actually be measured? What other variables might be used as independent or intervening variables with this variable? (Examples of variables might be alienation, urban segregation, age, social class, marital status, deviance, level of industrialization, social integration, or morale.)

2. By reviewing research articles, either in journals or as reprinted in workbook exercises for sociological statistics, you can find challenging ways to examine problems with variables.* First, you might pick out and classify variables in abstracts of research reports in the sociological literature. It usually helps to read the entire article, too. Secondly, you might at this point pick out vari-

*See H. J. Loether and D. G. McTavish, *Statistical Analysis For Sociologists: A Student Manual,* for useful exercises involving research variables.

ables and begin to specify how you would prefer to measure and relate them to hypotheses in a study of your own.

3. Identify the most likely scale of measurement (nominal, ordinal, interval, or ratio) and the continuity or discreteness of the following variables:
   a. Years of service on a job
   b. Occupational prestige
   c. Number of organizations to which a person belongs
   d. Population density of a city
   e. Home ownership (of a person)

4. Round the following numbers to whole numbers:
   a. 45.73
   b. 38.1
   c. 27.49
   d. 25.5
   e. 26.501
   f. 22.5

GENERAL REFERENCES

Lieberman, Bernhardt, *Contemporary Problems in Statistics: A Book of Readings for the Behavioral Sciences* (New York, Oxford University Press), 1971.
    A reader, this book includes in Section 1 a number of articles on measurement. The article by S. S. Stevens is included, as well as more detailed treatments of problems and possibilities in measuring.
Miller, Delbert C., *Handbook of Research Design and Social Measurement,* Third Edition (New York, David McKay), 1977.
    This book provides compact summaries of various facets of research and statistics. Part I includes an outline of steps in typical research projects and Part III includes comments on statistical analysis. The book also includes a sampling of scales used to measure sociological variables.
Mueller, John H., Karl F. Schuessler, Herbert L. Costner, *Statistical Reasoning in Sociology,* Third Edition (Boston, Houghton Mifflin), 1977.

OTHER REFERENCES

Allen, Michael Patrick, "Construction of Composite Measures by the Canonical-Factor-Regression Method," in Herbert L. Costner, editor, *Sociological Methodology 1973–1974* (San Francisco, Jossey-Bass Publishers), 1974, p 51–78.
Armor, David J., "Theta Reliability and Factor Scaling," in Herbert L. Costner, editor, *Sociological Methodology 1973–1974* (San Francisco, Jossey-Bass Publishers), 1974, p 17–50.
Bonjean, Charles M., Richard J. Hill, and S. Dale McLemore, "Continuities in Measurement, 1959–1963," *Social Forces,* 43 (March, 1965), p 532–536.

Bonjean, Charles M., Richard J. Hill, and S. Dale McLemore, *Sociological Measurement: An Inventory of Scales and Indices*, (San Francisco, Chandler Publishing Company), 1967.

Burke, Cletus J., "Measurement Scales and Statistical Models," Chapter 7 in Bernhardt Lieberman, *Contemporary Problems in Statistics: A Book of Readings for the Behavioral Sciences*, (New York, Oxford University Press), 1971.

Duncan, Otis Dudley, and Jay W. Artis, "Some Problems of Stratification Research," *Rural Sociology*, 16 (March, 1951), p 17–29.

Elder, Glen H., Jr., "Appearance and Education In Marriage Mobility," *American Sociological Review*, 34 (August, 1969), p 519–533.

Festinger, Leon, and Daniel Katz, *Research Methods in the Behavioral Sciences*, (New York, Holt, Rinehart and Winston), 1953. Chapter 11 by Clyde H. Coombs, is entitled, "Theory and Methods of Social Measurement."

Hamblin, Robert L., "Ratio Measurement for the Social Sciences," *Social Forces*, 50 (December, 1971), p 191–206.

Harman, H. H., *Modern Factor Analysis,* revised edition, (Chicago, University of Chicago Press), 1967.

Heise, David R., "Some Issues in Sociological Measurement," in Herbert L. Costner, editor, *Sociological Methodology 1973–1974* (San Francisco, Jossey-Bass Publishers), 1974, p 1–16.

Jones, Bryan D., and Richard Shorter, "The Ratio Measurement of Social Status: Some Cross-Cultural Comparisons," *Social Forces* 50 (June, 1972), p 499–511.

Labovitz, Sanford, "The Assignment of Numbers to Rank Order Categories," *American Sociological Review*, 35 (June, 1970), p 515–524.

Lazarsfeld, Paul F., "Problems in Methodology," in Paul F. Lazarsfeld, Anna K. Pasanella, and Morris Rosenberg, editors, *Continuities in the Language of Social Research* (New York, The Free Press), 1972, p 17–24.

Lazarsfeld, Paul F. and Wagner Thielens, Jr., *The Academic Mind: Social Scientists in a Time of Crisis* (New York, The Free Press), 1958.

Neal, Arthur G., and Melvin Seeman, "Organizations and Powerlessness: A Test of the Mediation Hypothesis," *American Sociological Review*, 29 (April, 1964), p 216–226.

Price, James L., *Handbook of Organizational Measurement,* (Lexington, Mass., D. C. Heath and Company), 1972.

Robinson, Richard, *Definition*, (London, Oxford University Press), 1954.

Shaw, Marvin E., and Jack M. Wright, *Scales for the Measurement of Attitudes*, (New York, McGraw-Hill Book Company), 1967.

Stevens, S. S., "On the Theory of Scales of Measurement," *Science*, 103 (1946), p 677–680.

Vargo, Louis G., Donald G. Schweitzer, Lawrence S. Mayer, and Sanford Labovitz, "Replies and Comments," *American Sociological Review*, 36 (June, 1971), p 517–522.

Webb, Eugene J., Donald T. Campbell, Richard D. Schwartz, and Lee Sechrest, *Unobtrusive Measures: Nonreactive Research In The Social Sciences*, (Chicago, Rand McNally), 1966.

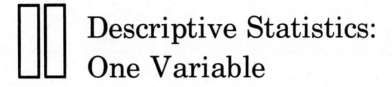

Descriptive Statistics: One Variable

3 *The Basic Logic of Valid Comparison*

Not long ago, an advertisment proclaimed that a major brand of cigarettes had 16 percent less tar and nicotine. While we probably could agree that less is better than more, the ad clearly goes on to suggest that because of this, the specific brand advertised is better than other cigarettes in this respect. Such an extension of the argument is, of course, inappropriate given only this evidence. The reason? Comparison information is not given. The 16 percent figure can not be interpreted. Does this brand have less tar and nicotine than it did formerly? Is this 16 percent less than the cigarette with the next least tar? Is it merely 16 percent less than the average of all tobacco products, or 16 percent less than the cigarette with the most tar? Even with the 16 percent drop, we do not know whether the levels of tar and nicotine are at a dangerous level or not, and there is no information which would permit one to judge the accuracy of the figure itself.

The problem is that a comparison is implied, but only half of the comparative information is given. Without some standard or something specified with which to compare the 16 percent, little can be concluded from this information.

This kind of problem confronts nearly everyone, whether sociologist or lawyer or layman. We are nearly always faced with comparative problems which call for some way to make a clear and valid contrast between things we want to compare. Is a figure of 9.9 percent of the population in the 65-and-older age bracket (as it was in the U.S. in 1970) a high or a low figure? A major university has a staff with 6.6 percent who are from minority groups. Does this indicate discrimination or not? In each case, comparative information is needed to reach a conclusion.

Appropriate comparisons are central to any inquiry, as the following illustrations suggest.

Do radio and television reports of National election results from the east coast influence voting behavior on the west coast where the polls close later? A sample of California voters was interviewed the day before and immediately after the 1964 Presidential elections. Among those voting after the eastern polls closed, [and] who were exposed to election broadcasts of eastern poll outcomes, 97 percent reported voting for Johnson, the eventual landslide winner. Among late voters who did not hear the broadcasts, 96 percent reported voting for Johnson; [this is] a comparison suggesting that eastern election victory reports had little overall effect on the later vote outcome. (Mendelsohn, 1966.)

Where does the after-tax, personal income go? In 1955, 27 percent of the total went to the tenth of the populaton with the highest incomes and 1 percent of the total went to the tenth of the population with the lowest incomes, an outcome generally predictable from social stratification theory. (Kolko, 1962.)

Presently, the birth rate in Africa is approximately 47 births per 1,000 population, compared to 38 for Asia and 18 for Europe. (Population Ref. Bur., 1971:6.)

"Do you think everyone should have the right to criticize the government, even if the criticism is damaging to our national interests?" In a 1970 national survey, 61 percent of the college graduates said "yes," 44 percent with a high school degree said "yes" and 28 percent with a grade school education said "yes." The higher the education, overall, the higher the percentage of "yes" responses. (Erskine, 1970:491.)

Among those 16 to 24 years old who were enrolled in high school or college in 1969, 6.5 percent were married. In this age bracket, 51 percent of those not enrolled in school were married. (U.S. Bur. Census, 1971: Series P-23.)

The percent illiterate in the U.S. dropped from 20 percent in 1870 to 11 percent in 1900 to 4 percent in 1930 and to 1 percent in 1969. (U.S. Bur. Census, 1971: Series P-20.)

A midwestern congressman who regularly polls some 150,000 households in his district revealed that the top-rank public concern was inflation and the cost of living, followed by concern for the Vietnam war and unemployment. The year before (1970) pollution was the number-one concern. (*Contact*, 1971.)

Does ordinal position in the family have an influence on potential scientific creativity? No, according to Lois-Ellin Datta, who studied a group of male high school seniors who competed in the 1963 Westinghouse Science Talent Search. Average creativity ratings for oldest boys was 6.1 on her scale and it was 6.0 for youngest boys. Other factors, however, influenced average rankings of both oldest and youngest boys. (Datta, 1968.)

Throughout these examples you will notice that there is a heavy emphasis upon comparison as the basis on which questions are posed and answered. These examples also indicate several ways in which comparisons can be made. At this point, two questions arise. First, for a camparison to be valid, *what* should be compared, and secondly, *how* should a comparison be made?

3.1   KINDS OF COMPARISONS

The answer to the question, "What should be compared?" depends largely on what one is interested in studying. A clearly stated question or research problem is basic to all comparisons. Without a clearly formulated problem, it is difficult to decide which of the myriad possible comparisons one should make and how it should be interpreted. Comparisons are aided if variables are carefully formulated and measured, and if objects to be measured are clearly and consistently identified so that comparable groups can be contrasted. Problems of theory, conceptualization, measurement, and research design cannot be over emphasized in statistics, although they are usually reserved as topics for a course in research methods. Here we will repeatedly call attention to these issues; you may want to read about some of them in a more systematic research methods text as well.*

In general, comparisons are made (a) *between groups* (group-group comparisons), either within the same study or in different studies, (b) *between a group and an individual* from that group (group-individual comparisons), or (c) *between study outcomes and standards* which have been established by previous research or are predictions derived from a formal model or theory the investigator is interested in testing.

3.1.1   Group-Group Comparisons

Group-group comparisons might be made between an experimental group that has been given some type of special treatment, such as a new instructional program, and a control group which has not had special treatment.† Comparisons could also be made between groups which have different characteristics but are from the same population, such as old versus young people or males versus females. Different contrasted groups might also represent different populations, such as a cross-cultural study where different societies are compared or an urban study where regions of a city (central city and suburbs) are contrasted. Which groups one compares depends to some degree, of course, on what problem one is studying. If the problem at hand is whether there is a difference in reactions to stress by men and women, then comparisons of stress-scores between sex groupings would clearly be the relevant contrast. The problem then becomes the statistical one of summarizing the stress-scores for each sex group so that the two summaries can be compared. In each instance, the groups that are compared are themselves composed of a number of scores measured on a series of definable cases or objects. Groupings could consist of measurements made on cases other than individual people, of course, such as groups of different kinds of organizations, or different kinds of dyads, or different types of encounters between people and organizations.

---

*An excellent research methods text which examines these issues is Orenstein and Phillips (1978). Also, see Riley (1963).

†Contrasts which are set up so that some subjects are assigned to an experimental group and some to control groups are usually discussed under the topic "experimental design." See Stouffer (1950) for a good reference to study design which introduces some alternatives.

3.1.2 Group-Individual Comparisons

Another kind of comparison is between a group and an individual who is a part of the group itself. Examples of individual-group comparisons would be a contrast of one person's grade in a course with the average grade for his class, or the increase in cost of living in one city compared to an overall, national average cost-of-living rise. Ranking states by the percentage of their population in institutions provides a somewhat different way to express where an individual unit (a state) falls in comparison to others in a group. The Coleman report cited in the first chapter compares test scores for specific minority groups with overall test outcomes, and in almost every grade studied, and for almost every minority group, average test scores were lower than overall, national test outcomes.

3.1.3 Study Outcome versus Standard Comparisons

Finally, there are comparisons in which data one has collected in a study are contrasted with some outside standard. Outside standards come from a number of sources. For example, one might have past data with which to compare some new information, such as comparisons between the percentage of males in a sample one has just drawn as against the percentage of males found in the latest census figures for that sampled area. Comparisons may be made between a study in a new setting as against a previous study in a similar group. Theory provides a powerful source of expectations or standards for comparison. Social stratification theory suggests that lower-class individuals will be relatively deprived (have poorer "life-chances") on many of the characteristics the society values, as compared to upper-class individuals. Population models of the transition of societies from non-industrial to industrial predict a change from high birth and death rates to low birth and death rates, and a resulting change in the age distribution and rate of growth of the society. One could compare changes in a society over time, with such model expectations to see if the theory of population transition works out as expected, or whether some modification or qualification in the theory needs to be made. Clearly, the more carefully thought out the standard and the more precisely it permits predictions to be made, the more interesting and useful it is to compare study data with the standard. This, in fact, is one of the main reasons why sociological theory is so important as a context for learning statistical reasoning.

Some time ago, Samuel Stouffer suggested the theory of "intervening opportunity," according to which the number of persons who make a move of a given distance is directly proportional to the number of opportunities at that distance, and inversely proportional to the number of intervening opportunities. Does this theory apply to the number of people who travel a given distance in the selection of marriage mates? A study could be designed in which the premarital residence of married partners is plotted on a map, and then the number of available partners of equivalent social class, ethnicity, education, and religion at that point and in intervening regions between the two partner's premarital residences is calculated. If the theory holds, the number of partners whose eventual spouses lived greater distances apart should decrease as intervening oppor-

TABLE 3.1 NUMBER OF INMATES OF CARE AND CUSTODIAL INSTITUTIONS FOR STATES CLASSIFIED AS RURAL OR URBAN, U.S. CENSUS, 1970 (IN THOUSANDS)*

Rural States	Urban States	
10	78	
8	~~1100~~	11
5	34	
124	219	
51	59	
60	106	
51	111	
37	86	
48	40	
9	9	
10	60	
22	113	
31	23	
5	12	
46	7	
14	3	
48	210	
22	4	
48		
26		
34		
31		
16	*Total Inmates*	
20	Rural States	933
33	Urban States	1185
37	All States	2118
8		
6		
3		
6		
38		
24		
2		

Source: U.S. Bur. Census, 1971: PC(V1)-1 and PC(V2)-1.
*An "urban" state is defined as one with 73.5 percent or more of its population residing in urban areas. This figure is the percent of the people in the U.S. who reside in urban areas. Other states are classified as "rural" in this example.

tunities for marital partners are increased, and there should be an increase in longer moves as the number of potential partners at a given distance increases relative to intervening marital opportunities. This is an example of comparison between a study outcome and outcomes which would be predicted from a theoretical statement.

3.1.4 An Example of Comparison Operations

Suppose we start with a small study in which we are interested in the institutionalized **population** of certain states. The institutional solution to problems of care and custody might be expected to be related to a more urbanized style of life with greater density of population, greater mobility of families, more specialization of jobs, and bureaucratization. Rural states might be expected to make less use of institutions as a solution to care and custody problems of the handicapped, homeless children, old people, and the various categories of deviants, because families and neighbors are more readily available alternatives to institutional custodial care. Given this formulation it is clear that rural-urban groups are to be *compared* on number of inmates in institutions.

Table 3.1 provides a list of 1970 data on the number of inmates of custodial and care institutions for all 50 states and the District of Columbia. Rural states are defined here as those states with less than 73.5 percent (the overall national percentage) of their people living in urban areas. The variable, "number of *inmates of institutions*" is defined as a discrete, ratio variable measured to the nearest one-thousand persons for each state on a scale that begins with zero.

First of all, before we proceed with our comparison example, we have to be alert to the possibility of error in the data. The number of inmates in the second score in the urban column is for Rhode Island, which has a total population of 946,725 so clearly it could not have over a million inmates of institutions! Upon checking, we would find the last two zeros were a clerical error. The number should have been 11 (thousands) rather than 1100 (thousands). We do not know if there was bias in the original data gathering operation. The data came from a source which is usually reliable and careful. One should keep in mind that the data are for 1970 and that this may have been an unusual year. Since the results of a comparison depend upon the quality of the original data, it is always worthwhile to check carefully for accuracy.

Although differences between rural and urban states are probably more accurately known when the actual figure for number of inmates is known than when one bases a comparison only on the impressions of an investigator who traveled around visiting institutions, the cause of comparison is still not greatly served by a list of scores such as those in Table 3.1. It is clear that without some type of ordering of even this small amount of data one could not say, in general, how much difference there is between rural and urban states. Certainly if more variables were measured or if there were more cases (*i.e.* if we measured counties rather than states), as there usually are in sociological studies, the problem of drawing a conclusion would be greatly compounded. Our first task, then, is to develop some of the basic ways to go about the task of making valid comparisons.

3.2   BASIC COMPARISON OPERATIONS

There are many procedures available for making comparisons. The field of descriptive statistics takes as one of its central problems that of the development of techniques for making valid comparisons among collections of quantitative data. In this chapter we will develop some of the basic ways data can be organized, and we will explore some of the summary indices which can be used in making comparisons. Chapter 4 will deal with graphic means for analyzing contrasts of interest to us, and later chapters will move toward more general and searching procedures for describing quantitative information in ways which facilitate valid comparisons the researcher or citizen wants to make.

There are two general operations which are usually involved in comparisons. One is the organization of scores into some convenient form or distribution, and the other is the arithmetic manipulation of scores, usually by means of subtraction and/or division. You will find that the division idea, the creation of a ratio of one number (numerator) to another (denominator), is one of the basic organizing themes of descriptive and inferential statistics. The problem is, of course, what to divide into what! The remaining part of this chapter is a discus-

TABLE 3.2 SCORES FROM TABLE ONE IN AN ARRAY, SEPARATELY FOR RURAL AND URBAN STATES

Rural States (N=33)		Urban States (N=18)
2	Lowest Score	3
3		4
5		7
5		9
6		11
6		12
8		23
8		34
9		40
10		59
10		60
14		78
16		86
20		106
22		111
22		113
24		210
26		219
31		
31		
33		
34		
37		
37		
38		
46		
48		
48		
48		
51		
51		
60		
124	Highest Score	

Source: U.S. Bur. Census, 1971: PC(V1)-1 and PC(V2)-1.

sion of these basic comparison operations, and you will find that they underlie most of the concepts in this book.

3.2.1   Organization of Scores

An unordered listing or accumulation of scores, as in Table 3.1, is relatively useless in itself for comparison purposes simply because it does not communicate the information that is there in a way that permits a focused contrast. Somewhat more helpful is an ordered set of observations (or scores), called an **array,** shown in Table 3.2. If scores are punched in hollerith cards,* an array would be created by sorting the cards into order from lowest number of inmates to the highest number of inmates. This ordering makes evident where one group starts and

---

BOX 3.1   Main Points in the Argument

The first chapter introduced some of the important concepts of the language of statistics such as "variable," "statistical observation," and "sample." Chapter 2 emphasized the importance of knowing something about the scores one has generated from a measurement process; also, the distinctions between "level of measurement," "scale continuity," and "independent" and "dependent" variables were discussed. At that point we were ready to make use of these ideas in sociological research.

The first part of this chapter emphasizes the point that almost all research involves some kind of comparison (three general kinds of comparison were noted). Now, the problem is how to set up a valid comparison. We have just introduced the first (relatively simple-minded) approach to comparison — make a list of scores. Next, we will talk about the very *important idea of creating distributions* of scores (three general kinds of distributions will be discussed plus some of the problems associated with their creation and meaning). Finally, in this chapter, we will discuss several kinds of ratios which are very valuable statistical tools for summarizing data and making valid comparisons.

The next two chapters will continue the theme started here. Chapter 4 will discuss graphic means for making comparisons, an all-too-frequently underrated approach, and Chapter 5 will discuss ways to summarize whole distributions of scores by using some selected index numbers. Chapter 5 depends rather heavily upon your understanding of the notion of "distributions," which we are about to begin here.

---

*A hollerith card is the standard punched card used in computer analysis. It has 80 columns, numbered from left to right, and each column has 12 possible punch positions. These are used for indicating the numbers 0 through 9, and letters are coded by punching two places in a single column. Some special characters are coded with three punches in a single column. The "hollerith" card got its name from Herman Hollerith (1860–1929) who developed a punch-card system used for vital statistics in Baltimore in 1887 and in the 1890 U.S. Census. Work in punch card processing can be pursued further in a statistics workbook, (such as Loether and McTavish, *Statistical Analysis for Sociologists: A Student Manual*).

BOX 3.2 SYMBOL NOTATION IN THIS BOOK

At this point a number of useful symbols are being introduced to simplify some often-used statistical expressions. Your attention is called to the following frequently used symbols.

N refers to sample size (the number of cases). An example of its use is in Table 3.3c.

X_i refers to a single score, the ith score. In grouped distributions it refers to the midpoint of a class, and these two uses will be clear from the context in which they are used. Sometimes, for simplicity, the subscript is dropped and only X is used.

f_i refers to the frequency in the ith category of a frequency distribution.

$\sum_{i=1}^{N}$ this is the *summation* operator. It instructs one to add up whatever is listed after the symbol. *Limits* of the summation operator are given below and above Σ (the capital Greek letter Sigma). In the instance shown above, the limits are where i equals one and where i equals N.

$\sum_{i=1}^{k} f_i = N$ merely instructs one to add up frequency of cases in each of the k categories of a distribution, starting with the first category $(i = 1)$ and stopping after all k category frequencies have been summed. This is a more elaborate way of symbolizing what is meant by N.

$\sum_{i=1}^{N} X_i$ would symbolize the operation of adding up raw scores (X_i), starting at the first score in a list $(i = 1)$ and ending when all N scores had been summed. Sometimes, in statistical notation, the limits of the summation operator are omitted because they almost always require addition over *all* the N scores. We will generally follow this practice, but we will include limits on the summation operator when they seem needed for clarity.

---

stops as compared to other groups, and by close examination one might be able to discern the general concentration of scores in the array. One could, for example, locate the score in the middle (later we will refer to this score as the 50th percentile or median) and gain some idea of the distance between the lowest and highest scores (later we will refer to this as the crude range). If the scores are on a nominal variable, simple clustering of like scores together would be analogous to an array of ordinal or interval/ratio scores.

At this point the process of organizing data for the purpose of comparison has to stop, *unless* we are prepared to lose or de-emphasize some detail in order to highlight or present some interesting characteristic. This is, in fact, usually a desirable step to take, since we usually want to make a comparison of scores in some particular respect rather than compare all of the raw scores with each other in various groups. In fact, one of the keys to valid comparison is to be selective in eliminating from the comparison those things which are distracting or contaminating as far as the comparison of interest is concerned. One of the important points beginning investigators sometimes forget is that irrelevant information, no matter how precise and accurate, is not only *not* positively worthwhile, but it is often a distinct handicap to making clear comparisons and drawing correct conclusions.

3.2.2   Grouping Scores into Classes

One of the procedures by which raw scores are further organized is that of creating a series of categories and then grouping the scores into these categories. Sometimes very little information is lost in the process. For example, if marital status were measured (Table 3.3(a) lists scores on marital status for 16 respondents) and a category system such as that in Table 3.3(b) were created, no information is lost by classifying the raw data in 3.3(a) into these categories. Clearly the "4" indicates that there were four subjects with the same "score"; that is, they

TABLE 3.3   CREATING A DISTRIBUTION OF DATA: AN EXAMPLE

(a) Raw Data on a marital status question asked of 16 adults.

Married	Single	No response
Single	Separated	Separated
Single	Divorced	Married
Divorced	Widowed	Married
Single	Single	Separated
Married		

(b) Frequency Distribution of the scores from 3.3(a) above.

Score Categories	Frequency of Cases in Each Score Category
Single	5
Married	4
Widowed	1
Separated	3
Divorced	2
No Response	1
Total Cases	16

(c) Grouped Frequency Distribution (Widowed, Separated and Divorced from Table 3.3(b) above have been grouped into a single category called "other").

X_i	f_i
Single	5
Married	4
Other	6
No Response	1
	$N = 16$

were each married. If we group the categories of marital status further, such as in Table 3.3(c), we have lost some information about whether some of the six respondents in the "other" category are widowed, divorced or separated. All we know from 3.3(c) is that there were six who were either divorced or widowed or separated. The organization of scores in (b) and (c) are called *frequency distributions*, a topic discussed in the next section. First, a note on "non-response."

The "non-response" category poses some problems. Clearly it is not a marital status, but just as clearly it is a type of result of the measurement process involved in measuring marital status. Usually, non-response is considered to reflect on the quality of the measurement procedures (and thus, most investigators take great pains to keep the rate of non-response very low and double-check possible biases which non-responses may create in the analysis of data), and non-responses would be reported in a footnote to a table rather than being included as a separate category of the variable being statistically summarized. On the other hand, one could be interested in studying non-responses, in which case this category would be included in statistical analysis. The fact that some groups (*e.g.* older people) are more likely than others to respond with "no response," is one of the factors a thoughtful investigator considers carefully in making statistical summaries.

3.2.3 Distributions

There are three types of distributions which are commonly used to organize data further. These are: (a) a frequency distribution, (b) a percentage distribution, and (c) a cumulative frequency (or percentage) distribution.

3.2.3a Frequency Distributions. In creating **frequency distributions,** we first list score categories and then tally up the number of cases which fall in each category.* Diagram 3.1 below lists some of the main rules for creating a frequency distribution. Any one of the frequencies in the "f" column is called a **cell frequency,** or category frequency, and is symbolized by the small letter f or f_i where the subscript i refers to the i-th class of the variable. The total number of cases is simply the sum of the frequency column and is symbolized by the letter N (or Σf_i).

While category labels for nominal variables are quite explicit (as in Table 3.3(b), for example) those for continuous variables need further explanation. Categories for the age distribution below, for example, are expressed in terms of a unit of measurement, in this case, "one-year" units. If age is measured to the

---

*A "tally" is created by putting a mark beside the category into which a score is to be classified, for each score, as one moves through a list of raw scores. It is usually a good idea to move systematically through a list of scores rather than jump around in an attempt to count all scores of a certain value, since this usually leads to fewer clerical errors. The following would be a tally of scores from which Table 3.3b was created.

    卌 Single
    IIII Married
    I Widowed
    III Separated
    II Divorced
    I No Response

AGE OF STUDENTS IN A CLASS

Age (Nearest Year)	Real Limits	f_i
11	10.5 – 11.5	3
12	11.5 – 12.5	6
13	12.5 – 13.5	9
14	13.5 – 14.5	4
		$N = 22$

nearest whole year, however, the three people in the 11-year-old class could have had ages between 10.5 and 11.5. We do not know what their exact age is but we know, if this has been accurately done, that they are nearer to 11 years old than they are to 12 or 10. The **real limits** (also called class boundaries and symbolized by L_i for lower real limit and U_i for upper real limit) of category 11 are 10.5 and 11.5, as discussed in Chapter 2. Eleven is, thus, the **midpoint** (symbolized by X_i) of a category which potentially covers a range of scores. For the purposes of further summary work, we use the value of the class midpoint to stand for the scores in that class. The **class width** (also called the **interval width** and symbolized by the small letter w_i) is one-year, the difference between the real limits of that class. Each category has real limits which extend half of one unit (here the unit is one-year) above and half a unit below the expressed limits of the class.

The problem of classification of age scores is handled for the investigator by rounding rules discussed in Chapter 2.

Sometimes a frequency distribution is *grouped* into fewer categories for ease of presentation. The distribution of age, above, might be grouped as follows:

Age in Years (Expressed Limits)	f_i	Class Midpoint X_i	Real Limits
11–12	9	11.5	10.5 – 12.5
13–14	13	13.5	12.5 – 14.5
	$N = 22$		

Here again, since the unit of measurement is one year of age, and measurement is to the *nearest* whole unit (year), the real limits extend one half of one unit (half a year in this case) on either side of the *expressed* category limits. The interval width is again the difference between real limits for a class, and in this example both classes have a width of 2 years.

(3.1)

$$w_i = U_i - L_i$$
$$2 = 12.5 - 10.5$$

Notice that the wider the categories become, the greater the loss of detailed information about specific scores. Again, the scores in each category are known by their midpoints, which can be found by adding half the interval width to the lower real limit of a class, as follows:

(3.2)
$$X_i = L_i + \frac{w_i}{2}$$

The midpoint of a class is referred to by the symbol X_i where i refers to the specific category being considered. This is the same symbol used for a raw score; but notice that, when data are grouped, the most reasonable single value to use for that raw score is the midpoint of the class into which it is classified. The X_i stands for "score" — raw scores where these are available; class midpoints where they are not. The usage is always clear in context. For the first class in the grouped frequency distribution above, the midpoint is $(10.5 + 2/2 = 10.5 + 1.0 = 11.5)$.

BOX 3.3 AN EXCEPTION IN CLASS LIMITS

The important exception to this class interval procedure is where measurement is to the *last* whole unit (rather than the *nearest* whole unit), a frequently used way of measuring age. The difference this makes in category midpoints and limits is discussed in Section 2.1.3c.

While grouping is frequently helpful to summarize information, it also poses some problems, or may pose problems, depending upon how the scores are distributed within categories, and upon the number of categories that are distinguished. The frequency distribution in Table 3.4a includes a set of 17 scores classified in a six-category distribution, each category width being one unit. The third column in Table 3.4(a) (headed fX_i or f_i times X_i) shows the total of the scores within each class. In the "1" category, for example, there are ten scores, each with the value of 1 (the midpoint of that class). The total or sum of these scores would be 1 times 10, or 10. There are two scores with the value 2 in the distribution, so the total of these scores would be $2 \times 2 = 4$, and so forth. If we add up the third column we would get the total of all of the 17 scores, and this value is 40 (also symbolized as ΣfX or the sum of the f times X column). Notice that the interval width is 1 for each category in this distribution.

Now, let us group these same scores into three categories instead of six. Each category will have a width of 2 rather than 1. The cell frequencies would be combined from cell frequencies in Table 3.4a, and they would still total to 17 cases, because we are not adding cases. The total value of the 17 scores (ΣfX) in Table 3.4b appears to differ from the same data classified into six categories in 3.4a.

We are again using the class midpoint as the value which stands for each of the scores in a category. The product of the midpoint and the category frequency (fX) yields the total of the scores which fall in that class. Twelve scores with an assumed value of 1.5 is a total of 18.0, for example. The sum of the fX column is the total of all scores. Notice that this figure is not equal to 40, the sum of the fX column in the distribution in 3.4a. Finally, let us group the original six categories into only two, with a width of three years of age, shown in Table 3.4c. Here the total of the scores appears to be 46.0 rather than 43.5 or 40.0. The difference is called **grouping error,** and it occurs because the midpoints of classes in this example do not adequately represent the value of the cases which fall in each class. The midpoint 2, for the 1–3 class in Table 3.4c, includes 10 scores of 1, 2 scores of

TABLE 3.4 AN EXAMPLE OF GROUPING ERROR

(a)

X_i	f_i	fX_i	$w=1$
1	10	10	
2	2	4	
3	1	3	
4	0	0	
5	1	5	
6	3	18	
	$N=17$	$40=\Sigma fX_i$	

(b)

Class	X_i	f_i	fX_i	$w=2$
1-2	1.5	12	18.0	
3-4	3.5	1	3.5	
5-6	5.5	4	22.0	
		$N=17$	$43.5=\Sigma fX_i$	

(c)

Class	X_i	f_i	fX_i	$w=3$
1-3	2	13	26	
4-6	5	4	20	
		$N=17$	$46=\Sigma fX_i$	

2, and 1 score of 3, and simply does not reflect the value of the scores in that class very well. Ideally, the cases in a class would be distributed rather uniformly across the interval. For this reason, especially where further computation is to occur, statisticians prefer a larger number of categories (such as 10, 15, or 20, depending on their purposes) or better yet, the raw scores themselves. Where grouping occurs, categories should be set up with some care and sensitivity to the way class midpoints reflect the value of cases in the class.* Other rules for creating frequency distributions are more a matter of convenience. Some useful guidelines are listed in Diagram 3.1.

The frequency distribution is exceedingly helpful in summarizing scores, partly because one can gain a sense of the way scores are spread over a scale, and partly because this distribution is often used as a basis for computing other summary measures. Later in this book we will avoid using grouped distributions as a basis for presenting other summary statistics, and instead will give only the raw-score form of these summary measures. It should be clear from the discussion above that we could readily convert raw-score formulas to formulas for grouped

---

*It probably goes without saying that statistical tools which have power to summarize data to facilitate comparison also are capable of summarizing errors and distorting comparisons if they are not appropriately used. The grouping error shown in Table 3.4, for example, comes about in spite of the fact that there are no technical errors in creating those frequency distributions. If one wanted to over-emphasize the total scores one could, of course, present only Table 3.4c, thereby misrepresenting the data and misleading the reader. Improper presentation of data can come out of inexperience as well as intention. A very nice presentation of some of the ways people inadvertently or intentionally give a distorted picture is in the book *How To Lie With Statistics* by Huff (1954).

data if we remember that class midpoints stand for each score in a class, and so instead of summing up the scores themselves we have to multiply class midpoints by class frequencies before summing. The reason we are going to drop grouped formulae for the most part is that the logic behind statistical reasoning can be presented quite nicely with raw scores, and most investigators and students have access to computers where programs typically accept raw-score rather than grouped data.

Table 3.5 presents frequency distributions of the data in our previous example in Table 3.1 on number of institutional inmates in various states. Notice that the frequency distributions for rural and urban states shown in Table 3.5

TABLE 3.5 DISTRIBUTIONS OF STATES ON NUMBER OF INMATES OF CARE AND CUSTODIAL INSTITUTIONS BY RURAL AND URBAN STATES, U.S. CENSUS, 1970, NUMBERS IN 1000's

(a) *Rural States*

Number of Inmates in Thousands	Distributions			
	Frequency	Percent	Cumulative Frequency	Cumulative Percent
0–24	17	51.5	17	51.5
25–49	12	36.4	29	87.9
50–74	3	9.1	32	97.0
75–99	0	0.0	32	97.0
100–124	1	3.0	33	100.0
125–149	0	0	33	100.0
150–174	0	0	33	100.0
175–199	0	0	33	100.0
200–224	0	0	33	100.0
Totals	33	100.0 (33)		

(b) *Urban States*

Number of Inmates in Thousands	Distributions			
	Frequency	Percent	Cumulative Frequency	Cumulative Percent
0–24	7	38.9	7	38.9
25–49	2	11.1	9	50.0
50–74	2	11.1	11	61.1
75–99	2	11.1	13	72.2
100–124	3	16.7	16	88.9
125–149	0	0	16	88.9
150–174	0	0	16	88.9
175–199	0	0	16	88.9
200–224	2	11.1	18	100.0
Totals	18	100.0 (18)		

Source: Data from Table 3.1.

DIAGRAM 3.1   Creating Distributions

THE PROBLEM

To organize raw data (separate scores) into a frequency distribution or percentage distribution which accurately reflects the data.

1. *Setting Up Classes.*
   If the data are to be used for *further statistical computation* it is best to either (a) use the raw data rather than create a distribution (computers are helpful in reducing the clerical problems in handling raw scores), or (b) preserve a relatively large number of classes (such as 10 to 15 or more). If the data are for *presentation* only, then fewer classes may be used and the problem is one of setting up classes which accurately portray main differences of importance in the data. Often this takes the form of an interest in grouping classes together so that minor differences are eliminated (the jaggedness of most distributions) and a more smoothed-looking distribution is shown.

   a.) *Select the number of categories* you want to use. Usually this should be more than two and probably no more than fifteen, depending on the purpose of the distribution (see point b, below). Five to ten is a useful range. Call this k.

   b.) *Find the interval size needed.* An approximation can be found by:

   $$w = \frac{(\text{highest score}) - (\text{lowest score})}{k}$$

   and take the nearest integer. This is the interval width, w. An odd number for w is preferred because then class midpoints can be arranged so that they are integers, but often this aid to hand computation is not necessary. Widths of multiples of 1, 2, 3, 5, or 10 are common.

   c.) *Set up categories.* Start with a lower limit for the lowest class which is (a) equal or less than the smallest score, and (b) preferably (traditionally) is some multiple of the selected class width, w. For example, if the class width was 5, one might begin the lowest class with some value such as zero, 5, 10, 15, etc. Thus the lower limit of each successive class is a multiple of the class width. It is generally preferred to have classes all of the same width although sometimes this rule is broken particularly where class frequencies are very small for large parts of the scale. It is also pru-

DIAGRAM 3.1 *(Continued)*

dent practice to create categories (especially if later statistical computation is to be carried out) so that there is a specified upper and lower limit for each class including the extreme ones. For presentation purposes sometimes the extreme classes are left "open" (*e.g.* 65 and over), but this practice makes it impossible to determine a class midpoint and thus makes the computation of some of the statistics discussed in Chapter 5 impossible.

d.) *Check the categories* to see if they accurately reflect the data they are to contain. That is, does the set of class midpoints fairly accurately reflect the balance of cases in the classes? Ideally, the sum of raw scores would equal the sum of class midpoints times class frequencies for interval-level variables, where this check is appropriate. If there is gross distortion, reset class limits, change (increase, usually) the number of classes, or change (decrease, usually) the class width.

2. *Group Cases Into Classes.*

If a computer or sorter is not available, the easiest and most accurate way to classify scores in categories is to create a "tally." Be sure to move systematically through the list of scores, tallying each score by making a tally-mark in the appropriate category rather than skipping around through the data to find identical scores. When you have tallied a score put a light line through it to show that it has already been tallied.

3. *Creating The Desired Kind of Distribution.*

Most kinds of distributions are created from an initial frequency distribution, so the frequency distribution should be totaled and checked for accuracy.

a.) *Percentage distribution.* Divide each class frequency in a frequency distribution by the total number of scores, and multiply each result by 100. When this kind of distribution is presented, it is good practice to include N, the total number of scores upon which the percentage is based, in brackets or parentheses below the total of the percentages. See Table 3.5.

b.) *Cumulative distributions.* Beginning with the lowest-value score category (traditionally), create a cumulative column by entering the class frequency of the lowest class next to that class, the total of the frequencies for the lowest two classes next to the second-lowest class, etc. In the case of a cumulative percentage distribution, the percentage distribution entries are accumulated. The last class (highest score category) should have N or 100 percent entered next to it. See Table 3.5.

allow us to get a much better picture of the distribution of scores than did the raw data in Table 3.1 or the array in Table 3.2. Here it is apparent that the urban states are less concentrated in the low-number-of-inmates categories, for example.*

3.2.3b Percentage Distributions. In creating **percentage distributions,** we simply divide each cell frequency by the total number of cases and multiply by 100.† Thus, the 0–24 category in Table 3.5(a) with a frequency of 17 out of 33 corresponds to a percentage:

(3.3)

$$\text{Percent} = \frac{f_i}{N}(100)$$

$$= \frac{17}{33}(100) = 51.5\%$$

The total of the percentage distribution should equal 100.0 percent except for very minor differences due to rounding.‡

While the frequency distribution permitted a somewhat clearer summary and contrast between urban and rural states than the original raw scores or the arrays, the percentage distribution contributes a great deal more to the ease and validity of comparison because it removes an important source of possible error. Notice that the percentage is a simple ratio, one which, incidentally, is easily understood in our culture since we tend to think in terms of parts of 100. Significantly, it takes account of the comparison problem of different totals, converting each total to 100. While 17 and 7 in Table 3.5a and 3.5b were not directly comparable because there were, after all, different numbers of rural and urban states, 51.5% and 38.9% are comparable because these are expressed as parts of the same total, namely 100 percent. One could, then, compare percentage distributions even though there were differing total frequencies, because this distorting feature of a distribution has been eliminated by computing percentages. Many comparisons that you will be making will be in terms of percentages, and you will get the feeling that they convey a good deal of useful information for making valid comparisons.

Table 3.6 presents data on reasons given by men aged 18 to 64 for a recent

*While the number of inmates is, as defined, a discrete variable, we will not treat these data differently from a continuous variable for two reasons. First, the difference in the kinds of descriptions we will be making is often of less importance than other distinctions we have to make, and, second, the results are generally identical. For example, if we treated the real limits of the first class in the distribution in Table 3.5 as extending from −0.5 to +24.5, it would be apparent that the category width is indeed 25. This is the same result one would get if one counted the interval widths in thousands. The class midpoint is −0.5 + ½(25) = 12.0, and so forth. There are points in statistics where different computations would be implied by differences between discrete and continuous variables. Here it is important because it is one aspect of the meaning of a score and it will be sufficient to keep its meaning in mind in interpreting computations.
†A *proportion* is simply a cell frequency divided by the total number of cases in a distribution. It is sometimes used instead of a percentage (which is the same ratio multiplied by 100). The chief virtues of percentages are cultural. We are familiar with parts of 100 and, in fact, these are more often whole numbers rather than small decimal fractions. Both proportions and percentages convey the same information about a distribution, however.
‡Some investigators make slight adjustments in the distribution so that the total actually comes out to 100.0 percent, and others let the total equal what it will, with a footnote indicating that the small difference (if any) is due to rounding errors. Rounding errors should be very small, otherwise one would suspect some computational error has been made.

TABLE 3.6 Percentage Distribution of Reasons Given by 6,292,000 Men Aged 18–64 for a Recent Change in Residence within the Same County, 1963

Reasons	Percentage
To take a job	2.7
To look for work	1.0
Job transfer	0.4
Commuting or Armed Forces	7.5
Housing	60.4
Change in Marital Status	11.0
Join or move with family	8.1
Health	1.1
Other reasons	7.9
	100.1
	(6,839,000)

Source: U.S. Bur. Census, 1966.

change of residence which was within the same county. It would probably not be hard to find someone who would claim, on the basis of these data, that 2.7 percent of the 6,292,000 men whose responses are summarized in the table moved to take a job. This is, of course, an error. The base of the percentages in Table 3.6 is not individual *men* who moved, but the 6,839,000 *reasons* that these 6,292,000 men gave for making the move. Some people gave more than one reason. The correct interpretation of the percentage would be that 2.7 percent of the reasons that were given for moving within the same county were to take a job. A good caution in reading a percentage table is to find out what 100 percent (the base or denominator of the percent) refers to. This total is supplied in parentheses below the total of the percentage column.

 3.2.3c Cumulative Distributions. Table 3.5 also presents two **cumulative distributions,** one for *frequencies* and one for *percentages.** A cumulative distribution is formed by indicating for each category, the number (or percentage) of cases which fall below the upper real limit of that class.† It is the sum of the frequency (percentage) in a given class and the frequencies (percentages) in all of the classes below that class. Thus for rural states in Table 3.5(a), the cumulative frequency of 29 means that by the upper real limit of the "25–49" class (that is, by the time one reaches the score of 49.5), 29 states have been categorized out of a total of 33 states. By the time one reaches the top of the "112" class, (using the midpoint to identify the class), all 33 cases have been classified and all of the cases have also been classified below the upper real limit of the 212 class. The same

*Just as frequency distributions are symbolized by the small letter *f*, percent and cumulative distributions have traditional labels. Percent distributions are labelled "%" or "Pct," cumulative distributions are labeled "cum %" or "cum *f*." Some authors use capital *F* to stand for a cumulative frequency distribution, but the more explicit "cum*f*" will be used in this text.

†In this book, we will adopt the convention of creating cumulative distributions by starting to accumulate from the low scores and accumulating so that *N* or 100% is by the high-score category. Here the low-score category is 0–24 (the smallest number of inmates) and the high-score category is 200–224 (the largest number of inmates). It has nothing to do with the geography of the printed page. In some cases the small scores will be printed at the bottom of the table and in other cases (as in Table 3.5) the small scores are printed at the top. This convention merely saves presentation of alternative formulae where accumulation is from high to low scores.

type of expression applies to the cumulative percentage distribution. By the upper real limit of the "25–49" class, 87.9 percent of the cases have been classified, or, 87.9 percent of the cases fall below the upper real limit of that class.

Cumulative distributions are useful for comparison if one wants to compare the way cases are spread across a scale. For example, for rural states, 97 percent have numbers of inmates below 99.5 (thousand) while for urban states, only 72.2 percent have numbers of inmates below 99.5 (thousand).

3.2.3d Percentiles. The smallest score below which a given percentage of cases falls is referred to as **percentile.** The smallest score below which 100 percent of the cases falls, for example, is known as the 100th percentile. For urban states this value is 224.5 (thousands) of inmates. One could be interested in any percentile, of course: for rural states the 51.5 percentile is the score 24.5 (thousands) of inmates; the 50th percentile for urban states is 49.5 (thousands). That means that half of the urban states have number-of-inmate scores below 49.5 (thousands) and half above this score. The 50th percentile is a frequently used percentile in statistics (later to be called the *median*). Frequent reference is also made in research to the 10th, 25th, 33rd, 66th, 75th, 90th, 95th and 99th percentiles.* Results of tests, such as the College Board Exam or the Graduate Record Exam, are reported in terms of percentiles—the percentage of individuals in some defined test-norm group, which have scores equal to or less than a particular person's score. Diagram 3.2 illustrates and explains how any percentile can be computed from a frequency distribution.

3.2.4 Other Uses of Ratios for Comparison

The percentage is a good example of the use of a ratio to create numbers which can be compared more validly. It takes account of the total number of cases, converting it to 100 for each distribution, and because of that, percentages can be directly compared without the distortion created by differing numbers of cases in the different distributions. There are other situations where ratios can be used and where other sources of distortion are controlled to provide a basis for a comparison. A classic example is the sex ratio.

The *sex ratio* is usually computed as a ratio of the number of men in a group divided by the number of women in that same group. Because the resulting number is a decimal, it is traditionally multiplied by 100 so that it reflects the number of males in the population per 100 females (rather than the number of males per one female).† Table 3.7 shows the sex ratio for the U.S. over three census years for the whole population and for various age groups. Although slightly more males than females are born, this ratio reduces to nearly an even split under 45 years of age, and decreases dramatically in the older years as women outlive men. From other information we know that the life expectancy at birth for males

---

*It should be noted that any "fractile" could be computed using the general logic shown here. Percentiles (100ths), deciles (10ths), and quartiles (4ths) are only the most frequently used.
†William Petersen, in his book *Population* (1961), notes that the European tradition is to define the sex ratio as the number of females per 100 males. More recently, the U.S. and European tradition is to express the sex ratio per 1000 rather than per 100, thus eliminating a decimal position. A number of other ratios useful in the field of demography are defined and used in this book.

TABLE 3.7 Sᴇx Rᴀᴛɪᴏs ʙʏ Aɢᴇ Gʀᴏᴜᴘ, U.S. 1950–1970

| | Males Per 100 Females | | |
Age	1950	1960	1970
All ages	98.7	97.1	94.8
Under 65	99.6	98.6	97.7
At Birth	105.8	105.5	104.2
Under 45	99.4	99.5	99.5
45–64	100.2	95.7	91.7
65 and Over	89.7	82.8	72.2
65–74	93.0	87.0	77.6
75 and over	82.7	75.1	64.0

Source: Brotman (1970).

was 68.2 years in 1974 and 75.9 years for women. By the year 2050 one estimate by the U.S. Census is a life expectancy at birth of 71.8 years for men and 81 for women.

Another interesting ratio is the *dependency ratio,* which is designed to show the relationship between the population in the middle years (21–65) and that at the young and older years. It is computed by using a numerator which is an esti-mate of the number of dependents (*i.e.* those "culturally defined" as younger than "adult" or "probably retired," namely below either 18 or 21 and above 65) and a

TABLE 3.8 Dᴇᴘᴇɴᴅᴇɴᴄʏ Rᴀᴛɪᴏs ꜰᴏʀ Sᴇʟᴇᴄᴛᴇᴅ Cᴏᴜɴᴛʀɪᴇs ꜰᴏʀ Cᴇɴsᴜs Dᴀᴛᴇs ɪɴ ᴛʜᴇ Eᴀʀʟʏ 1960s

Columbia	98.6
Mexico	96.6
Venezuela	92.5
Taiwan	91.5
Peru	89.9
Thailand	85.1
Chile	78.6
Canada	68.8
United States	67.8
Poland	64.6
Germany	61.4
Australia	61.2
Netherlands	60.8
Portugal	59.8
Yugoslavia	59.7
France	59.1
Japan	51.1
Hungary	50.7
Sweden	50.7
Switzerland	49.6
Bulgaria	48.0

Source: Metropolitan Life, 1968. Reprinted by permission of the publisher.

60

DIAGRAM 3.2 Computing Percentiles

The 10th percentile is the *score* value below which 10 percent of the scores fall. The *i*th percentile is the *score* value below which *i* percent of the scores fall. There are two typical kinds of percentile problems one might want to solve.

a.) *Find the percentile rank of a given score.*

$$\text{Percentile Rank of a Given Score} = (100)\frac{\text{cum} f \text{ up to and including that score}}{N}$$

The percentile rank of the score "7" in the example below is:

Raw Scores

1
4
4
6
7
8
12
14

$\begin{cases}\text{Percentile Rank}\\\text{of the Score 7}\end{cases} = (100)5/8 = 62.5$
Since $N=8$ and the score 7 is the 5th one in an array of these scores.

b.) *Find the score at a given percentile rank.*

Score at a given percentile is found by:

(1.) First, multiplying the percentile by N, and

(2.) Secondly, counting up an array of scores until that numbered score is found. The score value of that score is the desired value.

If one wanted the 62nd percentile in the above example, he would multiply .62 by N which in this example is 8. The result is 4.96 or, rounded to an inte-

DIAGRAM 3.2 *(Continued)*

ger, 5. We would then begin at the first score in the array above and count up to the 5th place. That fifth score is 7 and that would be the score at the 62nd percentile.

Grouped scores pose a slight problem. Usually in finding the score at a given percentile the class containing this score is identified and then a score value in that class is found by interpolation. The following formula may be used for this purpose and it is discussed later in Chapter 5 and in Diagram 5.1 in connection with the median, which is the 50th percentile.

$$\text{Score at a given percentile} = L_p + \left[\frac{(N)\,(P) - \text{cum}f_p}{f_p}\right]w_p$$

where L_p is the lower real limit of the class containing the Pth percentile score. This is determined by inspection of the cumulative frequency or cumulative percentage distribution;

w_p is the class width of the class containing the Pth percentile;

N is the number of cases (scores);

P is the percentile of interest, expressed as a proportion;

$\text{cum}f_p$ is the cumulative frequency up to but *not* including the frequency in the class containing the Pth percentile;

f_p is the frequency in the Pth percentile class.

Percentile rank of a score or score at a given percentile may be found as well by inspection of a graph of a cumulative percentage distribution as discussed in the next chapter.

denominator which is the number of people in the middle years (*i.e.* culturally defined as in their "productive" or "self-supportive" years), and multiplied by 100. Dependency ratios are shown in Table 3.8 for several different countries for the early to middle 1960's, again permitting a comparison across societies. The next question, of course, is why. Why do the ratios differ from country to country as they do?

Turning to our earlier example of the number of inmates for different states, suppose we wanted to compare New Hampshire's 7,917 inmates and Massachusetts' 77,752 inmates. It seems likely that a direct comparison of these two figures may be misleading if for no other reason than because Massachusetts has 5.7 million people within its borders and New Hampshire has only slightly less

TABLE 3.9 PERCENTAGE OF THE POPULATION OF STATES WHICH ARE INMATES OF CARE OR CUSTODIAL INSTITUTIONS FOR RURAL AND URBAN STATES, U.S. CENSUS, 1970

Rural States	Urban States
1.05	1.37
1.07	1.15
1.19	1.13
1.05	1.20
.98	.82
1.35	.99
1.35	1.00
1.32	.97
1.03	1.01
1.43	1.19
1.58	.88
1.45	1.01
1.38	1.05
.93	.69
.99	.67
.79	.62
.94	1.05
.86	.50
1.06	
.82	
.87	
.90	
.74	
1.04	
.90	
1.43	
1.09	
.77	
1.04	
.60	
1.12	
1.15	
.51	

Source: Data from Table 3.1, presented in the same order.

than one million. Other things being equal, one might expect a bigger state to have more inmates. We could create a ratio, then, which is a percent of the total population of a state who are inmates of care or custodial institutions. Table 3.9 shows these ratios in the same order for the rural and urban states as in Table 3.1. By inspection, it is hard to reach a definite conclusion about whether rural or urban states have a higher ratio (percentage) of their population as inmates in institutions, but it would appear that rural states have a higher percentage in institutions than urban states—a result, if true, which would be opposite to our earlier prediction. At this point we need some further means for statistical comparison between these two sets of percentages. These measures will be developed in Chapter 5.*

3.2.5 Time-Based Ratios

Some specialized ratios are especially useful for measuring the amount of some variable characteristic that occurs in a given time period or the change from one period to the next. Two of these time-based measures are the *rate* and *percentage change*.

3.2.5a Rates. One way of defining a **rate** is as a ratio of the number of events which actually occurred in a given time period to the number of such events which might potentially have occurred during the same period. Thus we have a formula which, stated in prose, looks like this:

(3.4)
$$\text{Rate} = \frac{\text{Number of events occurring during a time period}}{\substack{\text{Potential number of events which could have} \\ \text{occurred during the same time period}}}$$

The birth rate is one such use of a rate. The *crude birth rate* is simply the ratio of the number of births in a given year divided by the total number of people in a given society at the middle of that year. The midyear population is used because it is probably a good average of the number of people in the society during the year; (in times of increasing population, for example, this number would exceed the actual population early in the year but be below the population achieved by the end of the year). The total population is used as an estimate of the potential number of births in a society.

Unfortunately, the birth rate defined above does not take account of the fact that some societies have an unusually large number of pre-puberty females, males, or older people, which makes the total population a poor estimate of the potential number of births. A better measure of the extent to which a society is producing a maximum number of births would be the number of births divided by the number of women in that society, or, better yet, divided by the number of women aged 15 to, say, 45. In each instance, the improvement in the rate results from finding a more refined estimate of the number of *potential* births in a society

---

*By way of preface to Chapter 5, you could compute the summary ratio called the average (arithmetic mean). The average percentage of inmates for the 33 rural states is 1.05% and for the 18 urban states it is .96%. The average is but one of several simple ratios which greatly facilitate comparisons of groups of scores such as these.

in a given time period. Specialists in population use rates such as the *age-specific fertility rate* (defined as the number of births per 1000 women in a specific age group in a year), or the *fertility ratio* (defined as the number of children under 5 per 1000 women in the 15–45 childbearing period) for these reasons.*

Often rates, like proportions, result in small decimal values, and it is traditional to multiply the rate by some power of ten so that the result is a comfortable whole number. The death rate from cancer in the U.S. general population during the first half of 1971 was 162.4 deaths per 100,000 population. The suicide rate for the same period was 11.8 per 100,000. (Metropolitan Life, 1971).

A somewhat different use of the rate is illustrated in Table 3.10. Is there an increasing chance for people to go to college? Is the educational system retaining more people, longer than before? The *retention rate* shown in Table 3.10 is the ratio of the number of first-time college students in a given year to the number of students in the fifth grade eight years before. Of those who were in fifth grade eight years ago, how many entered college?

It should be clear by now that any summary measure is useful only if one is clear about the components which make it up. This is not so much a comment on the obscurity of summary measures as it is on the need for a clear understanding of what one wants to know and what the information permits him (her) to infer, and it is not a problem particularly unique to statistics. Some time ago, an investigator, T. N. Ferdinand, illustrated how a seeming contradiction between statements based on rates could arise if one is not rather clear about what the

TABLE 3.10 RETENTION RATE FOR FIRST-TIME COLLEGE STUDENTS PER 1000 STUDENTS STARTING AT GRADE 5 IN THE U.S.

1924	118
1926	129
1928	137
1930	148
1932	160
1934	129
1936	121
1938	—
1940	—
1942	205
1944	234
1946	283
1948	301
1950	308
1952	328
1954	343
1956	357

Source: U.S. Bur. Census, *Statistical Abstract,* 1966:112.

*An interesting example of the problem of defining and refining ratios in the area of crime statistics may be found in an article by Chiricos and Waldo, (1970). They develop ratios to measure certainty of punishment and severity of punishment plus measures of percent change, which are discussed later in this chapter.

rate is designed to represent (Ferdinand, 1967). Federal reports for some time had been reporting increases in the number and also the rate of crimes in the United States, but in Ferdinand's careful study of criminal patterns in Boston since 1849, he found declines in crime rates. How could this be so? Ferdinand explained it this way. Using a society with a population of some given size, say 1 million, for illustration, in one year, 800,000 people (80%) might live in villages and the rest might live in urban areas. It is known that villages have a lower crime rate than urban areas, so the following crime rates might hold:

RESIDENTIAL DISTRIBUTION AND CRIME RATE, YEAR ONE

Location of the Population	Percent	Crime Rate
Villages	80%	40 per 1000 population
Urban Areas	20%	100 per 1000 population

Overall crime rate for this year = 52 per 1000 population

In a following year two things could happen. First there might be a shift of population from low-rate areas (villages) to higher-rate areas (urban areas), but there might also be a decrease in crime rates in, say, the urban areas, as shown below.

RESIDENTIAL DISTRIBUTION AND CRIME RATE, YEAR TWO

Location of the Population	Percent	Crime Rate
Villages	10%	40 per 1000 population
Urban Areas	90%	60 per 1000 population

Overall crime rate = 58 per 1000 population

The result of these two shifts, a drop in rate in urban areas and a shift of population to the higher-rate areas, results in an overall increase in crime rate. Rates of sub-populations have gone down, but the overall rate has increased.

3.2.5b Percentage Change.   A further ratio which is sensitive to time and change, and the last of those we will discuss in this book, is called **percentage change.** It is defined as the ratio of the amount of change between two time periods to the amount at the start multiplied by 100.

(3.5)
$$\text{Percent change} = \frac{(\text{Amount at Time 2}) - (\text{Amount at Time 1})}{(\text{Amount at Time 1})}(100)$$

It is used to express the amount of change in a variable relative to the starting value of that variable. For example, the population of the United States changed (increased) by +13.3 percent between the 1960 and 1970 censuses. This is computed by using the 1960 population of 179,323,175 and the 1970 population of 203,184,772 as follows:

$$\text{Percent change} = \frac{203,184,772 - 179,323,175}{179,323,175}(100) = +13.3\%$$

The logic of this ratio is that the amount of change, the difference in the numerator, should take account of the size at the starting point. It is one thing for a society of 179,323,175 to increase in ten years by 23,861,597, but it would be much more startling if a society starting with only 40 million people made this amount of change in the same ten year period. If the change is a reduction, then the percentage change will turn out to be negative.

There are situations where it might make more sense to measure change in terms of an ending point rather than in terms of change from a starting point. Hans Zeisel points out some of these situations in his excellent book, *Say It With Figures* (1957). If, for example, we were interested in measuring percentage change in a student's grade-point average (GPA) from his freshman to senior year, we might reason that it is harder to make an increase of 1.0 overall if the student started out with a 3.0 as a freshman than if he started out with a 1.5 GPA as a freshman. In the latter case, several A's might make the change up to 2.5, but no amount of A work would be sufficient to raise the 3.0 to an overall GPA of 4.0. Students are approaching an upper limit, and the closer one approaches the limit the more difficult it is to make any given percentage change. In such a case one might use the ending GPA rather than the beginning GPA as the figure in the denominator of the percentage change formula. Of course such a computation would have to be clearly explained, since the percent change is usually computed with the beginning amount in the denominator.

Zeisel makes a similar point for change in sales volume from one year to the next for businesses that already control most of the market versus those who are just starting in business. It may be harder for one of the largest companies to double their sales volume than it is to double the sales of a very small business.

An important area of study in demography is the study of migration, both between states and within the boundaries of states. Percentage change in the population of states or regions provides the basis for many such studies.

3.3 SUMMARY

In this chapter we have considered the basic problem of making valid comparisons in order to reach important substantive conclusions. Almost any thoughtful inquiry involves a *comparison* of one group with another, an individual with the rest of his group, or the outcome of a study with some previously established standard or prediction. Just *what* should be compared and *how* it might be compared are the major topics we have introduced here. The following chapters in descriptive statistics have as one basic theme the formulation of procedures for making meaningful and valid comparisons. Often these comparative procedures involve some form of the simple idea of a *ratio*.

Some of the procedures, such as those for creating distributions, involve organizing data in such a way that certain information contained in the data is made more apparent in order to permit comparisons to be made more readily. *Frequency, percentage,* and *cumulative* distributions were discussed in this light, as well as *tallies, arrays,* and *unordered lists* of data. Where *grouped distributions* are used, detailed knowledge of individual scores is lost, and scores within a category are referred to by their *class midpoint.* Among the problems which confront a person making a comparison is the problem of irrelevant information. Loss of irrelevant detail is often helpful in exposing the desired information that the data contain.

BOX 3.4 Some Guidelines for Making Valid Comparisons

1. *Know what you want to compare,* what variable(s) should be compared for which groups. Is the comparison to be made between groups, between an individual and a group, or between the study group and some standard (previous study, predicted value, etc.)? A good line of reasoning or theory helps immensely at this point.
2. *Have comparable measures on comparable cases.* Training observers, providing standard interview schedules, questions that have been pre-tested, or special recording devices, etc. generally help assure comparability. The same scale of measurement should generally be used on all groups to be compared and the rules for "sameness" should be such that this claim can be checked. Furthermore, cases that are measured should be the same across groups. If individuals are measured in one group, individuals should be measured in other comparison groups. In general, the sweetness of apples and the sweetness of sweethearts are not comparable.
3. *Research design* is an important topic here, too. We will not go over design possibilities, but you should consult a research methods text on design *before* you become deeply involved in your own study.
4. Where there is a shifting base, consider taking account of this in terms of a *ratio* of some type.
5. Where the features to be compared are "buried" in accurate but irrelevant information, consider some of the *data summary* operations discussed in this chapter and in the rest of the book.

A second source of problems for those who would make comparisons is the *shifting base of comparison.* The number of inmates in rural states could not be directly compared with that for urban states, (a) because the totals involved more states classified as rural and fewer states classified as urban, and (b) because the urban states had larger populations to start with than did rural states, so that on this basis alone one might expect differences between states in the number of inmates. The solution was to create and examine a number of ratios which took out the effect of different kinds of incomparability. In one instance, we computed percentages of rural (and urban) states with lower or high numbers of inmates. In another, we computed the percentage of a state's population that were inmates of institutions so that we could compare these percentages for rural and urban states. The point is that a ratio was the mechanism by which we arrived at summaries which could be properly compared only when certain distorting irrelevancies were taken into account.

There are a number of different kinds of ratios. Two time-based ratios discussed in this chapter are the *rate* and the *percentage* change.

The next chapter will introduce graphic procedures for making comparisons. You will notice that often ratios, percentages, and rates are used in a graph, and graphic analysis provides some additional analytic insight which is useful in making valid comparisons.

CONCEPTS TO KNOW AND UNDERSTAND

types of comparisons

array

grouped frequency distribution

   class

   midpoint

   cell or class frequency

   interval width

   real limits

grouping error

distributions

   frequency

   percentage

   cumulative frequency

   cumulative percentage

percentile

ratio

rate

percentage change

QUESTIONS AND PROBLEMS

1. Find an interesting article from one of the sociology journals (*American Sociological Review, American Journal of Sociology, Social Forces, Sociometry, Sociological Quarterly,* etc.) and analyze the kinds of contrasts or comparisons its line of reasoning implies. What kind of comparisons are they? If standards are involved, where do the standards come from? What kinds of values might be computed to aid in the comparison? A statistical workbook is likely to contain reprints of several journal abstracts which might be used to locate interesting articles for this problem.*

2. Using a newspaper or magazine, find illustrations of individual-group, group-group, and study-standard types of comparisons. How are these contrasts made, and how would you make them if you were the investigator on the case?

3. Find illustrations of each of the summary measures we have discussed thus far: ratios, rates, distributions, percentiles, and percentage change.

4. The following set of scores are percentages of college enrollments of students who come from outside the region where their college is located (e.g., the first college whose score is listed had 15 percent of its student body coming from outside the region where the college is located). The scores represent a 20 percent sample of all of the colleges and universities in the United States for which data were available. Using these scores (a) set up a frequency distribution with class widths of 10, (b) compute the corresponding percentage distribution, (c) compute the cumulative frequency distribution, and (d) compute the cumulative percentage distribution. What do these distributions tell you about the variable classified? 15, 1, 14, 40, 20, 30, 10, 15, 20, 50, 45, 2, 10, 30, 3, 30, 10, 70, 73, 60, 25, 40, 10, 40, 3, 20, 3, 70, 5, 30, 2, 1, 1, 1, 3, 5, 15, 33, 15, 45, 5, 20, 55, 5, 5, 25, 35, 40, 25, 5, 30, 45, 15, 14, 10, 25, 2, 2, 3, 15, 25, 5, 1, 1, 30, 10, 18, 10, 30, 5, 1, 15, 4, 1, 5, 80, 15, 50, 1, 80, 20, 1, 11, 7, 15, 70, 10, 10, 35, 25, 5, 15, 10, 3, 25, 15, 5, 1, 5, 5, 5, 80, 25, 50, 29, 4, 2, 3, 1, 16, 1, 15, 15, 20, 8, 15, 38, 5, 1, 7, 4, 56, 4, 25, 11, 7, 1, 5, 5, 50, 1, 15, 27, 4, 5, 5, 30, 10, 10, 40, 65, 14, 20, 15, 12, 10, 75, 15, 19, 32, 2, 15, 3, 10, 2, 2, 40, 5, 20, 60, 6, 15, 9, 10, 10, 11, 10, 17,

*See H. J. Loether and D. G. McTavish, *Statistical Analysis for Sociologists: A Student Manual.*

1, 23, 50, 25, 13, 1, 80, 10, 1, 30, 10, 30, 5, 5, 32, 5, 13, 5, 5, 35, 5, 15, 3, 13, 35, 38, 2, 1, 2, 20, 15, 25, 1, 7, 15, 10, 10, 10, 10, 12, 1, 15, 8, 11, 1, 5, 2, 30, 15, 25, 34, 5, 40, 5, 13, 10, 24, 18, 35, 30, 30, 5, 3, 20, 37, 20, 5 34, 15, 2, 45, 5, 7, 10, 5, 1, 15, 48, 5, 10, 25, 25 (Source: *The CBS News Almanac 1977*)

GENERAL REFERENCES

Hagood, Margaret, and Daniel Price, *Statistics for Sociologists* (New York, Holt), 1952. Chapter 6 is a useful introduction and organization of the field of descriptive statistics. Chapter 7 deals with rates, ratios, proportions, and percentage change.

Hammond, Phillip E., *Sociologists at Work* (New York, Basic Books), 1964. This book is a collection of discussions about the background of some significant sociological research by a baker's dozen of sociologists. In addition to being a very readable and useful insight into research, the book presents the lines of reasoning which result in a need for data on specific kinds of contrasts between individuals, groups, and standards.

Zeisel, Hans, *Say It With Figures* (New York, Harper and Row), 1957. This book contains some clear discussion of the idea of comparison and some of the ways ratios and percentages may be used. Chapter 1 deals with percentages, another chapter illustrates how the "don't know" category might be handled, and Chapter 5 explains how index numbers might be constructed for comparison purposes.

OTHER REFERENCES

Brotman, Herman B., "Facts and Figures on Older Americans," Administration on Aging, U.S. Department of Health, Education and Welfare, Publication No. 182, 1970. (Data were originally from U.S. Census publications.)

Chiricos, Theodore G., and Gordon P. Waldo, "Punishment and Crime: An Examination of Some Empirical Evidence," *Social Problems,* 18 (Fall, 1970) p 200–217.

Contact, a newsletter published by Congressman Joseph E. Karth, (Summer, 1971).

Datta, Lois-Ellin, "Birth Order and Potential Scientific Creativity," *Sociometry,* 31 (March, 1968), p 76–88.

Erskine, Hazel, "The Polls: Freedom of Speech," *Public Opinion Quarterly,* 34 (Fall, 1970), p 491.

Ferdinand, Theodore N., "The Criminal Patterns of Boston Since 1849," *American Journal of Sociology,* 73 (July, 1967), p 84–99.

Huff, Darrell, *How To Lie With Statistics,* (New York, Norton), 1954.

Kolko, Gabriel, *Wealth and Power in America: An Analysis of Social Class and Income Distribution,* (New York, Praeger), 1962.

Mendelsohn, Harold, "Election-Day Broadcasts and Terminal Voting Decisions," *Public Opinion Quarterly,* 30 (Summer, 1966), p 212–225.

Metropolitan Life Insurance Company, *Statistical Bulletin,* December, 1968.

Metropolitan Life Insurance Company, *Statistical Bulletin,* September, 1971.

Orenstein, Alan, and William R. F. Phillips, *Understanding Social Research: An Introduction* (Boston, Allyn and Bacon), 1978.

Petersen, William, *Population,* (New York, Macmillan), 1961, (note 4, p 72).

Population Reference Bureau, *Population Bulletin,* 27 (April, 1971), p 6.

Riley, Matilda White, *Sociological Research: A Case Approach,* (New York, Harcourt, Brace and World), 1963.

Stouffer, Samuel A., "Some Observations on Study Design," *American Journal of Sociology,* 40 (January, 1950), p 355–361.

U.S. Bureau of the Census, *Statistical Abstract of the United States: 1966* (87th Edition), (Washington, D.C.), 1966.

————, *Current Population Reports,* No. 154, (Washington, D.C.).

————, *Current Population Reports,* Series P-23, No. 34, (Washington, D.C.), 1971.

————, *U.S. Census Report* PC (V1)–1 (*Final Population Counts for the 1970 U.S. Census*), (Washington, D.C.), 1971.

————, *U.S. Census Report* PC (V2)–1 (*Final Population Counts for the 1970 U.S. Census*), (Washington, D.C.), 1971.

————, *Population Characteristics,* (*Current Population Reports*), Series P-20, No. 217, (Washington, D.C.), 1971.

4 *Graphic Presentation and Analysis*

The sex ratio (number of males for every 100 females) declines steadily with increasing age not only in the United States, but in all other societies for which there is information. In the United States, slightly more males are born (104.2 males per 100 females in 1970), but by the early 20's the sex ratio begins to reverse, and for those over the age of 75, in 1970, there were only 64.0 males per 100 females.*

4.1 RATIOS—A GRAPHIC EXAMPLE

The line graph in Figure 4.1 shows sex ratios by age for each of three years, 1950, 1960 and 1966. In general the line for each of the three years shows a decline in the sex ratio with older age-groups, but past age 40 the difference between the three lines shows that there has been a rather marked drop in the sex ratio between 1950 and 1966. Why?

The graphic technique used in Figure 4.1, in this case a line graph, displays changes in the sex ratio very clearly, permitting both overall comparisons and comparisons for specific age-segments of the population. Such vivid contrasts help focus further research. Perhaps the greater drop in the mortality rate for females than for males is a factor, and this in turn might be traced to changes in medical techniques and social norms concerning births. The changing sex composition of the post-40 age brackets might also be related to the way sex roles are defined in the U.S. (*e.g.* male mortality from wars, or perhaps occupation-related

*See Table 3.7.

stress and disease). The consequences of such changes in composition would then begin to show up in data as an age cohort. A **cohort** is defined as those people experiencing a similar event in the same period or year — in this case, those *born* in the same year. (See Riley, 1973.) Such graphic comparisons also help focus research questions about possible *consequences* of an increasing sex imbalance with advancing years for such areas as housing, types of role relationships, and various types of medical and social services which may be used by older age-groups in our society. Each of these possibilities suggests further avenues of related research. Graphic presentation in this case probably communicates more relevant information more clearly and accurately than would be immediately evident if the same information were to be presented in a table.

An important aim of statistics is that of making clear the relevant kinds of information a collection of measures may contain. In a sense, statistical handling of data helps highlight the "information" contained in data so that an investigator and his (her) audience may examine it and draw appropriate conclusions in an efficient way. Graphic analysis has much to offer when it comes to clarifying complex relationships, and with the advent of computer-controlled plotting systems, there is an opportunity to exploit this means of analysis more than has been possible in the past. In this chapter we will discuss some of the basic techniques of graphic analysis and illustrate some of the more useful alternatives.*

4.2 BASIC TECHNIQUES FOR GRAPHING DISTRIBUTIONS

Histograms, polygons, ogives, and line graphs are basic graphic procedures used in statistics, and they provide a useful way to introduce some of the techniques and cautions of graphic analysis. The first three of these techniques illustrate how the distributions discussed in Chapter 3 may be handled graphically. A later section in this chapter will apply some of these techniques to other styles of graphic presentation and introduce some important and useful alternative approaches to graphic analysis.

Basic to any graphing is the idea of a **reference system** or **coordinate system.** The usual coordinate system for graphing consists of lines or "dimensions" at right angles to each other called a Cartesian Coordinate system.† The reference system in Figure 4.2 is two-dimensional but the idea can be generalized to three or more reference axes or dimensions. Traditionally, the vertical line is called the Y-axis or **ordinate** and the horizontal line is called the X-axis or **abscissa.** This divides a plane into four **quadrants.** The **origin** or zero point on both axes is the point where the axes cross and numerical scales extend outward from this point. Scores upward from the origin on Y are positive scores,

---

*For an excellent reference for graphic presentation, see Schmid, (1954), in the General References.
†René Descartes (1596–1650), the French philosopher and mathematician, is said to have devised the coordinate system named after him while he was in bed watching a fly. He could locate the fly's travels in terms of three lines perpendicular to the walls and floor. In two dimensions, on a page, points can be exactly located by two axes at right angles to each other and thus by a pair of numbers (X and Y) shown in the graph above. Descartes went on to combine graphing and algebra by expressing the location of points in terms of formulas or functions such as $Y = 3 + 2(X)$, ideas we will use in Chapter 7. Descartes published this important work as an appendix to a 1637 book on the solar system. For an interesting discussion of Descartes' work, see Asimov (1964:83–4). There are other kinds of reference systems, some of which will be introduced later in this chapter.

> **BOX 4.1** THE THEME OF COMPARISON—MAKING IT GRAPHIC
>
> Chapters 1, 2, and 3 introduced some of the language of statistics, the importance of comparisons, and some uses of the ratio as a means by which valid comparisons can be made. This chapter extends this theme to include graphic means for making comparisons, either within one collection of data or between several collections of data. Graphic techniques sometimes make use of ratios discussed in Chapter 3.
>
> There are two loose groupings of graphic techniques to be discussed here.
>
> *Group I* includes four basic statistical graphs—histogram, polygon, ogive and line graph. The first three of these are graphic representations of distributions discussed in Chapter 3 and we will use these as a basis for discussing some general techniques of graphing and some cautions.
>
> *Group II* includes a broader variety of specialized graphic techniques, each of which has important uses in graphic analysis of data in sociology. These techniques include population pyramids, pie charts, statistical maps, triangular graphs, scatter plots, and semi-logarithmic charts.

those downward are negative. Scores to the left on X are negative and to the right are positive. Since most social measurements are on scales that extend from zero in a positive direction only, quadrant one is frequently the only one needed, and the other quadrants are then simply omitted from a graph as a matter of aesthetics and convenience.

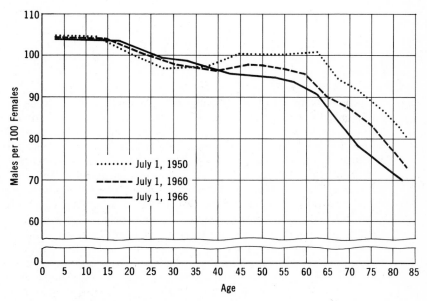

Sources: Riley and Foner, 1966; U.S. Bur. Census, 1966.

FIGURE 4.1 LINE GRAPH SHOWING THE SEX RATIO BY AGE FOR THREE YEARS, 1950, 1960 AND 1966 (Total population including Armed Forces overseas.)

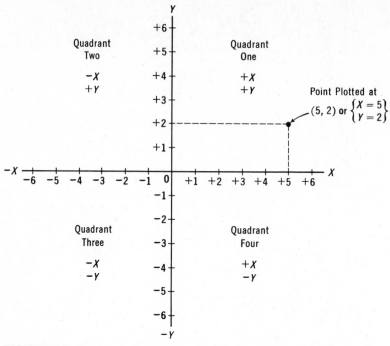

FIGURE 4.2 Cartesian Coordinate Reference System, Two Dimensional

4.2.1 Histogram

A histogram is a plot of a frequency or percentage distribution in which the frequency (or percentage) of cases in each category is represented by a bar as wide as the plotted score category on the X-axis and to a height which is proportional to the frequency (or percentage) of cases falling in that category as shown on the Y-axis. The scale of categories of the variable being plotted is laid out on the X-axis with equal physical distances corresponding to equal differences in scores. The frequency (or percentage) scale is arranged on the Y-axis, again with equal physical distances corresponding to equal frequency (or percentage) amounts. The Y-axis always begins with the origin or zero point to avoid distortions in comparing different columns, but the X-axis or score axis typically begins with any convenient low score. The objective is to create a figure with a total area corresponding to a total frequency, N (or total percent, 100%), which is distributed among the different categories of the variable on the X-axis in a way that correctly represents the relative number (percentage) of cases in each category or interval. A step-by-step discussion of the construction of a histogram is presented in Diagram 4.1.

Figure 4.3 illustrates the use of a histogram. In this case it shows the number of U.S. households in 1970 which have different numbers of persons in them. There are nearly eleven million households of size one, 18 million of size two, and so forth. The histogram clearly shows that most households are of size one, two or three. The average number of persons per household in the U.S. in 1970 is 3.17 persons.

Some variations of the basic histogram are used to reflect certain characteristics in the data. For example, if the variable is a *nominal* one, the bars in the histogram may be separated somewhat to visually carry the image that they are separate and distinct categories. This is usually called a "bar chart" or "bar graph," and the bars are drawn with equal widths.

For *ordinal* variables, where equal distances are not defined, one may choose to separate bars slightly to emphasize this fact (particularly if the variable is discrete rather than continuous). Alternatively, one might rely on the "stair-step" impression of the histogram to carry this message and still preserve the impression of rank order of categories by placing columns immediately next to each other (incorporating the lines between each column). For ordinal variables, it is customary to adopt a standard category width for plotting purposes, in spite of the fact that distances are not defined. This helps avoid distractions and apparent distortions from arbitrarily varying width. In this respect, one may want to treat ordinal variables by the same rules that apply to plotting interval-level variables.

For *interval* variables the basic procedures for plotting a histogram in Diagram 4.1 can be followed. A histogram may be particularly appropriate for discrete, interval variables. If the variable being plotted is continuous, one may prefer to turn to another type of basic graphing technique, the polygon.

A number of variations on the histogram are discussed in Section 4.3.

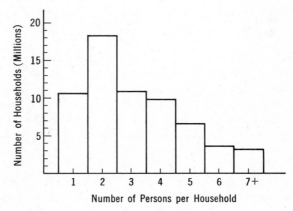

Source: U.S. Bur. Census, 1970, P-60: Table 6.

FIGURE 4.3 Histogram of Size of Household, U.S., 1970

DIAGRAM 4.1 How to Construct Basic Graphs

THE PROBLEM

Select an appropriate graphic technique and plot the following data on age of 223 grade-school children. Notice that we will illustrate all of the basic graphic techniques on the same set of data for comparative purposes. In practice, an investigator would choose one appropriate technique and *not* draw all possible graphs.

Age of Grade School Child	Frequency Distribution	Percentage Distribution	Cumulative Frequency Distribution	Cumulative Percentage Distribution
4 years	21	9.4	21	9.4
5	53	23.8	74	33.2
6	69	30.9	143	64.1
7	47	21.1	190	85.2
8	33	14.8	223	100.0
	$N = 223$	100.0%		

Step 1: Using graph paper, lay out the X- and Y-axes so that the Y-axis is three-fourths to equally as long as the X-axis. Use standard arithmetic graph paper (10 lines per inch is often a useful type of ruling to use).

Step 2: Lay out the score, or X-axis. Choose a standard size physical unit to represent a standard size score step. Start near the origin on the X-axis with a score that is one score interval lower than the lowest score category to be plotted. This permits one to leave a small space between the Y-axis and the left-hand side of the graph you plot. Notice in the examples that one year of age is represented by a fixed physical distance in a graph, and that age scores start with age 2 or 3, even though the lowest age category for which there are data is age 4. Label the X-axis and create a descriptive heading for the graph (and figure number for reference if there are several graphs).

The remaining steps depend upon the type of graph used.

HISTOGRAM

Step 3: After steps 1 and 2 above, lay off the frequency scale on the Y-axis. Begin with zero at the origin and lay off frequency intervals so that a fixed physical distance stands for a given jump in frequency. Make the height of the Y-axis correspond to somewhat more than the maximum frequency in the category with the greatest frequency of cases. Label the Y-axis clearly.

Step 4: Using a ruler, construct rectangles to a height which corresponds to the frequency in a given category as indicated on the frequency axis; (see note below). Sides of each rectangle should come down exactly at class boundaries. In this case, there are 21 cases in the age-four category, so

DIAGRAM 4.1 *(Continued)*

the height of that column is plotted at 21 and the sides of the column come down at 3.5 and 4.5, which are the class boundaries. This is done for each category, although the vertical lines may be omitted or erased, leaving only the outside lines of the enclosed figure.

Note: If a category is twice as wide as most of the others, its frequency and the height of the column should be reduced proportionately (to half) so that the area reflects the number of cases in that interval (*i.e.* so that the area of each column is proportional to the cases in that interval). For discrete, ordinal, or nominal data, one may want to separate bars somewhat to reinforce this impression of the data.

The *percentage* histogram is constructed in the same way, except that the Y-axis represents percentage points and is as tall as the largest cell percentage.

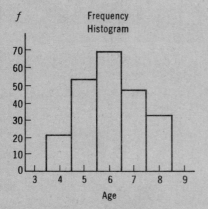
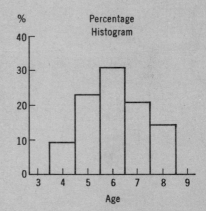

POLYGON

Step 3: After steps 1 and 2 above, lay off the Y-axis in frequency or percentage units in exactly the same way as discussed in step 3 under histograms.

Step 4: Plot the cell frequency (or percentage) as a dot at a height above the X-axis which corresponds to the frequency (or percentage) in that category as indicated on the Y-axis scale; (see note below). The point should be plotted directly above the *midpoint* of the class.

Step 5: Now plot two added points. Plot a zero at the next lower class midpoint below the lowest plot point on the X-axis. Then plot a zero at the midpoint of what would be the class just higher than the highest plotted point.

Step 6: Using a ruler, connect all of the dots so that there is a closed polygon — closed with the X-axis.

Note: If a category is wider than others, reduce the height of that

DIAGRAM 4.1 *(Continued)*

category proportionately. For example, if the category is twice as wide as others, plot the frequency (or percentage) above the midpoint of the class, but at a height which is equal to one half of the frequency (or percentage) in that class.

A percentage polygon is constructed in the same way as the frequency polygon, except that the *Y*-axis is laid out in percentages.

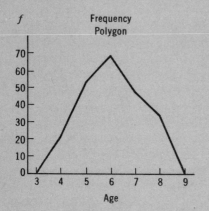

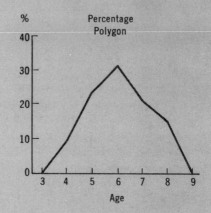

OGIVE

Step 3: After steps 1 and 2 above, lay off the frequency or percentage scale on the *Y*-axis in a way similar to that indicated in step 3 under histogram. The scale should extend up to *N*, the total number of cases (or 100 percent, in the case of a cumulative percentage scale).

Step 4: Plot a zero at the *lower* real limit of the first (lowest) score category. Then plot each successive cumulative frequency (or percentage) directly above the upper real limit of the class to which it corresponds; (see note below). A cumulative distribution shows the number (percentage) of cases below the upper real limit of each successive class up to *N* or 100 percent.

Step 5: Connect all dots with a straight line, ending at the last dot which is *N* or 100 percent. The graph should go up and to the right without any dips. A frequency in the next higher category can not be any lower than it was at the last category (although it may be the same height if there are no further cases in that next category).

Note: Always plot directly above the upper real limit of a category regardless of the widths of classes. The percentage ogive is the same as the frequency ogive, except that the *Y*-axis is laid off in percentages from zero to 100%. An added use of the percentage ogive is that it permits one readily to find any percentile (*i.e.* the score below which some percentage of the cases happen to fall).

DIAGRAM 4.1 *(Continued)*

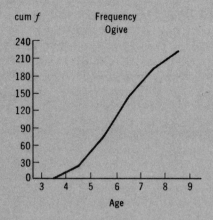

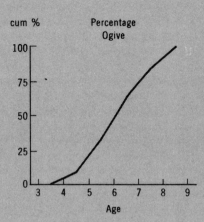

LINE GRAPHS

Step 3: The Y-axis is laid out in terms of the scale of the dependent variable that is being plotted, and the X-axis is laid out in terms of the scale of the independent variable. For example, the X-axis may represent time or age or regions of the country, etc. The Y-axis may be a sex-ratio, the number of crimes in a given year, etc. Label each axis. Again, equal physical distances correspond to equal score distances, and the Y-axis starts with zero.

Step 4: Plot with a dot, the value of the dependent variable above the appropriate category midpoint on the independent variable.

Step 5: Join the dots, using a ruler, without closing the figure with the X-axis. Where more than one line is plotted on the same graph, use a different type of line for each (dotted, solid, dot and dash, etc.). Clearly label each line.

An example of a line graph is as follows:

TOTAL BIRTHS EXPECTED PER 1000 WIVES 18–39 YEARS OLD, 1971*

Age	Births Expected per 1000 Wives
18–24	2,384
25–29	2,617
30–34	2,996
35–39	3,255

Source: U.S. Bur. Census, 1972:P-20, No. 232.

DIAGRAM 4.1 *(Continued)*

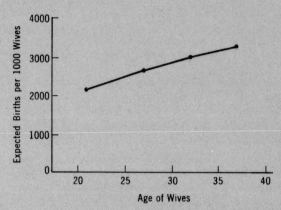

TOTAL BIRTHS EXPECTED PER 1000 WIVES 18–39 YEARS OLD, 1971*
*Data from the above table.

CAUTIONS

1. Use the three-fourths rule, creating the Y-axis about three-fourths the length of the X-axis (or equally long).
2. Always begin the frequency or percentage axis with zero at the origin. The score axis may begin with any convenient score to make an aesthetically balanced but accurate diagram.
3. Be sure that equal numeric differences are represented by equal physical distances on all scales.
4. Label the graph adequately to include all scales, the source of data, a heading explaining what is shown, and keys to explain different kinds of lines or symbols, etc.
5. To avoid confusion, do not attempt to show too many different graphs on the same axes.

4.2.2   Polygons

The frequency (or percentage) polygon is a closed figure connecting points plotted above the *midpoint* of each category at a height corresponding to the frequency (or percentage) of cases in the category. Figure 4.4 shows two percentage polygons of the age distribution of enrolled and not-enrolled people aged 3 to 34 years in the United States in October, 1971. Figure 4.4a is the age distribution for those enrolled in school and Figure 4.4b shows the age distribution for those not enrolled in school. Note that each figure is closed by extending the line down to the *X*-axis at the next midpoint beyond the extreme categories of the distribution. Diagram 4.1 presents step by step procedures for constructing a polygon.

A comparison of the two age distributions for those enrolled and those not enrolled, as in Figures 4.4a and 4.4b, shows marked differences in their form. Among those enrolled in school the distribution is concentrated around the 10–13 year ages, with smaller percentages enrolled shown on either side. For those not enrolled, the distribution is concentrated at both ends of the age range and the distribution forms a "U-shape."

The polygon gives the visual impression of more gradual shifts in frequency (or percentage) from category to category, while the histogram highlights shifts between categories. The frequency (or percentage) polygon also relies on distances between categories, and for this reason it is most appropriate in graphing interval-level variables.

Figure 4.5 shows an analytic use of the polygon to present the overlap between distributions of income for blacks and for whites in the United States in 1968. The area of overlap of the two distributions is used as one measure of *integration* which is computed by finding the percentage (or proportion as shown

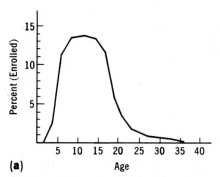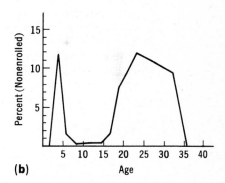

Source: U.S. Bur. Census, 1972, P-20, No. 234.

FIGURE 4.4

(a)   Percentage Polygon of the Age of 3 to 34 Year Olds Who Were Enrolled In School: U.S. October, 1971

(b)   Percentage Polygon of the Age of 3 to 34 Year Olds Who Were Not Enrolled In School: U.S. October, 1971

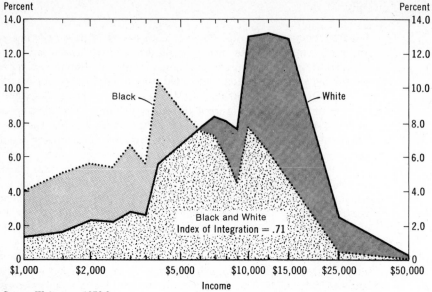

Source: Weitzman, 1970:9.

FIGURE 4.5 PERCENT DISTRIBUTION OF BLACK FAMILIES AND WHITE FAMILIES BY TOTAL MONEY INCOME; 1968.

in Figure 4.5) of overlap out of the total area under both curves (*i.e.*, the total shaded area). In this example, there is a 71 percent overlap, or the "index of integration" for blacks and whites on family income in 1968 is equal to .71. Complete segregation, where the index of integration is equal to 0.0, would be shown by a graph in which the income distributions of blacks and whites do not overlap at all. Complete integration, where the index of integration would equal 1.0, would be shown by two identical distributions exactly "overlapping" or corresponding to each other. Both polygons are plotted on the same graph for easier comparison, and essentially two bits of information are presented. Ultimately two measurements had to be made on each family, race, and income, for the graph to be prepared. The distribution of income was then plotted separately for the two racial groups.

4.2.3 Ogives

A third basic graphic technique is a plot of a cumulative distribution (either percentage or frequency). This kind of graph shows the frequency (or percentage) of cases falling below the upper boundary of each successive class.* Ogives are

---

*Note that we are accumulating from low scores to high scores for consistency throughout this book. Obviously one could accumulate from high to low, too, in which case statements like this one would have to be reversed or changed to the "lower" boundary. High to low accumulation would be useful in the analysis of waiting times in a queue, for example, where one would be interested in the percentage of people waiting a given number of minutes "or more."

used with ordinal or interval-level variables to aid in the examination of the overall shape of a distribution, and to aid in finding various percentiles. Diagram 4.1 illustrates the steps in drawing an ogive. Notice that the plot points are directly above the *upper boundary* of each class, so that the line connecting them extends from the X-axis up and to the right, ending at the total frequency (or total percentage, 100%, in the case of a percentage ogive).

Figure 4.6 illustrates one use of the ogive in sociology. Abbott Ferriss was interested in the rate of divorce and whether or not the percentage of couples who divorced within the first nine years of marriage was changing through time (Ferriss, 1970:363). He used a percentage ogive to plot the cumulative percentage of divorces among those married in the same year. That is, Ferriss gathered data on a given marriage cohort, calculating the percentage of that cohort who divorced each year for the first nine years of their marriage. These percentages were then accumulated and plotted above December 31st of each successive year of marriage. This was done for each of the annual marriage cohorts from 1949

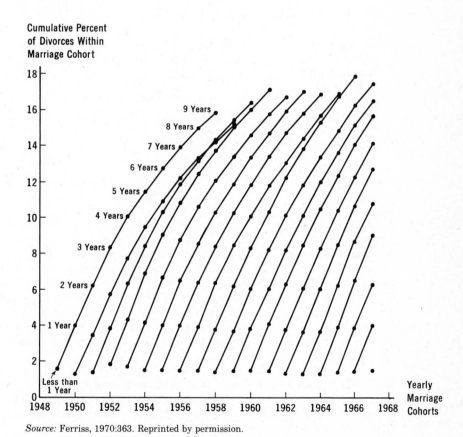

Source: Ferriss, 1970:363. Reprinted by permission.

FIGURE 4.6 Cumulative Percent of Divorces within Yearly Marriage Cohorts, by Duration of Marriage (through 9 years), 1949–1967.

to 1967. For couples married in 1949, for example, about 1.5 percent divorced in less than a year. By the end of the first year about 4 percent had divorced, and so on, until by nine years nearly 16 percent had divorced. The ogive in this case does not extend to 100 percent because the base of the percentage was all marriages of a given cohort, a more meaningful figure in this instance, rather than the final number of divorces of a given cohort (which would still be unknown at the time of Ferriss' study).

From the ogives in Figure 4.6 one can see that the percentage of married couples who divorce after a given number of years of marriage is relatively constant from cohort to cohort. There is some suggestion that the percentages might be rising for later-married cohorts (shown by taller ogives in later years) and the 1950 marriage cohort had a lower than usual percentage of divorces after the fifth year of marriage. This latter point is shown by the ogive bending to the right rather than extending upward as steeply as other cohort ogives.

BOX 4.2   THE DISTRIBUTIONS BEHIND THE GRAPHS

The basic graphic techniques discussed so far — histograms, polygons and ogives — are used to present the distributions which were discussed in Chapter 3. Do you understand the following?

Frequency distribution   (if not, see Section 3.2.3a).

Percentage distribution   (if not, see Section 3.2.3b).

Cumulative distributions   (if not, see Section 3.2.3c).

Percentiles   (if not, see Section 3.2.3d).

4.2.4   Line Graph

A fourth type of basic graph is the line graph, which shows the value of some dependent variable (scaled along the Y-axis) for each of several categories of another variable, usually an independent variable, (scaled along the X-axis). The plotted points are connected by a straight line, and the figure is not closed with the X-axis, since the area under the curve has no particular meaning as it does in the case of the frequency or percentage histogram or the polygon. Diagram 4.1 provides a step-by-step discussion of the way these graphs are constructed.

Figure 4.7 is an example of a line graph, where time is represented on the X-axis. The dependent variable, rate of marital dissolution through death, is plotted on the Y-axis, and the line extends from 1860 to 1964. This is often called a **trend line** because it is a plot of some characteristic through time. The graph shows rather clearly the overall decline in marital dissolutions through death during the century, with minor fluctuations from this trend for most years. The year 1918 stands out as a major departure from the trend. It was during 1918 that

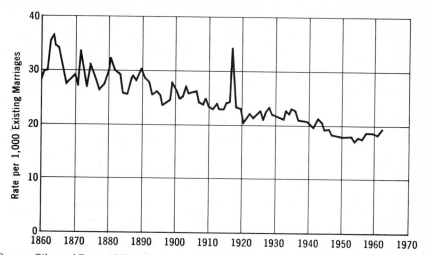

Sources: *Sources:* Riley and Foner, 1968:19; Jacobson, 1959:118; and Jacobson, 1966: Fig. 1. Used by permission.

Note: Includes Armed Forces overseas during 1917 to 1919 and 1940 to 1964; excludes civilian population of Alaska and Hawaii prior to 1960.

FIGURE 4.7 MARITAL DISSOLUTIONS BY DEATH, U.S., 1860–1964

there was a major epidemic of influenza, an epidemic which also claimed German sociologist Max Weber in 1920.

Line graphs are particularly appropriate where one of the variables (on the *X*-axis) is a continuous, interval-level variable, such as age or time. A bar graph could be used instead of a line graph for situations where the variable on the *X*-axis is nominal or ordinal. Figure 4.8 provides an example where columns represent the number of children desired for two groups of countries, those with relatively low birth rates and those with higher birth rates, as recorded in a study by W. Parker Mauldin (1965). The graphic technique clearly shows the author's conclusion that "the number of children desired by people in countries with low birth rates is generally lower than in countries with higher birth rates." Interesting, too, is the difference between the two collections of countries. Those with lower birth rates in this graph are generally considered more industrialized than those with higher birth rates.

4.2.5 DISTORTIONS IN GRAPHING

Whichever form of graph is selected, the key criterion should be the accuracy and clarity with which it communicates the aspects of the data of interest to an audience or to the investigator himself (herself). The investigator has the opportunity and a wide range of techniques for selecting the specific part of the data which is of interest in his (her) study. To communicate properly requires clear knowledge

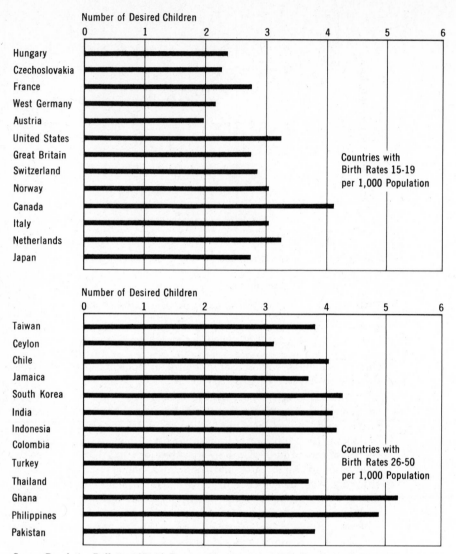

Source: Population Bulletin, 1971:20. Reprinted by permission of the Population Reference Bureau, Inc.

FIGURE 4.8   DESIRED FAMILY SIZE, SELECTED COUNTRIES

of what is there to communicate, and of possible misinterpretations and qualifications which should be taken into account, along with a healthy sensitivity for the ways people interpret and understand evidence. Clearly, too, there are vast possibilities for doing mathematically or technically correct manipulations which may intentionally or inadvertently misrepresent the state of affairs being investigated. Some cautions and rules for careful representation are given in this book, but there are always exceptional situations, and the responsibility for clear communication is not solved by mere rules for technical accuracy.

There are several ways of graphing which are numerically accurate but which give a distorted impression of the data. These are illustrated in Figure 4.9. The series of graphs, Figure 4.9a through 4.9c, illustrates the distorting effects of extending or shortening one of the axes. Although the data are accurately plotted in each graph, the impression of the steepness of change from category to category is greatly altered. A partial corrective is the three-quarters rule discussed in Diagram 4.1.

Similar in effect to the distortion involved in changing scale length is the distortion involved in not starting with zero on the frequency (or percentage) scale on the *Y*-axis in a histogram or polygon. Figure 4.1 illustrates a helpful

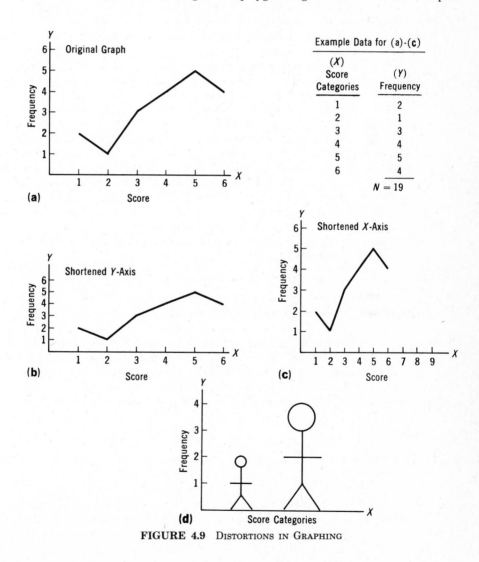

FIGURE 4.9   DISTORTIONS IN GRAPHING

procedure to use where a graph would be awkwardly tall if all values from zero were included. In Figure 4.1 sex ratios were plotted in the 70 to 110 range, and in order to have the graph large enough so that values could be read off the Y-axis, physical units on the Y-axis were made relatively large. In this case, Riley did not include all score values from zero to 70, because this would have resulted in a graph longer than the printed page (Riley and Foner, 1966). Rather, she skipped some values between zero and 60, but clearly indicated in her presentation that this had been done and the reader should beware. This warning is expressed by "breaking" the graph and the Y-axis to show that part of the table is missing. *In general, it is a poor practice to omit values on the frequency or percentage axis.*

A further potential distortion of graphs is the use of figures which differ in area or volume as well as height. In Figure 4.9, graph (d) illustrates the problem. While the stick figures are drawn so that the taller is twice as high as the shorter to indicate a doubling in frequency, the figures, if drawn aesthetically, also cover an area, and the ratio of these areas is likely to be about 1 to 4 rather than 1 to 2 as intended. It is often not clear to a reader which of these possible comparisons correctly represents the data. If a figure is a square or circle, a doubling of its height or diameter results in an area that is four times larger. If a figure in a graph is illustrated as a cube or sphere, a doubling of its diameter or height results in an eight-fold increase in volume. To solve this possible distortion, investigators generally prefer to adopt a standard-sized figure to represent a given number or percentage of cases and merely use as many of these figures as is necessary to indicate different category values. Figure 4.10 illustrates this more careful procedure. Called a **pictograph** because of its use of small pictures to represent a standard quantity of the variable being studied, Figure 4.10 shows the relationship between birth and death rates for seven world areas and for the world as a whole. Notice that each figure represents two births or deaths per 1000 population, and that fractional figures are used to represent fractional amounts.

4.3   VARIATIONS ON BASIC GRAPHIC TECHNIQUES

There are a number of other graphic techniques which are useful either for analysis of data or presentation of results or both. In this section of the chapter we will illustrate some useful variations of the basic graphing techniques we have been discussing thus far. In the following sections we will discuss techniques involving somewhat different approaches, and illustrate new applications of analytic graphing in sociology.

4.3.1   The Bar-Chart

The **bar-chart** is a variation of the histogram often used for nominal variables or when emphasis is on categories of a variable (see Figure 4.8). Figure 4.11 shows bar-charts of "place of physician visits" for four age categories (0–4 years, 5–14,

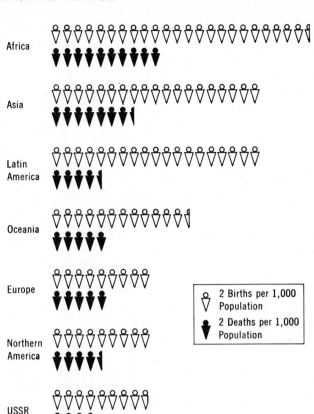

Source: Population Bulletin, 1971:6. Reprinted by permission of the Population Reference Bureau.

FIGURE 4.10 BIRTH AND DEATH RATES BY AREA, 1971

15–24 and 25 and over). Data from monthly interviews by the U.S. Public Health Service are shown for a year ending June, 1958, in the top graph and for the year 1968 in the bottom graph. In this application each bar represents 100 percent of the physician visits made by a given age group in a given year, but the bars are divided differently to show the percentage breakdown of visits by where they occurred—in the physician's office, a hospital, home, or "other" places. For ease in interpretation, percentages are also entered on each of the shaded portions of each bar. Notice that in both graphs, older age groups are more likely to visit physicians in their office, and between 1958 and 1968 there was a shift to greater relative use of the physician's office than alternative sites. Between the two years, home visits nearly disappeared, while hospital visits remained relatively con-

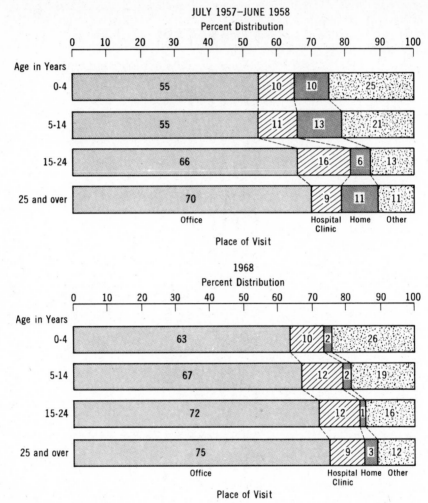

FIGURE 4.11   PERCENT DISTRIBUTION OF PHYSICIAN VISITS BY PLACE OF VISIT, ACCORDING TO AGE: JULY 1957–JUNE 1958 AND 1968.

stant both over age groups within a given year and between 1958 and 1968. Generally this form of graph is adopted to facilitate comparison of the size of separate components within several bars.

Figure 4.12 provides another illustration of a bar-chart, where again each bar is the same length, representing 100 percent of the cases in a category (*e.g.* number of crimes of a given type known to the police). In this instance, however, the bars are adjusted so that the part extending to the right of a vertical line represents the percentage of that type of crime which has been "cleared" by an

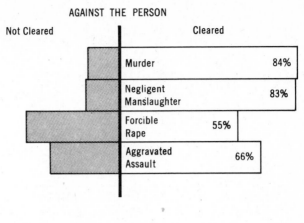

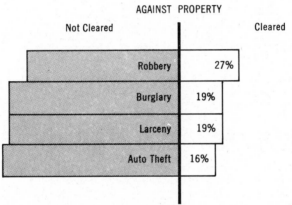

Source: U.S. Dept. Justice, 1971:32.

FIGURE 4.12   Crimes Cleared by Arrest, U.S., 1971

arrest having been made. The top graph presents a group of crimes "against persons" and the bottom graph presents similar data for crimes "against property."
It is obvious in these graphs that crimes against persons are much more likely
to be cleared by an arrest than are crimes against property — at least among those
crimes known to police.

4.3.2   The Population Pyramid

The **population pyramid** is somewhat more complex in design, while again
illustrating some simple graphing procedures discussed in connection with the
histogram. Figure 4.13 provides an illustration. Each bar in the graph represents
the percentage of the population in a specific age-sex category. The bars are

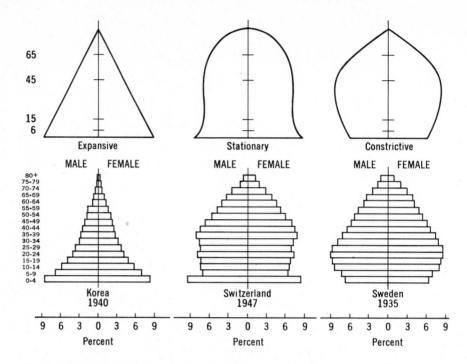

Source: Petersen, 1961:628. Reprinted by permission of the publishers, The Macmillan Company. Copyright © William Petersen, 1961.

FIGURE 4.13 THREE ANALYTIC POPULATION PYRAMIDS SHOWING DIFFERENT POPULATION STRUCTURES AND THREE ACTUAL POPULATION PYRAMIDS.

organized with those for males on one side and those for females on the other and are stacked from youngest to oldest ages. The population pyramid is used chiefly by demographers to show the overall distribution of a population and to make comparisons of the populations of different countries. Figure 4.14 shows the population pyramid for the United States in 1969 (shaded pyramid) and, on the same axis, a pyramid which is a projection of the population of the U.S. in the year 2037 under the assumption of a stable population just producing enough children to replace itself. Compared with the stable population pyramid, the 1969 pyramid shows a relatively greater percentage of younger people and a relatively smaller percentage of older people. Notice too that the imbalance of males and females in the older age brackets is clearly shown, as are the effects of the depression of the 1930's (shown by the "notches" in the 30- and 35-year bars for the 1969 pyramid), and the "baby boom" is shown by the percentage of population in the 20- and-younger age brackets. Sometimes one can see traces of the effects of disease or war on a cohort.

A similar idea has been used in the study of organizational hierarchies. Figure 4.15 shows several hypothetical and actual "pyramids" drawn symmetrically so that ratios between the number of people at different levels in an organi-

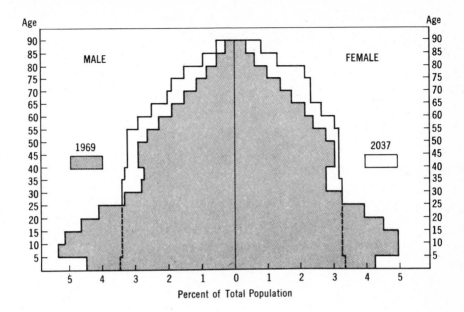

Source: U.S. Bur. Census, 1970, P-25.

FIGURE 4.14 POPULATION PYRAMIDS FOR THE U.S. IN 1969 AND FOR A STATIONARY 2037 PROJECTION; by percent of total population.

zation are accurately shown. Kaufman and Seidman (1970), whose data are shown in Figure 4.15, conclude from their study of 25 federal agencies and 4 county health departments that "line-personnel" tend to form pyramid structures with a broad base and a narrow top but "staff-personnel" tend to form an inverse pyramid-shaped organization with the base narrower than the top.

4.3.3 The Pie Chart

A **pie chart** is shown in Figure 4.16, where an entire circle represents the total of some characteristic such as the total world population. In this example, the world population at different dates is shown by larger and larger sized circles (the area representing the changing total size). The "pie" is divided into segments with different kinds of shading, where each segment represents a component of the total. In this figure, the components are four general continental population areas of the world. Each wedge has an angle at the center which represents its relative share of the 360 degrees in a circle. Each 3.6 degrees represents one percent of the total. In 1970, the Americas held about 14 percent of the world's population and its "wedge" of the pie is represented by 50.4 degrees of the circle, $(14\% \times 3.6° = 50.4°)$.

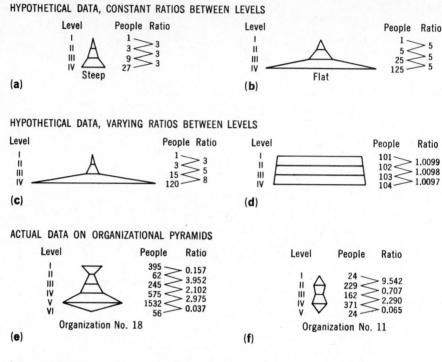

Source: Kaufman and Seidman, 1970:440. Reprinted by permission.

FIGURE 4.15 Organization Hierarchy Pyramids with Varying Ratios between Levels

4.4 OTHER APPROACHES TO GRAPHIC ANALYSIS IN SOCIOLOGY

In addition to the methods discussed above for graphing distributions, there are four further graphic techniques which we will introduce here because they offer some special advantages for the examination of data and because they suggest some of the types of contrasts an investigator may be interested in making. These techniques include statistical maps, scatter diagrams, triangular charts, and semilogarithmic graphs.

4.4.1 Statistical Maps

Statistical maps have been widely used in ecological studies in sociology and three somewhat different approaches will be presented here. A statistical map shows the distribution of a variable over a geographic area. In Figure 4.17 a base map (the outline map of some area) of Chicago was divided into community areas for which data on suicide rates had been gathered for the 1959–1963 period. Often

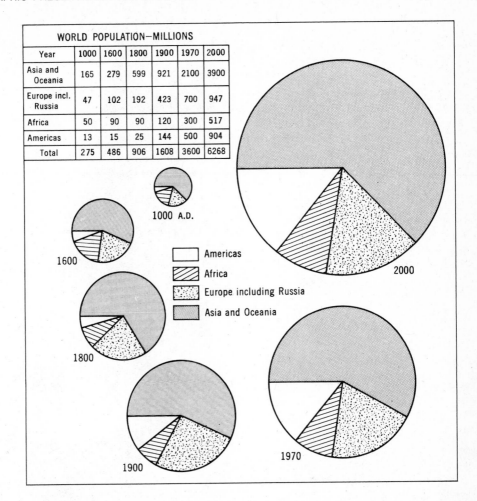

WORLD POPULATION—MILLIONS

Year	1000	1600	1800	1900	1970	2000
Asia and Oceania	165	279	599	921	2100	3900
Europe incl. Russia	47	102	192	423	700	947
Africa	50	90	90	120	300	517
Americas	13	15	25	144	500	904
Total	275	486	906	1608	3600	6268

Americas
Africa
Europe including Russia
Asia and Oceania

Source: Population Bulletin, 1971:9. Reprinted by permission of the Population Reference Bureau.

FIGURE 4.16 ONE THOUSAND YEARS OF POPULATION GROWTH

statistical maps report data by states, counties, or census tracts. Each area is shaded or "cross-hatched" in a way which illustrates the value of the variable in that area. In Figure 4.17, the more dense shading is used for higher suicide rates (12 to 23.1 per 100,000 population) and a *key* explains what each type of shading means. Usually variables with only a few categories are plotted, and shading or coloring is selected to reflect gradations of a variable by brightness or density. The suicide rate map prepared by Maris shows that rates tend to increase toward the center of the downtown area (Maris, 1969). Where other characteristics of community areas are known, clues to the pattern of suicide rates can be found, and changes over time can be studied by comparing similar maps for successive years.

96

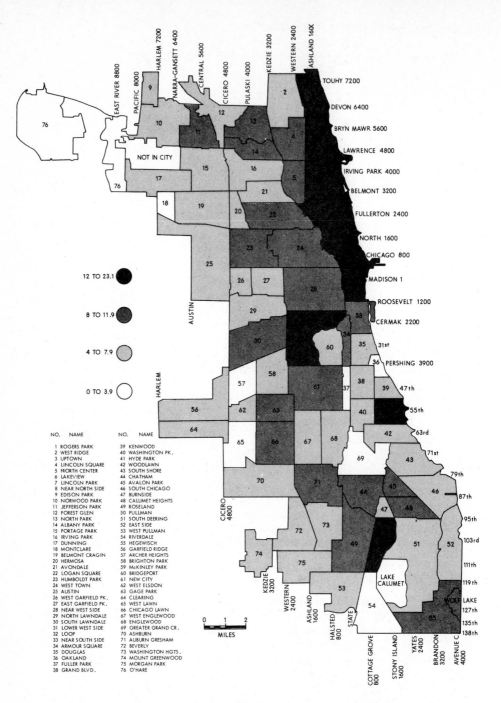

Source: Maris, 1969:139. Reprinted by permission of the author and publishers.

FIGURE 4.17 Suicide Rates by Community Areas, City of Chicago, 1959–1963

There are now a number of computer programs which permit an investigator to have his (her) data automatically plotted in some graphic form, either by using printed characters or by creating instructions which can be used on a separate plotting device or shown on a television-like computer terminal. This greatly facilitates visual inspection of data and permits an investigator to examine even more complex plots which show several variables simultaneously.* Figure 4.18 shows a statistical map of the percentage of blacks in community areas of Chicago in 1964; in this case, the map was prepared on a computer. The concentration of blacks in three areas on this map shows the geographic segregation of this group. It is interesting to note that areas with a high percentage of blacks in 1964 are generally not the areas with a high suicide rate shown in Figure 4.17.†

A second illustration of statistical maps is shown in Figures 4.19 and 4.20, which were created in a study by Thomas Donnelly, F. Stuart Chapin, Jr., and Shirley F. Weiss, showing land use in Greensboro, North Carolina (Chapin and Weiss, 1965). Instead of using existing and often irregular community areas, they divided Greensboro into arbitrary but equal-sized land areas and created a "land development" index which was divided into ninths and used to shade in the grid map. The darkest areas are the areas of greatest development as of 1960. The authors also developed a statistical model (*i.e.*, a set of equations) using scores on some variables for each of the land areas in 1948 and "simulated" the process of development of land use. That is, their equations helped them project the land use patterns in 1960 on the basis of earlier development patterns. Figure 4.20 shows this "expected" or "computed" 1960 outcome. Differences between the two statistical maps help identify factors which perhaps should be taken into account in developing a better predictive model. Models of this sort (*e.g.* voting prediction models) will not be discussed in this book, but they represent some exciting applications of both sociological theory and statistical technique. A computer-controlled plotter could have produced maps such as those in Figures 4.19 and 4.20, although in this example the maps were prepared by hand.

The third statistical mapping technique is illustrated in a study by Hoiberg and Cloyd (1971) in which they show the distribution of social class throughout a community by "isolines." These are lines drawn in such a way that they connect geographic areas with the same values of a variable, creating closed figures which do not cross or meet. Lines are drawn for each of several values of a variable and the values are usually picked at equal intervals along a scale of that

---

*Some computer-driven devices will create a three-dimensional plot by pushing small wires through a board (which represents two of the dimensions) to a height above the board which represents the value of a third dimension. The result is similar to a patch of grass that has been clipped in an uneven pattern to represent the distribution of some characteristic. Such plots might be useful in showing the distribution of social class in areas throughout a city, for example, a graphic alternative to the two-dimensional plots in Figure 4.17 and 4.18.

†This is probably a good place to caution the reader about making direct inferences about the relationship between two characteristics which describe groups, such as percentage black *vs.* suicide rates in "community areas." It would be an error (called the "ecological fallacy") to make any direct simplistic inference from these patterns for community areas such as, for example, that white people rather than black people may be more likely to commit suicide. It is entirely possible, for example, that in largely white areas it is blacks who commit suicide and in largely black areas it is the whites who commit suicide (or vice versa). The general point is that conclusions about *community areas,* generally cannot be inferred from data collected only on *individual people* (or subgroups), and conversely conclusions on individuals are not warranted from community (or group) data.

B. VARIATIONS IN THE PERCENTAGE OF POPULATION BLACK IN CHICAGO, 1964[a]

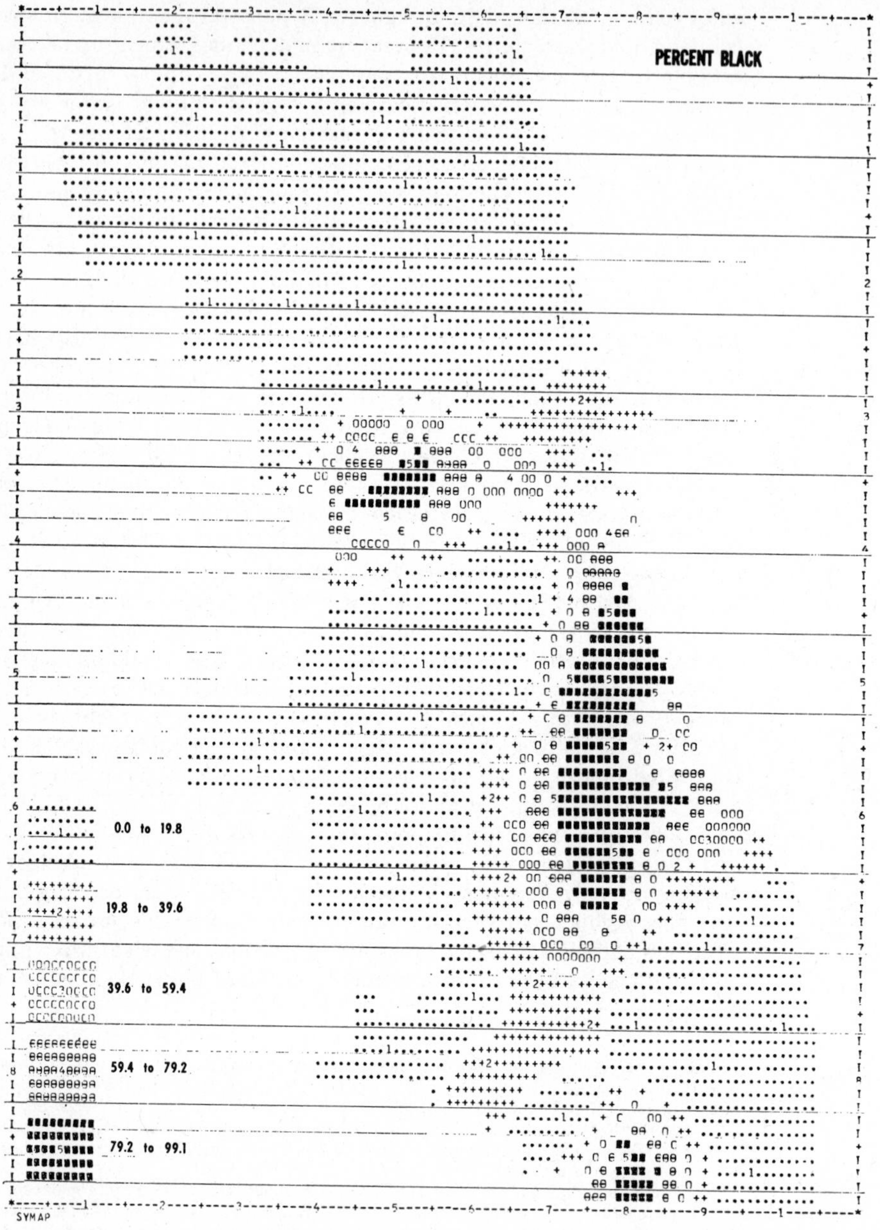

PERCENT BLACK

0.0 to 19.8

19.8 to 39.6

39.6 to 59.4

59.4 to 79.2

79.2 to 99.1

[a] Percentages were computed for each community area, and the sequence of mapping decisions then was identical to that used in preparing Figure A.

Source: Berry, 1971. Reprinted by permission.

FIGURE 4.18 VARIATIONS IN THE PERCENTAGE BLACK IN COMMUNITY AREAS OF CHICAGO IN 1964, A COMPUTER-CREATED STATISTICAL MAP variable. Isolines are like lines on a contour map showing the elevation of hills and valleys (or pressure and temperature lines on a weather map) except, in this case, they reflect not elevation but social class (or some other selected social characteristic).

Figure 4.21 was created by laying a grid over a base map of a midwestern community of some 150,000 population. Values of social class were determined for residences in the grid areas and the isolines were then drawn. The value of the social class measure through which a given line is drawn is indicated on the line itself and, in this case, larger numbers represent higher social class. Notice that the "peak" of social class appears to be in the central and southern edge of the town. Where isolines are very close together this indicates a rather steep gradient (like a cliff in physical characteristics). Again, as with other statistical maps, this map helps an investigator consider a pattern of geographic distribution for some characteristic, and his (her) comparison of this pattern with maps of other characteristics is often useful in producing insight into reasons underlying the pattern. When isolines (a more general term) refer to similar values of *ratios* or *indices* they are generally called "isopleths." More precisely, the isoline map shown in Figure 4.21 would be a map where a social class index provided the information for drawing "isopleths."

BOX 4.3 A PAUSE FOR REFLECTION

The expressed purpose of this chapter has been a discussion of different ways to present and analyze data graphically. The "hidden" purpose, which is also important, has been to provide exercise in the idea of scores and distributions of scores on a variable. You have probably seen examples of most of the graphic techniques presented thus far in the newspaper or in other texts. Most of these techniques are rather direct expressions of ideas contained in some kind of distribution. If you feel somewhat confused at this point, we suggest that you "loop back" for a quick review of distributions in Section 3.2.3 ff. All of the techniques described there are useful because they permit some of the kinds of comparisons and contrasts discussed in Section 3.1 ff. The following three analytic graphs are also part of this same theme.

4.4.2 The Scatter Diagram

A particularly useful form of analytic graph is the scatter diagram, where individual cases (cities, states, people, organizations, etc.) are plotted as points on Cartesian coordinates. In this graph there are as many dots as there are cases, and the dots are *not* joined by lines. Each dot is placed on the plot in a way that reflects the standing of an observed unit on two variables, one scaled on the Y-axis and the other scaled on the X-axis. We will make extensive use of scatter plots later on, but for now we will illustrate their use in detecting deviant cases and in seeing how cases bunch together in various ways.

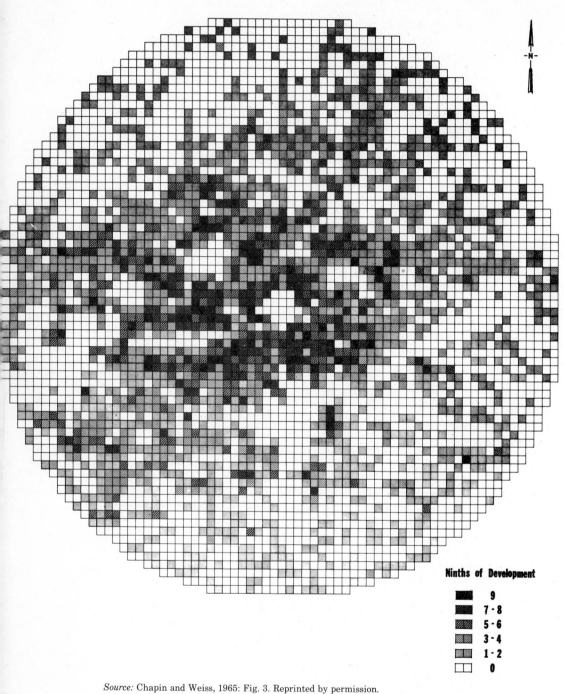

Ninths of Development

9
7 - 8
5 - 6
3 - 4
1 - 2
0

Source: Chapin and Weiss, 1965: Fig. 3. Reprinted by permission.

FIGURE 4.19 LAND IN RESIDENTIAL USE, GREENSBORO, NORTH CAROLINA, 1960

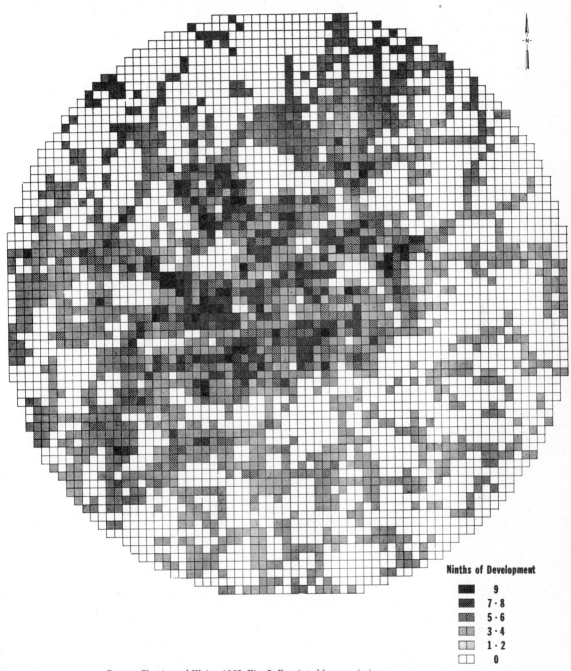

Ninths of Development

- 9
- 7·8
- 5·6
- 3·4
- 1·2
- 0

Source: Chapin and Weiss, 1965: Fig. 5. Reprinted by permission.

FIGURE 4.20 EXPECTED RESIDENTIAL LAND USE, GREENSBORO, NORTH CAROLINA, 1960, BASED ON USE OF PROBABILISTIC MODEL — MEDIAN OUTCOME OF 50 RUNS

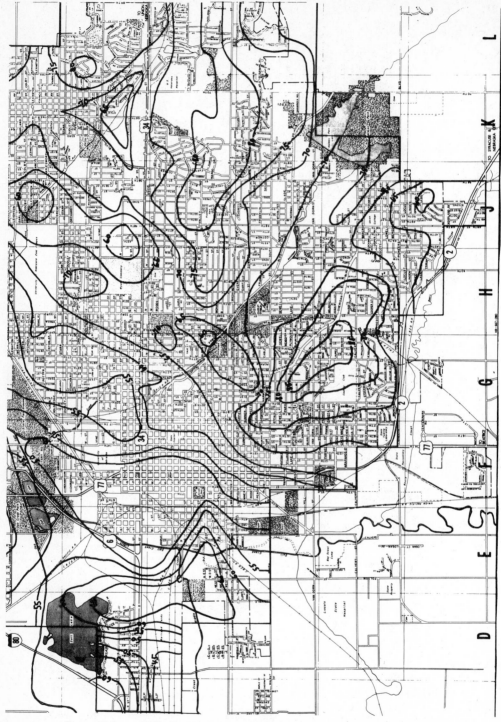

Source: Hoiberg and Cloyd, 1971. Reprinted by permission.

FIGURE 4.21 AN ISOLINE MAP OF SOCIAL CLASS VARIATION IN A MIDWESTERN CITY

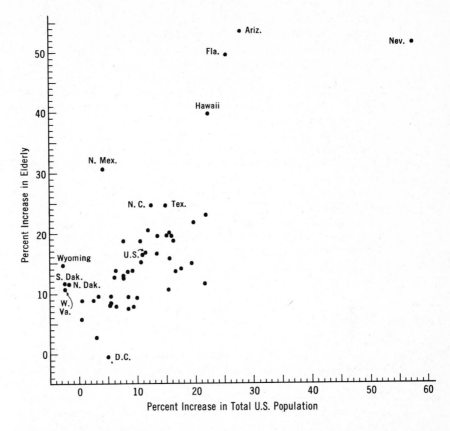

Source: U.S. Senate: 1969:326.

FIGURE 4.22 SCATTER PLOT OF PERCENT INCREASE IN TOTAL POPULATION AND IN POPULATION AGED 65 AND OVER BETWEEN 1960 AND 1968 FOR STATES AND THE DISTRICT OF COLUMBIA

Figure 4.22 shows a scatter plot of the fifty states plus the District of Columbia. Each dot represents one of these areas. The dot is placed to show on the X-axis the percentage change in its population overall between 1960 and 1968, and on the Y-axis the percentage change in the state's elderly population aged 65 and older between 1960 and 1968. As shown on the scatter diagram, most states show a total population increase of 10 to 20 percent, and most likewise show a 10 to 20 percent increase in the elderly population. The dots tend to bunch around the overall U.S. average of an 11 percent population increase and a 16 percent increase in the elderly. Yet several states show an unusually large increase in the elderly, or in both elderly and total population. Some of the deviant states are labelled on Figure 4.22. Nevada, Arizona, Florida, Hawaii, and New Mexico show a larger percentage increase in the elderly population than in the total population. Wyoming, North and South Dakota, and West Virginia show a total population decline coupled with a 10 to 15 percent increase in the elderly population. Only the District of Columbia showed a decline in the elderly and an increase in total population between these two years.

An examination of these deviant cases which are made evident in an examination of a scatter diagram may lead to other important insights into migration trends. Likely population processes are relatively obvious for the first two groupings of states noted above. The first group consists of states which are often defined as "retirement areas" with large villages developed mainly for the elderly. The second group of states are those where other population groups are moving out, leaving the elderly behind. The District of Columbia presents an interesting reversal of this pattern of increase in the 65-plus population. Isolation of such deviant cases often serves as the focus for productive future research to better understand and explain the sociological phenomenon of migration, in this instance. Other examples of the use of the scatter diagram will be presented in Chapter 7.

Scatter diagrams are useful to show how individual cases are spread out in a "space" formed by the two dimensions or variables scaled along the X- and Y-axes. Diagram 4.2 provides a further explanation of the mechanics of creating scatter diagrams.

4.4.3 Triangular Plot

A triangular plot is like a scatter diagram in that each case or individual is located by a dot in the space of the graph so that clusterings and distances between the dots on measured variables may be examined. The triangular plot, however, locates dots on a graph which is the shape of an equilateral triangle. It is used in situations where a variable has three meaningful categories (or can be regrouped to have three), and where a case can be characterized in terms of a percentage in each category totalling to 100 percent for each case (*i.e.* it is composed of three parts that include all the possibilities being considered in the case).* Usually the "cases" plotted are groups, like "male graduate students," "members of a certain religious denomination," "the leading crowd in a high school," etc., where each case can be characterized in terms of a total percentage (100%) made up of those for example, who have a "favorable," "unfavorable" or "no opinion" attitude on some issue. Other variables with three categories, in addition to some opinion questions, might be marital status (single, currently married, other); residence (in the central city, in suburbs, in rural areas); judgments (other jobs are better, the same, or worse than mine), and so forth.

The graph paper, which is available commercially, is an equilateral triangle with scales extending from each side to the opposite corner. Each scale is marked from zero percent at the opposite side to 100 percent in the corner. A point that is one third of the way toward a corner from each of the three sides falls exactly in the middle of the graph at the point which is $33\frac{1}{3}\%$ on each of the scales. Thus a dot's location reflects three percentages which total to 100 percent. Figure 4.23 illustrates this type of graph and Diagram 4.2 discusses the mechanics of construction.

*A triangular plot may also be used to plot values on three variables, given that their values add up to 1 or 100 percent for each case.

DIAGRAM 4.2 How to Construct Scatter Plots, Triangular Plots and Semi-Log Graphs

SCATTER DIAGRAMS

Step 1. Data on at least two variables for each case are needed to construct a scatter diagram. For a 2-dimensional scatter plot, two scores per case are needed. In the following example, large cities have been measured using a health facility measure (number of hospital beds per 100,000 population in 1970) and an economic measure (that city's per capita personal income as a percentage of the national average per capita income).*

	Selected Cities	No Health Facilities	Economic Measure (%)
a	Youngstown, Ohio	366.8	100
b	Trenton, N.J.	525.1	115
c	Springfield, Mass.	437.5	97
d	St. Louis, Mo.	496.4	108
e	Newark, N.J.	436.9	129
f	New Orleans, La.	507.2	97
g	Huntington, W. Va.	593.5	87
h	Erie, Pa.	397.5	96
i	Cleveland, Ohio	418.3	119
j	Cincinnati, Ohio	370.7	106
k	Chicago, Ill.	436.9	127
l	Charleston, W. Va.	619.1	93
m	Birmingham, Ala.	493.3	87

Step 2. Set up coordinates and mark the scale of each variable on one of the axes. Axes should be about the same physical length, and the scales should be marked so that the highest and lowest scores are nearly as far apart as the axis is long. Equal physical units on the scale, of course, should correspond to equal distances on the variable. Label the axes clearly.

Graph paper is usually used. It is commercially available with different numbers of lines per inch, but for much of the graphing in sociology a 10-line-per-inch graph paper is most useful.

Step 3. Plot each case as a dot, located so that it is just opposite the proper point on both axes correctly showing its score on both variables. One dot is plotted per case. Often, in this descriptive usage, the dots will be identified by some sort of coding system such as the small letters used below.

Step 4. Analysis. The scatter diagram can then be examined to see to what extent the dots tend to cluster. By looking at other characteristics of cases in a cluster, often additional insight into possible reasons for the cluster-

---

*These cities all are Standard Metropolitan Statistical Areas with 200,000 population or more in 1970. All areas chosen have a death rate of 10.0–10.2 (the national death rate is 9.7). The U.S. average on the health facilities measure is 414.6 and for the economic measure it is 100%. Both variables are interval level. *Source:* U.S. Bur. Census, 1971; Stat. Abstract.

DIAGRAM 4.2 *(Continued)*

ing can be found. Some investigators used this technique to identify types of cities, for example.

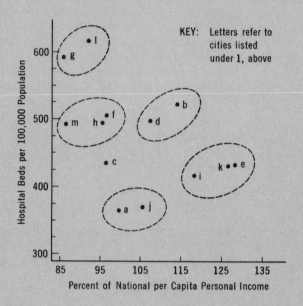

TRIANGULAR PLOTS

Step 1. Data for a triangular plot must have three parts for each case that is to be plotted, and, for each case, the sum of the three parts must equal 100 percent. Thus we could plot a series of questions or *items* if they all were characterized by the same three-part response scale such as "agree," "disagree," and "undecided," and each item could be scored in terms of the percent of people who gave these three responses so the sum of percentages for any item would be 100 percent. Alternatively, one could measure *cases* on a specific variable where, again, that variable has three categories and each case is measured in such a way that the sum of the measures for a case is always 100 percent. Cities, for example, might be characterized in terms of percentage of their population which falls in three inclusive general ethnic categories.

For this illustration we will use data from a Gallup poll in which the question was asked for several years, "Do you think Communist China should or should not be admitted as a member of the United Nations?" The answers were either "favor," "oppose," or "no opinion." For each poll the percentage responding in each category is given below,

107

DIAGRAM 4.2 *(Continued)*

and the total of these percents is 100 for any given poll. Here we can examine change through time.

Poll Date	% Favor	% Oppose	% No Opinion	Total
1954, July	7	78	15	100
1954, Sept.	8	79	13	100
1955, June	10	67	23	100
1955, Sept.	17	71	12	100
1956, July	11	74	15	100
1957, Feb.	13	70	17	100
1958, Feb.	17	66	17	100
1958, Sept.	20	63	17	100
1961, March	20	64	16	100
1961, Oct.	18	65	17	100
1964, Feb.	15	71	14	100
1964, Nov.	20	57	23	100
1965, March	22	64	14	100
1966, Jan.	22	67	11	100
1966, April	25	55	20	100
1966, Oct.	25	56	19	100
1969, Feb.	33	54	13	100
1970, Oct.	35	49	16	100

Source: Erskine, 1971. Used by permission of George Gallup.

Step 2. Set Up Coordinates and mark one of the corners of the triangular chart with each response alternative. The perpendicular from the opposite side to a corner is laid off on a scale of percents with 100 percent the value in a corner and 0 percent the value at the opposite side. Label the graph clearly.

Graph paper is available from most engineering supply stores. It is called "triangular coordinate" paper.

Step 3. Plot each case as a single dot. The dot is located in terms of percentages. For example, the July, 1954 poll outcome would be represented by a dot at the 7% point on the scale that ends in the corner labeled "favor" where that line crosses the line at 78% out toward the corner marked "oppose." This will be at a point which is 15% out from the side opposite the corner marked "no opinion." (See graph on following page.)

Step 4. Analysis. Two comments are relevant. First, it would be a mistake to examine only percentages in favor without also taking account of whether the rest of the cases were "oppose" or whether there was merely a shift from favor to "no opinion," for example. Secondly, by connecting dots with a line, we can show the direction of the shift through time. It appears, overall, that there has been a shift toward the "favor" corner.

DIAGRAM 4.2 *(Continued)*

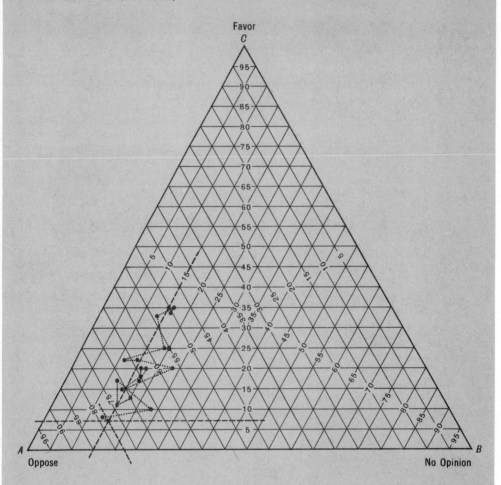

SEMI-LOGARITHMIC CHARTS

Step 1. Data used in semi-logarithmic charts are usually measures for a series
of different years made on one case, such as a society. There are exceptions, of course, but the usual use of semi-logarithmic charts is to show
rate of change through time. The following data are median (a kind of
average which is discussed in the next chapter) income for families in
the United States and these measurements were made for a number of
different years. What is the rate of change in median income of families?

109

DIAGRAM 4.2 *(Continued)*

Year	Median Income
1947	$3,031
1950	3,319
1958	5,087
1959	5,417
1960	5,620
1961	5,737
1962	5,956
1963	6,249
1964	6,569
1965	6,957
1966	7,500
1967	7,974
1968	8,632
1969	9,433
1970	9,867

Source: U.S. Bur. Census, 1971:P-60.

Step 2. Set Up Coordinates. A semi-logarithmic chart could be created by marking an arithmetic scale with the logarithms of numbers. Instead, semi-log graph paper is available, where the rulings are in terms of original scores rather than their logs, so that scores need not be converted to logs prior to plotting. This is the reason for the unequally spaced appearance of semi-log graph paper; it has lines more closely spaced together toward the top of the paper.

Paper is available with several tiers or "cycles" of lines numbered from 1 to 10. One-cycle paper is used to plot scores where the largest is no more than 10 times the smallest. If the largest number is 10 to 100 times the smallest, two-cycle paper is needed, and so on. The X-axis is a standard arithmetic scale (although double-logarithmic paper is available and used for some specialized purposes not discussed in this text).

In the example above, the largest median income figure is less than 10 times the smallest so one-cycle paper is needed. The log-scale is labeled in score units, in this case $1000's of dollars. Notice that the starting point is some convenient power of ten (.01, .1, 10, 100, 1000, etc.) rather than zero because the rate of increase from zero can not be determined (recall that percentage change is the difference between a starting and ending figure divided by a starting figure and division by zero is not a defined operation). We chose $1,000. and proceeded upward in $1,000. increments.

The X-axis is labeled in terms of time, in this case years between 1947 and 1970.

Step 3. Plot the value of the score above the proper X-axis score. Here, we are plotting each median income figure above the year to which it refers.

DIAGRAM 4.2 *(Continued)*

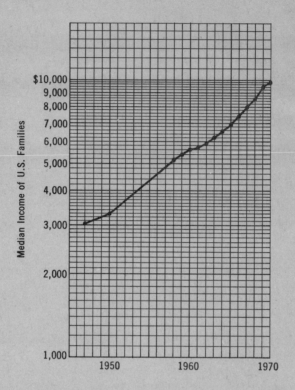

These dots are connected with straight lines and the figure is *not* closed with the X-axis.

Step 4. Analysis. A straight line indicates a constant rate of change. The more steeply the line goes up or down, the more rapid the rate of change, and by inspection of these graphs one can determine whether there is a change at an increasing rate or a decreasing rate of change.

In the example, the rate of increase has been relatively constant, overall, but there have been periods of decreasing-rates-of-increase prior to 1950, just after 1960, and just before 1970. In the mid-60's there was a slightly increasing rate of increase, as shown by the arching upward of the line into a steeper ascent.

Figure 4.23 is from a study of adolescent society by James Coleman (1961) in which he studied ten high schools selected from northern Illinois in 1957 and 1958. Data for the triangular graph came from a question in his study which asked boys what they most wanted to be — jet pilot, nationally famous athlete, or atomic scientist. For each school two sets of percentages were computed: the percentage of all boys selecting each of the three "ideals," and the percentage of boys who were among the "elite" or leading crowd who selected each of the three

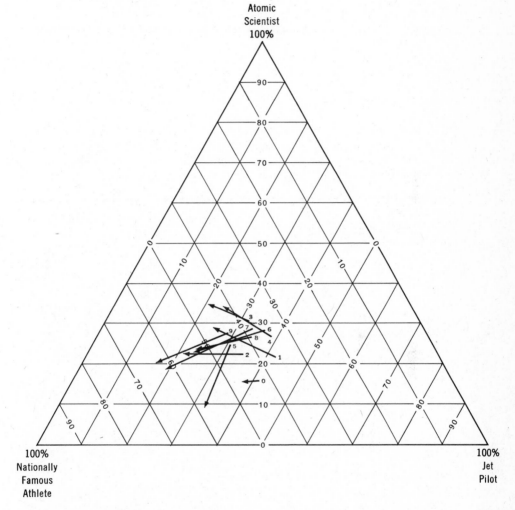

Source: Coleman, 1961:132. Reprinted by permission of The Macmillan Company. Copyright © The Free Press, A Corporation, 1961.

FIGURE 4.23 TRIANGULAR PLOT OF THE RELATIVE ATTRACTIVENESS OF "NATIONALLY FAMOUS ATHLETE," "ATOMIC SCIENTIST," AND "JET PILOT" TO BOYS IN EACH SCHOOL (tail end of arrow) AND LEADING CROWD OF BOYS IN EACH SCHOOL (point of arrow)

ideals. Most triangular plots stop with this display of results, but Coleman took an additional step which highlights the relative attractiveness of the "nationally famous athlete" category among the leading crowd in each school. He connected each overall percentage for each school with the dot representing the elite's position in that school with the arrow pointing toward the elite's dot. The length of the arrow reflects the difference between boys in general and the leading crowd and the direction of the arrow indicates the category toward which this difference is directed. Notice that most leading crowds are "further out" than boys in general, and that they are different in the direction of finding "nationally famous athlete" a more attractive ideal category. This triangular plot, and others that Coleman presents, greatly aid in his analysis of the social trends and pressures of high schools he studied.

4.4.4 The Semi-Logarithmic Chart

The last graphic technique we will discuss here is the semi-logarithmic chart. This is a graph plotted on rectangular coordinates which are similar to the Cartesian coordinates we have used before. The X-axis is a numeric scale often showing time or age. The Y-axis, however, is laid off in terms of the logarithms of

TABLE 4.1 AN ILLUSTRATION OF RATE OF CHANGE AS REFLECTED IN DIFFERENCES BETWEEN RAW SCORES AND DIFFERENCES BETWEEN LOGS OF SCORES

(a) Data Illustrating a *Constant Rate* of Increase

Year	Raw Scores	Amt. of Change	Percent Change	Log of Scores	Amt. of Change in Logs
1930	12.00			1.079	
1935	15.00	+3.00	+25%	1.176	+.097
1940	18.75	+3.75	+25%	1.273	+.097
1945	23.44	+4.69	+25%	1.370	+.097

(b) Data Illustrating an *Increasing Rate* of Decrease

Year	Raw Scores	Amt. of Change	Percent Change	Log of Scores	Amt. of Change in Logs
1950	8.00			0.903	
1955	6.00	−2.00	−25%	0.778	−.125
1960	4.00	−2.00	−33%	0.602	−.176
1965	2.00	−2.00	−50%	0.301	−.301

(c) Data Illustrating a *Decreasing Rate* of Increase

Year	Raw Scores	Amt. of Change	Percent Change	Log of Scores	Amt. of Change in Logs
1970	2.00			0.301	
1975	6.00	+4.00	+200%	0.778	+.477
1980	10.00	+4.00	+66%	1.000	+.222
1985	14.00	+4.00	+29%	1.146	+.146

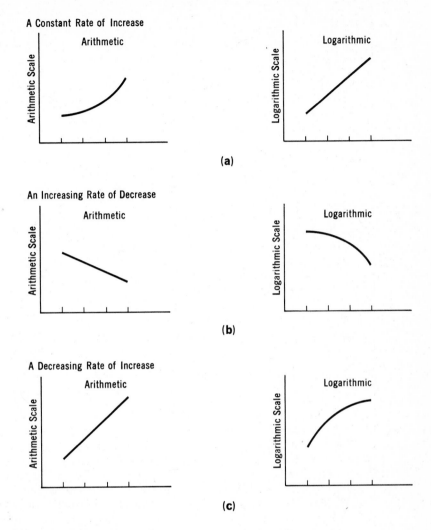

Source: Data from Table 4.1.

FIGURE 4.24 Graphic Illustration of the Difference between an Arithmetic and Logarithmic Chart to Portray Rates of Change

numbers rather than their arithmetic value. The semi-log chart is especially useful in an analysis of *trend lines,* because equal numeric differences between logarithms indicate constant *rates of change.** Table 4.1 and Figure 4.24 illustrate this useful feature of logarithms. Notice that a constant rate of change shown in Figure 4.24 can be quickly detected on a semi-logarithmic chart by a straight line rather than a curved line.

*There are charts with logarithmic scales on both axes, and these are called "log-log" charts. They are used for more complex plots of rates against each other and will not be included in this discussion.

BOX 4.4 LOGARITHMS AND THEIR USE IN STATISTICS

Common logarithms are a system of exponent numbers for the base-number 10, that is, exponential powers to which 10 may be raised to produce any given original number. For example, the common logarithm (base 10) of 100 is 2.0 because $10^{2.0}$ is 100; the $\log_{10}$ of 1000 is 3.0, and so on, with the majority of log numbers being a fractional power of 10. Logarithms are convenient for calculation because the process of *adding* these logarithmic exponents amounts to a process of *multiplying* the original numbers produced by the logarithms (and subtracting amounts to dividing). Thus the relatively simple process of adding or subtracting can be used to show more complex multiplying and dividing variations. In statistical graphing, logarithms can thus be used to exaggerate certain effects.

The usual fractional common logarithm has two parts: a whole number to the left of the decimal point called the *characteristic,* and a fractional number to the right of the decimal point called the *mantissa.* The value of the characteristic may be positive or negative and depends on the location of the decimal point in the original number; the value of the mantissa is normally read from a table of common logarithms (usually in reference sections of mathematics texts or in some statistics workbooks). The table below has been developed to show both the characteristic and the mantissa.

Original Number	Logarithm	Original Number	Logarithm
0.01	−2.0000	50	1.6990
0.1	−1.0000	100	2.0000
1	0.0000	200	2.3010
2	0.3010	500	2.6990
5	0.6990	1000	3.0000
10	1.0000	2000	3.3010
20	1.3010	5000	3.6990

Notice that the characteristic for original numbers less than one is negative and is equal to the number of zeros between decimal point and first digit, plus one. For original numbers greater than one, the characteristic is a positive number which is one less than the number of integers in the original number (*i.e.,* zero for ones, 1 for tens, 3 for thousands, etc.). Thus the complete common logarithm is formulated with the appropriate characteristic for the original number plus its mantissa as shown in a standard reference table. *Anti-logarithms* are found by reversing the procedure to find the original number from a logarithm.

Since commercial logarithmic graph paper is available, conversion of statistical scores or frequencies to logarithms by calculation is not normally necessary, and the scale of the original scores can be placed within a labeled sequence in the unequally (*i.e.,* logarithmically) spaced Y-axis scale points. Original scores are then plotted directly on the scale.

Logarithms are quite useful in statistics in a number of ways. They simplify the computation of the specialized mean called the geometric mean (see Chapter 5); they greatly aid in the study of trends, as indicated in this chapter; and they are sometimes used to transform scores to simplify a study of relationships between variables (see Chapters 7 and 8).

Two studies will serve to illustrate uses of the semi-logarithmic chart and Diagram 4.2 provides details about their construction. Figure 4.25 presents data on infant mortality for the United States during the period from 1915 to 1967. The Y-axis is a logarithmic scale labeled in terms of death rates for infants. The trend line shows a general decline which is almost a straight line but not quite. The rate of decline from year to year is nearly but not quite constant. In years prior to 1945 the rate of decline slowed up slightly and the rate of decline dropped more rapidly again in the few years after about 1945. Again around 1960 there was a slowing in the rate of decrease in infant mortality shown by a leveling off of the trend line. More recently the steep rate of decline has been resumed. Demographers sometimes use infant mortality figures as an indicator for changes in the quality of health care in a country. Research could be focused on explanations for the increasing or decreasing rate of decrease in mortality in the U.S. over this half-century period. The semi-logarithmic chart makes these changes in rates more apparent. In Figure 4.25, the fact that this is a semi-logarithmic chart can be readily detected by the characteristically unequal spacing on the Y-axis and the constant size of the unit on the X-axis.

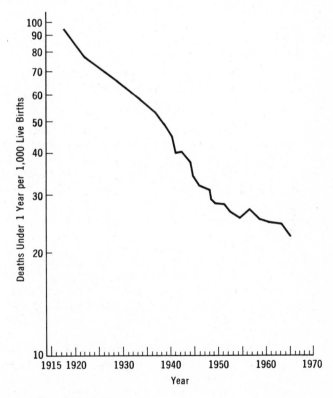

Source: Chase and Byrnes, 1972: Fig. 1.

FIGURE 4.25 Infant Mortality Rates, U.S. 1915–1967 (a semi-logarithmic chart)

The second example of the use of a semi-logarithmic chart is from a study of population growth by Wray (1971) which matched infant death certificates with the corresponding birth certificates for the 1.3 million births in England and Wales in 1949–1950 (an example of the use of recorded data). The graph in Figure 4.26 shows infant mortality ratios on the logarithmic scale on the Y-axis. A mortality ratio is the ratio between the mortality rate for infants in a given population sub-group divided by the overall mortality rate for all infants. The value 100 on the Y-axis indicates that the sub-group mortality rate and the overall mortality rate are the same; higher values indicate that the sub-group mortality rate is higher. In this graph, the sub-groups which are examined reflect different social classes (I and II are upper classes, III is middle class, and IV and V are lower classes in these data) of the infant's mother. Within social class groupings mortality ratios are computed for infants of different birth order. That is, infants who are first-born, second- or third-born, and fourth- or later born are handled separately in the graphic. presentation. Thus there is a separate line on the graph for each birth order within each of three social classes. Each line shows the mortality ratio of infants in a given birth-order and social class category by the age of its mother at the time of its birth. From this semi-logarithmic chart a large number of conclusions may be drawn. In virtually all cases the mortality ratio drops with increases in the mother's age until some time around the 30's where the rate of decline slows up and gradually changes direction. Toward the 40's the mortality ratio shows an increasing rate of increase, in most instances with the increasing age of the mother. The mortality ratio (and, by the way, the infant mortality rate as well) is higher for later-born infants than for early-born infants, and this finding holds up within each of the social class categories. Furthermore, looking across the three social class categories for the latest-born category of infants, the mortality ratio is higher among infants of upper-class mothers of a given age than it is among infants of lower-class mothers of that age. The investigators conclude that high mortality rates among infants of young mothers with large families occurs in all social classes and the concentration of young mothers with large families in lower social classes can not totally account for higher mortality rates among infants in lower classes.

BOX 4.5 TRENDS AND TREND LINES

The study of *trends,* or change in some characteristic through a period of time, is one of the important areas of sociology. It is an area which poses many problems for research design, for measurement, and for statistical analysis. (For example, see Harris, 1967, for a discussion of several methods for handling change with techniques which are referred to in the chapters of this present volume.)

The study of change via *trend lines* was first discussed in Section 4.2.4. Up to this point we have discussed four methods for describing change: (a) rates (Section 3.2.5 a), (b) percentage change (Section 3.2.5 b), (c) time-line graphs (Section 4.2.4), and semi-logarithmic graphs (Section 4.4.4). In Chapter 5 some methods for "smoothing" trend lines will be discussed.

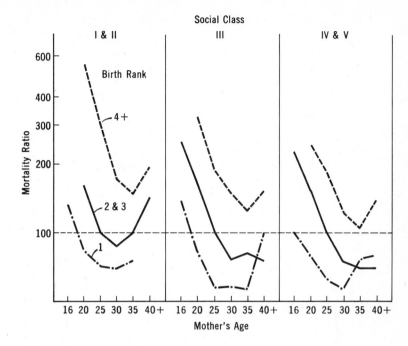

Source: Wray, 1971:414. Reprinted by permission.

FIGURE 4.26 MORTALITY RATIOS BY BIRTH RANK, MOTHER'S AGE AND SOCIAL CLASS, ENGLAND AND WALES, 1949–1950; A SEMI-LOGARITHMIC CHART

4.5 SUMMARY

In this chapter several basic illustrative and analytic graphing techniques have been discussed and their principles analyzed. Graphic techniques, properly used, constitute a very useful and powerful means for communicating more involved comparisons in data and for analyzing complex relationships. They are particularly useful in showing the overall form of a distribution (a feature to be discussed at greater length in Chapter 5), in showing the relationships between variables, and in showing trends through time.

It seems to be characteristic of more useful procedures that they also have a potential for distortion as well as a potential for more accurate and helpful displays of interesting features of data. A number of possibly distorting practices have been pointed out in this chapter.

The next chapter returns us again to the basic univariate distribution discussed in Chapter 3, which is graphed by techniques such as the histogram and polygon as shown in this chapter. Chapter 5, however, will approach the description of whole distributions of scores in terms of a few summary index numbers which will communicate key features of a univariate distribution. Later chapters will address similar problems for bivariate and multivariate distributions.

CONCEPTS TO KNOW AND UNDERSTAND

reference system
  ordinate
  abscissa
  origin
  quadrants
histogram
polygon
ogive
line graph
bar chart
logarithm

cohort
population pyramid
pie chart
pictograph
triangle plot
scatter diagram
statistical map
semi-logarithmic chart
trends
cautions about graphic distortions

QUESTIONS AND PROBLEMS

1. Find several examples of graphic presentation of data in popular magazines and newspapers. For each graph describe its strengths and weaknesses in presenting relevant features of the data for the reader. Are there flaws in the graphic presentation? If so, what corrective measures would you suggest, and what possible distortions might your changes promote?

2. If you are assigned a project or piece of research in this course, now is a good time to decide upon the kinds of phenomena you are going to study. This often means selecting an interesting dependent variable. After you have selected your dependent variable and found relevant data, begin your analysis by creating a distribution of the dependent variable. Select an appropriate graphic technique and graph the distribution.

3. Using data in a journal research article or from a statistics workbook, select two scores on each of several cases. Create a scatter diagram with scales for the two variables along the X and Y axes. Explain in your own words what it is that the graph shows.

4. Using the data given in problem 4 of Chapter 3 (page 68) and graph paper, construct (a) a histogram, (b) a frequency polygon, and (c) an ogive distribution. What do each of these distributions tell you about the data?

5. The data on page 119 give the labor force participation rates of married women in the United States for the years 1948 through 1974. Construct a line graph from these data. What does the line graph tell you about the labor force participation of women over the time period covered by the data?

Year	Labor Force Participation Rate
1948	22.0
1949	22.5
1950	23.8
1951	25.2
1952	25.3
1953	26.3
1954	26.6
1955	27.7
1956	29.0
1957	29.6
1958	30.2
1959	30.9
1960	30.5
1961	32.7
1962	32.7
1963	33.7
1964	34.4
1965	34.7
1966	35.4
1967	36.8
1968	38.3
1969	39.6
1970	40.8
1971	40.8
1972	41.5
1973	42.2
1974	43.0

Source: U.S. Bureau of Labor Statistics (1975)

6. Construct bar graphs from the data provided in the table below. Do the graphs reveal any differences in job satisfaction of blacks and whites?

TABLE 7.9 PROPORTION OF MEN HIGHLY SATISFIED WITH JOB, BY TYPE OF OCCUPATION AND SELF-RATING OF HEALTH: EMPLOYED MEN 45–59 YEARS OF AGE, BY COLOR

Type of occupation and self-rating of health	WHITES		BLACKS	
	Total number (thousands)	Percent who like their job very much	Total number (thousands)	Percent who like their job very much
White collar	5,065	69	190	68
Excellent	2,094	77	84	66
Good	2,162	64	55	80
Fair or poor	634	57	41	54

continued

TABLE 7.9–*continued*

Blue collar	5,824	50	738	49
Excellent	1,963	56	254	56
Good	2,557	48	295	46
Fair or poor	1,094	39	164	42
Service	643	51	176	55
Excellent	228	51	68	57
Good	281	60	69	53
Fair or poor	88	34	35	59
Farm	1,097	54	133	36
Excellent	239	68	20	42
Good	468	51	50	37
Fair or poor	352	48	60	29
Total	12,655	58	1,240	51
Excellent	4,543	66	425	58
Good	5,476	55	472	50
Fair or poor	2,168	46	301	43

Source: U.S. Department of Labor (1979)

GENERAL REFERENCE

Schmid, Calvin F., *Handbook of Graphic Presentation,* (New York, Ronald Press), 1954. This is a carefully explained and illustrated book on different types of graphic presentation and how to solve some of the presentational problems an analyst encounters.

OTHER REFERENCES

Asimov, Isaac, *Asimov's Biographic Encyclopedia of Science and Technology* (Garden City, N.Y., Doubleday and Company), rev. ed. 1972.

Berry, Brian J. L., "Problems of Data Organization and Analytical Methods in Geography," *Journal of the American Statistical Association,* 66 (September, 1971), p 510–523.

Chapin, F. Stuart, Jr., and Shirley F. Weiss in collaboration with Thomas G. Donnelly, *Some Input Refinements For A Residential Model,* a project of the Center for Urban and Regional Studies, University of North Carolina at Chapel Hill, in cooperation with the U.S. Bureau of Public Roads, July, 1965, p 28.

Chase, Helen C., and Mary E. Byrnes, "Trends In 'Prematurity': United States: 1950–1967," Vital and Health Statistics Series 3, Number 15, (U.S. Department of Health, Education and Welfare, Public Health Service, Washington, D.C.), 1972.

Coleman, James S., *The Adolescent Society: The Social Life of the Teenager and its Impact on Education* (New York, The Free Press of Glencoe), 1961.

Erskine, Hazel, "The Polls: Red China and the U.N." *Public Opinion Quarterly,* 35 (Spring, 1971), p 125.

Ferriss, Abbott L., "An Indicator of Marriage Dissolution By Marriage Cohort," *Social Forces,* 48 (March, 1970).

Harris, Chester W., (ed), *Problems in Measuring Change,* (Madison, Wisc., University of Wisconsin Press), 1967.

Hoiberg, Eric O., and Jerry S. Cloyd, "Definition and Measurement of Continuous Variation in Ecological Analysis," *American Sociological Review,* 36 (February, 1971) p 65–74.

Jacobson, Paul H., *American Marriage and Divorce,* (New York, Rinehart), 1959, p. 118.

————, "The Changing Role of Mortality in American Family Life," *Lex et Scientia,* vol 3, no. 2, 1966.

Kaufman, Herbert, and David Seidman, "The Morphology of Organizations," *Administrative Science Quarterly,* 15 (December, 1970).

Maris, Ronald W., *Social Forces in Urban Suicide* (Homewood, Ill., Dorsey Press), 1969.

Mauldin, W. Parker, "Fertility Studies: Knowledge, Attitude, and Practice," *Studies in Family Planning* #7 (New York, The Population Council), June, 1965.

Petersen, William, *Population,* Second Edition (New York, Macmillan), 1969.

[Population Reference Bureau] "Man's Population Predicament," *Population Bulletin,* vol. 27, no. 2, April 1971.

Riley, Matilda White, and Anne Foner, *Aging and Society: An Inventory of Social Research Findings,* (New York, Russell Sage Foundation), 1968.

Riley, Matilda White, "Aging and Cohort Succession: Interpositions and Misinterpretations," *Public Opinion Quarterly,* 37 (Spring, 1973), p 35–49.

U.S. Bureau of Labor Statistics, *Handbook of Labor Statistics 1975 – Reference Edition,* Bulletin 1865 (Washington, D.C.), 1975, Table 14, p 57.

U.S. Bureau of the Census, *Current Population Reports,* P–25, No. 352, (Washington, D.C.), 1966, p 4.

————, "Projections of the Population of the United States by Age and Sex (interim projections): 1970 to 2020," *Current Population Reports,* P–25, No. 448, (Washington, D.C.), 1970.

————, "Consumer Income," *Current Population Reports,* P–60, No. 72, (Washington, D.C.), 1970, Table 6.

————, "Consumer Income," *Current Population Reports,* P–60, No. 78, (Washington, D.C.), 1971, Table 1.

————, "Metropolitan Area Statistics," *Statistical Abstract of the United States, 1971,* (Washington, D.C.), 1971.

————, "Birth Expectation Data: June 1971," *Current Population Reports,* P–20, No. 232, (Washington, D.C.), 1972.

————, "School Enrollment in the United States, 1971," *Current Population Reports,* P–20, No. 234, (Washington, D.C.), 1971.

U.S. Department of Health, Education, Welfare: Public Health Service, "Children and Youth: Selected Health Characteristics," *Vital and Health Statistics,* Series 10, No. 62 (Rockville, Md.), 1971, p 31.

U.S. Department of Justice, *Uniform Crime Reports for the United States,* (Washington, D.C.), 1971, p 32.

U.S. Department of Labor, *The Pre-Retirement Years,* Vol. 1, Manpower Research Monograph N. 15 (Washington, D.C.), 1970, Table 7.9, p 216.

U.S. Senate, Committee on Aging, "Developments in Aging," (Washington, D.C.), 1969, p 326.

Weitzman, Murray S., "Measures of Overlap of Income Distributions of White and Negro Families in the United States," *Technical Paper* No. 22, U.S. Bureau of the Census, (Washington, D.C.), 1970, p 9.

Wray, Joe D., "Population Pressure on Families: Family Size and Child Spacing," Chapter 11, *Rapid Population Growth: Consequences and Policy Implications,* vol. 2, (Baltimore, Md., Johns Hopkins Press), 1971, p 414. (Prepared under direction of Dr. Roger Revelle.)

Summary Measures of the Location, Variation and Form of Univariate Distributions

5

A very basic type of research continually pursued by all quantitative disciplines is research on measurement instruments. Some of this research is to develop new or better measurement procedures for some interesting characteristic. Other research is for the purpose of identifying possible biases which may enter into the use of a measurement procedure, and also for identifying those conditions under which the measurement will produce valid and reliable scores.

5.1 UNIVARIATE DISTRIBUTIONS—AN EXAMPLE AND OVERVIEW

Cloud and Vaughn conducted research of the kind described above, using the Wilson-Patterson "Conservatism Scale," (1970). The Wilson-Patterson scale includes 50 words or catch phrases which, in many Western contexts, reflect ideas held by more "liberal" or more "conservative" people. A respondent is asked to indicate which words represent ideas he favors or believes in. Ten of the 50 items are given below, and agreement with the odd-numbered items is thought to indicate a conservative attitude.

Which of the following do you favor or believe in?

1	death penalty	Yes	?	No
4	striptease shows	Yes	?	No
7	patriotism	Yes	?	No
12	birth control	Yes	?	No
22	legalized abortion	Yes	?	No
29	royalty	Yes	?	No
33	apartheid	Yes	?	No

35	church authority	Yes	?	No
36	disarmament	Yes	?	No
50	pyjama parties	Yes	?	No

An individual's "conservatism" score is created first by assigning numeric weights to the answer alternatives, and then by summing up the weights for a respondent's answer alternatives for all 50 items. A weight of 2 was assigned to the most conservative response to each item, which means that the "Yes" response on the 25 conservative words, or the "No" response on the 25 liberal words, were each assigned the weight 2. The undecided, question-mark response was assigned a weight of 1, and the liberal response was assigned a score of zero. Thus, individual respondents could get a total score for all 50 items from zero (for those giving all "liberal" responses) to 100 (for those giving all "conservative" responses). This scale is "balanced" because there were an equal number of conservative and liberal items, and by using the scoring procedure discussed above, the final score should be one which eliminates acquiescence bias and produces a "true" content score. To say the same thing somewhat differently, any tendency an individual may have to give a "yes" answer regardless of the content of an item will occur equally often both for liberal items (and be scored zero) and for conservative items (and be scored 2), thus balancing any bias toward conservativism or toward liberalism in the overall score.

Cloud and Vaughn then developed two other ways of scoring the same kind of scale which would represent acquiescence bias. One scoring procedure, called a "yeasaying" score, is simply the sum of weighted responses as before, except that the "yes" response always was weighted 2 regardless of the content of the item. The score zero would be for those who never checked "yes" to any of the 50 items and the score 100 would go to those who always checked "yes." The other score, called a "style" score, is somewhat more complicated. It is a ratio created by taking the difference between the proportion of agreements to conservative items and the proportion of agreements to liberal items, divided by the proportion of all items the subject agrees with.* Both the yeasaying scores and the style scores are measures of response style of the kind called acquiescence.

*As an example of the three scores, suppose that there were only a scale of six (rather than 50) items, alternating liberal and conservative, as follows:

	One Person's Answers
Item 1 (conservative item)	Yes
Item 2 (liberal item)	No
Item 3 (conservative item)	Yes
Item 4 (liberal item)	Yes
Item 5 (conservative item)	Yes
Item 6 (liberal item)	No

One person's answers to these six items are given above and his three scores are computed as follows:
(a) Conservativism Score (C), in this example would count 2 for each conservative item answered "yes" or each liberal item answered "no." Here, the total would equal 10, two points for the answer to items 1,2,3,5, and 6.
(b) Yeasaying Score (Y), counts two points for each yes regardless of whether the item is a liberal or conservative item. Here there are four yes answers for a score of 8.
(c) Style Score (S) is a ratio made up of the following proportions:

Proportion of conservative items that the subject agreed with. Here he agreed with all three of the conservative items for a score of 1.00.

5.1a Features of a Distribution. Figure 5.1 shows frequency polygons for content (*C*), yeasaying (*Y*), and style (*S*) scores from the Wilson-Patterson

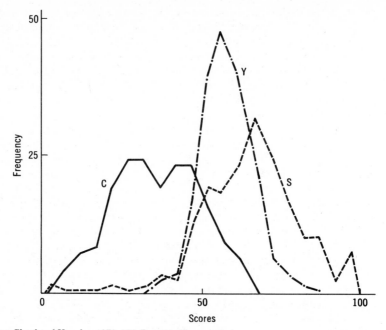

Source: Cloud and Vaughn, 1970: 199. Reprinted by permission.

FIGURE 5.1 FREQUENCY DISTRIBUTIONS OF CONTENT (*C*), STYLE (*S*), AND YEASAYING (*Y*) SCORES FOR 183 NEW ZEALAND COLLEGE STUDENTS ON THE WILSON-PATTERSON CONSERVATISM SCALE.

scale for 183 New Zealand college students. The three distributions are clearly different in a number of respects. First, they differ in their *location* or concentration along the score scale. The "lowest" score distribution is the *C*, next the *Y*, and the highest is the *S* distribution. Secondly, the distributions differ in their relative *concentration*. The *Y* distribution is much more "piled up" in the center of the distribution, with fewer cases at the extremes, than is the case for the

---

Proportion of liberal items that the subject agreed with. Here, he agreed with one out of the three liberal items for a score of .33.

Overall proportion of items agreed with. Here four of the six items were agreed with for a score of .67.

Combining these three components, the Style score is:

$$\text{Style} = \frac{1.00 - .33}{.67} = \frac{.67}{.67} = 1.00$$

While the construction of this type of score need not be understood to handle material in this chapter, it does provide an illustration of a use of statistical summaries in combination to create an index which may get at the phenomena one wants in a useful fashion. Index formation and "scaling" techniques are topics usually reserved for a course in research methods.

other two distributions. Both C and S distributions are much more spread out, with lower frequencies in any one category than is true of Y, but greater frequencies in more categories than is true of Y. Thirdly, the *form* of the distributions differ in several respects, such as the number of peaks, extent of "lopsidedness" etc. These three features of distributions are referred to as **central tendency** (or, sometimes, *location*), **variation,** and **form.**

In the case of Cloud and Vaughn's research (1970), the C distribution represents acquiescence-free conservatism content scores. The Y and S distributions represent somewhat different approaches to relatively pure "acquiescence" score distributions. Their point is that unbalanced tests, those which do not include equal numbers of "opposite" items, would result in a distribution of some type of combination of the C and Y (or C and S) distributions. The effect of not eliminating acquiescence bias would not only be a change in the *location* or *central tendency* of the distribution of conservatism scores, but it would likely also be a change in the the amount of *variation* among scores and the *form* of the distribution of conservatism scores—rather complex biasing of scores from a measurement procedure. They naturally recommend that attitude scales, such as the Wilson-Patterson scale, be carefully "balanced" as they were able to do in their research.

In this case, differences between three distributions were shown by the graphic technique of the frequency polygon discussed in Chapter 4. In this chapter we will discuss the possibilities of an even more compact characterization of distributions than the graphic technique permits. We will do this by using a very few summary "index numbers" which will indicate central tendency, variation and form of a distribution. We will then be in a position to compare distributions both more efficiently and in terms of some of the specific ways in which distributions differ from each other.

5.1b. On the Need to Drop Detail. Cloud and Vaughn (1970) could have compared each of the three scores (*i.e.* the C, the Y and the S scores) for each of the 183 students in their study and based a judgment of differences on this type of detailed search. This would be tedious, unnecessary, and probably misleading in their research. The myriad differences between *individual* scores would probably serve to obscure differences between the three **sets** of scores, and the researchers were interested in differences between the three whole distributions. Furthermore, examining individual score differences for all combinations of scores would probably focus attention on differences person-by-person, and

---

BOX 5.1

The idea of a **distribution** of scores—in frequency or percentage form—is basic to the discussion in this chapter. The objective of this chapter is to describe whole distributions in compact ways. The idea of distribution is introduced in Section 3.2.3 and graphic presentations of distributions are shown in Figures 4.3 and 4.4.

would neglect statistical comparisons between sets of different scores. Their research question is at the level of differences in the overall **distribution** of a variable, and the possible effects one kind of bias may have.

The typical interest of investigators is in distributions rather than individual scores. Thus it is very helpful to have ways to describe whole distributions and drop real but nonetheless distracting information about specific individual scores. As we noted in our discussion of distributions of raw scores, one of the helpful procedures in the analysis of data may be selective elimination of distracting detail in order to permit a clearer focus on differences which are important to the research question at hand.

5.1c.   What Is to Come.   Thus it turns out that there are three main ways that distributions differ from each other, and the point of this chapter is to develop accurate ways to measure the features of a univariate distribution for various kinds of data. There are several different approaches to each feature, and the approach one selects in any specific instance will depend on such things as the level of measurement of the variable, the research question in which one is interested, and, sometimes, on the way the variable happens to be distributed. Separate sections of this chapter will deal with approaches to measuring *central tendency* (Section 5.3 ff), *variation* (Section 5.4 ff.) and *form* (Sections 5.2 ff. and 5.5 ff.) of distributions.

5.2   THE FORM OF A DISTRIBUTION—I

The feature of a distribution most readily apparent from a histogram or polygon is its form—the overall shape of the distribution. For many purposes a simple verbal description is sufficient, but single summary "indices" can be developed to reflect certain aspects of the form of a distribution. These numeric indices will be presented in Section 5.5 at the end of this chapter after some of the basic notions of location and variation have been developed. For now, we will call attention to aspects of the form of distributions and introduce some of the terminology by which these features can be expressed.

5.2.1   Number of Modes

A first characteristic of the form of a distribution readily picked up from a histogram or frequency polygon is the number of high points or peaks (modes) the distribution has. In Figure 5.1, for example, the distribution of conservatism scores (C) has two high points with a low between them, and it would be called a **bimodal** distribution. The other two distributions (S) and (Y) in Figure 5.1 are **unimodal** since they have only one main high-point. Notice that the determination of the number of peaks depends to some extent upon one's judgment of the importance of differences in the frequency in categories. For example, the curve (S) has some slight dips and peaks on each side of the main peak, but these are clearly minor when compared to the main peak in the distribution. If they were judged important, comparatively and absolutely, one would refer to a many-moded distribution or "multimodal" distribution. It seems to be true that very many distributions actually encountered in research are unimodal, overall.

5.2.2 Symmetry

A second aspect of the form of a distribution is the extent to which it is symmetrical rather than lopsided or "skewed." The "skew" in a distribution refers to the trailing off of frequencies toward extreme scores in one direction, away from the bulk of cases. Figure 5.2 shows curves skewed to the left (negatively) in Figure 5.2f, and to the right (positively) in Figure 5.2d.

A symmetrical curve is one in which the two sides of the distribution would exactly correspond, if the figure were to be folded over at its central point. Income is a variable that is usually positively skewed, there are a few extremely high scores (incomes) which are not "balanced" by equally extreme low scores and most individuals have modest to moderate incomes. Ability or experience test scores very often have a nearly symmetrical shape.*

5.2.3 Kurtosis

Kurtosis refers to the extent to which cases are piled up closely around a point in the distribution or are distributed rather widely among categories. Or alternatively, it is interpreted as the relative concentration of scores in the tails of the distribution (relative to the normal distribution). As indicated in Figure 5.2, an unusually concentrated distribution of scores is called a **leptokurtic** distribution, one which is flatter than "normal" is called a **platykurtic** distribution. A "normally" concentrated distribution of scores is called **mesokurtic.** Here "normal" has a technical and precise meaning which will be discussed later on. For the time being it is sufficient to know the terminology and be able to compare curves in an approximate way.

There are some other terms which refer to the overall shape of a distribution and are handy to know. A **"J-shaped"** distribution has almost all cases piled up at one end of the scale and then tails off uniformly in one direction from the concentration as indicated in Figure 5.2j. **Rectangular** distributions have equal frequencies in all categories so the distribution has a flat top (Figure 5.2l).†
A **"bell-shaped"** distribution is unimodal and slopes gently off in both directions from the mode to form the general shape of a bell (Figure 5.2h). Finally, a **"U-shaped"** distribution is bimodal, with modes at both extremes and the low-frequency area in the center of the distribution (Figure 5.2k).

---

*The symmetry is generally a function of the way scores are defined.
†It is interesting to note that pseudo-random numbers are a good example of a *rectangular distribution,* where the likelihood of any of the ten digits from 0 to 9 appearing is equal. Computers sometimes are used to create pseudo-random numbers for random sampling purposes, and one check on such computer programs is to see if they indeed produce a rectangular distribution of numbers. Sampling and the use of random numbers is a topic in the area of inferential statistics (see Chapter 11).

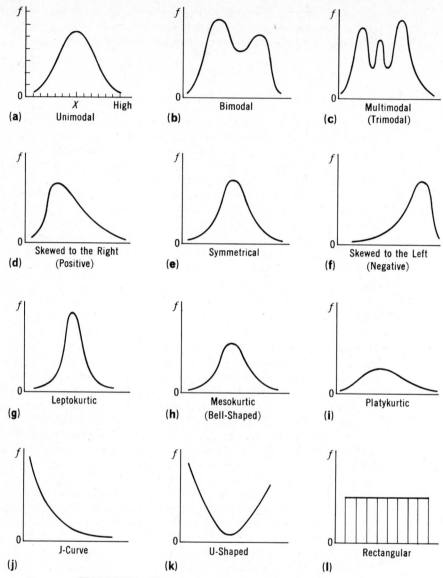

FIGURE 5.2   VARIETIES OF OVERALL FORM OF DISTRIBUTION

5.3   LOCATION OF A DISTRIBUTION

The location or "central tendency" of a distribution refers to the place on the scale of score values where a particular distribution is centered. Consider the following five sets of scores on height of people, measured in feet. The five individuals in group (c) generally have lower heights than those in (a) or (b), while individuals in (d) and (e) are generally taller. Notice this is generally true even

group a	2	3	4	5	6	N = 5
group b	2	4	4	4	6	N = 5
group c	2	2	2	4	6	N = 5
group d	2	4	6	6	6	N = 5
group e	5	6	7	8	9	N = 5

though some individuals in the group with the tallest heights (e) have heights that are no taller than the tallest individuals in the group with the shortest individuals (c). Location is usually measured in terms of some center point or "typical" score in the distribution around which the scores tend to center or cluster in some defined way. An appropriate measure of central tendency would bring out clearly and succinctly the differences in these five distributions. There are three main location measures we will use frequently throughout this book.

5.3.1 Mode (Mo)

The **mode** is the value of the score which occurs most frequently in a collection of scores. It would be represented by the tallest column on a histogram or peak on a frequency polygon. For grouped data, the mode is the midpoint of the *class* which has the largest frequency of cases in it. The mode of group (b), above, is the score "4," for example, and the modal category of education for 20–21 year old whites in Table 5.1 is "one-year-of-college-or-more." One may choose to designate several modes, as explained in the previous section on the form of a distribution (Section 5.2.1), even though not all "modes" are of exactly equal frequency.

The mode has the virtue of being relatively easily discovered by inspection, and therefore it is often used as a first, quick index to the location of a distribution. In fact, the mode is almost never "computed" by any formula.

The mode can be used with nominal variables (*e.g.* at birth, the modal sex is "male") since it does not depend upon the ordering of scores. Stated differently, the mode ignores score information about rank order and interval size, even if it is available in the data. Distributions (f) and (g), below, have the same mode even though they are located at seemingly quite different points on a score scale if the distances between scores are taken to be meaningful.

group f	1	3	3	3	3	4	N = 6
group g	3	3	50	75	100	119	N = 6

If the variable is ordinal or interval, important locational information will be ignored by using the mode, and this loss of information is usually not desirable. The mode does not require a closed-ended, grouped distribution, provided that the modal category is not the open-ended category (so that a midpoint can be computed). As in (g), above, there may be (and, generally there are) more scores that are *not* equal to the mode than those that *are* equal to the mode. The mode is merely the particular score that most frequently occurs in the data.

TABLE 5.1 PERCENT DISTRIBUTION OF YEARS OF SCHOOL COMPLETED FOR PERSONS 20 YEARS OLD AND OVER, BY AGE AND RACE: U.S. 1970

Age and Race	Total	Less Than 4 Years High School	High School, 4 Years	College 1 Year or More	Median Years of School Completed
WHITE					
20 and 21 years old	100.0	17.7	39.5	42.8	12.8
22 to 24 years old	100.0	17.1	45.9	37.1	12.7
25 to 29 years old	100.0	22.2	45.0	32.8	12.6
30 to 35 years old	100.0	25.8	44.8	29.4	12.5
35 to 44 years old	100.0	33.1	41.9	25.0	12.4
45 to 54 years old	100.0	38.8	40.0	21.2	12.3
55 to 64 years old	100.0	53.6	28.2	18.2	11.2
65 to 74 years old	100.0	67.0	18.4	14.6	8.9
75 years old and over	100.0	75.0	13.8	11.2	8.6
NEGRO					
20 and 21 years old	100.0	32.5	43.5	24.1	12.4
22 to 24 years old	100.0	36.9	41.6	21.4	12.3
25 to 29 years old	100.0	43.9	39.0	17.1	12.2
30 to 34 years old	100.0	50.0	37.6	12.4	12.0
35 to 44 years old	100.0	58.5	29.4	12.2	11.2
45 to 54 years old	100.0	70.9	19.8	9.3	9.3
55 to 64 years old	100.0	83.1	10.7	6.3	7.9
65 to 74 years old	100.0	90.0	4.7	5.4	6.1
75 years old and over	100.0	92.5	4.4	3.1	4.6

Source: U.S. Bur. Census, 1970, P-60:9, Table 7.

5.3.2 Median (Mdn)

The median is the score value below which (and above which) half the scores in an ordered distribution fall. It is the 50th percentile. In group (e), above (in Section 5.3), the median is 7, since this score value exactly divides the odd number of scores ($N = 5$) into high and low halves. Group (g) has a median falling between the two center scores, since there are an even number of scores ($N = 6$). In this simple case, the usual procedure is to take the average of the two center scores as the median. In this case, the median would be:

$$\text{Mdn} = \frac{50 + 75}{2} = 62.5$$

In a **grouped distribution,** the median is usually computed on the assumption that cases in the category containing the median are evenly distributed throughout that interval (and, in fact, on the assumption that the interval width is a meaningful value). Diagram 5.1 below illustrates the steps involved in computing the median from grouped data by use of the following formula:

(5.1)
$$\text{Mdn} = L_{\text{mdn}} + \left(\frac{\frac{1}{2}N - \text{cum} f_{\text{mdn}}}{f_{\text{mdn}}} \right) w$$

> **BOX 5.2** PERCENTILES
>
> The 50th percentile is the *score* below which 50 percent of the scores in a distribution fall. In general, the *n*th percentile is the *score* below which *n* percent of the scores in a distribution fall. You may want to review percentiles by looking back at Section 3.2.3d. Diagram 3.2 illustrates how percentiles may be computed from grouped data. Diagram 5.1 also explains in somewhat more detail the computation of the median itself.

where: L_{mdn} is the lower limit of the category containing the median, N is the total number of cases, $\text{cum}f_{\text{mdn}}$ is the cumulative frequency up to but *not* including the frequency in the median category, f_{mdn} is the frequency in the median category, and w is the width of the median category.

This formula is identical to the formula for computing percentiles since, after all, the median is merely the 50th percentile, although symbols appropriate to the median have been inserted. We will postpone explaining the rationale behind this simple interpolation until Diagram 5.1, since most computations of the median are made by computer directly from raw data. Grouped-data formulas like the one above provide a very good approximation if the number of categories is relatively large (15 or more, for example).

From this definition, it is clear that scores must be at least ordinal before the median can be computed, since the concept of a median implies direction (ranking above and below the median). The median is an index of location that does not imply knowledge of distance, however.* This means that it, like the mode, loses some information contained in interval-level scores. This is sometimes a strength since the median, for example, is little influenced by wild and erratic extreme scores; it is simply the point on the scale of scores dividing all cases into an upper and lower half. Like the mode, the median does not require that extreme classes in a grouped distribution be closed, provided that the median does not fall in one of the open, extreme classes (in which case a category width could not be identified for use in Formula 5.1, above).

To illustrate the median, we have taken data from Reiss concerning the variable, "attitude toward premarital sexual permissiveness"—clearly an ordinal variable (Reiss, 1967:36). Reiss used five attitude questions reflecting permissiveness attitudes, and asked a sample of college students and a national sample of adults to indicate the extent to which they agreed with each of the five questions. An individual's answers were combined, and he (she) was assigned a "permissiveness score" using a scaling technique called Guttman scaling. The scores ranged from a conservative or low permissiveness score of zero to a high permissiveness score of 6. The percentage distribution of students and of adults is given in Table 5.2, below.

From an examination of the cumulative percentage distributions it is clear that slightly over half (51%) of the students have permissiveness scores of types 0, 1, or 2. The median, therefore, could be some value between 2 and 3 on the

*The exception is the width of the median interval when the grouped-data formula for the median is used.

TABLE 5.2 Percentage Distribution of Attitudes Toward Premarital Sexual Permissiveness for a College Student and Adult Sample

College Student Sample

	Permissiveness Scale Types		Percent	Cumulative Percent
	0	(low)	15	15
	1		17	32
Median →	2		19	51
(2.4)	3		18	69 ←
	4		16	85
	5	(high)	15	100
			100	
			(844)	

National Sample of Adults

	Permissiveness Scale Types		Percent	Cumulative Percent
	0	(low)	37	37
Median →	1		11	48 ←
(1.6)	2		31	79
	3		5	84
	4		6	90
	5	(high)	10	100
			100	
			(1399)	

Source: Reiss, 1967:36. Reprinted by permission of Holt, Rinehart, and Winston, Inc. Copyright © 1967 by Holt, Rinehart, and Winston, Inc.

ordinal scale of types of permissiveness, below which half of the scores fall. If the grouped formula were used, the median for these grouped data would turn out to be a value of 2.4.* For the national sample, however, it is clear that subjects are much more concentrated toward the low-permissiveness scale types, and in fact nearly half (48%) have scale type scores of 0 or 1. Again, we could use a convenient value between 1 and 2 as the point on the scale below which (about) half of the cases fell. If this were computed by the grouped formula, the value would be 1.6 for the median. These two values can be compared; the national sample of adults are "located" at a lower point on the permissiveness scale than are the college students. The median is a useful measure of location for a distribution of cases on an ordinal variable.

---

*It should be noted that the Guttman scaling procedure can result only in integers as score values. Thus the score 2.4 is not defined within the operations of that procedure. Nevertheless, the variable "permissiveness attitudes" is defined as a continuous variable, so that fractional values are conceptually meaningful. Thus the score of 2 would be assumed to have a lower limit of 1.5 and an upper limit of 2.5. Use of the grouped data formula also implies that the width of the median interval is a meaningful concept. This is not defined in an ordinal variable, so the median computed by this formula should be considered an approximation. Investigators generally are willing to make this approximation, and they use fractional values for the median of ordinal variables or of interval variables which are defined as discrete. Other than the assumptions about the category containing the median, the median uses only the ranked aspect of scores.

Table 5.3 shows another use of the median, in this instance, used on an interval variable — income — which usually has a skewed distribution with extreme or erratic scores. Table 5.3 is the income distribution for the U.S. as a whole and for household heads who have attained differing amounts of formal education. Notice that the *median* income for all households in 1969 was $8,389 and that the median increases markedly as the amount of formal education increases. In fact, for households whose head had four or more years of college, the median income ($13,362) was about three times the median income for families whose head had completed no more than seven years of elementary school. The median greatly aids in making these kinds of contrasts between the location of different income distributions. In each case, half the individuals have scores below the median income for that group, and half above.

5.3.3 Arithmetic Mean ($\overline{X}$)

The common average or **arithmetic mean** is simply the sum of all scores divided by their number, as in Formula 5.2, below.* You have used this average for some time in computing your grade point average, batting averages, etc.

(5.2)
$$\overline{X} = \frac{\Sigma X_i}{N}$$

It has some other interesting and useful characteristics beyond its value as the most commonly known and widely used measure of central tendency.

To start with, it is another example of the statistical use of the ratio as an aid to valid comparison. The aggregate or total of scores is "standardized," so to speak, in terms of the number of scores that are included in the sum. This permits one to compare "averages" for groups of different sizes, whereas a direct comparison of totaled-up scores would be misleading. Sometimes, however, the number of scores contributing to a sum is not the only source of distracting differences one may want to take into account in making comparisons of central tendency. In arriving at a sum, each score contributes a different amount depending upon its numerical value. Big scores count more than small ones in the sum, and this means that extremely large (or small) scores will tend to have a stronger influence on the mean than more modest, single scores. The arithmetic

---

*Two comments are relevant. First, the symbol for the arithmetic mean is traditionally $\overline{X}$ or μ (the Greek letter *mu*). The X-bar symbol is used for the statistic — the arithmetic mean computed on sample data — while the symbol μ is used for the parameter — the arithmetic mean computed for the population. Since sample data are generally used in sociology, we will use symbols appropriate for sample statistics. In the section of the book on inferential statistics, where the distinction becomes critical, the two different symbol usages will be made plain and observed carefully. Secondly, we should note that the summation operator, sigma, correctly includes limits of summation (see Box 3.2) written below and above the symbol as follows:

$$\overline{X} = \frac{\sum_{i=1}^{N} X_i}{N}$$

However, in statistics, summations are nearly always over all N cases, so we will simplify our notation by omitting the limits where this is true. Limits will be introduced when they are needed for clarity.

TABLE 5.3 PERCENTAGE DISTRIBUTION OF HOUSEHOLD HEADS ON 1969 INCOME, BY EDUCATION, IN THE UNITED STATES (Households as of March, 1970)

Educational Attainment of Head	All Households (Thousands)	Total	Total Household Income									
			Under $1,000	$1,000 to $1,499	$1,500 to $1,999	$2,000 to $2,499	$2,500 to $2,999	$3,000 to $3,499	$3,500 to $3,999	$4,000 to $4,999	$5,000 to $5,999	$6,000 to $6,999
ALL HOUSEHOLDS												
Total	62 874	100.0	3.2	3.4	3.4	3.4	2.8	3.1	3.1	5.7	6.0	6.4
Elementary: 0 to 7 years	9 203	100.0	7.7	9.2	8.6	6.9	5.8	5.8	4.9	8.3	7.5	6.1
8 years	8 274	100.0	4.5	6.0	5.4	5.3	4.8	4.3	4.7	8.3	7.3	6.7
High School: 1 to 3 years	10 505	100.0	3.1	3.1	3.5	3.9	3.1	3.2	3.3	6.7	7.1	7.4
4 years	19 522	100.0	2.0	1.7	1.7	1.9	1.7	2.2	2.6	4.8	5.9	7.0
College: 1 to 3 years	7 146	100.0	1.7	1.6	1.6	2.2	1.4	2.2	2.2	4.5	4.9	6.3
4 or more	8 225	100.0	1.1	0.6	1.0	1.3	0.9	1.3	1.4	2.3	3.1	3.6
HOUSEHOLDS WITH HEAD AGED 25 OR OVER												
Total	58 570	100.0	3.2	3.5	3.5	3.3	2.7	3.0	3.0	5.5	5.7	6.1
Elementary: 0 to 7 years	9 086	100.0	7.7	9.3	8.7	6.9	5.7	5.9	4.9	8.2	7.4	6.0
8 years	8 123	100.0	4.5	6.0	5.4	5.3	4.8	4.3	4.6	8.3	7.2	6.5
High School: 1 to 3 years	9 737	100.0	2.9	3.1	3.5	3.6	3.0	3.0	3.3	6.4	6.6	7.2
4 years	17 605	100.0	1.9	1.7	1.7	1.9	1.5	2.1	2.3	4.4	5.4	6.7
College: 1 to 3 years	6 249	100.0	1.5	1.5	1.6	1.8	1.3	1.7	1.7	3.8	4.0	5.9
4 or more	7 770	100.0	1.0	0.5	0.9	1.0	0.7	1.1	1.2	2.0	2.7	3.2

TABLE 5.3 (Continued)

Total Household Income

Educational Attainment of Head	$7,000 to $7,999	$8,000 to $8,999	$9,000 to $9,999	$10,000 to $11,999	$12,000 to $14,999	$15,000 to $24,999	$25,000 to $49,999	$50,000 and Over	Median Income (Dollars)	Mean Income (Dollars)
ALL HOUSEHOLDS										
Total.	6.8	6.7	6.2	11.3	11.8	13.4	2.8	0.4	8 389	9 544
Elementary: 0 to 7 years .	5.3	4.8	4.1	5.5	4.9	3.8	0.6	(Z)	4 108	5 494
8 years	6.9	6.3	5.7	8.7	7.8	6.5	0.7	(Z)	5 928	6 951
High School: 1 to 3 years .	8.1	7.0	6.3	11.4	11.3	10.3	1.1	0.1	7 687	8 454
4 years	8.0	8.3	7.4	14.2	14.2	14.0	2.1	0.1	9 275	10 003
College: 1 to 3 years .	6.7	6.8	7.0	12.7	14.7	18.8	4.2	0.5	10 145	11 314
4 or more . . .	4.2	4.9	5.0	12.1	16.1	29.0	10.4	1.8	13 362	15 452
HOUSEHOLDS WITH HEAD AGED 25 OR OVER										
Total.	6.6	6.6	6.1	11.5	12.2	14.2	3.0	0.4	8 614	9 755
Elementary: 0 to 7 years .	5.3	4.8	4.1	5.5	5.0	3.9	0.6	(Z)	4 097	5 501
8 years	6.8	6.4	5.7	8.8	7.9	6.6	0.7	(Z)	5 936	6 979
High School: 1 to 3 years .	8.1	7.0	6.5	11.7	11.8	11.0	1.1	0.1	7 919	8 667
4 years	7.6	8.2	7.4	14.6	15.0	15.3	2.3	0.2	9 640	10 329
College: 1 to 3 years .	6.4	6.4	7.0	13.5	15.9	21.0	4.7	0.6	10 836	11 960
4 or more . . .	4.0	4.9	4.9	12.2	16.5	30.2	11.0	1.9	13 756	15 916

Source: U.S. Bur. Census, 1970: Table 2.

135

mean is "pulled" toward unbalanced, extreme scores in a distribution. To state it differently, the mean is pulled toward the tail of skewed distributions; it is pulled higher in positively skewed distributions and lower in negatively skewed distributions. The median, on the other hand, uses the numerical value of the score merely to establish the rank order of a point so that the "upper" and "lower" half of the scores can be identified. Extreme scores would have a minimal effect on the median (each score has the same weight in determining ranking) but a larger effect on the arithmetic mean (where each score contributes different amounts according to its magnitude).

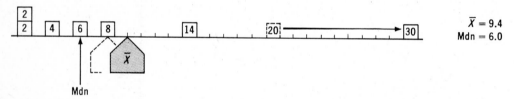

The arithmetic mean is like a balancing point or fulcrum of a lever where uniform blocks are placed on the lever at distances representing their score value. Extreme scores are farther out in both directions on the lever. If the top score in the diagram above were larger, say 30 instead of 20, the balancing point would move to a higher score point on the scale, as shown below.

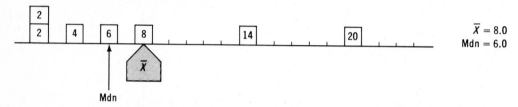

If the extreme score were 200 instead of 20, the fulcrum would have to move to the right even farther as illustrated below, a value for the mean which is quite different from the bulk of scores and perhaps gives too much weight to the single extreme score, 200. The median, on the other hand, would be the same (Mdn = 6), in all three of these situations, since the center of the distribution in the 50%–50% sense has not changed.

If one were to compute the difference between the arithmetic mean and each of the scores in the group upon which it was based as in Table 5.4, and sum these differences algebraically, another interesting and useful property of this measure of location would be evident. The algebraic sum of deviations of scores from their arithmetic mean is zero, and this is always the case, since the mean is

TABLE 5.4 DEVIATION OF SCORES FROM THE ARITHMETIC MEAN AND FROM THE MEDIAN OF A DISTRIBUTION

Raw Scores X_i	Difference $(X_i - \overline{X})$	Absolute Difference $\|X_i - \overline{X}\|$	Difference $(X_i - \text{Mdn})$	Absolute Difference $\|X_i - \text{Mdn}\|$
2	$2-8= -6$	6	$2-6= -4$	4
2	$2-8= -6$	6	$2-6= -4$	4
$N=7$ 4	$4-8= -4$	4	$4-6= -2$	2
6	$6-8= -2$	2	$6-6=\ \ 0$	0
8	$8-8=\ \ 0$	0	$8-6= +2$	2
14	$14-8= +6$	6	$14-6= +8$	8
20	$20-8=+12$	12	$20-6=+14$	14
$\Sigma X_i = 56$	0	36	+12	34

$$\overline{X} = \frac{56}{7} = 8.0$$

$$\text{Mdn} = 6$$

defined as that value around which scores balance or cancel each other's effect. Notice that the sum of deviations from the median bears out this rule. It is not zero. On the other hand, the median has an interesting property of being the point on the scale for which the sum of absolute deviations is a minimum value. In Table 5.4, the sum, 34, is smaller than the sum of absolute deviations from the mean, which is 36 in this case.

It is also a property of the arithmetic mean that it is the value for which the sum of the squared deviations of the scores from the mean is a minimum value. The sum of the squared deviations of the scores from any other point will be larger.*

For grouped distributions, the arithmetic mean can be computed only if the distribution is "closed" so that midpoints can be computed for each and every class (so that the total of all scores can be computed). This is because the midpoint is used to stand for the value of each of the scores in a class, since information on the exact value of scores is lost when a grouped frequency distribution is created. Formula (5.3) is for the mean for grouped data, and computing Diagram 5.1 explains the computational procedures involved. Note that, in this case, X_i is the class midpoint and f_i is the frequency in the ith class.

(5.3)
$$\overline{X} = \frac{\Sigma f_i X_i}{N}$$

The arithmetic mean for household income is shown in Table 5.3 in addition to the median income. Notice that in each case, the mean is higher than the median, even though it, too, increases as one moves toward higher educational attainment categories. This difference between the mean and median indicates a positive skew in the distribution of income; some extremely high scores are evident.

*The significance of this property for measures of variability will be discussed in Chapter 12.

DIAGRAM 5.1   Measures of Central Tendency

MODE (Mo)

The mode is almost always found simply by inspecting for the most frequent single raw score or, in the case of grouped distributions, the midpoint of the class which has the highest frequency of cases in it.

MEDIAN (Mdn)

Raw Scores:

If N, the number of cases, is *odd:* the median is the value of the middle score. This would be the Kth score in the array of scores, where:

$$K = \frac{N+1}{2}$$

For the following seven scores, the median is 8, which is the value of the $K = (7+1)/2 = 4$th score in the array.

$$2, 3, 5, 8, 8, 10, 12$$
$$\uparrow$$

If N is *even:* the value of the median is the average of the two center scores (or, said differently, the score which is exactly mid-way between them). This would be the point on the score scale between the Kth and $K + 1$st scores in the array, where this time $K = N/2$.

For the following six scores, the median is $(5 + 8)/2 = 6.5$. This is the value between the $K = 6/2 = 3$rd and the $K + 1 = 3 + 1 = 4$th scores in the array.

$$2, 3, 5, 8, 8, 10$$
$$\uparrow$$

Grouped Scores:

Step 1: Create a cumulative frequency distribution, starting with the lowest score category as shown in the illustration below.

Scores	f	cum f	Real Limits	Class Width
22–26	18	88	21.5–26.5	5
17–21	21	70	16.5–21.5	5
12–16	26	49	11.5–16.5	5
7–11	15	23	6.5–11.5	5
2–6	8	8	1.5–6.5	5
	$N = 88$			

Step 2: Find $N/2$, the *number of cases* that fall below the median score. Here this is 44 cases (88/2).

DIAGRAM 5.1 *(Continued)*

Step 3: Using the cumulative frequency distribution of step 1 and the results of step 2, find the category within which the median falls. This is the category with a cumulative frequency equal to or bigger than (but closest to) $N/2$.

In this example, the class 12-16 is the median category because the 44th case falls in that class. The "cum f" of 49 is just bigger than $N/2 = 44$.

Step 4: If the cumulative frequency in a class exactly equals $N/2$, then the upper class boundary is the value of the median.

If the median falls some place within a category as is usually the case, it is traditional to "interpolate" to find a median value within the median class on the assumption that cases are distributed in an even (rectangular) fashion within the median interval.

The logic is that we want to locate a median score within the median class, which is a certain distance into that class. The distance depends upon the proportion of the frequency in the median class that needs to be added to the cumulative frequency below the median class, in order to equal $N/2$, or the number of cases that should fall below the median score. This proportion is found by asking how many additional cases are needed:

$$\frac{N}{2} - \text{cum} f_{mdn}$$

(cumf_{mdn} is the cumulative frequency up to but *not* including the frequency in the median category.)

or, $$\frac{88}{2} - 23 = 21$$

This is .81 or 81 percent of the cases in the median category (21 divided by 26, which is the number of cases in the median category itself, f_{mdn}).

Since the median class is 5 score units wide, 81 percent of the way through that width is:

$$.81 \times 5 = 4.0$$

Adding this 4.0 points to the lower real limit of the median class yields a median of 15.5,

$$11.5 + 4.0 = 15.5$$

a score below which 50% or 44 of the 88 cases should fall. These relationships are shown in the "off-set" bar chart diagram below. In practice, the formula for computing the median of grouped data is used directly. It is a specific case of percentiles discussed in Section 3.2.3d.

140

DIAGRAM 5.1 (Continued)

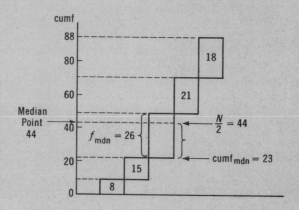

$$\text{Mdn} = L_{\text{mdn}} + \left(\frac{\frac{1}{2}N - \text{cumf}_{\text{mdn}}}{f_{\text{mdn}}}\right)w$$

where: L_{mdn} is the lower limit of the median class; N is the number of cases; cumf_{mdn} is the cumulative frequency up to *but not including* the frequency in the median class; f_{mdn} is the frequency in the median class; w is the width of the median class interval.

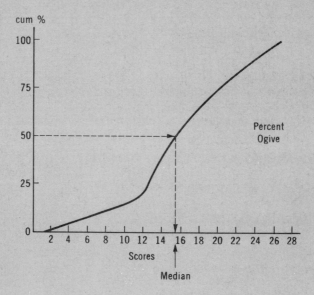

DIAGRAM 5.1 *(Continued)*

Graphically, the median could be found from a percentage ogive (plot of a cumulative percent distribution) by extending a line from the 50 percent point to the ogive and dropping down to the score axis to read off the 50th percentile or median score.

ARITHMETIC MEAN ($\overline{X}$)

Raw Scores:

Simply add up the scores, algebraically, and divide by N, the number of cases.

$$\overline{X} = \frac{\Sigma X_i}{N}$$

For example, the mean of the 8 scores at the right is:

$$\overline{X} = \frac{23}{8} = 2.88$$

X_i
0
2
3
3
−4 $N=8$
6
10
3
$\Sigma X_i = 23$

Grouped Scores:

Step 1: Create a column of class midpoints. Scores in a class will be treated as if they have the value of the midpoint of a class since no other information on their specific values is known.

Scores	X_i	f_i	$f_i(X_i)$
22–26	24	18	432
17–21	19	21	399
12–16	14	26	364
7–11	9	15	135
2–6	4	8	32
		$N=88$	1362

Step 2: Multiply, for each class, the class midpoint times the frequency in that class, $f_i(X_i)$.

Step 3: Sum the products in Step 2, $\Sigma f_i X_i$, and compute the mean by dividing by N.

$$\frac{\Sigma f_i X_i}{N}$$

The mean is, thus, 15.5 since $1362/88 = 15.5$.

5.3.4   Special Kinds of Means

There are other measures of central tendency designed for special situations, although most of these measures do not find frequent use in the social sciences. One might, for example, weight scores unequally, before combining them into a mean, by multiplying each score times some previously arranged weighting factor. If, for example, a score were simply too wild or unusual, it might be considered to be due to errors in recording information or making measurements and, accordingly, weighted zero (*i.e.* dropped). If this were done, the base of the ratio, the number of scores, would usually be reduced accordingly. Other weighting schemes take account of over- or under-represented segments of samples.

Two somewhat different kinds of means are the **geometric mean** and the **harmonic mean,** neither of which is widely used as a measure of central tendency, but both of which have specific uses. The geometric mean is the *n*th root of the product of all scores (or, using logarithms, the antilog of the average logarithm of the scores). It is used in averaging rates or scores where one expects a constant rate of change.

(5.4)
$$\text{Geometric Mean} = \sqrt[N]{(X_1)(X_2)(X_3) \cdots (X_N)}$$

We will use this average in Chapter Seven to combine two summary statistics. The harmonic mean is the reciprocal of the average reciprocal of scores and it is used to average ratios in which the numerators are constant but the denominators vary.

(5.5)
$$\text{Harmonic Mean} = \frac{N}{\Sigma(1/X_i)}$$

Both the geometric and the harmonic mean turn out to be somewhat smaller than the arithmetic mean.*

5.3.5   Selecting the Most Appropriate Measure of Central Tendency

Generally speaking, in analyzing the distribution of a variable only one of the possible measures of central tendency would be used. Its selection is largely a matter of judgment based upon the kind of data, aspect of the data to be examined, and the research question. Some of the points that might be considered are the following.

Central tendency for interval-level data is generally represented by the arithmetic mean, which takes into account the available information about distances between scores. For ranked data, the median is generally most appropriate, and for nominal data, the mode or modes.

If there are several modes, then the mode may be useful alone, or in addition to one of the other two measures, even for ordinal or interval data. If the

---

*Wallis and Roberts (1956) discuss these means and provide several suggestive examples of their application. A somewhat more detailed, illustrated discussion of the geometric and harmonic means is presented in Newman (1956: v 3, 1489–1493).

BOX 5.3   Features of a Distribution

At this point you have made it through essentially two thirds of the different kinds of features of a univariate distribution one might describe: form and central tendency. The third feature, variation, will be discussed next, followed by a bit more on form. Check yourself at this point to see if you are ready to go on. The idea of *variables* and *level of measurement* is discussed in Sections 2.1 and 2.1.1. The different kinds of distributions are defined and illustrated in Section 3.2.3 and the following few pages. The graph in Figure 4.5 illustrates differences in form and location, and Figure 5.1 illustrates differences in form, location and the degree of concentration of scores. After a comment on the use of the arithmetic mean to "smooth" curves, our next main topic will be measures of the variability of scores.

---

distribution is badly skewed, one may prefer the median to the mean, because the median would not be affected as much by unusual extreme scores. For this reason, for example, the median income of people is usually reported rather than the arithmetic mean.

If one is interested in prediction, the mode is the best value to predict if an *exact* score in a group has to be picked. More cases occur at the value of the mode than at any other *single* score value one could pick. The median has the property of exceeding and being exceeded by half of the scores, so the result of guessing the median would be an over-prediction as frequently as an under-prediction. Overall, the median produces the smallest absolute error (*i.e.* the sum of absolute deviations is less from the median than from any other point). The arithmetic mean is the score around which the aggregate of deviations (algebraically considered) is zero, so the result of guessing or predicting the value of the mean would always be an algebraic deviation which balances out. In each case, a certain kind of predictive error is minimized, and one would pick a measure of central tendency to use in prediction which depends upon the kinds of predictive error one wants to minimize.

Finally, it should be noted that it is possible to have median or arithmetic-mean measures which do not correspond to any specific score in a distribution. There may be a gap at that point where the median or mean happens to fall, but the score at that gap would still be the result of the computation. This is, of course, quite appropriate, since we are trying to develop an index which characterizes the *distribution* of scores — not a single number which will be close to each individual case. The mode, of course, is an actually observed score, and the most frequent one, at that.

The arithmetic mean is probably the most important measure of central tendency since it appears repeatedly as a part of the logic of other statistical techniques you will encounter in this book; techniques such as the variance and standard deviation, standard scores, correlation and regression. Furthermore, the sample mean is important as an estimator of the population mean, and the population mean is a key parameter in many common population distributions (especially the normal distribution).

5.3.6   Smoothing Trend Lines: The Moving Average

In Chapter 4 we introduced the idea of a line graph which might be used to plot the value of some variable through time. This would result in a trend line, but often the trend line shows a confusing picture. In addition to the general trend, most plots of real data show a "saw-toothed" pattern of cyclic or minor variations which tend to obscure the overall trend. This is shown in Figure 5.3 by the dotted lines, showing components of population change by month for the 1968 to 1972 period in the United States. If an investigator wants to show the general trend, he may want to "smooth out" some of the minor ups and downs. The arithmetic mean is sometimes used to smooth trend lines, and in this application it is called a **"moving average."**

A moving average is an average of a fixed number of scores over successive periods of time. In Figure 5.3, the heavy line represents the result of using successive 12-month periods. In our example in Table 5.5, annual net growth rate figures are given, one for each year between 1935 and 1966. We would create a moving average by deciding upon a fixed time period over which to average, say, five years. We would then add together and average the rates for the first five years (*i.e.* for 1935 through 1939), and place this value, 8.2, at the midpoint of that time span, the year 1937. This is the first of our five-year averages.

The second five-year average is created by dropping the earliest year's rate and adding in the following year's rate. Thus the rate for 1935 would be dropped, and the rate for 1940 would be added, and again these five figures would be averaged; then the resulting average would be placed by the year which is mid-way in the five-year period. Successive sets of five years are averaged in this way until the last period, 1966, is reached, as shown in Table 5.5. Notice that the two years at the extreme ends of the distribution do not have five-year average figures, because they are not the middle year for an average. The moving averages are then plotted, and the result is a smoothed trend line.

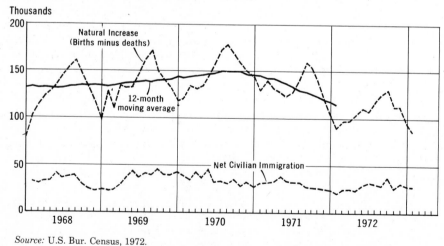

Source: U.S. Bur. Census, 1972.

FIGURE 5.3 COMPONENTS OF POPULATION CHANGE BY MONTH, U.S. 1968 TO 1972.

TABLE 5.5 ILLUSTRATION OF A FIVE-YEAR MOVING AVERAGE FOR ANNUAL RATES OF NET GROWTH, U.S. 1935–1966

Year	Net Growth Rate*	5-Year Moving Average
1935	8.0	— —
1936	7.1	— —
1937	7.9	8.2
1938	9.1	8.4
1939	8.9	9.1
1940	9.2	10.0
1941	10.3	10.8
1942	12.7	11.4
1943	13.1	
1944	11.5	
1962	15.1	
1963	14.3	
1964	13.6	13.3
1965	12.2	— —
1966	11.4	— —

Source: U.S. Bur. Census, 1971.

The longer the period over which the average is taken, the smoother the trend line. By careful choice of the time period over which averages are taken, an investigator can remove or smooth out cyclic or seasonal variations in data. This is very clearly indicated in Figure 5.3, where the 12-month moving average removes the seasonal fluctuation in the rate of natural increase over a year. The result shows the trend as it changes year by year, with seasonal variation removed.

Moving averages tend to "anticipate" marked changes in the direction of the un-smoothed graph; that is, the moving average begins to drop or rise somewhat before the point where a marked shift occurs. You will recognize this as one of the properties of the arithmetic mean; it tends to be influenced in the direction of extreme scores. This method of smoothing trend lines is useful in analytic graphing.

5.4 VARIATION

Adams (1953), in a study of the occupational origins of physicians, sent mailed questionnaires and did interviews with physicians in four northeastern cities. Their "occupational origin" was measured by using the North-Hatt occupation prestige scale on their reported father's occupation. Needless to say, physicians tend to come from families where the father also has a relatively highly prestigious occupation (lawyer, banker, physician, etc.). Separating his data by the date of birth (age) of sampled physicians, he noted that the prestige of father's occupation tended to decrease. For physicians born between 1875 and 1895 the mean of father's prestige was 79.8, and it dropped to 74.6 for the later period, 1895 to 1920. He also found an interesting second feature in the data, namely that there seemed to be more variation in the backgrounds of physicians born more recently. That is, their mean prestige rating not only dropped somewhat through time, but there was more variation in father's occupational prestige scores; they were not as similar to each other at the end of the time period as they were before. How can this feature be measured?

The third of the three characteristics of a univariate distribution, for which we are developing simple measures, is the characteristic of spread, or dispersion of scores. Some groups of scores are wide-ranging; others are more bunched together, as the following illustrations show, quite apart from their location or the distinctive features of the distribution's form.

group a	2	3	4	5	6	N = 5
group b	5	6	7	8	9	N = 5
group c	51	52	53	54	55	N = 5
group d	52	53	53	53	54	N = 5
group e	47	50	53	56	59	N = 5

In the first three groups, (a), (b), (c), each has the same amount of variability between scores. Notice that the scores are all consecutive integers. Group (d), however, has less variability since the scores are more closely grouped, many of them identical in value. Group (e) is even more dispersed than group (c), yet (c), (d), and (e) all have the same mean and median, 53. (As far as the other measure of central tendency is concerned, only group (d) could be usefully described as having a mode which is 53.)

There are several different approaches to the measurement of variability in a group of scores, and the basic distinction between measures appropriate for nominal (categorical) data and interval data is needed here as well. Variation in ordinal data is often treated by techniques used for interval-level data even though the distance idea is not defined. Ordinal data could also be treated by measures of variation appropriate to nominal data.* Diagram 5.2 (below, near Section 5.4.6.) provides the step-by-step procedures for computing each measure. We will begin with some of the measures for interval-level data.

For interval-level scores where the idea of distance on a score scale has

---

*To the authors' knowledge there is no measure of variation designed specifically for ordinal data. If each score in a set of ordinal data is unique—that is, there are no tied scores—then the spread of the set is dependent upon the number of scores there are. Thus, as many ranks are represented as there are cases or observations. If there are ties and the ordinal data are grouped into a frequency distribution, then the use of the usual measures of variability leads to the implicit assumption that the ordinal categories are separated by equal distances. And of course, ordinal measurement does not justify such an assumption.

meaning, there are two general approaches to the measurement of variation in a distribution. First, one could think in terms of the range of the scale over which the scores in a distribution actually fall. Secondly, one could think of measures of variation describing the extent to which scores in a distribution differ from some single score, such as a central-tendency score for the distribution.

5.4.1 Range

The range is simply a number representing the difference in value of the highest and the lowest score in a distribution. More accurately, it is the difference between the lower real limit of the lowest score category and the upper real limit of the upper score category in a distribution.

(5.6)
$$\text{Range} = U - L$$

In Formula (5.6), U refers to the upper real limit of the highest class in the distribution, and L refers to the lower real limit of the lowest class. Usually the range, like the mode, is quickly computed by inspection of a distribution, and this is one of its chief claims to fame.

Unfortunately, the range only depends upon the two extreme scores in a distribution, and it ignores completely the way the data between these extremes are distributed. In practical research and everyday life we often expect that extreme scores will be undependable. They may be due to some gross measurement error or, perhaps, a very rare and unusual score.

For example, if one were interested in studying multiple births he (she) might examine birth records. In most modest-sized samples of these records he (she) would undoubtedly find scores from one to three (single-birth to triplets) or a range of 3. With much larger samples, or by unusual luck in small samples, an investigator might observe a range of 5 (from single-birth to a quintuplet-birth). Only seven quintuplet births (where all survived infancy) are known to have occurred in the world in the decade of the 60's.

If sample size is increased, the crude range can only stay the same or increase, regardless of the concentration or dispersion of the rest of the scores. For these reasons, a more refined measure of variation is generally preferred.*

5.4.2 Interquartile Range (Q)

The **interquartile range** is defined as the range which includes the center fifty percent of cases in a distribution or the distance between the first and third quartiles.

(5.7)
$$Q = Q_3 - Q_1$$

Although, again, this range measure is based on the difference between two points in a distribution, these points are determined in a way that is sensitive to the concentration in the data themselves.

---

*The range is often a useful measure of variability for small samples because it is easy to compute and, in the case of small samples, is an efficient estimate of the standard deviation (discussed in section 5.4.4 below).

The first quartile is the point on the score scale below which 25 percent of the cases fall, and the third quartile is the point on the score scale below which 75 percent of the cases fall. These are determined in ways quite similar to the way the median (point below which 50 percent of the cases fall) is computed. (See Diagram 5.2 and the discussion of percentiles in Chapter 3.)

The interquartile range avoids the exclusive use of the two extreme scores, and it is thus less subject to the erratic variation in extreme scores. Distance between other points on a distribution may be used instead of quartiles. For example, the interdecile range would be the difference between the 9th and the 1st decile (the difference between the point below which 10 percent of the cases fall and the point below which 90 percent fall). Sometimes the semi-interquartile range is used, which is simply half the interquartile range. Its virtues and properties are the same as those of the interquartile range itself.

5.4.3   Average Absolute Deviation (AD)

Another kind of measure of variation is that in which the deviation of scores is measured from a central point in a distribution, usually the arithmetic mean. Although seldom used, the **average absolute deviation** illustrates nicely the principles involved in this approach.

A portion of Table 5.4 provides a good illustration of the logic of this dispersion measure.

Raw Scores X_i	Absolute Difference $\lvert X_i - \overline{X} \rvert$	
2	6	$N = 7$
2	6	
4	4	$\overline{X} = \dfrac{56}{7} = 8.0$
6	2	
8	0	
14	6	
20	12	
56	36	

You will recall from our discussion of the arithmetic mean that one of its properties is that the sum of algebraic deviations of scores from it is always zero. Thus, as a measure of dispersion, the average deviation shifts to the use of the absolute value of the deviation of each score from the mean, as shown above.* The average absolute deviation is simply the arithmetic mean of absolute deviations, another ratio.

(5.8)
$$AD = \frac{\Sigma \lvert X_i - \overline{X} \rvert}{N}$$

$$AD = \frac{36}{7} = 5.1$$

*Absolute deviations ignore the direction and thus, the sign of the deviations of scores from the mean.

On the average, these seven scores deviate 5.1 points from the arithmetic mean of this distribution. The larger the average deviation, the more variability there is between scores in the distribution. If all the scores are identical to the mean, the numerator becomes zero and the average absolute deviation is, appropriately, zero. The average deviation can become quite large; in fact, its upper limit depends upon the unit of measurement and the magnitude of variation in the data themselves.

While the average deviation is easily interpreted and relatively easily computed, another measure is generally preferred, simply because of its mathematical uses in other areas of statistics. This measure is the variance, or its square root, the standard deviation.

5.4.4 Variance and Standard Deviation (s^2 and s, respectively)*

The **variance** and **standard deviation** are similar to the average deviation in that differences between the mean and each score are used, but instead of taking the absolute value of these deviations, the square of the deviation is used. This has a similar effect to side-stepping the zero-sum-of-deviations property of the arithmetic mean, and results in a measure of dispersion for interval data that has wide applicability and some interesting connections with other topics in statistics.

The variance is simply an average of squared deviations of scores from the arithmetic mean, and the standard deviation is the square root of the variance.†

(5.9)
$$s^2 = \frac{\Sigma (X_i - \overline{X})^2}{N}$$

(5.10)
$$s = \sqrt{s^2}$$

Using the illustrative data below from a portion of Table 5.4, the variance and standard deviation can be computed. Computing diagram 5.2 (below, near Section 5.4.6) presents the step-by-step procedures and more convenient formulas for computation.

---

*Again, we should note that there are, conventionally, two sets of symbols for the variance and for the standard deviation, one for the statistic (s^2 and s) and one for the parameter (the lower-case Greek letter sigma, σ^2 and σ). In this part of the book we will use the symbols for sample statistics (s^2 and s), whereas in the parts on inferential statistics, both sets are introduced when contrasts between them are important.

†Although the variance is defined as the simple "average" of squared deviations from the mean (and thus it has N in the denominator as in Formula 5.9, above), this is a biased estimate of the population standard deviation if Formula 5.9 is used on sample data from small samples. Since we are dealing in the area of description rather than inference at this point, we will use the formula with N in the denominator. It simplifies some later computations in this book, and the "unbiasing" can be better explained and treated in inferential statistics. (See Chapter 12.) The square root of the variance is called the "standard deviation," a term coined by Karl Pearson (1894). As an aside, the sample variance as computed in Formula 5.9 yields a slight *under*-estimate of the population variance, especially for small, probability samples. This under-estimate bias is usually corrected by using $N-1$ in the denominator for the sample variance rather than N. In most sociological studies the sample size is sufficiently large so that the effect of this correction is nil.

	Raw Scores X_i	Differences $(X_i - \overline{X})$	Squared Differences $(X_1 - \overline{X})^2$
	2	$(2-8)= -6$	$(-6)^2 = 36$
	2	$(2-8)= -6$	$(-6)^2 = 36$
$N=7$	4	$(4-8)= -4$	$(-4)^2 = 16$
	6	$(6-8)= -2$	$(-2)^2 = 4$
$\overline{X}=8.0$	8	$(8-8)= 0$	$(0)^2 = 0$
	14	$(14-8)= +6$	$(6)^2 = 36$
	20	$(20-8)=\underline{+12}$	$(12)^2 =\underline{144}$
	56	0	272

$$s^2 = \frac{272}{7} = 38.9$$

$$s = \sqrt{38.9} = 6.2$$

In the study of the occupational origins of physicians discussed at the beginning of this section, Adams actually chose the standard deviation as a measure of variation to show what has been happening from 1875 to 1920 in terms of variation. Table 5.6 shows his data. The standard deviation shows the variation in physician's father's occupational prestige in units of the North-Hatt occupation prestige scale. In the earliest period the standard deviation was 7.5 North-Hatt units, while in the last period, 1915–1920, the standard deviation was nearly double, 13.6 North-Hatt score points. The average standard deviation prior to 1895 was 9.2 and the average standard deviation from 1895 on was 11.3.* There is clearly an increased variation in the prestige backgrounds of phy-

---

TABLE 5.6 TREND IN THE NORTH-HATT OCCUPATION PRESTIGE SCORES FOR THE OCCUPATION OF FATHERS OF PHYSICIANS BY DATE OF BIRTH OF PHYSICIANS.

Date of Birth	Number	Mean Prestige Rating	Standard Deviation of Prestige Rating
1875–1879	9	78.7	7.5
1880–1884	9	77.9	6.7
1885–1889	7	81.6	8.8
1890–1894	15	80.9	7.0
1895–1899	19	74.8	10.4
1900–1904	22	74.6	8.7
1905–1909	24	74.2	14.2
1910–1914	23	74.0	10.6
1915–1920	9	76.4	13.6
Overall	137	76.1	10.7

Source: Adams, 1953: 406, Table 2. Reprinted by permission.

---

*These averages cannot simply be computed by adding up standard deviations and dividing by their number, because each standard deviation is based on a different number of cases. The averages quoted in the text are weighted averages.

sicians in addition to a slightly declining average level of father's occupational prestige. This finding is backed up by other data Adams collected. Physicians have a 93 on the 100-point North-Hatt rating scale, and if many physicians had fathers who were also physicians, the average prestige score would be higher (and the standard deviation smaller). There has been, however, a decline in the percentage of physicians who come from families where their father is also a physician. For physicians born in the 1870–1879 period, 22 percent came from physician-families, while the figure drops to 9.4 percent from physician-families among those born in the 1910 to 1920 period.

Apart from its interpretation in terms of sheer difference in magnitude, the standard deviation has a number of interesting properties and may be interpreted in at least two different ways. To start with, the standard deviation and variance will both be zero where all scores have the same value (*i.e.*, the value of the mean), and they reach a maximum magnitude for a given set of data when scores are divided between the extreme ends of the scale. These conditions are illustrated in Figure 5.4. Given some difference between scores, the value of the variance can be interpreted in terms of a scale extending from a minimum possible value which equals the range divided by the square root of twice the sample size, to a maximum which is the square of half the range. The standard deviation could be interpreted in terms of a scale extending between values which are simply the square root of the values given above.

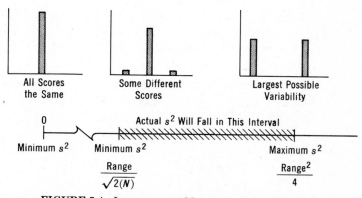

FIGURE 5.4 Limits on the Magnitude of the Variance

For most curves, the range is approximately equal to six times the standard deviation as a rough rule of thumb.

The variance and standard deviation can be interpreted in a somewhat different way too; in terms of the area under a curve within z standard deviations of the mean of a distribution.* This central area, between z standard de-

---

*The number of standard deviations a score is from the mean of its distribution is called a "standard score" or z-score. If all scores in a distribution were expressed in terms of the number of standard deviations they are from their mean, the distribution of z-scores which would result would have a mean of zero and a standard deviation of one. This simply follows from the definition of the standard score or z-score. We will discuss standard scores somewhat further later in this chapter (Section 5.4.7).

viations below the mean and z standard deviations above the mean, is always, and for any shape of distribution, at least as follows:

$$\text{Minimum percent of cases within } z \text{ standard deviations above and below the mean of a distribution of any shape} = 100\left[1 - \left(\frac{1}{z^2}\right)\right]$$

Thus at least 75 percent of the area under a curve will fall within 2 standard deviation units of the mean: $100\left[1 - (1/2^2)\right]$. This is called Chebyshev's Theorem (or Tchebycheff's inequality), and it has applicability especially in inferential statistics. In descriptive statistics it provides a useful interpretation of the standard deviation. For many curves, particularly those more concentrated around the mean, the percentage of cases within a given number of standard deviation units is much larger than the minimum given by Chebyshev's Theorem. For curves which are unimodal and symmetrical, for example, the minimum percent within $(\overline{X}+z)$'s and $(\overline{X}-z)$'s is at least as follows:

$$\text{Minimum percent of cases within } z \text{ standard deviations above and below the mean of a unimodal, symmetrical distribution} = 100\left[1 - \left(\frac{4}{9}\right)\left(\frac{1}{z^2}\right)\right]$$

The minimum percentage of cases within two standard deviations above and below a unimodal, symmetrical distribution would be 89 percent rather than 75 percent, which is the general figure for any shaped distribution.

TABLE 5.7 Percent of Area Under a Curve between the Mean Plus z and the Mean Minus z Standard Deviations

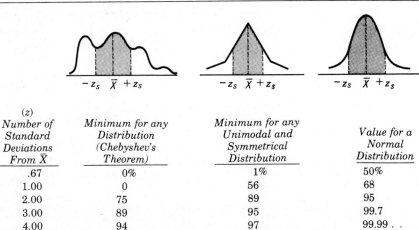

(z) Number of Standard Deviations From $\overline{X}$	Minimum for any Distribution (Chebyshev's Theorem)	Minimum for any Unimodal and Symmetrical Distribution	Value for a Normal Distribution
.67	0%	1%	50%
1.00	0	56	68
2.00	75	89	95
3.00	89	95	99.7
4.00	94	97	99.99 . .
10.00	99	99.6	99.999 . . .

In fact, the special bell-shaped curve called the "normal curve" has considerably more area within z standard deviations of the mean, as is shown in Table 5.7. The normal curve has 95 percent of its area within two standard deviations above and below the mean. The normal curve will be discussed further in Section 5.5 on the form of a distribution, since it is an important shape which is used in other topics in statistics.

5.4.5 Index of Dispersion (D)

The measures of variation discussed above all imply the knowledge of distance and thus are most appropriate for interval-level measures. The idea of variation in scores is not limited, however, to interval-level data. The index of dispersion is suggested as a measure of variation for nominal (or ordinal) variables by Hammond and Householder (1962: 136–142).*

The **index of dispersion** is a ratio whose numerator and denominator are counts of the number of pairs of scores. The denominator is the maximum number of unique pairs that could be created out of the scores such that each member of the pair is different. It turns out that this condition is met where the N scores are evenly distributed among the k categories into which the scores are grouped. This corresponds exactly with our idea of maximum variability in a nominal (and ordinal) variable.

To illustrate the logic behind this measure, suppose we consider the following nine scores on marital status, classified into three categories ($k=3$):

	f
Married	5
Single	3
Other	1
Total	9

If variability among the 9 scores were a maximum, there would be three scores in each of the three categories. The number of pairs of cases where the members of a pair came from *different* categories would be 27. That is, three pairs could be formed by pairing each of the three married people with the single people. Since there are three married people this means $3 \times 3 = 9$ pairs. The same pairing could occur between married and "other," and between single and "other" categories, yielding a total of $9 + 9 + 9 = 27$. In this example, however, cases are not evenly distributed over categories; they are more concentrated than that. The 5 married people could be paired with the 3 single (15 pairs) and with the 1 "other" person (5 pairs), and the single people could be paired with the "others" (yielding 3 pairs), or a total of 23 pairs. The D statistic is simply a ratio of the number of different pairs that could be made out of the data at hand, compared with the maximum number of unique pairings that could be created if cases were evenly spread over all available categories. Here $D = .85$.

If all scores were in a single category of a variable that has several possi-

*Another expression of this measure is called the "index of qualitative variation" (IQV), in Mueller and Schuessler (1961: 177-179).

ble categories, then there is maximum concentration or minimum variability, and D would equal zero. On the other hand, if cases were evenly distributed among the possible categories, there would be maximum variability, and the numerator and denominator of the D ratio would be the same, and D would equal 1. D varies, then, from zero to one and is a useful measure of variation for nominal or ordinal variables.

The computational formula for D is given below and the step by step procedures for computing D are discussed below in Diagram 5.2.*

(5.11)
$$D = \frac{k(N^2 - \Sigma f_i^2)}{N^2(k-1)}$$

where: N is the number of scores; k is the number of categories of the variable into which data might be classified (apart from whether all categories are used or not); f_i is the frequency of cases in the ith category; and the summation is over the squared category frequency of all categories.

An application of the index of dispersion might be made on data presented by Linn (1971), who studied the career patterns and preferences of 785 women dentists in the United States. Among the questions asked in the 1968 mail questionnaire was one requesting the subject to check items which were found to be most satisfying, second most satisfying, etc. from a list Linn provided. He subsequently compared the distribution of sources of satisfaction for women dentists who were married and those who were separated, single, divorced, or widowed.

Table 5.8 shows that the modal activity giving the most satisfaction shifted from family relationships for married dentists to career or occupation for the dentists who were not married, and the percentage checking the mode in each case was 68 and 64 percent, suggesting a good deal of agreement on priorities. As D, the index of dispersion, indicates, the variation in sources of satisfaction is also greater among those not currently married as compared to currently married women dentists, suggesting a focusing of sources of satisfaction around fewer areas for the latter.

The D statistic (with slightly different symbols used to express it) was proposed by Rushing and Davies (1970) as a measure of the concept, "division of labor." The concept of division of labor refers to the difference or variability there is among individuals in their sustenance activity. Generally this can be thought of in terms of the different kinds of occupations there are in a society, and the extent to which individuals are spread out among them rather than all concentrated in, say, laboring or farming or housewifing. Division of labor, then, could be measured by D, using information about the number of individuals in different occupational categories. The more evenly spread they are among the

---

*Maurice G. Kendall defined a measure of variation which examined all pairs of scores in a set of data. The measure is equal to one half of the average squared difference between all possible pairs of scores. This one can be computed as:

$$\text{Variation} = (N^2 - f_1^2 - f_2^2 \cdots - f_k^2)/2N^2$$

where f_1^2 is the frequency squared in category one of the variable and the square of the frequency in each of the "k" categories is subtracted in the numerator. This is a slightly different expression of the identical formula, above (number 5.11), and it has several useful implications for sociological analysis, as shown in Hawkes (1971).

DIAGRAM 5.2 Computing Measures of Variation and Form

As is true of central tendency, measures of variation and form may be computed on raw scores or on grouped frequency distributions. Both procedures are illustrated below using the same set of scores on number of items correct on a 20-item exam in arithmetic. The 32 scores are listed and the same scores are shown in a grouped distribution below:

Raw scores:

2 3 4 4 5 6 6 7 | 7 7 8 8 8 8 9 9 9 9 9 10 10 10 11 11 | 11 11 12 12 12 12 15 16

Q_1 Q_3

$\Sigma X_i = 281$ $\Sigma X_i^2 = 2791$

Grouped distribution: (Using the data, above)

	Class	X_i Midpoint	f_i	cum f_i	f_iX_i	$f_iX_i^2$
	2–3	2.5	2	2	5.0	12.50
Class	4–5	4.5	3	5	13.5	60.75
Containing Q_1 →	6–7	6.5	5	10	32.5	211.25
Class	8–9	8.5	9	19	76.5	650.25
Containing Q_3 →	10–11	10.5	7	26	73.5	771.75
	12–13	12.5	4	30	50.0	625.00
	14–15	14.5	1	31	14.5	210.25
	16–17	16.5	1	32	16.5	272.25
	Total		$\Sigma f_i = 32$		$\Sigma f_iX_i = 282.0$	$\Sigma f_iX_i^2 = 2814.00$

1. RANGE (Formula 5.6)

$$\text{Range} = U - L$$

where: U is the upper real limit of the highest score (or class), and L is the lower real limit of the lowest score (or class).

For raw data: Range $= 16.5 - 1.5 = 15$
For grouped data: Range $= 17.5 - 1.5 = 16$

The difference between the two values for raw and grouped data is due to the way data were grouped. The U value of the grouped data is 17.5.

2. INTERQUARTILE RANGE (Formula 5.7)

$$Q = Q_3 - Q_1$$

where: Q_3 is the *score value* below which 75 percent of the cases fall, and Q_1 is the score value below which 25 percent fall.

DIAGRAM 5.2 *(Continued)*

For raw scores, Q_3 and Q_1 may be found by creating an array as shown in the raw data above, and counting up the array to the respective score values. If a quartile falls between two actual scores, it is traditional to simply take as the quartile score value the average of the two adjacent scores. In this example:

$$Q_3 = 11$$

$$Q_1 = 7$$

$$Q = 11 - 7 = 4$$

For grouped data, quartiles are found in the same way that any percentile is found. This is discussed with respect to the median (the second quartile) in Diagram 5.1 and in Diagram 3.2.

(D1)
$$Q_3 = L_{Q_3} + \left(\frac{\tfrac{3}{4}N - \operatorname{cum} f_{Q_3}}{f_{Q_3}} \right) w$$

(D2)
$$Q_1 = L_{Q_1} + \left(\frac{\tfrac{1}{4}N - \operatorname{cum} f_{Q_1}}{f_{Q_1}} \right) w$$

where L_{Q_3} and L_{Q_1} are lower limits of the classes containing the third and first quartiles; N is the number of cases; w is the width of the class containing the quartile of interest; *cum* f_{Q_3} and *cum* f_{Q_1} are cumulative frequency of cases *up to but not including* cases in the class containing the quartile being computed; f_{Q_3} and f_{Q_1} are frequency of cases in the quartile class being computed. In these data:

Step 1: Find the class containing the quartile of interest by using a cumulative frequency distribution and finding the class below which the $3N/4$-th case would fall for the third quartile, or the $N/4$-th case for the first quartile. For the first quartile that class is the $6-7$ score class (since the $\tfrac{1}{4}$ of the 32 scores is 8, and the 8th score falls in the $6-7$ class). The third quartile is the $10-11$ score class because it contains the 24th score out of 32.

Step 2: Substitute values in the formulas for Q_3 and Q_1 and compute the estimated third and first quartiles.

$$Q_3 = 9.5 + \left(\frac{\tfrac{3}{4}(32) - 19}{7} \right)(2) = 10.9$$

$$Q_1 = 5.5 + \left(\frac{\tfrac{1}{4}\,32 - 5}{5} \right)(2) = 6.7$$

Step 3: Compute the interquartile range.

$$Q = 10.9 - 6.7 = 4.2$$

A similar procedure could be used to compute other percentiles (or other "fractiles"), and other ranges, such as the interdecile range, could be used.

DIAGRAM 5.2 *(Continued)*

3.3 VARIANCE AND STANDARD DEVIATION

   The definitional formulae for variance and standard deviation are as follows:

$$s^2 = \frac{\Sigma(X_i - \bar{X})^2}{N} \qquad \text{Variance* (Formula 5.9)}$$

$$s = \sqrt{s^2} \qquad \text{Standard Deviation (Formula 5.10)}$$

These formulae are rarely used in computational work because they require the extra step of computing the arithmetic mean and finding the deviation of each score from the mean—steps which may result in extensive rounding error. Computational formulae given below make use of raw scores (or category midpoints and frequencies in the case of grouped data) and are mathematically identical to the definitional formulae above.

For raw data:

(D3)
$$s^2 = \frac{\Sigma X_i^2 - (\Sigma X_i)^2/N}{N}.$$

where: ΣX_i^2 is the sum of *squared scores;* $(\Sigma X_i)^2$ is the square of the *sum of scores;* N is the number of cases.

   As before, the standard deviation is the square root of the variance.
   In the example above, where $\Sigma X_i = 281$, $\Sigma X_i^2 = 2791$, and $N = 32$

$$s^2 = \frac{2791 - (281)^2/32}{32} = \frac{323.47}{32} = 10.1$$

$$s = \sqrt{10.1} = 3.2$$

For grouped data:

   The formula for the variance of grouped data is similar to that given above, except that category midpoints and frequencies are used rather than raw scores. The formula for the variance is this:

(D4)
$$s^2 = \frac{\Sigma f_i X_i^2 - (\Sigma f_i X_i)^2/N}{N}$$

where $\Sigma f_i X_i^2$ is the square of the category midpoint times the category frequency summed up over all categories; $(\Sigma f_i X_i)^2$ is the sum of category midpoint and category frequency products, squared.

---

*See the first two footnotes in Section 5.4.4 in this chapter. The variance given here is a "biased" estimate of a population variance. Where inferences to population variances on the basis of sample variances is of interest, the unbiased sample variance is used. It simply substitutes $N-1$ in place of N in the denominator of the variance. Where N is large, this makes little numeric difference, but the resulting variance is referred to as "unbiased." This topic is explored in the field of inferential statistics (see Chapter 12). Here we will use the somewhat simpler formula given above.

DIAGRAM 5.2 *(Continued)*

In this example, these two sums can be found by creating two columns of products as shown above. The result is this:

$$s^2 = \frac{2814 - (282)^2/32}{32} = \frac{328.88}{32} = 10.3$$

$$s = \sqrt{10.3} = 3.2$$

4. INDEX OF DISPERSION (Formula 5.11)

$$D = \frac{k(N^2 - \Sigma f_i^2)}{N^2(k-1)}$$

where k is the number of categories of the variable; Σf_i^2 is the sum of squared frequencies; N is the number of cases.

In the example, above, the scores would be grouped into a frequency distribution prior to computing D, and a new column of f_i^2 values, not shown above, would be created and summed as follows:

f_i	f_i^2
2	4
3	9
5	25
9	81
7	49
4	16
1	1
1	1
32	186

D is then computed as:

$$D = \frac{8(32^2 - 186)}{32^2(8-1)} = \frac{6704}{7168} = .94$$

5. Sk, A MEASURE OF SKEWNESS (Formula 5.14)

$$Sk = \frac{m_3}{m_2^{3/2}}$$

where m_3 is the third moment, and m_2 is the variance or second moment, defined under part 3, above. Since the square root of the variance is the standard deviation (s), the denominator may be expressed more simply as the cube of the standard deviation or s^3. Thus, the formula may be expressed alternatively as follows:

$$Sk = \frac{m_3}{s^3}$$

DIAGRAM 5.2 *(Continued)*

The third moment may be computed from raw scores by the following formula.

$$m_3 = \frac{\Sigma X_i^3 - [3\Sigma X_i \Sigma X_i^2/N] + [2(\Sigma X_i)^3/N^2]}{N}$$

The grouped-data formula is similar with the substitution of category midpoints and category frequencies for raw scores. In the example, above, where $\Sigma X_i^3 = 30149$

$$m_3 = \frac{30149 - [3(281)(2791)/32] + [2(281)^3/32^2]}{32} = \frac{-40.4}{32} = -1.26$$

where 3.2 is the standard deviation computed above in section 3.3 of this Diagram.

$$Sk = \frac{-1.26}{(3.2)^3} = \frac{-1.26}{32.77} = -.038$$

6. Ku, A Measure of Kurtosis (Formula 5.15)

$$Ku = \frac{m_4}{m_2^2}$$

where m_4 is the fourth moment, and m_2^2 is the square of the variance.

The fourth moment may be computed from raw scores by the following formula.

$$m_4 = \frac{\Sigma X_i^4 - [4\Sigma X_i \Sigma X_i^3/N] + [6(\Sigma X_i)^2 \Sigma X_i^2/N^2] - [3(\Sigma X_i)^4/N^3]}{N}$$

In this example, where $\Sigma X_i^4 = 347887$

$$m_4 = \frac{347887 - [4(281)(30149)/32] + [6(281)^2(2791)/32^2] - [3(281)^4/32^3]}{32}$$

$$m_4 = \frac{9376.72}{32} = 293.02$$

$$Ku = \frac{293.02}{(10.6)^2} = 2.61$$

TABLE 5.8 ACTIVITIES INDICATED BY WOMEN DENTISTS AS THOSE WHICH PROVIDE THE MOST SATISFACTION

Activities Listed	Married Women Dentists		Separated, Single, Divorced, or Widowed Women Dentists	
Your career or occupation	27%	(101)	64%	(133)
Family relationships	68%	(253)	24%	(50)
Leisure time recreational activities	3%	(11)	4%	(8)
Religious activities	2%	(7)	5%	(10)
Participation in the public affairs of your community	0%	(0)	1%	(2)
Others (written in)	0%	(0)	2%	(4)
Totals	100%	(372)	100%	(207)
	$D=.56$		$D=.63$	

Source: Linn, 1971: 401, Table 4. Reprinted by permission.

different possible occupations, the greater the "division of labor." The amount of division of labor could be expressed for societies, cities, states, and within organizations. It could also be expressed for different groups of people. Undoubtedly the division of labor for women in the United States is less than it is for men. Besides noting the relative concentration of individuals within occupational categories, it would be useful to note the number of different categories (k) that exist for given organizations or societies, as Rushing and Davies note. The Rushing and Davies suggestion indicates a rather important usage of statistics: to measure an interesting characteristic of social organization. In addition, of course, statistics are used to summarize collections of such scores.

5.4.6 Selecting the Most Appropriate Measure of Variability

Several criteria might enter into the selection of a measure of variability. One, of course, is the meaning of the scores one has. For interval-level scores, the interquartile range or, more likely, the standard deviation (or variance) would be chosen. Since range-based measures tend to be sensitive to only two scores or points on the distribution, they would probably not be selected if the standard deviation could be computed. If a distribution is severely skewed, so that the mean is thought not to give an appropriate indication of central tendency, the range-based measures may then be preferred.

Ordinal data present something of a problem. The index of dispersion, D, is not sensitive to the ordering of categories implied in ordinal variables; thus D loses some information. On the other hand, measures which rely on distances, such as the interquartile range or the standard deviation, imply information in data which are not defined into the scores. The usually recommended halfway house is to use the median for central tendency and the interquartile range for variation of ordinal data, interpreting the interquartile range as the range of ranked categories which include the middle fifty percent of cases.

For nominal, and perhaps ordinal variables, the index of dispersion provides a nice solution to the problem of measuring the variation in scores, D has another feature which recommends its use. It can be interpreted more readily because it varies on a scale from zero (for no variation) to 1.0 (which corresponds to the maximum amount of variation possible in the data at hand). Although the other measures of variation indicate greater amounts of variation by larger magnitudes, the maximum possible magnitude of the variation measures, other than for D, varies depending upon the size of the score units used (*e.g.* years or decades) and the spread of the scores. Zero may indicate no variation (for the range and for the variance and standard deviation), but beyond that the magnitude of the number has little meaning in an absolute sense.

5.4.6a The Coefficient of Variation. This state of affairs has led to the development of some relative measures of variation. For example, the coefficient of variation is the standard deviation divided by the arithmetic mean.

(5.12)
$$V = \frac{s}{\overline{X}}$$

This serves to take into account the units of measurement (*e.g.* years or decades) and the location of the distribution. This measure is not used very frequently, in part because it also does not have defined limits (unlike the D statistic). Re-

---

BOX 5.4 REVIEWING FEATURES OF A DISTRIBUTION

Form, central tendency, and variation, the three features of a univariate distribution, may each be measured in several different ways. Thus far only verbal descriptions of the form — kurtosis (Section 5.2.3), symmetry (Section 5.2.2) and number of modes (Section 5.2.1) — of a distribution have been discussed, but the next section will introduce some indices for these features as well. For the central tendency of a distribution we have discussed the mode (Section 5.3.1), the median (Section 5.3.2) and the arithmetic mean (Section 5.3.3), plus a couple of special types of means which are not too frequently seen. Finally, variation measures included the range (Section 5.4.1), the interquartile range (Section 5.4.2), the average absolute deviation (Section 5.4.3), the variance and standard deviation (Section 5.4.4), the index of dispersion, D (Section 5.4.5) and the coefficient of variation (Section 5.4.6a).

An important point you should be clear about is the three different ways that the standard deviation (and variance) can be interpreted: (a) in terms of sheer magnitude (Section 5.4), (b) in terms of the minimum and maximum values it could take for a given set of data (Section 5.4.1), and (c) in terms of the proportion of the area under curves between points defined in terms of the number of standard deviations there are on either side of the mean (Section 5.4.4).

cently two sociologists developed a relative measure of variation which does vary between the limits 0.0 and 1.0 (this is known as a "normed" measure of variation, since its maximum and minimum possible values are fixed at zero and 1.0). (See Martin and Gray, 1971.*)

5.4.7 Standard Scores

Early in Chapter 3 we discussed the various types of comparisons that one might well want to make, statistically. Many of our procedures in Chapters 3, 4, and 5 were designed to facilitate the group-group or group-standard types of comparison. One could also make use of what we have described up to this point to indicate the relative standing of an individual in his group. One of these ways is to compute an individual's *percentile rank* (*i.e.* the percentage of all scores equal to or less than his score). Another way to make this individual-group comparison is to create **standard scores** which are also called **z-scores.** A standard score is merely the number of standard deviation units an individual falls above (or below) the mean of his group.

(5.13)
$$z = \frac{(X_i - \overline{X})}{s}$$

The standard score takes out the effect of the mean (by subtraction) and expresses the difference in standard deviation units (by dividing by the standard deviation). It is a simple re-scaling of scores so that the mean of z-scores is zero and the standard deviation is one—a change from dollars and years, for example, to "standard" units.† Because of this, one could compare z-scores of an individual on different distributions (*i.e.*, he may be 2 standard deviations above the mean on one score distribution and only 1 standard deviation above on another). More will be said about standard scores later in this chapter.

5.5 FORM OF A DISTRIBUTION—II

Thus far, we have described the form of a distribution only in terms of some general verbal labels which pointed to symmetry, kurtosis, and number of peaks in a general way. In this section we will introduce some approaches to summary measures of form that might be expressed as single numbers, which are similar in purpose to the summary index numbers used to described central tendency

---

*Their measure, called $s(d)$, is simply a ratio of the actual value of V, the coefficient of variation computed in Formula 5.12, above, divided by the maximum possible V.

$$s(d) = \frac{V}{V_{max}}$$

where $V = s/\overline{X}$ and $V_{max} = \sqrt{N-1}$. They also propose a variation ratio which is the average absolute deviation divided by the arithmetic mean. Its maximum, and the denominator of $s(d)$ where it is used is $2(1 - 1/N)$.

†An interesting property of z-scores, which will be used later on in Chapter 7 when we discuss a correlation coefficient, is that the sum of squared z-scores always equals the number of cases, N. $\Sigma z^2 = N$

and variation in a distribution. In the process, we will discuss the idea of standard scores (z-scores) and special kinds of curves such as the standard normal curve.

As a matter of fact, we already have a number of tools at hand which would be useful in creating an index of skewness. If the mean and median are different, for example, we would conclude that the distribution is skewed in the direction of the mean and, for the same range of scores, bigger discrepancies would indicate more skewness. We could also examine differences between quartiles in a distribution. If the distance between the first and second quartiles is greater than the distance between the second and third, we would conclude that the distribution is negatively skewed. In this section, however, we will develop more useful and interpretable measures of skewness and kurtosis.

5.5.1 The Central Moment System

When one is dealing with interval-level data, it is often useful to describe the data in terms of its balance around some central point. The arithmetic mean, for example, is the point around which the algebraic "balance" of scores is perfect in the sense that the algebraic sum of deviations of scores is zero. Deviation of scores from the mean of a distribution is often expressed by the small letter x.

$$x = (X_i - \bar{X})$$

The **first moment** about the arithmetic mean or first central moment, then, is simply the average of the first power of deviations from the mean, or:

$$m_1 = \frac{\Sigma x}{N}$$

Since the sum of deviations around the mean is always zero, the first moment is always zero, a defining characteristic of the mean. If higher powers of deviations about the mean are examined, however, additional information about a distribution is revealed. The **second central moment,** for example, is the variance:

$$m_2 = \frac{\Sigma x^2}{N}$$

Statisticians generally consider only the first four moments. The **third** and **fourth central moments** are simply the average of the third and fourth power of deviations from the mean.

$$m_3 = \frac{\Sigma x^3}{N}$$

$$m_4 = \frac{\Sigma x^4}{N}$$

The advantage of each of the moments stems from two factors: (a) the fact that even powers have the effect of eliminating negative signs but odd powers preserve negative signs in the numerator of the moments, above, and (b) the fact that higher powers tend to emphasize larger deviations from the mean, as illustrated below:

(x)	Power To Which x Is Raised			
Deviation	1st	2nd	3rd	4th
-9	-9	$+81$	-729	$+6561$
$+1$	$+1$	$+ 1$	$+ 1$	$+ 1$
$+1$	$+1$	$+ 1$	$+ 1$	$+ 1$
$+2$	$+2$	$+ 4$	$+ 8$	$+ 16$
$+5$	$+5$	$+25$	$+125$	$+ 625$
Totals	0	$+90$	-594	$+7204$

The third moment provides an index of skewness because it is an odd moment; thus if the high and low scores do not balance out around the mean (as they do not in the example above), it will not be equal to zero. Also, it is a higher moment, and thus it emphasizes the extreme deviations there may be from the arithmetic mean. The fourth moment is an even moment, thus it does not distinguish between deviations above or below the mean. It is a higher moment, so that it emphasizes the deviation of scores which fall in the tails or extremes of a distribution. Thus the fourth moment is useful in measuring the degree of kurtosis in a distribution.

None of the moments are relative measures. That is, they range in magnitude over a scale which starts at zero and extends upward to a value which depends upon the units of measurement and the variability of scores. For this reason, skewness and kurtosis measures are sometimes created in such a way that the range of index values is over a defined scale. This would for example, permit one to compare skewness and kurtosis scores regardless of units of measurement. Two such measures, Sk and $Ku,$ are sometimes created out of the moment system:

(5.14)
$$Sk = \frac{m_3}{m_2^{3/2}}$$
(Skewness)

(5.15)
$$Ku = \frac{m_4}{m_2^2}$$
(Kurtosis)

5.5.1a Skewness (Sk). The *skewness* measure, $Sk,$ is the ratio of the third central moment divided by the cube of the square root of the second central moment. Since the second central moment is the variance, its square root is the standard deviation (s); therefore, the denominator of formula 5.14 can be expressed more simply as s^3. Thus, the formula becomes

$$Sk = \frac{m_3}{s^3}$$

If the distribution is symmetrical, the third moment will be zero and thus the Sk measure will be zero. If the distribution is skewed to the right, Sk will be a positive value, and if it is skewed to the left, Sk will be negative. The magnitude of Sk expresses the relative amount of skewness and can be compared between distributions of different units of measurement. An Sk value of $+1$ represents rather extreme positive skewness and a value of -1, rather extreme negative skewness.

5.5.1b   Kurtosis (Ku).   The *kurtosis* measure, Ku, is the ratio of the fourth moment to the square of the second moment, and it too is a relative measure. Small values of Ku indicate a platykurtic (flatter than normal) distribution, and high values indicate a leptokurtic distribution. The normal or mesokurtic distribution has a Ku value of three.

5.5.2   The Standard Normal Distribution

The so-called "normal curve" is only one possible shape of a distribution, but it is a frequently used shape in statistics because it usefully describes a large number of chance distributions which are explored at great length in the field of inferential statistics (see Chapter 12 and the chapters following). Here it is useful to mention for two reasons, first, because it is the "mesokurtic" normal curve, and secondly, because it is helpful in explaining some of the usefulness of the standard deviation as a measure of variation. Platykurtic and leptokurtic distributions are, quite aptly, less peaked and more peaked than this normal one. As noted above, the normal curve has a Ku value of 3.0 and, because it is a symmetrical curve, it has a Sk value of zero. It is a unimodal, bell-shaped distribution which can be precisely described in terms of the following formula:

(5.16)
$$Y = \frac{1}{s\sqrt{2\pi}} e^{\frac{1}{2}\left(\frac{X-\overline{X}}{S}\right)^2}$$

where Y is the value on the ordinate of a graph; X is the score value on the abscissa, e is the base of the natural logarithms (2.71828), raised to the power indicated by the somewhat complicated exponent attached to it; and π is pi (3.14159).

Notice that the formula includes s, the standard deviation, as well as the mean. It is not at all necessary to remember this formula; it is only introduced to indicate that the normal curve is a specific form of curve which is a function of the mean and standard deviation.

5.5.2a   The Standard Normal Curve.   There is a family of normal distributions each having different means and standard deviations and with scores expressed in original units (*e.g.* tons, pounds). In statistics, however, it is often useful to think of a normal curve whose scores are expressed as standard scores (z-scores) rather than in terms of original units. Such a normal curve is called a *standardized normal curve*. It has a mean of zero, a standard deviation of 1.0, and an area which is set as equal to 1.0. Figure 5.5 shows a standard normal curve.

Notice that for any normal distribution the percentage of area under the curve falling between a point which is one standard deviation below the mean ($z = -1$) and one standard deviation above the mean ($z = +1$), is about 68 percent of the total area under the curve. Since 50 percent of the area under the curve is on either side of the mean (it is a symmetrical curve), there is only about 16 percent of the area under a curve falling above a z-score of $+1.0$.

5.5.2b   Area under a Normal Curve.   Since the normal curve is used rather frequently in inferential statistics, tables of areas under the normal

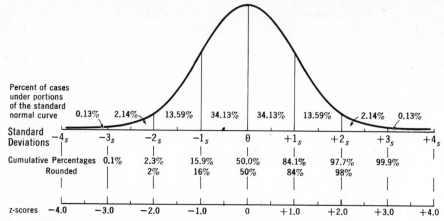

FIGURE 5.5 AREAS AND PERCENTAGES UNDER THE STANDARD NORMAL CURVE

curve have been prepared. Table B in the Appendix of this book is such a table. The first column in Table B is the z-score value. Notice that it is necessary to list only positive z-scores, since we know that the normal curve is symmetrical and thus identical values would be given for negative z-scores. The second column in Table B shows the proportion of the area under the entire normal curve which falls between the mean and a point which is a given number of standard deviation units away in a given direction. Column 3 in Table B gives the proportion of the area under the curve falling beyond a given z-score in one direction. With this table, it is possible to compute quickly the proportion of the area under a normal curve between any two z-scores.* The following table lists some of the more frequently used z-score values from Table B and corresponding percentages of the area under a normal curve.

	Percentage Of The Total Area In:	
z-score	*One-Tail*	*Both Tails Combined*
1.64	5%	10%
1.96	2.5%	5%
2.33	1%	2%
2.58	0.5%	1%

To illustrate the use of the table of areas under a normal curve and z-scores, let us consider the case of an achievement test with normally distributed scores, with a mean of 50, and with a standard deviation of 15. What proportion of the scores in the distribution will fall above the score of 66? To solve this we compute the z-score for the score 66:

$$z = \frac{66 - 50}{15} = \frac{16}{15} = +1.07$$

Looking up this value in the z-score column of Table B in the Appendix, we will take the value in column C, which is the proportion beyond a given z-score. For

*Tables are also available which give the height of the ordinate on a normal curve at various points represented by z-scores. See, for example, Hagood and Price, 1952:558.

the z-score of +1.07, this table value is .1423. Slightly more than 14% of the area under the curve is above this z-score or, in terms of the problem, slightly more than 14% of the scores in this test distribution are above the score 66. Using similar procedures, we could compute the percent of the area under a curve "below" a given z-score, between z-scores of different values, or we could compute the scores at the first and third quartiles.*

Notice in Figure 5.5 that the tails of the distribution do not touch the baseline. No matter how far the tails of the distribution are extended to the right or the left, although they get closer and closer to the baseline, they never touch— they go to infinity in each direction without touching. This property of the standard normal distribution is called *asymptosis*. It is obvious from this property that the standard normal distribution is a theoretical distribution rather than an empirical one; nevertheless, it is a very useful one as you will see. Because of this property of asymptosis the entries in column B of Table B in the Appendix (areas between the mean and z) never quite reach .5 and the entries in column C (areas beyond z) never quite reach zero.

5.5.3 Other Mathematical Descriptions of Form

Although the standard normal curve is a useful curve in statistics and one which can be described mathematically, it is but one form of curve. It was once thought that many characteristics were distributed in this way, but it is clear that this is not particularly true. The main use of the normal curve will be as a description of a certain kind of probability of sampling outcomes. For most other distributions, a description in terms of central tendency, form, and variation adequately expresses what the distribution is like. Where a mathematical description of a curve is possible, a clearly powerful tool for description is available. Examples in the social sciences include a curve describing the likelihood of the end of wars and strikes, the Lorenz curve in demography, and the J-curve hypothesis concerning adherence to social norms. Some examples of these may be found in various sociology texts, or illustrated in the data section of a workbook (such as the one designed for use with this text: *Study Guide to Accompany Descriptive and Inferential Statistics: An Introduction,* 2nd ed., by H. J. Loether and D. G. McTavish, prepared by Sally Gorelnik).

5.6 SUMMARY

In this chapter we have discussed ways of describing distributions of univariate data. Three overall features of distributions were described— *form, central tendency* and *variation* —and several indices of each feature were presented. The form of a distribution varies in terms of *kurtosis, skewness*, and number of *modes* or peaks, and it is often simply described in terms that are broad and descriptive such as a "bell-shaped" distribution. Measures of kurtosis and skewness were

---

*Note that the relationship between standard scores and areas in Table B holds only for normal distributions. These areas will not be accurate for non-normal distributions, although the amount of error involved will depend upon how far a distribution departs from the normal distribution.

developed, and mathematical formulas can be used to describe precisely the shapes of distributions. Measures of central tendency and variation were also discussed, and some of the reasons one might prefer one measure over another were presented. Typically, an investigator would describe his (her) data in terms of one index of variation, one index of central tendency, and measures of form (kurtosis, skewness and modes). The specific measures used depend on the type of data and research interests.

Comparisons of individuals and between individuals and their group are facilitated by using standard scores, and, if the curve happens to be a normal curve, tables of areas under the normal curve would permit one to readily convert a z-score into an expression about the proportion of cases in a group falling below a given z-score. In any case, measures presented in this chapter are highly useful in making valid comparisons between individuals, groups, or outside standards as dictated by one's problem.

Thus, looking back at the Cloud and Vaughan study of the control of acquiescence bias in the measurement of attitudes (Figure 5.1), we can describe the content curve (C) as bimodal, unlike the other two which are unimodal, and we can describe it as located lower on the score scale than the other two. The "yeasayer" curve (Y) is unimodal, leptokurtic, and has much less variability than the other two; also it is located between the others on the score scale. The highest curve (S) for style, is unimodal and of an average amount of variation and kurtosis. Numeric indices for these features of the distributions would have greatly aided our overall comparisons and would have aided us in drawing clearer conclusions about the differences between these distributions.

The next section of this book also deals with description, but instead of describing and contrasting univariate distributions, it will confront the problem of describing sets of univariate distributions where each individual is simultaneously characterized by scores on two different variables. This is called a bivariate distribution. Although specific tools of description vary from those we have covered, the essential principles remain: the selective losing of information with more pointed and powerful summary measures; the utility of the idea of a ratio; the description of location and variation in a dependent variable across subgroups defined in terms of the second variable; and, finally, the importance of clearly identifying the contrasts and comparisons one wants to make.

CONCEPTS TO KNOW AND UNDERSTAND

three features of a distribution
  central tendency (location)
  variation (dispersion)
  form
kurtosis
  platykurtic
  mesokurtic
  leptokurtic
unimodal, bimodal, multimodal

median
other specialized means
range
interquartile range
average absolute deviation
variance, standard deviation
index of dispersion
variation ratio
moving average

skewness
  negative
  positive
  symmetry
arithmetic mean
mode

standard scores (z-scores)
standard normal distribution
area under a normal curve
central moment system
asymptosis

QUESTIONS AND PROBLEMS

1. Select data on an interesting variable from the U.S. Census sources available to you, or from the data section of a workbook (such as the one designed for use with this text: *Study Guide to Accompany Descriptive and Inferential Statistics: An Introduction,* 2nd ed., by H. J. Loether and D. G. McTavish, prepared by Sally Goralnik). Compute appropriate measures of central tendency, variation, and form. Justify your selections and explain in your own words what each of the measures means.

2. Sometimes it is helpful at this point to set up a chart which organizes the main differences between the measures included in this chapter. Computing formulae and interpretations could be entered on such a chart for each measure. See if you can set up a chart which helps you select and interpret measures.

3. Sociology often deals with the idea of "normal" or "average" and with the idea of similarity and difference. Find one research article in a sociology journal which uses statistical measures of central tendency and variation to discuss these concepts. Explain how the authors use and interpret these measures in their work.

4. Using the scores from problem 4 of Chapter 3, pages 68–69, compute the arithmetic mean, the median, the mode, and the standard deviation. What do the three measures of central tendency tell you about the form of the distribution? Does the standard deviation adequately represent the variability of the distribution? If so, why? If not, why not? Compute the interquartile range. How does it compare with the standard deviation as a measure of variability for this distribution?

5. The table on the following page shows annual international suicide rates (per 100,000 population) for 33 countries. Compute the appropriate measures of central tendency and variability for these data. What do these measures tell you about the central tendency, dispersion, and form of the distribution?

6. Using the formula for a standard score and Table B in the Appendix find how many of the suicide rates in problem 5 above would be expected to be as high as or higher than 19.2 if the distribution of suicide rates were normal. How many in the actual distribution of scores are as high as or higher than 19.2? How closely do the two numbers correspond? Now find out how many suicide rates would be expected to be as low as or lower than 6.8 if the distribution were normal. How many in the actual distribution are as low as or lower than 6.8? How closely do the two numbers correspond? What can you conclude about the distribution from the analysis you have just performed?

ANNUAL INTERNATIONAL SUICIDE RATES (per 100,000 population)

Nation	Rate	Nation	Rate
Hungary	36.9	Australia	11.6
E. Germany	30.5	Bulgaria	11.6
Czechoslovakia	24.7	Singapore	10.9
Finland	24.0	Hong Kong	10.6
Denmark	23.8	New Zealand	9.0
Austria	21.9	Norway	9.0
Sweden	20.3	Portugal	8.6
W. Germany	19.9	Scotland	8.4
Switzerland	19.2	Netherlands	8.2
France	15.4	England & Wales	7.7
Belgium	15.4	Israel (Jews only)	6.8
Yugoslavia	13.8	Italy	5.8
Luxembourg	13.4	N. Ireland	4.5
Iceland	13.2	Spain	4.4
Canada	12.2	Greece	3.0
Poland	11.7	Ireland	3.0
United States	11.7		

Source: World Health Organization (Data for 1970, 1971, 1972, or 1973)

GENERAL REFERENCE

Hagood, Margaret, Jarman, and Daniel O. Price, *Statistics for Sociologists,* Revised Edition (New York, Henry Holt and Company), 1952
See especially Chapters 9 and 14.

OTHER REFERENCES

Adams, Stuart, "Trends in Occupational Origins of Physicians," *American Sociological Review,* 18 (1953), p 404–409.

Cloud, Jonathan, and Graham M. Vaughn, "Using Balance Scales to Control Acquiescence," *Sociometry,* 33 (June, 1970), p 193–202.

Hammond, Kenneth R., and James E. Householder, *Introduction to the Statistical Method* (New York, Alfred A. Knopf), 1962, p 136–142.

Hawkes, Roland K., "Multivariate Analysis of Ordinal Measures," *American Journal of Sociology* 76 (March, 1971), p 908–926.

Linn, Erwin L., "Women Dentists: Career and Family," *Social Problems,* 18 (Winter, 1971), p 393–404.

Martin, J. David, and Louis N. Gray, "Measurement of Relative Variation: Sociological Examples," *American Sociological Review,* 36 (June, 1971), p 496–502.

Mueller, John H., and Karl F. Schuessler, *Statistical Reasoning in Sociology,* (Boston, Houghton Mifflin), 1961, p 177–179.

Newman, James R., *The World of Mathematics,* vol. 3 (New York, Simon and Schuster), 1956, p 1489–1493.

Reiss, Ira L., *The Social Context of Premarital Sexual Permissiveness* (New York, Holt, Rinehart, and Winston), 1967, p 36.

Rushing, William A., and Vernon Davies, "Note on the Mathematical Formalization of a Measure of Division of Labor," *Social Forces,* 48 (March, 1970), p 394–396.

U.S. Bureau of the Census, "Consumer Income," *Current Population Reports,* P-60, No. 72, (Washington, D.C.), 1970, Table 2.

———, "Estimates of the Population of the United States and Components of Change: 1940 to 1971," *Current Population Reports,* P-25, No. 465, (Washington, D.C.), 1971.

———, "Selected Characteristics of Persons and Families, March, 1970," *Current Population Reports,* P-20, No. 204, (Washington, D.C.), 1970, p 9, Table 7.

———, "Estimates of the Population of the United States to September 1, 1972," *Current Population Reports,* P-25, No. 491, (Washington, D.C.), 1972.

Wallis, W. Allen, and Harry V. Roberts, *Statistics: A New Approach,* (New York, The Free Press of Glencoe), 1956, p 226–230.

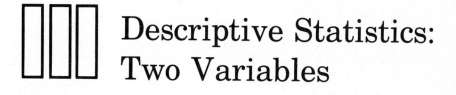 Descriptive Statistics:
Two Variables

6 Cross Classification of Variables

Sociologists concerned with urban styles of life have theorized that people living in larger cities will have greater tolerance of racial and ethnic differences, for example, than people living in smaller cities or rural areas. This is thought to be due to a number of features of the urban environment such as the greater impersonality, greater variety of contacts between people, greater concern for utility of others, and the greater use of universalistic standards for making judgments and decisions. Does the evidence support this prediction? To examine this question, Fischer (1971) used data from five Gallup polls taken between 1958 and 1965 in the U.S.

6.1 BIVARIATE DISTRIBUTIONS: AN EXAMPLE

In these polls, the question was asked, "If your party nominated a generally well-qualified man for president and he happened to be a (Negro or Jew or Catholic), would you vote for him?" Tolerance was measured by counting the number of these three minority groups for which the respondent would agree to vote (*i.e.* none, one, two, or all three). Table 6.1 presents the overall frequency and percentage distribution of tolerance scores for the 7,714 cases in the combined polls.

It is interesting to note that about 41 percent of the respondents would vote for all three minority groups—that is, they show high tolerance for ethnic and racial minorities as measured by this type of question.* It is clear, however,

---

*It is important to distinguish between tolerance views expressed in polls and behavior indicative of tolerance which may be expressed in the privacy of an election booth or in other situations.

TABLE 6.1   Tolerance Scale for 7,714 Respondents of Five Gallup Polls Taken between 1958 and 1965 in National Samples of the U.S.

Tolerance Level	f	%
3 (High)	3,147	40.8
2	2,317	30.1
1	1,229	15.9
0 (Low)	1,021	13.2
Totals	7,714	100.0%

Source: Fischer, 1971: 849. Used by permission.

that this overall distribution does not permit one to say whether the theory is supported, simply because it does not provide necessary information which would permit us to make a comparison between cities of different sizes. We need to know whether tolerance scores are distributed differently for people residing in places of different sizes, and if they are, we also need to see whether the percent of respondents in the high tolerance categories of the tolerance scale is higher in larger cities compared to smaller ones. We need to compare this univariate distribution of tolerance scores for small and medium and large cities to see if the differences are in the predicted direction.

Table 6.2 provides the detail we need. Here four separate univariate distributions (both frequency and percentage) are shown: one for only those respondents who live in rural areas; one for small places; one for medium places,

TABLE 6.2   Tolerance Scales for Respondents in Table 6.1

6.2A   TOLERANCE SCALE FOR FARM OR COUNTRY RESIDENTS

Tolerance Level	f	%
High	597	26.2
2	700	30.7
1	497	21.8
Low	485	21.3
Total	2,279	100.0%

6.2B   TOLERANCE SCALE FOR SUBJECTS LIVING IN TOWNS UNDER 25,000

Tolerance Level	f	%
High	483	36.0
2	396	29.5
1	272	20.3
Low	191	14.2
Total	1,342	100.0%

6.2C   TOLERANCE SCALE FOR RESPONDENTS LIVING IN PLACES 25,000 TO 500,000

Tolerance Level	f	%
High	738	43.4
2	536	31.5
1	242	14.2
Low	185	10.9
Total	1,701	100.0%

6.2D   TOLERANCE SCALE FOR SUBJECTS LIVING IN CITIES OVER 500,000

Tolerance Level	f	%
High	1,328	55.5
2	684	28.6
1	220	9.2
Low	160	6.7
Total	2,392	100.0%

Source: Fischer, 1971: 849, Table 2. Used by permission.

and one for large cities. In each case, however, we are looking at the distribution of the dependent variable, tolerance scores.* Comparing the "high-tolerance" end of the scale of tolerance, for example, we see that the predictions seem to be borne out: 26.2% have "high-tolerance" among rural and country people; this percentage increases to 36.0% for people living in places of under 25,000 population; it increases again to 43.4% for people living in the next larger category of cities; and reaches 55.5% scored as "high-tolerance" among people living in the largest cities. A comparison of the "low-tolerance" category reflects this trend also. Smaller places have a bigger percentage of people with "low-tolerance," and the percentages decrease as city size increases. Thus far, then, it seems to be the case that the theory holds for these data.† Notice that we arrived at this conclusion by comparing percentages rather than absolute frequencies, since it is clear that the number of respondents differs between tables we want to compare. Chapter 3 discussed this use of percentages as one means by which valid comparisons could be made in just such circumstances.‡

Up to this point we have discussed the problem of comparing tolerance scores for each category of city size as a comparison of several univariate distributions. The dependent variable is common to each of the distributions, and separate tables are distinguished in terms of categories of the independent variable, size of place. It happens to have four categories, and thus there are four separate distributions of the dependent variable to examine. A more efficient way to reach conclusions under these conditions would be to combine all of the

TABLE 6.3 PERCENTAGE DISTRIBUTION OF TOLERANCE SCORES FOR RESPONDENTS IN A COMBINED SAMPLE CONSISTING OF FIVE GALLUP POLLS OF U.S. NATIONAL SAMPLES CONDUCTED BETWEEN 1958 AND 1965, BY SIZE OF PLACE OF RESIDENCE

Tolerance Level	Size of Place of Residence				
	Farm & Country	Under 25,000	25,000 to 500,000	Over 500,000	Total
High	26.2%	36.0%	43.4%	55.5%	40.8%
2	30.7	29.5	31.5	28.6	30.1
1	21.8	20.3	14.2	9.2	15.9
Low	21.3	14.2	10.9	6.7	13.2
Totals	100.0%	100.0%	100.0%	100.0%	100.0%
	(2279)	(1342)	(1701)	(2392)	(7714)

Source: See Tables 6.1 and 6.2

*Recall from Chapter 2 that a dependent variable is either the primary variable whose variation one is interested in, or the effect or outcome of variation in an independent or causal variable. The role variables play in a study depends upon the investigator's theoretical reasoning about what influences what.
†Actually Fischer presents a more exciting analysis, tracing out differences in the distribution of tolerance by such other variables as occupation, race, region, and religion. These appear to be important conditions which govern the distribution of tolerance, and this leads to important questions about the explanatory importance of city size alone.
‡We might also have made the comparisons from table to table in terms of some set of measures of central tendency, form, and variation, as discussed in Chapter 5. Certainly a comparison of medians, for example, would have reduced some of the detail involved in comparing all of the percentages shown in Table 6.2 above.

separate tables (*i.e.* Tables 6.2A, 6.2B, 6.2C, 6.2D and Table 6.1) into one overall table as shown in Table 6.3. This table permits us to compare the separate groups more readily and it leads to a type of summary of the whole table in terms of a relationship between the two variables in general. Table 6.3 is known as a **bivariate percentage distribution** because it permits one to examine the percentage distribution of one variable (the dependent variable) within the different categories of the other variable. The comparison ideas behind such cross-classifications are the basis of analysis in sociology, because the science attempts to develop theoretical statements about the relationship of variables and the conditions under which these occur.

6.2   CONDITIONAL DISTRIBUTIONS

A bivariate distribution, such as that shown in Table 6.3, permits one to examine not only the overall distribution of some dependent variable, but also some of the conditions which may be thought to influence how that variable is distributed. The theory suggested that under certain conditions tolerance would be higher than under other conditions. In Table 6.3 these conditions corresponded to different sizes of place of residence, but one could think of other conditions which may have a bearing upon the level of tolerance. In fact, one could think of a scientific discipline as one which is searching for the kinds of conditions which help predict and explain the level of some kind of phenomenon. A table merely puts together a related set of **conditional distributions** and an overall total distribution of some dependent variable.

Our objective in this chapter is to explore some of the characteristics of bivariate distributions or cross-classifications of two variables. The following chapter will discuss some of the more useful index numbers which can be used to summarize variables that are related to each other. This is a road we have already traveled once before in previous chapters. In univariate descriptive statistics we started with a set of raw scores or a distribution, and then we asked if there were ways it could be summarized in terms of a few overall index numbers. We developed several measures for each of the three features of a distribution: central tendency, form, and variation. Our objective in bivariate description is quite similar, and, in many respects, simpler, more interesting, and more useful in sociological inquiry.

6.3   HOW TO SET UP AND EXAMINE TABLES

A cross-classification of two variables requires a table with rows and columns. The categories of one variable are labels for the rows, and the categories of the other variable are used to label the columns. Usually, where there is a dependent variable, it is used as the row variable, and the independent variable is used as the column variable, but this tradition is sometimes broken.

6.3.1 Creating a Bivariate Frequency Distribution

To illustrate how a table is constructed, we could think of 13 sets of scores. Figure 6.4 shows a tolerance score and a sex score for each of 13 individuals. The bivariate distribution is set up as shown, with the categories of tolerance (here two categories, high and low) down the **stub** or side of the table, and the categories of sex across the **heading** at the top.

This table is a **2 by 2 table,** or **fourfold table,** because it has two rows and two columns (or four cells in the **body** of the table where the rows and columns intersect). Tables can, of course, have any number of rows or columns (in general we refer to an r by c table where r refers to the number of rows and c refers to the number of columns); r and c depend upon the number of categories which are distinguished for row and column variables.

The problem now is to count the number of cases which have various possible combinations of values on the two variables, and to enter these totals into the table to form a bivariate frequency distribution. Notice that three of the 13 cases in this sample are males who are "high" on tolerance. Three are males who are "low" on tolerance, five are females who are "high" on tolerance, and two are females who are "low" on tolerance. These numbers are written in the

TABLE 6.4 A FREQUENCY TABLE

Score Key	Case I.D.	Tolerance Score	Sex Score
M = male	A	L	M
F = female	B	L	M
H = high	C	H	M
L = low	D	H	M
	E	H	F
	F	L	F
	G	H	F
	H	H	M
	I	L	F
	J	L	M
	K	H	F
	L	H	F
	M	H	F

TABLE OF TOLERANCE LEVEL BY SEX

Tolerance Level	Sex		Row Totals
	Male	Female	
High	III 3	ЈЖ 5	8
Low	III 3	II 2	5
Column Totals	6	7	13

Source: Hypothetical data.

boxes or cells in the body of the table corresponding to the appropriate row and column labels as shown in Table 6.4. Sometimes it is helpful to create a tally within each cell as a workmanlike way to assure accuracy.

Each of the boxes in the table is called a **cell** and the frequency in a cell is called a **cell frequency.** Cell frequencies are sometimes symbolized by the small letter n_{ij}, where the first subscript indicates the number of the row and the second subscript indicates the number of the column, as follows:

	column 1	column 2	row totals
row 1	n_{11}	n_{12}	$\sum_j n_{1j}$
row 2	n_{21}	n_{22}	$\sum_j n_{2j}$
column totals	$\sum_i n_{i1}$	$\sum_i n_{i2}$	N

They indicate the number of cases in the total sample which fall in a certain category of the row and column variable as indicated by the row and column labels.* The cell frequencies indicate the number of cases with two characteristics simultaneously. Row and column totals each add up to 13, which is the total number of cases there were. The cell frequencies constitute the conditional distributions, and the row and column totals reflect the marginals or univariate distribution of each variable.

6.3.2 Traditions of Table Layout

Usually tables are set up so that the dependent variable is the one with categories listed down the *stub,* or left side of the table, and the independent variable is listed across the top in the *heading.* This convention, of course, is not always kept, but it does tend to aid in the examination of conditional distributions in each column to have it set up this way. Table 6.3 illustrates proper labeling of a table. Notice that low categories of the independent variable, where there are low categories on that variable, are listed at the left and the high category is at the right. For the dependent variable the high categories are at the top of the table and the low categories are at the bottom. This is similar to the labeling of other graphs, although in the case of tables the convention is not as rigidly adhered to, and the investigator would do well to double-check the table layout before proceeding to make any interpretation.

A table usually has a title which lists the dependent variable, whether the table contains frequencies or percentages (or some other measure), the independent variable(s), and the kind of case upon which the measurements were taken. Table 6.3 contains data on 7,714 individuals. If the table is a percentaged table,

---

*Some authors use a different set of symbols for row and column totals, where a dot is used in place of a subscript for totals. Thus $n_{.1}$ would be the sum of column 1 over all of the rows in the table (the row subscript is replaced by a dot) instead of $\sum_i n_{i1}$ and $n_{1.}$ would symbolize the total of row one, instead of $\sum_j n_{1j}$. One could use n rather than N or $\sum\sum n_{ij}$ to indicate the grand total number of cases.

it is important to indicate the base upon which the percentage was computed in brackets, at the bottom by the column total percentages*; when this is done cell frequencies may be omitted from a percentaged table. The source of data is indicated, typically, in a footnote to the table, and both the stub and heading are clearly marked with the variable and the name of each of the categories of each variable.

Table 6.4 is a frequency table. It has categories of the tolerance variable down the side (the stub), and categories of the variable, sex, across the top. In this case, tolerance played the role of dependent variable.

Notice that cell frequencies in columns are summed, and the sums are put at the bottom of the table. Rows are also summed, and the totals put at the right hand side. These row and column totals are called **marginals,** or simply row totals and column totals, and they are merely the univariate distribution of each variable separately.

If the table shows percentages it is called a bivariate percentage distribution, and if frequencies are shown, it is called a bivariate frequency distribution.

6.3.3   Percentaged Tables

Probably the most often used type of table is the percentaged table. Its value lies in the way it helps one to make comparisons across the conditional distributions one wants to compare. The basic rule for computing percentages on a table is as follows:

Compute percentages in the direction of the independent variable.

This means that percentages should sum up to 100% for each category of the independent variable. For tables set up such as Table 6.3, the percentaging rule leads to computation with column totals as the base of the percent; thus column percents add up to 100% for each column. If the independent variable and the dependent variable were switched around, the percentages would have to be run in the other direction. There are three ways that a table can be percentaged, as shown in Table 6.5, using the hypothetical data from Table 6.4. Tables could be percentaged with *column totals* as the base of percentages, with *row totals* as the base of percentages, and with the *grand total* as the base of percentages. Since the dependent variable is down the stub of Table 6.4, the proper table to examine to see what differences there may be between categories of the sex variable would be to percentage with column totals (the number of males or the number of females) as the base of the percentages. One wants to contrast the distribution of the dependent variable between men and women, and the only way to do this is to take out the effect of different numbers of men and women by percentaging down (in the direction of the independent variable). This type of operation permits one to make comparisons in the *other direction. Comparisons are made* (made possible) in the *opposite direction from the way percentages are run.*

---

*This is true if column totals are the bases of percentages. If rows sum to 100% then row total frequencies are given.

TABLE 6.5 Illustration of Different Ways Percentages Can Be Computed on Tables

ORIGINAL FREQUENCY DISTRIBUTION FROM TABLE 6.4

Tolerance Level	Sex Male	Female	Total
High	3	5	8
Low	3	2	5
Total	6	7	13

6.5A PERCENTAGING TO COLUMN TOTALS AS THE BASE

Tolerance Level	Sex Male	Female	Total
High	50%	71%	62%
Low	50	29	38
Total	100%	100%	100%

6.5B PERCENTAGING TO ROW TOTALS AS THE BASE

Tolerance Level	Sex Male	Female	Total
High	38%	62	100%
Low	60%	40	100%
Total	46%	54	100%

6.5C PERCENTAGING TO OVERALL GRAND TOTAL AS THE BASE

Tolerance Level	Sex Male	Female	Total
High	23%	39%	62%
Low	23%	15%	38
Total	46%	54	100%

Independent Variable

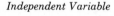

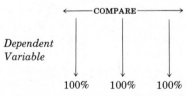

Dependent Variable

100%   100%   100%

Comparisons are made in a percentaged table by examining differences between percentages. In Table 6.5a, for example, the difference between percentage "high" on tolerance among men and women is 21% (71% − 50% = 21%). This value is called **epsilon,** the percentage difference in a table, and it is symbolized by the Greek letter ϵ. For tables larger than a two by two table, there are a number of percentage contrasts or epsilons which may be computed and used in interpretation. Epsilon will be discussed further later on in this chapter.

Sometimes an investigator will compute percentages, as in Table 6.5c, with the total number of cases (N) as *the base for all cell percentages*. Where this is done, we no longer can compare conditional distributions, but we can express the percentage of cases which have each of the different combinations of characteristics labeled by the rows and columns.

If it is not clear which variable is dependent or independent, or if we could think of the data in both ways, we might compute percentages to *both row and column totals* (as in Tables 6.5a and 6.5b) and *examine each table*. Table 6.5a would permit us to say that females are more likely to be higher on tolerance than are males. Table 6.5b would permit us to say that high-tolerance people are more likely to be female than are low-tolerance people—a subtle shift with worlds of import, as we shall soon see.

As shown in Table 6.5, percentaging *down* permits an examination of any influence sex may have on the distribution of tolerance; percentaging *across* shows the possible recruitment pattern into tolerance levels from each sex, and percentaging to the *grand total* permits us to examine the joint percentage distribution of tolerance levels and sex.

6.3.3a   THE EFFECT OF SAMPLING DECISIONS ON PERCENTAGING TABLES

Sometimes an investigator will decide to draw a sample in such a way that he (she) guarantees, say, equal numbers of cases in each study category or condition he (she) wants to examine. This means, of course, that the marginals for that variable do not necessarily reflect or represent the way that variable is distributed in the population from which the sample was drawn. This happens, for example, in situations where the phenomenon being studied is quite rare. Only a small percentage of the population of the U.S. is classified as alcoholic, and it would be easy to draw a sample and completely miss this kind of person. If one were interested in comparing alcoholics and non-alcoholics on some characteristic, the whole study would collapse if alcoholics were not sampled in sufficient numbers to analyze. To protect himself (herself), an investigator may sample in a way that is called *stratified sampling* (sampling techniques are introduced and discussed in Chapter 11). That is, the investigator may divide a population list up into alcoholic and non-alcoholic and deliberately draw an equal (or specified) number of cases from each condition into the sample.

This procedure results in the distribution of alcoholism which may not reflect the way that variable is distributed in the population as a whole (*i.e.* alcoholics would be over-represented in the sample). This procedure would mean that alcoholism would be an "unrepresentative" variable. Because of this essentially arbitrary changing of the distribution of a variable in a sample, we must follow procedures which take this into account; namely, the table must be percentaged so that the category totals of the unrepresentative variable are used as the base of percentages. This is called **percentaging in the direction of the non-representative variable.** If the non-representative factor happens also to be an independent variable, then the percentages we have to examine are indeed the percentages we need to use in drawing conclusions. If, however, we stratify or make non-representative in some way a dependent variable, then the percentages have to be run the "wrong" way for this analysis, and we are unable to make the contrasts we need to make.

6.3.3b ILLUSTRATION OF PERCENTAGED TABLES

Table 6.6 provides data on the bivariate distribution of income and ethnicity in the U.S. in 1970. The table clearly shows differences in the distribution of income for the different ethnic groups. Whites are less likely to be in the lowest income categories as compared to those of Spanish origin or blacks. Epsilon, the difference between percentages, is 3.2% between whites and those of Spanish origin, and 14.1% between whites and blacks. This table would be quite useful to answer the question: What is the consequence of ethnicity for the distribution of income? Had the percentages been run the other way, we could answer the question: What is the consequence of different income levels on the ethnic distribution of the United States? The first question is probably much more sensible and useful to answer than the second; thus percentaging down, as has been done in Table 6.6, would be the appropriate way to percentage this table.

Table 6.7 provides a somewhat different example of percentaged tables which also highlights one of the problems involved in comparison of percentages. These data are from a summary of questions asked in a number of public opinion polls concerning the role of women. One of the questions asked over several years by the Gallup Poll is: "If your party nominated a woman for President, would you vote for her if she were qualified for the job?" The response categories are, basically, "yes" and "no," and the percentage giving these two responses out of the total number of men and women sampled is shown in the table.

Notice that for men and to some extent for women, the percentage saying that they would vote for a woman increases over the years between 1937 and 1969. The percentage saying they would not vote for a woman also tends to decline. Using these data, we could answer the question: "What is the difference in percent agreeing to vote for a woman from year to year?"

Notice too that the comparison is contaminated by the troublesome "no opinion" or "no response" category. Where percentages include these cases, the

TABLE 6.6 PERCENT DISTRIBUTION OF INCOME IN 1970 FOR MALES WITH INCOME, AGED 25 YEARS OLD AND OVER, BY ETHNIC ORIGIN, U.S., MARCH 1971

| | Ethnic Origin | | |
Income	White	Spanish	Black
Less than $3000	14.0	17.2	28.1
$3000 to $5999	18.2	30.3	31.9
$6000 to $9999	32.0	35.1	29.6
$10000 and over	35.9	17.3	10.4
Total	100.0	100.0	100.0
Number in millions	(45.9)	(1.8)	(4.4)

Source: U.S. Bur. Census, 1971.

TABLE 6.7 Percentage Indicating Support for a Woman President in Gallup Polls in the United States between 1937 and 1969, by Sex

"If your party nominated a woman for President, would you vote for her if she were qualified for the job?"

| | FOR MEN | | | | | FOR WOMEN | | | |
Time	Would	Would Not	No Opinion	Total*	Time	Would	Would Not	No Opinion	Total
1937	27%	69%	4%	100%	1937	40%	57%	3%	100%
1945	29	58	13	100	1945	37	51	12	100
1949	45	50	5	100	1949	51	46	3	100
1955	47	48	5	100	1955	57	40	3	100
1963	58	37	5	100	1963	51	45	4	100
1967	61	34	5	100	1967	53	44	3	100
1969	58	35	7	100	1969	49	44	7	100

Epsilons between Men and Women on Percentage Saying They *Would* Vote for a Woman for President

Time	ε†
1937	+13%
1945	+ 8
1949	+ 6
1955	+10
1963	− 7
1967	− 8
1969	− 9

Source: Data from Erskine, 1971:278. Used by permission of George Gallup.
*Total number of cases was not given in the source.
†Epsilons here are computed by subtracting the percentage who would vote for a woman president among men from that among women in any given year. Minus signs indicate that the male percentage is higher and plus signs indicate that the female percentage is higher.

percentage comparisons of voting intention between one year and another are distorted because they also reflect differences in the level of "no opinion." This is particularly important to watch for in comparisons involving the year 1945, where the percentage of men and women giving no opinion increased a very great deal. Among men, there is little difference in percentage who would vote for a woman between 1937 and 1945, but there is a big drop between these years in percent saying they would not vote for a woman. Apparently, overall, there was a shift to "no opinion" from the "would not vote for a woman" category. Between 1945 and 1949, however the percent "no opinion" drops back to a modest level, and there is a corresponding increase in the percentage who would vote for a woman. In this case, then, epsilons computed between years on one response category of the dependent variable (*e.g.* willingness to vote for a woman as president) would be misleading without taking account at the same time of the shift in the "would not" and "no opinion" categories.* We need some descriptive device which will permit an overall examination of a table rather than

*Other procedures for handling the "no response" or "no opinion" categories in research of this sort are discussed by Davis and Jacobs, 1968.

only a contrast of individual pairs of percents, (although these too are interesting for some purposes).

BOX 6.1

Thus far in Chapter 6 we have discussed the way in which bivariate distributions may be created (Sections 6.1, and 6.2, and 6.3.1) and the use of percentages to examine the distribution of a variable within categories of some other variable (Section 6.3.3. ff). Percentaging rules which are useful in this process are: (a) percentage in the direction of the independent variable and compare in the other direction (Section 6.3.3), (b) percentage in the direction of the non-representative variable, if any (Section 6.3.3a), and (c) percentage to the grand total to examine the joint percentage distribution (Section 6.3.3). These uses of percentages should be understood before you go on into some further applications of these rules under somewhat more detailed conditions, and before we discuss the measures of association which follow.

6.3.4   More Complex Conditional Distributions

It is often the case that investigators look at percentage tables which are a good deal more complex than those we have just been examining. That is, they may be interested in conditional distributions which are distinguished by more than one independent variable. We will have more systematic things to say about such tables later on in this book, but for now we should indicate how we might begin to approach the interpretation of such a table.

Often a table is set up so that the focus is on either the *dependent variable alone* under various kinds of conditions, or on some basic set of conditional distributions—*a basic table*—which is examined in another set of conditional situations.

Table 6.8 illustrates a more complex table of the first kind. Four variables, education of mother, education of father, I.Q. of the child, and sex of the child, define the conditions within which a student's plans regarding college can be examined. The dependent variable is, as the title suggests, the percentage *planning on going to college*. Actually, there could be two tables presented here instead of one. One table would show the percentage *planning* on college and the other would show the percentage *not planning* on college. Since these two percentages would total to 100 percent, it is sufficient to look at only one of the two possible tables. The other table would be a mirror image of the first. There is a relatively orderly way to go about reading such a table, and we will discuss these procedures here.

The *first step* in examining a table as complex as this one is to *identify clearly the dependent variable* and the kind of *unit* (*i.e.* family, person, parent) upon which it has been measured. In this table it is "expressed plans to go to college,"

TABLE 6.8 Percentage Who Planned on College by Sex, Intelligence, Father's Education, and Mother's Education[†]

Father's Education	MALES				FEMALES				TOTAL			
	Mother's Education				Mother's Education				Mother's Education			
	High	Middle	Low	Total	High	Middle	Low	Total	High	Middle	Low	Total
a. High Intelligence												
High	89.3(186)	87.8(115)	55.6(36)	85.2(337)	86.0(186)	74.5(102)	68.6(35)	80.5(323)	87.6(372)	81.6(217)	62.0(71)	82.9(660)
Middle	85.7(63)	66.0(241)	53.5(144)	64.7(448)	62.7(75)	50.0(238)	40.5(131)	49.3(444)	73.2(138)	58.0(479)	47.3(275)	57.1(892)
Low	55.1(78)	54.8(210)	45.3(415)	49.2(703)	62.0(100)	37.3(220)	28.1(438)	35.2(758)	59.0(178)	45.8(430)	36.5(853)	42.0(1461)
Total	80.4(327)	66.3(566)	47.9(595)	62.0(1488)	74.5(361)	49.5(560)	33.1(604)	48.9(1525)	77.3(688)	57.9(1126)	40.5(1199)	55.4(3013)
b. Middle Intelligence												
High	71.0(69)	50.0(84)	36.1(36)	55.0(189)	79.5(78)	56.9(58)	38.2(34)	63.5(170)	75.5(147)	52.8(142)	37.1(70)	59.1(359)
Middle	51.1(47)	39.5(218)	30.9(139)	37.9(404)	49.0(49)	31.7(205)	17.1(152)	28.3(406)	50.0(96)	35.7(423)	23.7(291)	33.1(810)
Low	36.0(50)	32.7(211)	26.8(598)	28.8(859)	37.2(78)	24.1(232)	17.1(679)	20.3(989)	36.7(128)	28.2(443)	21.6(1277)	24.2(1848)
Total	54.8(166)	38.4(513)	27.9(773)	34.7(1452)	56.1(205)	31.1(495)	17.9(865)	27.1(1565)	55.5(371)	34.8(1008)	22.7(1638)	30.8(3017)
c. Low Intelligence												
High	48.6(35)	32.1(53)	21.4(28)	34.5(116)	50.0(34)	26.2(42)	30.8(26)	35.3(102)	49.3(69)	29.5(75)	25.9(54)	34.9(218)
Middle	43.3(30)	16.6(181)	15.8(101)	18.9(312)	43.3(30)	20.6(136)	13.4(142)	19.5(308)	43.3(60)	18.3(317)	14.4(243)	19.2(620)
Low	27.3(44)	14.2(211)	0.5(765)	11.3(1020)	30.8(39)	8.3(193)	7.6(887)	8.5(1119)	28.9(83)	11.4(404)	8.5(1652)	9.8(2139)
Total	38.5(109)	17.3(445)	10.6(894)	14.8(1448)	40.8(103)	14.8(371)	8.9(1055)	12.5(1529)	39.6(212)	16.2(816)	9.7(1949)	13.6(2977)
d. Total												
High	80.0(290)	63.5(252)	39.0(100)	67.1(642)	80.2(298)	59.4(202)	47.4(95)	67.9(595)	80.1(588)	61.7(454)	43.1(195)	67.5(1237)
Middle	65.0(140)	43.0(640)	35.4(384)	43.1(1164)	54.6(154)	36.6(579)	23.1(425)	34.0(1158)	59.5(294)	40.0(1219)	28.9(809)	38.6(2322)
Low	42.4(172)	33.9(632)	23.7(1778)	27.4(2582)	47.5(217)	23.9(645)	15.3(2004)	19.6(2866)	45.2(389)	28.8(1277)	19.2(3782)	23.3(5448)
Total	65.8(602)	42.6(1524)	26.4(2262)	37.4(4388)	63.7(669)	34.1(1426)	17.8(2524)	29.5(4619)	64.7(1271)	38.5(2950)	21.8(4786)	33.3(9007)

Source: Data from Sewell and Shah, 1968:198, Table 2. Used by permission.

[†]The base of percentages is shown in parentheses by each percentage. Data are based on a large, randomly selected cohort of Wisconsin high school seniors who were followed for a seven-year period.

which is the dependent variable asked of individual people who are the school-aged sons and daughters of parents who have differing amounts of formal education.

The *second* step is to identify the variables involved in the table and the categories of each. In the education variables, three categories are used in this table, and they are labeled "low," "medium," and "high." One education variable has to do with father's education and one has to do with mother's education. There is also an intelligence variable which has three values: "high," "middle," and "low." Finally, female students and male students are distinguished.

Thirdly, the structure of the table should be clarified. In the case of Table 6.8, categories of sex are reflected in the left two strips of percentages down the table. The right strip down the table shows percentages for all cases, regardless of sex, in the total sample. Father's education appears in rows of the table; three rows for father's education within the "high intelligence" category, three in the "middle intelligence" category and three within the "low intelligence" category. The bottom general row of numbers combines together students of all intelligence categories but distinguishes (in rows, again) the father's education. Mother's education is indicated across the top of the two sex and total strips down the table, and there are "total" columns which summarize over father's education or over mother's education, etc. In Table 6.8, rows of figures around the bottom and right hand sides of the table indicate totals, and more detailed breakdowns are concentrated in the rows and columns in the middle and upper left areas of the table. Having identified the structure of the table, we can examine percentages.

The percentage 89.3% in the upper left corner of the table means that 89.3 percent of students plan on going to college among those students who are (a) male, (b) of high intelligence, have (c) a father with "high" education, and (d) a mother with "high" education. Pick another percentage and express its meaning in your own words.

At this point, it is a good idea to find the overall percentage planning on college regardless of education of parents, I.Q., or sex; that is, find the *overall* distribution of the dependent variable without any conditional distribution being considered separately. That figure can be found in Table 6.8 in the lower right hand corner—the total column which summarizes all of the conditional distributions. This figure is 33.3%. Then we can look at the distribution of this variable by each of the other variables one at a time, making contrasts between the percentages, perhaps in terms of epsilon. For example, 67.5% of those with high father's education plan on college, but only 23.3% of those students with low father's education have similar plans. For mother's education the figures are 64.7% and 21.8% respectively. For males, overall, the percentage is 37.4% and for all females it is 29.5%. Finally, among those with high measured intelligence, 55.4% plan on college, but only 13.6% of those with low measured intelligence plan to attend college.

The next step would be to look at the various combinations of variables two at a time, three at a time, and four at a time to see what the effect of various conditional distributions is. The highest percentage planning on going to college is the 89.3% in the upper left corner of the table, and this corresponds to high I.Q. males with parents each of whom has high education.

At the other extreme, 7.6% of the female students with low intelligence and with parents who have low educational backgrounds plan on going to college. Notice that percentage planning on college increases uniformly as one moves from the less advantaged conditions to the more advantaged. Each improvement in advantage seems to be related to an increase in the percentage planning on college.

We could draw several conclusions from these data. First, it seems to be the case that the higher the educational background of the mother or of the father, the higher the percentage of students who plan on college becomes. This will later be called a *positive* relationship, since increases in one variable tend to go along with increases in the other. Males are more likely to plan on college, and the higher the measured intelligence, the higher the percentage planning on college. These findings seem to hold up even within various categories of other variables shown in the table. Furthermore, all of these conditions seem to be cumulative in the sense that individuals with two sources of "advantage" do better than those with only one, and so forth.

At this point we should recall our initial purpose. We wanted to examine the distribution of a *dependent* variable under conditions which are defined in terms of other variables singly or in combinations.

While many investigations call for the examination of conditions based on one variable at a time (*i.e.* by sex or separately by education, etc.), many of our core interests are posed in terms of more complex conditions defined in terms of more than one variable, as was the case in Table 6.8. In many instances the examination of percentaged tables provides the kind of summary and the kinds of contrasts we need for the problem at hand. They are easily understood and direct. On the other hand, it is clear that even simple percentaged tables may contain more details than we want, and some of these may interfere with the analysis. The solution, as it was in the case of univariate summary statistics, is to try to create overall indices which summarize the aspect of the distributions we are most interested in examining.

6.4 FOUR CHARACTERISTICS OF AN ASSOCIATION

Going back to a bivariate distribution such as that shown in Table 6.3, we can think of that distribution as a relationship between two variables, and we want to know how the distribution in the dependent variable varies as we move from category to category of the other variable. The way two variables relate to each other is called an **association** between the two variables. In Table 6.3, as city size increased, the percentage of individuals showing higher tolerance began to increase. The two variables were associated in that particular fashion.

We can speak of the association of any two variables and describe that association in terms of a percentaged table, as we have shown. There are other ways to summarize the association between two variables, however, and, in fact, there are four characteristics of an association which we will single out for summary, just as there are three characteristics of a univariate distribution which we summarized in terms of different index numbers (*i.e.* central tendency, variation, and form). The four aspects of a bivariate association are:

(a) Whether or not an association *exists*.
(b) The *strength* of that association.
(c) The *direction* of the association.
(d) The *nature* of the association.

Each of these characteristics will be discussed in turn, and in the next chapter we will develop several alternative measures of them. In fact, we will create a single number which will be used to describe the first three features of an association listed above and in some cases a simple formula can be used as an efficient description of the last of the features above.

6.4.1 The Existence of an Association

An association is said to exist between two variables if the distribution of one variable differs in some respect between at least some of the categories of the other variable. This rather general statement can be pinned down in a number of ways, the first of which we have already discussed. If, after computing percentages in the appropriate direction in a table, there is *any* difference between percentage distributions, we would say that an association exists in these data. In the first table, below, the distribution of education is slightly different for men compared with women. We know this by percentaging in the direction of the

Education	Men	Women	Total
High	40%	38%	38%
Low	60	62	62
Total	100%	100%	100%
	(43)	(56)	(99)

(fictitious data)

independent variable and comparing across. In the second table, below, there is *no* association between "toe-nail length" and "education," and this is shown by

| Education | Toe Nail Length | | Total |
	Short	Long	
High	33%	33%	33%
Low	67	67	67
Total	100%	100%	100%
	(521)	(1756)	(2277)

(fictitious data)

the fact that there is no difference in the percentage distribution of education (the dependent variable) regardless of the category of the independent variable within which we examine the dependent variable. In the following table it is clear that there *is* an association between social class and the number of arrests, because the percentage distributions, comparing across the way percentages were run, are different.

Number of Arrests	Social Class		
	Low	Medium	High
None	11%	28%	45%
Few	11	18	35
Many	66	54	20
	100%	100%	100%
	(129)	(260)	(73)

(fictitious data)

There is a name for these comparisons: **epsilon** (ϵ), which is the percentage difference computed across the way percentages were run in a table. In a table where *all* of the epsilons are *zero*, there is *no* association. If any epsilon is non-zero, there is an association in the data even though we may not choose to consider the very small differences important enough to talk about.

The second way to tell whether or not there is an association in a table is to compare the **actual observed table frequencies** with the frequencies we would expect if there were no association, or **expected frequencies.** If the match between actual data and our model of no association is perfect, then there is no association in the actual data between the two variables which were cross tabulated in the table.

6.4.1a   No Association Models.   A **model of no association** can be set up for a specific table as follows. Usually in setting up a model of the way frequencies in a table should look if there were no association, we assume that the marginal distribution of each variable is the way it is in the observed data table, and that the total number of cases is the same. The problem is to specify the pattern of cell frequencies in the body of the table in a way which shows no association. As an example, suppose the marginals on variables X and Y are as follows:

(Y)	(X)		
	Low	High	Total
High	a	b	57
Low	c	d	50
Total	34	73	107

The problem is to find a pattern of frequencies for cells a, b, c and d such that they exhibit no association between X and Y. The reasoning goes like this. If there is no association in the table, then the ratio of "high" cell frequencies for variable Y as related to the corresponding column totals should be the same throughout the table, as it is in the overall distribution of Y itself, namely 57 to 107. In the table above, we would expect 57/107ths of the 34 cases in the "low" category of X to be in the "high" category of Y. Furthermore, we would expect the same ratio, 57/107ths of the 73 cases in the high column of X to be in the top row. This would mean that, relatively speaking, there is no difference between the proportion of cases in the top row for any column of the table.

$$\frac{57}{107}(34) = .533(34) = 18.1 \text{ cases } \textit{expected} \text{ in cell } a$$

$$\frac{57}{107}(73) = .533(73) = 38.9 \text{ cases } \textit{expected} \text{ for cell } b$$

Given that one of the above cell frequencies in a 2×2 table is computed, the other expected cell frequencies could be determined by subtraction. The resulting table of expected cell frequencies (expected if there were no association between the two variables, X and Y for these 107 cases) is shown below.

"EXPECTED" CELL FREQUENCIES			
		(X)	
(Y)	Low	High	Total
High	18.1	38.9	57.0
Low	15.9	34.1	50.0
Total	34.0	73.0	107.0

This is a hypothetical tabulation showing no association and thus fractional frequencies are appropriate.

Expected cell frequencies (f_e) can be computed for a given cell by multiplying the row total for that cell by the column total for that cell and dividing by N, which is the operation explained above.

(6.1)
$$f_{e_{ij}} = \frac{(n_{i.})(n_{.j})}{N}$$

Where $f_{e_{ij}}$ refers here to the expected cell frequency for the cell in the ith row and jth column of the table, $n_{i.}$ is the total for the ith row and $n_{.j}$ is the total for the jth column, and N is the total number of cases. An expected cell frequency is computed for each cell in the table.

Now, the difference between the table of observed data and the model we could construct of how this table would look if there were no association can be compared. This comparison is made by subtracting an expected cell frequency, f_e, from an observed cell frequency, f_o. The difference is called **delta,** and in this text we will symbolize delta with the upper case Greek letter Delta (Δ).

For a given cell,

(6.2)
$$\Delta = f_o - f_e$$

A delta value can be computed for each cell in a table, regardless of the size of the table. If any of the deltas are *not* zero, then there is at least some association shown in the table. Whenever all deltas are zero, all epsilons will also be zero. Later we will discuss summary measures of association based on these ideas.

In summary, whether or not an association exists in a table of observed frequencies can be exactly determined in two ways. One way is to compute percantages in one direction and compare across in the other direction, using epsilon. The other way we discussed is to create a table of expected cell frequencies and compare the observed and expected cell frequencies, cell by corresponding cell, using delta. If all of the epsilons that can be computed in a table, or if all of the delta values for a table amount to zero, then there is no association between the two variables cross-tabulated in the bivariate distribution. This is called **statistical independence.** If, on the other hand, there is any epsilon or any delta which is not zero, then to that extent there is an association in the observed frequency table, however slight or large that association might be.

6.4.2 Degree (Strength) of Association

Where the differences between percentages (epsilons) are large, or where the deltas are large, we speak of a strong **degree of association** between the two variables; that is, the dependent variable is distributed quite differently within the different conditional distributions defined by the independent variable. This can be contrasted with a weak association where there is very little difference or where the epsilons and deltas are very small, approaching or equalling zero.

Often investigators use epsilon (or delta) as a crude first indicator of the strength of association. The problem with both delta and epsilon is that it is difficult to determine what a given sized delta or epsilon means, other than that there is some association in the table. The reason for this is that both delta and epsilon values for any cell(s) can vary from zero or near-zero up to a magnitude which is not, in general, fixed. They are not "normed" or standardized. Later, in this chapter and the next, the problem of creating good standardized measures of the strength of association will be discussed and several alternative measures will be described. Suffice it to say here that some tables show a strong relationship between independent and dependent variables, and some show a weak association or no association at all.

6.4.3 Direction of Association

Where the dependent and independent variables in a table are at least ordinal variables, it makes sense to speak about the **direction** of an association that may exist in a table. If the tendency in the table as shown by the percentage distribution is for the higher values of one variable to be associated with the higher values of the other variable (and the lower values of each variable also tend to go together), then the association is called a *positive* association. Height and weight tend to have a positive association, since the taller a person is, the heavier he (she) tends to be, in general, across the people in a general population.

On the other hand, if the higher values of one variable are associated with the lower values of the other (and the lower values of the first with the higher values of the second), the association is said to be *negative*. Sociologists generally expect that the higher the educational level of people, the lower their degree of normlessness will be — a negative association.

The association between city size and tolerance scores (see Table 6.3) is positive because the larger the city, in general, the higher the tolerance level becomes (*i.e.* the higher the percentage of people who have high tolerance scores). The older a person's age, in general, the fewer the years left until retirement, a negative association.

6.4.4 The Nature of Association

Finally, the **nature** of an association is a feature of a bivariate distribution referring to the general *pattern* of the data in the table. This is often discovered by examining the pattern of percentages in a properly percentaged table. Often the pattern is irregular, and an investigator would cite many epsilons in describing where the various concentrations of cases are in the different categories of the independent variable. Sometimes there is a rather uniform progression in concentration of cases on the dependent variable as we move toward higher values of the independent variable. If, with an increase of one step in one variable, cases tend to move up (or down) a certain number of steps on the other variable we might call the nature of the association "linear." That is, the concentrations of cases on the dependent variable (the mode, for example) tend to fall along a straight line that could be drawn through the table.

The nature of association will be discussed at length later, in the next chapter. Simple linear associations have an intrinsic interest to investigators as one of the simplest natures of association, but some associations are curvilinear in nature, or of some more complex patterning. In most cases the nature of association will be determined from a percentage table or a scatter plot, but in some cases nature can be described in terms of an equation.

At this point we should pause to examine several tables and describe them in terms of these four features of an association. Table 6.9 presents a series of examples together with brief summary statements.

TABLE 6.9 Illustrations of the Existence, Strength, Direction, and Nature of Association

6.9A PERCENTAGE DISTRIBUTION OF SEVERITY OF DISABILITY BY EDUCATION FOR DISABLED, NON-INSTITUTIONALIZED ADULTS AGED 18–64, UNITED STATES, SPRING, 1966

Severity of Disability	Education				Total
	No School	Elementary	High school	College	
Severe	65.6%	43.8%	28.0%	21.2%	34.4%
Occupational	11.2	29.1	29.6	24.3	28.2
Secondary Work Limitations	23.2	27.1	42.4	54.4	37.4
Total	100.0%	100.0%	100.0%	100.0%	100.0%
Number in millions	(0.5)	(6.9)	(7.5)	(2.5)	(17.8)

Source: Treitel, 1972:Table 5.

(1) *Existence of Association.* This table is percentaged down in the direction of the independent variable, education, so that comparisons may be made across. The percentages across are different so that there is an association evident in the table. Compare any

TABLE 6.9 *(Continued)*

row, say the row for "Severe" disability; percentages range from a high of 65.6% down to 21.2%, all different from 34.4%, the total percentage for that category.

(2) *Strength of Association.* Overall, if education makes any difference in severity of disability, we should find the largest percentage difference (epsilon) between the "no-school" and "college" groups for the extreme categories of disability – namely the "severe" category or the "secondary work limitation" category. The epsilon for the "severe" disability comparison is 44.4%, and for the "secondary work limitation" category it is 31.2%. Both epsilons are substantial percentage differences, although they are not equal to 100%, which in some tables is the largest that epsilon could be.

(3) *Direction of Association.* As an aid in finding the direction of an association between variables each of which is at least ordinal, a useful procedure is to make comparisons across the way percentages are run, underlining the highest percentage for each comparison. In this table, we could make three comparisons, one for each row. In the "severe" row, 65.6% is clearly the largest percentage. In the "occupational" row, 29.6% is largest, but it is virtually tied by 29.1%, so we will underline both of these largest percentages. In the "secondary work limitations" row, 54.4% is again clearly the largest. One could draw a line through the underlined percentages in the table. In this case, the line would extend from the "severe/no school" corner down to the "secondary work limitation/college" corner. This is the major diagonal along which cases tend to concentrate. Notice that the more extreme disabilities tend to go with less education, and the more mild types of disabilities tend to go with higher educational categories, with intermediate stages on both variables falling in between. This is a *negative* direction of association: the higher the education the less severe the disability tends to be.

(4) *Nature of Association.* In this table, the nature of association, that is, the pattern of concentration in the table, tends to be almost linear. There is a relatively uniform shift toward less education as one shifts toward the more severe disability categories. There are no "reversals" in this general trend of concentration. Because the variables are ordinal, it would be more appropriate to speak of this nature of association as "*monotonic*" rather than linear. If distances were defined then one could determine whether in fact there is a *constant* amount of shift in values of one variable, given a fixed amount of change in the other. In ordinal variables one can only say that the value of one variable remained the same or shifted in a fixed *direction* with increases in the other variable – a monotonic nature of association. Contrasted with this type of nature are those shown in later tables here.

6.9B PERCENTAGE DISTRIBUTION OF EARNINGS OF U.S. WORKERS IN 1967, BY AGE

Annual Earnings	Age of Worker				
	Under 25	*25–44*	*45–64*	*65 and over*	*Total*
Under $2,000	60.7%	21.4%	20.6%	55.9%	32.7%
2,000–3,499	18.3	13.8	15.1	12.2	15.3
3,500 and over	21.0	64.8	64.3	31.9	52.0
Totals	100.0	100.0	100.0	100.0	100.0
(in millions)	(22.2)	(33.5)	(26.4)	(3.6)	(85.5)

Source: Data from U.S. Dept. H.E.W., 1971:Table 1. Data are for earnings of covered employment in the old-age, survivors, disability and health insurance program which covers virtually all work categories. Income refers to taxable wages or salaries.

(1) *Existence of Association.* After percentaging down, comparisons across reveal differences in percentage; thus there is an association.

TABLE 6.9   *(Continued)*

(2) *Strength of Association.* Although we do not have formal ways to measure the strength of association yet (they will be discussed later), it is clear that some of the percentage differences are relatively large. Take, for example, the 40.1 percentage point difference between the "under 25" and the "45–64 year old" workers on percentage earning under $2,000 per year.

(3) *Direction of Association.* In this table the direction of association is somewhat confused. Among those aged 44 or less, the association is positive – the older the person the more likely his income will be higher. Among those aged 45 and over, however, the association is negative – the older the age of worker, the more likely he is to earn less.

(4) *Nature of Assocation.* This association is curvilinear in nature since both extremes of age tend to earn less, while those in the middle tend to earn more.

6.9C   PERCENTAGE DISTRIBUTION OF COLLEGE ENROLLMENT OF PERSONS AGED 18–21 YEARS OLD, BY RACE, OCTOBER, 1970

College Enrollment	Race			
	White	Black	Other	Total
Enrolled	36%	21%	44%	34%
Not Enrolled	64	79	56	66
Total	100%	100%	100%	100%
(Number in Millions)	(11.3)	(1.6)	(0.1)	(13.1)

Source: Data from U.S. Bur. Census, 1972:Table 15.

(1) *Existence of Association.* In this table, percentages enrolled and not enrolled are computed down, within race categories. Comparing across, there is a difference in percentages and thus, an association exists between race and college enrollment.

(2) *Strength of Association.* Using epsilon, we might note that $\epsilon = 8\%$ between "white" and "other," and it is 15% between "white" and "black," and 23% between "black" and "other." In general, other races are more likely to be enrolled in college, but the percentage differences are not very large. Notice that at this point we do not have tools for saying how strong some kind of difference in percentages is. Clearly, however, these are not zero, nor are they 100 percent differences.

(3) *Direction of Association.* Since race (and enrollment) are generally considered nominal variables, it would not make sense to talk about the direction of association, so this feature can be ignored.

(4) *Nature of Association.* The pattern of percentages in the table can be described at length. Here the percentage differences cited under #2, above, seem to cover all we can say about a 2×3 table, but in larger tables or for tables with ordinal or interval variables, much more description of patterning is generally possible.

6.5   CREATING MEASURES OF ASSOCIATION

It is possible to create single summary index numbers which will indicate the existence, degree, and direction of association all in the same number. Typically these measures are set up so that they will vary along a scale from a minus value, indicating a negative relationship, to zero, indicating no association, to a positive value, indicating a positive association. The larger the magnitude of the index number, the stronger the association.

Ideally a measure of association would also have the characteristic of varying between two *fixed limits* such as −1.00 and +1.00, where 1 indicates a **perfect association** or a maximally strong one in some sense, and zero indicates statistical independence.

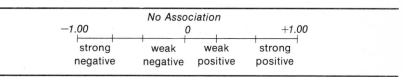

SCALE OF STRENGTH OF ASSOCIATION WITH FIXED LIMITS OF −1.00 AND +1.00'

A measure of association which varied over a scale with fixed limits, such as that shown above, is called a **normed** or **standardized measure of association.** Its value lies in the fact that normed measures can be validly compared. If a measure of association computed on one table was −.56, and the same kind of measure computed on another table was −.13, both could be interpreted as falling between the highest possible negative value of −1.00 and zero (no association). They could be compared. The first table has a stronger negative association (about four times stronger) than the second, and the difference between the two measures is substantial. If the limits on a scale were not fixed or known, then an index number such as −.56 would be very difficult to interpret and it would be essentially impossible to use the index in making comparisons of strength of association between two different tables. Since comparison and interpretation are important to sociologists, normed measures of association are important.

Table 6.10 illustrates the problem of norming with epsilon, an unnormed measure of association. Notice that the marginals of a table may be such that an epsilon of 100% could not be achieved regardless of the way cell frequencies are rearranged.

TABLE 6.10 Epsilon, an Illustration of the Norming Problem in Developing Measures of Association

Two illustrative tables are given below. The first shows a 2 × 2 table with balanced marginals (both row and column totals are the same) and it is possible in this particular table to fix cell frequencies so that, computing percentages down, an epsilon could equal zero or 100 for this table. Thus, the achieved epsilon of 17% can be interpreted as falling on the scale from 0% to 100%. In the second illustration, however, it is *not* possible to adjust cell frequencies (given fixed marginals, as before) so that zero or 100 could be achieved for epsilon. The epsilon in this second table, 19%, thus has to be interpreted in terms of a more limited scale extending from 3% to 83%. It would not be appropriate to compare the two epsilons because they not only reflect the pattern of frequencies in a table, but they are also constrained by the way the marginals are distributed, and this confounds the intended comparison. About all one is able to say is that epsilon indicates that there is an association in both original tables.

TABLE 6.10 *(Continued)*
ILLUSTRATION 1

	Weakest Possible Association				*Original Table*				*Strongest Possible Association*		
	X				*X*				*X*		
	Low	High			Low	High			Low	High	
Y High	9	6	15	Y High	10	5	15	Y High	15	0	15
Low	6	4	10	Low	5	5	10	Low	0	10	10
	15	10	25		15	10	25		15	10	25
	$\epsilon = 0\%$				$\epsilon = 17\%$				$\epsilon = 100\%$		

ILLUSTRATION 2

	Weakest Possible Association				*Original Table*				*Strongest Possible Association*		
	X				*X*				*X*		
	Low	High			Low	High			Low	High	
Y High	8	7	15	Y High	9	6	15	Y High	13	2	15
Low	5	5	10	Low	4	6	10	Low	0	10	10
	13	12	25		13	12	25		13	12	25
	$\epsilon = 3\%$				$\epsilon = 19\%$				$\epsilon = 83\%$		

In order to create a measure of association which is normed and interpretable, it turns out that we need to define 0 and 1, that is, we need to define what no association and perfect association are supposed to mean. We have already discussed two ways to identify the situation of no association, or statistical independence, in a table, using epsilon or delta. The definition of the opposite extreme is less clear cut.

6.5.1 In Pursuit of a Normed Measure of Association

Measures of association turn out to be rather simple ratios designed to be sensitive to changes in the strength of association and, in some cases, to the direction, nature, and role played by independent and dependent variables. Often the ratios suffer from the usual ailments of ratios: the numerator or denominator may fluctuate because of some irrelevancy, such as the number of cases in the table or the number of rows and columns — things which are generally of no interest when one wants to compare strength of association, and things which prevent the valid comparisons needed in a study. As in the other approaches to refining ratios, better measures of association become better by excluding irrelevant influences on the numerator and denominator. Actually you should be able to think up some of your own measures of association on the basis of the kinds of univariate and bivariate statistics we have discussed thus far.

In the following paragraphs we will illustrate this pursuit with a set of measures based on delta, the difference between observed and expected cell frequencies in a table. Most of the delta-based measures are not used widely because of the problems they *do not* adequately take into account. On the other hand, there are not very many measures of association for nominal variables, so certain of these measures are familiar to and used by investigators in certain situations. Here they are presented for the purpose of showing the straightforward way measures of association are developed.

To start with, it appears intuitively useful to measure the "distance" between an *observed* set of frequencies in a bivariate table (f_o) and an *expected* set of frequencies (f_e) which serve as a model for statistical independence or no association. Clearly, if there are no differences between observed and expected tables, the association in the observed table is zero, as defined earlier. The larger the difference in general, the stronger the association. In order for one to characterize a whole table rather than a single cell in a table, the delta values are traditionally summed up over all cells in the table. Thus:

$$\Sigma\Delta = \Sigma(f_o - f_e)$$

where the summation is over all cells in a table.

Such a measure is deficient for a number of reasons. First, the importance of any $f_o - f_e$ difference clearly depends upon the size of f_e. After all, a delta of 6 is large if only three cases were expected in that cell, but a delta of 6 is not very impressive if 300 cases were expected. Secondly, it will always turn out to be zero, because positive differences between f_o and f_e will be balanced out by negative differences elsewhere in the table. This occurs because row and column totals in the observed and expected tables are identical, so that excesses in one place have to be offset by deficiencies elsewhere.

To help take account of problems of the total delta above, it is traditional to compute a different measure. The difference between observed and expected cell frequencies is squared to eliminate negative differences (and skirt the problem of always totaling up to zero), and this squared difference is divided by the expected cell frequency to help take out the effects of different numbers of cases in a category of the row or column variable. This division of the squared difference for each cell by f_e takes out the effect of different marginal distributions in different tables and results in what is called a "margin-free" measure of association. The measure which sums over all of these cell by cell calculations is called **chi-square** (χ^2).

$$\chi^2 = \Sigma\left(\frac{\Delta^2}{f_e}\right)$$

where $\Delta = (f_o - f_e)$, and the summation is over all cells.

Chi-square (χ^2) is used in inferential statistics as a basis for a test of significance called the "chi-square test." It is useful in that role, but it is *not* useful as a measure of degree of association in general, because of norming problems. To repeat, *chi-square is* not *used as a measure of association*, it is a test used and discussed in inferential statistics.*

*See Chapter 16.

χ^2 is always a positive number, and it will equal zero if there is no associa-tion in the table. However, the upper limit on the magnitude of χ^2 is:

$$N(k-1) = \text{upper limit of } \chi^2$$

where N is the sample size and k is the number of rows or columns in a table, whichever is the smaller number. For a 2×2 table, for example, the upper limit of χ^2 is N. Given two tables with equal association (*i.e.*, in percentaged form, they are identical), where one table has twice the number of cases as the other, the table with twice the number of cases will have a χ^2 value twice as large as the other table.

The advantages of χ^2 (advantages useful in inferential statistics) lie in the fact that it works for nominal variables. Categories of rows or columns can be re-ordered without any effect on delta, and χ^2 will always be zero if there is indeed a perfect match between observed and expected cell frequencies in a whole table.

To begin to take account of the maximum possible χ^2, at least for the table with 2 rows (or 2 columns), another frequently mentioned coefficient has been designed called the "mean square contingency" or **phi squared, ϕ^2**. It is simply the χ^2 value divided by N, the maximum possible χ^2 for a table with two rows (or columns).

(6.3)
$$\phi^2 = \frac{\chi^2}{N} \quad \text{or} \quad \phi = \sqrt{\frac{\chi^2}{N}}$$

Phi varies from 0 (for independence) to a maximum of $+1$ (perfect associa-tion) for any $2 \times k$ table (where k is the number of categories in the other vari-able), and its magnitude can be interpreted as a measure of the strength of associ-ation. It also retains the margin-free, nominal variable features of χ^2 itself. So far, so good. The problem with this measure is that its maximum value exceeds 1.0 in tables which have more than 2 categories in both variables, since the upper limit of χ^2, $N(k-1)$ can be larger than N. The maximum value for $\phi^2 = k-1$, where k is the smaller of the number of rows or columns.

A partial solution to this problem was developed by Karl Pearson in his "coefficient of contingency," **Pearson's C**. C can not exceed 1.0 in magnitude, regardless of the size of the table.

(6.4)
$$C = \sqrt{\frac{\chi^2}{\chi^2 + N}}$$

This is so simply because χ^2 appears in both the numerator and the denominator of the ratio. It can always reach zero, which means, again, that there is no associ-ation, but its upper limit, while it can not exceed 1.0, likewise can never reach 1.0. In fact, its maximum value varies depending upon the number of rows and columns there are in a table. For a square table (where $r = c$), the maximum value of C can be computed as follows:

$$\text{Maximum } C = \sqrt{\frac{k-1}{k}}$$

where k is the number of rows (or columns) in a square table.

For a 2×2 table, for example, the maximum C is .707 and for a 5×5 table the maximum C is .894. Again, comparisons of this measure of association could not be made between tables of different sizes.

A further refinement on measures of association is called **Tschruprow's T,** which attempts to correct the upper limit problem of C by changing the denominator slightly so that it includes a value reflecting the number of cells in a table instead of χ^2 itself, as in Pearson's C. This new element in the denominator of T is called **degrees of freedom** (df) and it is computed thus:

$$df = (r-1)(c-1)$$

It is the product of the number of rows minus one times the number of columns minus 1.* Tschruprow's T is defined as:

(6.5)
$$T = \sqrt{\frac{\chi^2}{N(df)}}$$

As before, this is an advance in our pursuit of a properly normed measure of strength of association appropriate to any table, in that the upper limit of T is 1.0 and that value can always be achieved regardless of the size of the table — so long as the table is square (*i.e.* $r=c$). For tables which are not square, T cannot possibly reach 1.0, although if tables to be compared have the same number of degrees of freedom, the maximum possible T is constant, and therefore a valid comparison of T values could be made.

Finally, a different measure of association, called **Cramer's V,** handles the norming problems of T by a slightly different figure, t, in the denominator, in place of degrees of freedom.

(6.6)
$$V = \sqrt{\frac{\chi^2}{Nt}}$$

Here, t is defined as the smaller of the two quantities, $(r-1)$ or $(c-1)$. Cramer's V can always attain a value of +1.0 even if the table is not square, and it equals zero if there is no association. It is, so to speak, a properly normed measure of association for bivariate distributions of nominal variables, it is "margin free" in that the number or distribution of cases in row or column totals does not influence its value, nor is it influenced by the number of categories of either variable.

But norming is only part of the problem of developing a summary index number to stand for the association in a table. There are at least two other desirable characteristics for such measures. *First,* the number should be interpretable in some intuitively useful sense, and, *secondly,* the meaning of 1.0, the "perfect association" norm, should be definable. In the first instance, Cramer's V (and, in fact, most of the other delta-based measures of association) can only be thought of as a magnitude on a scale between zero and 1.0; the bigger the number the stronger the association. It can *not* be interpreted, for example, as the

---

*Degrees of freedom (df) is a concept used heavily in inferential statistics. (See Chapter 12.) Here it can be grasped intuitively by noting that in a 2×2 table, where $df = (2-1)(2-1) = 1$, we could set up an expected frequency table by merely supplying one cell frequency. The rest could then be obtained by subtraction, since the marginals were taken as fixed. We had, so to speak, one degree of freedom in choosing a cell frequency before all other cell frequencies were determined. In a 2×3 table, two cell frequencies can be chosen before others are determined — there are two degrees of freedom, etc.

percentage of variation in one variable explained by another, nor can it be interpreted as the proportion of predictive errors which may be reduced by prior knowledge of one of the variables—interpretations which are especially useful for substantive investigators. More will be said about this criterion, and the next chapter will develop a group of measures called "proportionate reduction in error" ("PRE") measures of association, which have as one of their strengths this kind of ready operational interpretation.

The other criterion for a good measure of degree of association is that the limits be defined in a meaningful way. We have already established the idea of *statistical independence,* or no association. In fact, all of the measures of association discussed here are always zero when there is no association in a table.* What about the upper extreme? What does a *perfect association* mean? What is the pattern in a table when a normed measure of association equals 1.0?

6.5.2   Meanings of Perfect Association

In this chapter we will limit our attention to the 2×2 table in considering alternative meanings of **perfect association.** In the next chapter, this idea will be extended to larger tables. For now, however, we can focus on two somewhat different notions of perfect association: a more restrictive or stringent definition, and a less restrictive one.

6.5.2a   The Restrictive Meaning of Perfect Association. The first model, the *more restrictive* meaning of perfect association, is one where all of the cases in a table are in one diagonal of the table. Stated differently, each value of one variable is associated with only one single value of another variable, so that for any category of the independent variable, only one cell of the dependent variable is non-zero. In a 2×2 table, this means that two diagonal cells have frequencies, and the other two cells do not. This condition for a positive and a negative association is indicated by X's (for some frequency) and zeros (for no frequency) in Figure 6.1. The left-hand tables in Figure 6.1 indicate a perfect negative association, and the right hand tables indicate a perfect positive association. The middle table shows symbols for cell frequencies which are traditionally used in expressing formulas for measures of association for a 2×2 table.

If we were testing a theory that stated that the larger the city size the higher the degree of tolerance, this model of perfect association would seem to reflect a situation where this hypothesis held exactly. All individuals living in large cities would have "high" tolerance scores, and all individuals living in smaller cities would have "lower" tolerance scores. Any deviation from this pat-

---

*While it is the case that measures we have discussed will be zero when a table shows no association (when one variable is statistically independent of another), it is not always the case that they will be zero only when two variables are statistically independent of each other. As we shall see later, measures of association are defined to be sensitive to certain features of an association, often depending upon the level of measurement, for example. Variables may be statistically independent with respect to categorical or ranking features, but not independent with respect to the distance features defined into interval-level variables. For this reason selection of an appropriate measure of association again depends critically upon meanings defined into variables and variation in data which an investigator is interested in examining. This will be discussed later, in this chapter and the next, as measures of association are presented and contrasted.

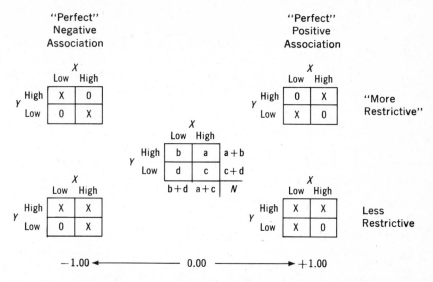

In the notation shown in the middle table, cell *a* is in the high-high corner and *d* is in the low-low corner; thus cells *a* and *d* are on the positive diagonal. This is important simply because it assures the correct sign in computations which follow.

FIGURE 6.1   Two Models of Perfect Association for 2 × 2 Tables

tern would be less than a perfect association. The *more restrictive* definition of a perfect association would require that all of the data be concentrated in one diagonal of the table for there to be a perfect association, as in Figure 6.1.

Phi, one of the delta-based coefficients discussed above, is a normed measure of association for a 2 × 2 table in this more restrictive sense. The *phi coefficient* on a 2 × 2 table can be computed as follows:

(6.7)
$$\phi = \frac{ad - bc}{\sqrt{(a+b)\ (c+d)\ (b+d)\ (a+c)}}$$

where a, b, c and d refer to cell frequencies in a 2 × 2 table as indicated in Figure 6.1, and the denominator of ϕ is the square root of the product of the marginals.

ϕ can be interpreted as the degree of diagonal concentration, and it will have values between −1.0 and +1.0. The values of 1.0 will be reached only if all cases are concentrated in the main diagonal of the table, and the magnitude of this coefficient indicates the extent to which this condition is exhibited in the table. For nominal variables, of course, one would use only the magnitude of the coefficient and drop the sign.*

6.5.2b   The Second Model of Perfect Association.   For a 2 × 2 table, the second model of perfect association, the less restrictive one, is illustrated by the set of tables in Figure 6.1. Here the requirement is that only one of the four cells need have a zero frequency and the other three, including the two on the

---

*As we will show in the next chapter, the 2 × 2 table is an interesting situation because many measures of association which are different for larger tables turn out to have the same numeric value for a 2 × 2 table. Thus $\phi = T = V = r$. Pearson's r is a measure of association to be developed in the next chapter.

main diagonal in the table, may have any frequency. Where might this definition be useful? Consider two illustrations.

Social science investigators often measure attitudes in terms of a series of individual items which are carefully graded to indicate different degrees of that attitude. For example, in a study of permissiveness attitudes regarding sex, an individual might be asked to indicate his agreement with statements such as these:

a) I believe that petting is acceptable for the female before marriage when she is engaged to be married.
     ( ) Agree
     ( ) Disagree

b) I believe that petting is acceptable for the female before marriage even if she does not feel particularly affectionate toward her partner.
     ( ) Agree
     ( ) Disagree

Now, if these questions do indeed reflect differences in degree of permissiveness attitudes, and if people realize this and respond accordingly, we would not expect anyone to disagree with the first question and agree with the second. If a person's permissiveness attitudes are sufficient to accept or agree with the second statement, then that person should also agree with the first. This expected state of perfect association between answers to these two attitude items would be shown in a table, as follows:

"More Permissive" *Attitude Item (b)*	*"Less Permissive"* *Attitude Question (a)*	
	Disagree	*Agree*
Agree	O	X
Disagree	X	X

where X indicates some cell frequency.

The scale items may not be graded as we thought, or people responding to the items may not understand them in this light, so any given set of responses may not match this idea of a perfect logical relationship between these "strong" and "weak" attitude items. A measure of association which used this less restrictive definition of perfect association as its definition of 1.0 would be the measure we would need to measure the distance between the data we observe and this idea of perfect association.

Another example is more of a philosophical distinction between "cause" and "effect" conditions. If one is interested in factors which influence the state of other variables, one may think of two kinds of effect. One situation is where the effect does not occur unless and until a given factor is present in a certain amount (called a *necessary* condition). The other situation is where the outcome may occur for other reasons, but it will never be the case that the outcome will not occur if a given factor is present to a certain extent (called a *sufficient* condition).

These states can be illustrated in terms of the two tables below. The first would be the model of a sufficient condition and the second a model of a necessary condition. The third table below is a model of a condition that is *both* necessary and sufficient. To what extent does the real world behave the way we reason that it should? A measure of association using the less restrictive model of perfect association would permit one to contrast actual data with the models for either a necessary or a sufficient condition to help provide the answer.

MODEL OF A "NECESSARY" CONDITION

Effect	*Cause*	
	Absent	*Present*
Present	O	X
Absent	X	X

where X indicates some cell frequency

MODEL OF A "SUFFICIENT" CONDITION

Effect	*Cause*	
	Absent	*Present*
Present	X	X
Absent	X	O

where X indicates some cell frequency

MODEL OF A CONDITION WHICH IS BOTH NECESSARY AND SUFFICIENT

Effect	*Cause*	
	Absent	*Present*
Present	O	X
Absent	X	O

where X indicates some cell frequency

6.5.2c Yule's Q. Yule's Q is one measure of association for a 2×2 table which uses the less restrictive definition of perfect association.* In a 2×2 table, Yule's Q is computed by finding the cross products, the product of the cell frequencies on one diagonal, and the product of the cell frequencies on the other diagonal in the table.

(6.8)
$$Q = \frac{ad - bc}{ad + bc}$$

where $a, b, c,$ and d are cell frequencies as shown in Figure 6.1.

If one of the cells on the minor diagonal is zero, then the Q value will be

*Actually Yule's Q is identical to gamma, a coefficient to be introduced in the next chapter. Gamma is suitable for any size of table, including a 2×2 table. The different name introduced here is again tradition and a usage frequently found in the literature. The formula with its particular notation is not useful for larger tables, but here it nicely indicates one important difference between measures of degree, direction and existence of association.

BOX 6.2 THE THEME

Thus far in this chapter we have (a) discussed the idea of association (end of Section 6.1; Section 6.2; Section 6.4) and how percentages may be computed to examine an association (Section 6.3.3); (b) we then introduced four features of an association — *existence, degree, direction,* and *nature* (Section 6.4ff.); and (c), we considered how to define no association (6.4.1a) and the more and less restrictive models of perfect association (Section 6.5.2ff). Measures of association which, in one index number, can summarize the existence/degree/direction of association are simple ratios. The problem is to select numerator and denominator in such a way that (only) those features of an association which are of interest are revealed. Several delta-based measures were presented to illustrate the pursuit of a properly normed measure of association. Finally, and most importantly, we ended up with two properly normed measures for 2×2 tables, each "normed" at a different but useful definition of perfect association. Other normed measures will be presented in Chapter 7, but two other topics — one caution and one distinction — will be discussed next.

---

either $+1.0$ or -1.0, depending upon the direction of association. In any event, the value 1.0 means that the data meet the less restrictive criteria for patterns of frequencies required for a perfect association. Notice that the Q and ϕ coefficients have the same numerator, but the denominators differ and produce measures which norm at different definitions of perfect association. If ϕ is 1.0, Q will also be 1.0, since the data will at least show the less restrictive pattern of perfect association. On the other hand, if $Q = 1.0$, it may be the case that $\phi \neq 1.0$, because the criterion for perfect association for ϕ is more stringent, calling for a different pattern of concentration of cases in the 2×2 matrix. As in the case of ϕ, Q can be used for nominal variables, but the sign of Q would not be used, only its absolute magnitude.* Q can reach 1.0 for any 2×2 table which is identical in pattern with its model of perfect association.

6.6 GROUPING ERROR

Before we proceed, we should update our attention to grouping error. You will recall that grouping error was one possibly distorting factor involved when we wished to group data into a smaller number of categories which might be handy for presentation purposes (see Section 3.2.3b). In univariate statistics, if the midpoint of a grouped class did not reflect the character of scores in that class,

---

*It is true, of course, that the sign of Q or ϕ resulting from their computation on the association of nominal variables would indicate which categories of one variable go with which categories of the other variable. Direction of association would not have meaning, however, and one would have to examine the table layout carefully to determine and express what the sign means. For this reason, Q and ϕ computed on nominal variables are expressed on a scale extending in magnitude from 0 to 1.0.

some grouping error would result. The same type of problem may happen where a bivariate distribution is grouped into fewer categories. Table 6.11 illustrates this possibility. Notice that the original 2×3 table is grouped in two ways to create a 2×2 table. We might want to group data if we preferred to use 2×2 measures of association (usually we would be well advised to pick a measure that works on the table we have), or we might have relatively few cases in some rows or columns of a larger table which might suggest that less precision is called for. In any event, as Table 6.11 shows, different groupings are not necessarily equivalent.

In the first 2×2 table (6.11b) the degree of association is rather strong and negative; in Table 6.11c, the association is zero. Which accurately represents the

TABLE 6.11 AN ILLUSTRATION OF GROUPING ERROR IN BIVARIATE DESCRIPTION

(a) ORIGINAL 2×3 TABLE OF FREQUENCIES

Income	Education High	Medium	Low	Total
High	5	2	8	15
Low	2	3	1	6
Total	7	5	9	21

(b) THE SAME DATA WITH "HIGH" AND "MEDIUM" CATEGORIES OF EDUCATION GROUPED TOGETHER TO CREATE A 2×2 TABLE

Income	Education Hi/Med	Low	Total
High	7	8	15
Low	5	1	6
Total	12	9	21

$$Q = -.70$$
$$\phi = -.33$$

(c) THE SAME DATA WITH "MEDIUM" AND "LOW" CATEGORIES OF EDUCATION GROUPED TOGETHER TO CREATE A 2×2 TABLE

Income	Education High	Med/Low	Total
High	5	10	15
Low	2	4	6
Total	7	14	21

$$Q = 0$$
$$\phi = 0$$

relationship between income and education in the original table? Neither one does, and the distortion is called *grouping error*. Generally speaking, it is a useful practice to retain as much precision as possible and to be reluctant to group more precise data.*

6.7 SYMMETRIC AND ASYMMETRIC MEASURES OF ASSOCIATION

One of the distinctions between different measures of association which we have discussed is whether or not they are properly normed. Another distinction was between measures which define perfect association in a more rather than less restrictive fashion. There is a third distinction between measures of association which should be mentioned here. That is the distinction based upon whether or not the measure distinguishes between independent and dependent variables.

Symmetric measures of association do not distinguish between dependent and independent variables, but merely address themselves to the strength (and direction) of relationship between pairs of variables. The role of variables in a bivariate distribution does not matter as far as the computation of the measures is concerned. Measures such as Yule's Q, the phi coefficient (as well as C, T, V, and some others) are all symmetric measures of association.

Asymmetric measures of association, by contrast, do require a distinction between independent and dependent variables for their computation. In general, they are oriented toward measuring the usefulness of the independent variable in predicting values of the dependent variable. In general, two different asymmetric coefficients can be computed on a single bivariate table. One of these coefficients measures the value of predicting Y from a knowledge of variable X and the other measures the value of predicting X from a knowledge of Y. Most of the measures of association of this type will be discussed in the next chapter (*e.g.* Lambda, Somers' d, Eta, etc.) but we have used a measure in this chapter which illustrates the asymmetric idea. That measure, crude (not normed) though it may be, is called **epsilon** (ϵ), and it is simply the difference between percentages. If percentages are computed down, epsilon (ϵ) is simply the difference between percentages compared across in the other direction. Traditionally ϵ is computed as the difference between the extreme corners of a table, as shown below, but it is clear that many ϵ's can be computed to compare cells of particular interest. Here ϵ would equal +25%, the difference between 40% in the upper right corner (the high-high corner of this table), and 15% in the upper left corner (the high-low corner of this table).† The reason for taking this particular difference is that if one expects cases to pile up on the main diagonal as the association becomes stronger, the smallest cell should be in one corner and the biggest in the other, so that this difference is expected to be the biggest in the table.

---

*Even where it is decided to group data into a smaller table, it is advisable to preserve at least three categories in each variable. The reason for this is that one is able to identify curvilinear natures of association if at least three categories are preserved. With only two categories, if there is an association, it will always appear to be linear.

†We could have taken the overall difference between 60% and 29% in the bottom row of the table, following the same logic. The ϵ is likely to be different, of course, and this is one reason why other measures of the overall association in a table are generally preferred.

ELECTORAL ACTIVITY OF RESPONDENTS IN A CROSS-SECTIONAL SAMPLE OF THE U.S., BY SOCIAL CLASS

Electoral Activity	Social Class			Total
	Low	Medium	High	
High	15%	24%	40%	26%
Medium	25	31	31	29
Low	60	45	29	45
	100%	100%	100%	100%
	(857)	(871)	(871)	(2599)

Overall $\epsilon = 40\% - 15\% = 25\%$

Source: Kim, 1971:900, Table 1. Data are from a cross-sectional study of the United States. Used by permission.

If percentages were computed in the opposite direction (by rows rather than columns as shown here) the percentage comparison would be made the other way, and it would be a different value as shown below.

SOCIAL CLASS OF RESPONDENTS IN A CROSS-SECTIONAL SAMPLE OF THE U.S., BY ELECTORAL ACTIVITY

Electoral Activity	Social Class			Total
	Low	Medium	High	
High	18%	31	51	100% (687)
Medium	28%	36	36	100% (756)
Low	45%	34	21	100% (1156)
Total	32%	34	34	100% (2599)

Overall epsilon $= 45\% - 18\% = 27\%$

ϵ, then, is an asymmetric measure because it is a comparison made after we percentage in the usual manner—in the direction of the independent variable. As we noted earlier, ϵ suffers from norming problems, but it will serve here as a rough and ready measure of overall differences and as an example of an asymmetric measure. Normed asymmetric measures will be presented in the next chapter.

6.8 SUMMARY

A number of key steps in statistical description have been taken in this chapter, and these will be elaborated in the next chapter and later in this book.

One of the first ideas was that of *association* between two variables. This amounted to a systematic way of examining the differences in the distribution of one variable which are associated with differences in a second variable. As values of one variable increase, there is a tendency for there to be an increase (or decrease) in the scores cases have on a second variable. Stated more generally, a table is said to show some association if a variable is distributed differently within the various categories of some other variable.

An association can be described in terms of four features — *existence* of an association, *degree, direction,* and *nature.* These may be examined by percentaging a table properly (and three rules for percentaging were given). Alternatively, the first three features — existence, degree, and direction — may be measured in terms of a variety of different ratios called *measures of association.* A single measure of association may be computed which (a) will indicate whether or not there is an association in the table (measures are generally zero if there is no association); (b) will indicate the degree or strength of association (by a magnitude which is usually between zero and 1.0), and (c) will show direction (by a plus or minus sign), although this is used only for bivariate distributions of ordinal or interval variables.

To permit valid comparison of measures of association computed on different tables, and to aid in the interpretation of these measures, investigators generally prefer *"normed"* measures of association. This means that the numerator and denominator of the measure of association are selected so that the measure could be −1.0 or 0 or +1.0 for any table, regardless of the number of cases it contains, the number of rows or columns, or the way the marginals happen to be distributed. Various delta-based measures of degree of association were discussed to illustrate the problem of norming.

Distinctions among measures of association can be made in terms of (a) whether or not they are normed, (b) the model of perfect association and no association the measure assumes, (c) whether the measure is asymmetric or symmetric, (d) whether it is most appropriate for cross-classifications of nominal, ordinal, or interval-level variables, and (e) the size of table for which it may be computed. Chapter 7 will discuss measures of association for nominal, ordinal, and interval-level cross-classifications.

The *nature* of association in a table refers to the patterning of concentration. It is often examined using a properly percentaged table, but Chapter 7 will discuss the use of a regression equation to describe the nature of association between interval-level variables. Sometimes this patterning is monotonic or linear or curvilinear, but often the pattern is more complex. In any event, the description of the existence, degree, direction, and nature of association between pairs of variables provides a powerful means for describing how variables relate to each other and this, after all, is the central activity of any scientific inquiry.

CONCEPTS TO KNOW AND UNDERSTAND

bivariate distribution; conditional distribution
  heading
  stub

body
cell
cell frequency
2×2 or fourfold table
$r \times c$ tables
marginals
percentaging rules
epsilon (percentage difference)
how to examine complex conditional distributions
association
  existence
  degree
  direction
  nature
expected cell frequencies
delta
statistical independence
norming measures of association
definition of perfect association
  more restrictive
  less restrictive
phi coefficient
Yule's Q coefficient
grouping error in tables
asymmetric and symmetric measures of association
Pearson's C
Tschruprow's T
Cramer's V

QUESTIONS AND PROBLEMS

1. Using data from a statistics workbook or from a piece of research you have been asked to conduct, select a series of cases which have been measured on two variables of interest to you. Construct a bivariate frequency distribution and compute percentages in the appropriate direction. Set up the table with labeling in a form suitable for presentation. Write a paragraph describing the relationship between the pairs of variables shown in this table.

2. Select an interesting percentaged table from the back of a statistics workbook or from a journal article. Compute percentages in each of the three possible ways (to row totals, column totals, and the grand total). Then, in your own words, express what it is that percentages in each of these three tables tells about the existence, degree, direction, and nature of association in the table.

3. Look through one of the professional journals for an example of a properly percentaged table and for a table which was not properly percentaged for the kinds of conclusions the author wanted to make. Discuss the use of percentaged tables in both articles.

4. Find one example of the use of Yule's Q and phi in a statistics workbook or in journals or texts. Examine each usage and discuss the merits of selecting Q or ϕ in each case.

5. The data in Tables A and B below are from a study of the relationship between economic strain and the consumption of alcohol to relieve distress (Pearlin and Radabaugh, 1976:657–658). Compute percentages in the appropriate direction for each table and use epsilons to interpret the data. What conclusions do the data warrant about the relationship between economic strain and the disposition to use alcohol to relieve distress?

TABLE A: ECONOMIC STRAIN AND LEVEL OF ANXIETY (Frequencies)

| | | Anxiety | |
	Intense	Moderate	Low
Economic Strain:			
Severe	41	27	32
Moderate	51	46	48
Little	56	80	98
None	190	310	691

TABLE B: LEVEL OF ANXIETY AND USE OF ALCOHOL FOR CONTROL OF DISTRESS (Frequencies)

| | | Anxiety | |
	Intense	Moderate	Low
Disposition to Use Alcohol for Distress Control:			
Strong	77	66	104
Weak	61	75	138
Minimal	198	329	623

6. Compute the percentages in Tables A and B in the opposite direction from that which you computed them in problem 5. Again use epsilons to interpret the data. Now what conclusions do the data warrant?

7. The following data are from a study of political leadership in Chile (Zeitlin, Neuman, and Ratcliff, 1976:1019). Compute phi and Yule's Q for these data. How do they compare? Which is more appropriate for the analysis of these data? Why?

TABLE C: FREQUENCIES OF LARGE CORPORATION EXECUTIVES WHO HELD NATIONAL POLITICAL OFFICE, BY POLITICAL OFFICEHOLDING IN IMMEDIATE FAMILY

Political Officeholders in Immediate Family	Number of Political Offices Held	
	None	*One or More*
Yes	39	29
No	126	35

GENERAL REFERENCES.

Davis, James A., and Ann M. Jacobs, "Tabular Presentation," in volume 15, p 497–509, David L. Sills (ed), *International Encyclopedia of the Social Sciences* (New York, Macmillan Co. and The Free Press), 1968.
This article discusses appropriate ways to set up tables, how to handle "don't know"s, and how to use percentages in examining tables.

Riley, Matilda White, *Sociological Research I: A Case Approach* (New York, Harcourt, Brace and World), 1963.
Unit 8 presents brief summaries of several important sociological studies, and commentary discusses their examination of the relationship among variables.

Rosenberg, Morris, *The Logic of Survey Analysis* (New York, Basic Books), 1968.
Chapter 1 discusses the meaning of relationships.

Weiss, Robert S, *Statistics in Social Research: An Introduction* (New York, John Wiley and Sons), 1968.
See especially Chapter 4 on tables and Chapter 9 on association.

Zeisel, Hans, *Say It With Figures* (New York, Harper and Row), 1957.
Chapter 1 discusses the use of percentages, Chapter 2 presents examples of percentaging rules and Chapter 3 handles the problem of "don't know"s or "no response"s. More complex tables are discussed in Chapter 4. This is a very thoughtful and readable treatment of the use of percentages in tables.

OTHER REFERENCES

Erskine, Hazel, "The Polls: Women's Role," *Public Opinion Quarterly,* 35 (Summer, 1971) p 278.

Fischer, Claude S., "A Research Note on Urbanism and Tolerance," *American Journal of Sociology,* 76 (March, 1971) p 847–856.

Kim, Jae-On, "Predictive Measures of Ordinal Association," *American Journal of Sociology*, 76 (March, 1971) p 900.

Pearlin, Leonard I., and Clarice W. Radabaugh, "Economic Strains and the Coping Functions of Alcohol," *American Journal of Sociology,* 82 (November, 1976). Copyright 1976 by the University of Chicago and Published by the University of Chicago Press.

Sewell, William H., and Vimal P. Shah, "Parents' Education and Children's Educational Aspirations and Achievements," *American Sociological Review,* 33 (April, 1968).

Treitel, Ralph, "Onset of Disability," Report 18, (U.S. Department of Health, Education and Welfare, Social Security Administration, Office of Research and Statistics, Washington, D.C.), 1972.

U.S. Bureau of the Census, "Selected Characteristics of Persons and Families of Mexican, Puerto Rican, and Other Spanish Origin; March, 1971," *Current Population Reports* P-20, No. 224, (Washington, D.C.), 1971.

————, "Characteristics of American Youth: 1971," *Special Studies*, P-23, No. 40, (Washington, D.C.), 1971.

U.S. Department of H.E.W., "Earnings Distributions In The United States: 1967," (U.S. Department of Health, Education and Welfare, Social Security Administration, Office of Research and Statistics, Washington, D.C.), 1971.

Zeitlin, Maurice, W. Lawrence Neuman, and Richard Earl Ratcliff, "Class Segments: Agrarian Property and Political Leadership in the Capitalist Class of Chile," *American Sociological Review*, 41 (December, 1976).

7

Measures of Association for Nominal, Ordinal and Interval Variables

The ecology of Chicago has long been a source of attention and study by sociologists since Park, Burgess and other early investigators at the University of Chicago began to study the patterning of urban phenomena (*e.g.*, social class, ethnic organization, etc.). A recent study by Albert Hunter (1971) examines ecological trends in Chicago, using census data from the 1930's to the 1960's based on 75 community social areas into which the city had been divided by sociologists in the 1920's. Among the interests Hunter had in this longitudinal study was whether there has been increasing or decreasing segregation in the city over these years. One would expect greater assimilation to occur the longer a group has been in an area, so that the area could no longer be distinguished in terms of the kinds of jobs or income level or family status of people living there compared to others in the city in general.

Ecological research in the U.S. has found that there are three main differences between urban areas, namely: social rank, family status, and segregation or ethnic status of the people living in an area. Hunter used percent black and percent foreign-born in each of the 75 Chicago areas to measure segregation or ethnic status. Percent females employed was used as a measure of family status, and median dollar value of homes was used as a social rank measure. Four other variables were measured but are not reported here. The basic statistical measure he used to describe the relationship between ethnic status and each of the other variables was a correlation coefficient or measure of the existence, strength, and direction of association between pairs of variables.

Hunter's reasoning is that if there is low segregation (*i.e.* high assimilation) then there should be little or no correlation between the percentage black in an area, for example, and the percentage of females who are employed—one of the family status measures he uses. Ethnic status should not make a difference in the distribution of that particular variable, nor should it be correlated with the median value of homes if there is no segregation. If correlations become smaller in magnitude over time, there is evidence of decreasing segregation or increasing assimilation. If, on the other hand, there has been no drop or if there is a shift from a lower to a higher correlation through time, then there is evidence of either no change in segregation or increasing segregation in urban Chicago. The data in Table 7.1 indicate some interesting increases and decreases through time, and these changes may be due in part to recent migration from urban to suburban areas which has been occurring across the nation in recent years.

Notice, in Table 7.1, that the correlation measure for the bivariate relationship between percent black and percent females employed in these 75 Chicago areas was +.46 in 1930, and that the correlation drops steadily in magnitude over time to +.42 in 1940, +.13 in 1950, and −.03 (smaller in magnitude but a shift to a negative rather than positive direction of association) in 1960. Com-

---

TABLE 7.1 MEASURES OF THE BIVARIATE ASSOCIATION BETWEEN EACH OF TWO SEGREGATION VARIABLES (PERCENT BLACK AND PERCENT FOREIGN-BORN) AND MEASURES OF STATUS IN 75 COMMUNITY AREAS OF CHICAGO FOR THE YEARS 1930–1960.

A. CORRELATIONS BETWEEN PERCENT BLACK AND THE PERCENT FEMALES EMPLOYED.†

1930	.46
1940	.42
1950	.13
1960	−.03

B. CORRELATIONS BETWEEN PERCENT BLACK AND THE MEDIAN VALUE OF HOMES.†

1930	−.09
1940	−.17
1950	−.22
1960	−.18

C. CORRELATIONS BETWEEN PERCENT FOREIGN-BORN AND PERCENT FEMALES EMPLOYED.†

1930	−.28
1940	−.13
1950	.06
1960	.12

D. CORRELATIONS BETWEEN PERCENT FOREIGN-BORN AND MEDIAN VALUE OF HOMES.†

1930	−.31
1940	−.41
1950	−.15
1960	−.01

E. CORRELATIONS BETWEEN PERCENT BLACK AND PERCENT FOREIGN-BORN.†

1930	−.52
1940	−.63
1950	−.71
1960	−.77

Source: Data from Hunter, 1971: 437, Table 5. Used by permission.
†Correlation coefficients are Pearson product-moment correlations (r), described at the end of this chapter. These coefficients vary between −1.00 and +1.00 and, in general, they can be interpreted in terms of their absolute magnitude—the larger the absolute value of the coefficient, the stronger the association or the more closely related are the two variables being compared.

BOX 7.1 ASSOCIATION

Double-check yourself. If you are not really clear about the following ideas, go back for a review.

Association of two variables (Section 6.4)
Direction of association (Section 6.4.3)
Degree of association (Section 6.4.2)
Nature of association (Section 6.4.4)
Normed measure of association (Sections 6.5 and 6.5.1)
Statistical independence (end Section 6.4.1a)
Perfect association (Section 6.5.2ff)

The purpose of the current chapter is to develop several measures of association which are normed, which are interpretable in an intuitively meaningful way, and which are appropriate for variables defined at various levels of measurement and related in certain characteristic ways.

---

parison of these correlation coefficients indicates that percent black in an area is becoming less associated with percentage of women employed, and this can be interpreted to mean that assimilation has occurred in this aspect of family status for this ethnic group. Notice that just the opposite pattern characterizes the association of percent black and median value of homes. The small negative association in 1930 becomes larger through time, although there is a reduction to −.18 in 1960. A decline in correlations from 1930 to 1960 is more characteristic of foreign-born than of black, indicating a more general pattern of assimilation for the foreign-born group. Notice too that the association between the two segregation measures, percent black and percent foreign-born for these 75 areas is getting stronger through time, and that it is a negative association. Where one group is concentrated in an area, the other group, increasingly, is absent. This is what one would expect if the foreign-born were moving to suburban areas, leaving higher percentages of black families behind in the central city neighborhoods.

Now for some comments on the coefficients we have been examining. No comparison such as those Hunter made would be possible unless the measures of association he used were comparable, that is, properly normed so that correlations of a given magnitude have the same meaning. This is an important point, because we want to compare across time and across different combinations of variables to reach a conclusion. Secondly, notice that the numbers may be positive or negative, showing direction of association (*e.g.* the *higher* the percentage black in an area, the *higher* the percentage of employed women in the area—a positive association—or, the *higher* the percentage of foreign-born in an area the *lower* the percentage black in the area—a negative association). Magnitude of the number varied from a small value, indicating a weak association between the cross-classified variables (a value of zero would mean no association), to a larger number, approaching 1.0, indicating a strong association between the two variables.

This chapter will introduce three families of measures of association which Hunter might have used for the purpose of validly comparing strength and direction of association between variables. There are several other types of measures we could find and use.* In the last chapter, for example, we discussed some of the so-called "delta-based" measures of association. In fact, you could devise your own measures of association after you see what is involved in measures discussed here. The purpose of this chapter is to present some of those measures used more frequently by sociologists and to illustrate the straightforward logic underlying a normed measure of association. We will present measures suitable for the association of nominal, ordinal, and interval variables.

7.1   PROPORTIONATE REDUCTION IN ERROR MEASURES ("PRE")

All of the measures we will discuss in this chapter are of a type called "proportionate reduction in error" measures, or **"PRE"** measures. They are all relatively simple ratios of the amount of error made in predicting under two situations: first, the situation where there is no more information than simply the distribution of the dependent variable itself, and, secondly, a situation where there is additional knowledge about an independent variable and the way the dependent variable is distributed within the categories of that independent variable. PRE measures simply state the proportion by which one can reduce errors made in the first situation by using information from the second situation, above.

$$PRE = \frac{\text{Reduction in Errors with More Information}}{\text{Original Amount of Error}}$$

The problem of prediction is a common one to the sciences, so it makes some sense to focus a measure of association on the idea of making accurate predictions of the values of some dependent variable. If theoretical knowledge leads us to say that people with a higher social class standing in a society will feel less alienated than those with lower class standing, we are saying, in effect, that knowledge of social class score differences will permit us to make more accurate predictions of differences in alienation scores. If *all* of the errors of prediction can be eliminated by basing predictions on social class, then there is a perfect association between these two variables, and our theoretical basis for expecting this outcome is supported. On the other hand, if the association between social class and alienation is poor or non-existent, then that fact would be indicated in a measure of association which expresses the proportion of the original predictive errors that can be avoided by virtue of the additional knowledge about social class — in this case little or none.

There are three things we might be interested in predicting, depending, essentially, upon the definition of the variables involved in a problem. For *nominal* variables, we are usually interested in predicting the category or exact score

---

*Among the variety of measures for special situations, a good sampling is presented in an elementary statistics book by Freeman (1965).

of the dependent variable. Often it is sufficient to focus prediction on the most typical or *modal* value of the dependent variable. If the dependent variable is *ordinal,* then we are probably interested in predicting *rank order* of pairs of scores on the dependent variable (although we could think of other possibilities, such as predicting the median or some other percentile). Finally, if the dependent variable is an *interval* level variable, we would probably be interested in predicting the arithmetic *mean* of that dependent variable. These alternatives are shown in Figure 7.1.

Generally speaking, we make errors in predicting modes or rank orders or

The Feature of the Dependent Variable to Be Predicted.	Rules Used in Making Predictions	
	Rule 1: Minimum Guessing Rule	Rule 2: Improved Guessing Rule
A. Predicting Modes (Nominal variables)	For each case to be predicted, predict the modal category of the variable, overall. (Alternatively, predict the category placement at random.)	For each case, first determine the category of the independent variable into which the case falls; then predict the mode of the dependent variable for that category. (Alternatively, predict category placement within categories of the independent variable.)
B. Predicting Rank Order (Ordinal variables)	For each pair of cases for which rank order on a variable is to be predicted, determine which is ranked higher by a flip of a coin (random selection).	For each pair of cases for which rank order on a dependent variable is to be predicted, first determine the ordering on an independent variable, and then, if the overall association is positive, predict "same ordering" on the dependent variable. If the overall association is negative, predict "reverse ordering" on the dependent variable.
C. Predicting the Mean (Interval variables)	For each case to be predicted, predict the overall mean of the dependent variable.	For each case to be predicted, first determine either (a) the category of the independent variable it falls in and predict the mean of that category, or (b) develop a regression equation which permits you to compute a predicted value for the dependent variable, given the score on the independent variable.

FIGURE 7.1 DEVELOPMENT OF PRE MEASURES OF ASSOCIATION*

*There are a number of good references for further discussions of PRE measures of association. See, for example, Costner, (1965) and Kim, (1971).

means, but with the proper theory and proper information we can cut down the amount of predictive error we would otherwise make.* Figure 7.1 indicates the way in which we might make predictions both with and without information. The PRE measures of association we will discuss below are simply a contrast between the errors made in using Rule 1 and those made in using Rule 2 to predict the mode or rank order or mean of a variable that interests us. Usually, the contrast will be formed as follows:

(7.1) $$\text{PRE} = \frac{(\text{Errors made using Rule 1}) - (\text{Errors made using Rule 2})}{(\text{Errors made using Rule 1})}$$

7.2   MEASURES OF ASSOCIATION FOR NOMINAL VARIABLES

PRE measures for each of the situations illustrated in Figure 7.1 will be discussed in turn in the following sections.

7.2.1   Lambda

The **Lambda** measure of association (also called "Guttman's Coefficient of Predictability"), λ_{yx}, is an asymmetric measure of association especially suited to bivariate distributions where both variables are interpreted to be nominal variables. It is a measure which very nicely illustrates the logic of PRE measures.

Suppose that we are interested in marital status as a dependent variable, and we are interested in making predictions about the marital status of individuals who are household heads in the U.S.† Given information in Table 7.2 about how marital status is distributed in the U.S., we would do best to predict that household heads are married. That is, if we know that the modal marital status is "married," then the most rational single score we could predict for a head of household would be that he (she) is married. We will be correct more frequently than we would be if we picked any other single category of marital status for our prediction. Information on the overall distribution of marital status for household heads in the U.S. in 1970 is shown in the total column at the right of Table 7.2.

If we were to guess that the head of the household is married before

---

*In the case of nominal variables, we will also discuss a measure of association which focuses upon predicting the category into which a case may fall, whether or not this category is the modal category. This measure focuses, in a sense, upon the *distribution* across categories of the dependent variable rather than upon an *optimal* single category. The various coefficients are introduced here, not only because they find important uses in sociology, but because they highlight the notion that there are many things about a dependent variable which one might be interested in predicting. The choice among available measures of association involves not only technical or computational matters, but also a basic understanding of what one wants to predict, and this in turn is tied directly to the substantive logic of the research problem itself.

†The U.S. Census defines a head of family as one person, usually the person regarded as the head by family members, but women are not classified as heads if their husbands are resident members of the family.

TABLE 7.2 Frequency Distribution of Marital Status of Household Heads in the U.S. in 1970 by Type of Household (in thousands).

| Marital Status of Head | Type of Household | | | | |
| | Male Headed | | Female Headed | | |
	Related Children Under 18	No Related Children Under 18	Related Children Under 18	No Related Children Under 18	Total
Married	25,776	19,214	313	198	45,501
Separated	79	502	998	425	2,004
Divorced	74	946	1,135	1,105	3,260
Widowed	181	1,199	942	6,457	8,779
Single	64	2,302	349	2,113	4,828
Total	26,174	24,163	3,737	10,298	64,372

Source: U.S. Bur. Census, 1971: 17, Table 6.

knocking on the door for an interview with each of the 64,372 (thousand)* households in the U.S., we would be right 45,501 times and wrong 18,871 times (*i.e.* 64,372 − 45,501 = 18,871). This is the total number of errors of prediction we would make if we merely predicted the overall mode of the marital status of the head of household variable. How many of these errors could be eliminated if we had more information to start with? (See Rule 2 information listed in Figure 7.1.)

The body of Table 7.2 shows the distribution of marital status of the head of household for separate categories of the "type of household" variable. A quick glance at this table suggests that indeed, marital status is distributed differently depending upon whether the household is headed by a male and includes related children 18 years of age, is male-headed with no related children under 18, is female-headed with related children under 18, or is female-headed with no children under 18. How much is this added information worth? If we knew the "type of household" before making our prediction of modal marital status, would we be able to refine our prediction and make fewer errors in predicting scores on the dependent variable?

In this example, the answer is "yes." If we knew that the head of household is a male and that the household contained related children under 18 years of age, we would certainly predict the modal marital status for the head of that kind of household — that he would be "married." We would be right 25,776 times out of 26,174 households of this type. We would also predict "married" for the male-headed-no-young-children households — again the mode for that category — and we would be right 19,214 times out of 24,163. If we knew, however, that the household head was female and that there were under-18, related children in the household, we would predict that the household head is "divorced," the mode of marital status for that type of household. We would be right 1,135 times out of 3,737 households of this type. Finally, if the household head were female and if there were no under-18, related children, we would do best to predict "wid-

---

*In Table 7.2, the number of households is expressed in terms of thousands of households — the unit of measurement. It is important, of course, to be consistent in using the same unit throughout our computations and to interpret the result with this unit in mind.

Within-Category Modal Frequency m_y	Category of the Independent Variable(s)
25,776	Households with male heads and related children under 18 years of age.
19,214	Households with male heads and no related children under 18 years of age.
1,135	Households with female heads and related children under 18 years of age.
6,457	Households with female heads and no related children under 18 years of age.
$\Sigma m_y = 52{,}582$	Total of within-category modal frequencies

owed" as the household head's marital status. Here we would be right 6,457 times out of 10,298.

Have we improved our predictive power by virtue of this added information? We can determine the answer by adding up the correct predictions resulting from within-category prediction (*i.e.* using Rule 2 to predict) and contrast that with the overall frequency in the modal category of the marital status variable.

The more refined prediction would lead to 52,582 correct predictions, which is some 7,081 fewer errors than would be made if we had merely predicted the overall mode of marital status (in which case we would be correct only 45,501 times out of 64,372). That amounts to a reduction of 37.5% in the errors made in predicting marital status of households. This value is called **lambda** (λ_{yx}), and it is a simple substitution of total errors and reduction in errors in the earlier generalized PRE formula. More specifically, λ_{yx} is computed as follows:*

(7.2)
$$\lambda_{yx} = \frac{\Sigma m_y - M_y}{N - M_y}$$

where N is the total sample size, M_y is the *overall* modal frequency of the dependent variable, Y, (45,501 in this case), and Σm_y is the sum of modal frequencies on the dependent variable, Y, *within* separate categories of the independent variable, X, (52,582 in this example). Applied to the current example, the computation is as follows:

$$\lambda_{yx} = \frac{52{,}582 - 45{,}501}{64{,}372 - 45{,}501} = \frac{7{,}081}{18{,}871} = .375$$

The numerator expresses the reduction in error with improved information and

*Lambda could be expressed in terms of the format given earlier for PRE measures as:
$$\lambda_{yx} = \frac{(N - M_y) - (N - \Sigma m_y)}{N - M_y}$$

where the first term in the numerator is the number of errors made in using rule 1, and the second term is the number of errors made in using rule 2. Simplifying the numerator, as shown in the text above, yields a numerator which is the number of *non-errors* under rule 2 (Σm_y) minus the number of *non-errors* under rule 1 (M_y). The denominator is the number of errors under rule 1.

the denominator expresses the error with minimum information. Notice that the symbol for lambda is a lower case Greek letter, and that it has two subscripts: λ_{yx}. The first of these subscripts indicates which variable is the *dependent* variable (variable Y, traditionally), and the second subscript indicates which variable is the *independent* variable (variable X, traditionally).

λ_{yx} is the asymmetric measure of degree of association which expresses the proportionate reduction in errors made in predicting modal values of variable-Y when prior information about variable-X is available to use in refining predictions of modal scores on the dependent variable. λ_{xy} simply reverses the role of the two variables, predicting X from information about Y. Asymmetric measures of association, unlike symmetric measures, always must be labeled in this way because the value of the measure depends upon which variable is predicted from which.

Suppose, to turn the predictive problem in Table 7.2 around, we were interested in predicting scores on "type of household." How much improvement in prediction would result from using marital status as the predictor variable? The formula for lambda is the same as before, with the exception that the x subscript is substituted for the y subscript (and vice versa) in the previous example. We are interested in predicting modes on type of household, overall, and then within categories of the marital status variable. The computations are as follows:

$$M_x = 26,174 \qquad \begin{array}{c} \overline{m_x} \\ 25,776 \\ 998 \\ 1,135 \\ 6,457 \\ 2,302 \\ \hline \end{array}$$

$$\Sigma m_x = 36,668$$

$$\lambda_{xy} = \frac{36,668 - 26,174}{64,372 - 26,174} = \frac{10,494}{27,704} = .379$$

Marital status permits one to reduce errors in predicting type of household by 37.9%. The difference between λ_{yx} and λ_{xy} in this example is rather slim. In general, the two different coefficients need *not* be the same for any given table, and they are interpreted quite differently. Using lambda, one could ask which variable(s) permits the greatest reduction in errors in predicting modes of some specific dependent variable, and lambda is a measure of association which helps one assess the utility of certain additional information.* Note that some of the difference between the two lambdas in a table may result from the precision of measurement of the predictor variable. If one wants to predict a dependent variable which has five categories by using a predictor variable which has only four categories, then one could predict only four different modes, not five. Where the predictor variable is more precise than the dependent variable, then more exact-

---

*Some investigators compute a "symmetrical" lambda coefficient, which is a kind of average of the two asymmetric lambda coefficients shown here. This is believed to be of limited value and its computation will not be discussed. The formula for the symmetrical lambda, however, is:

$$\lambda = \frac{\Sigma m_y + \Sigma m_x - M_x - M_y}{2N - M_x - M_y}$$

TABLE 7.3 PERCENTAGE DISTRIBUTION OF TYPE OF IMPAIRMENT BY AGE, U.S., JULY 1966–JUNE 1967 FOR CHILDREN AND YOUTH AGED 25 OR LESS

(Y) Type of Impairment	Age (X)			
	Under 15	15–24	25 and over	Total
Visual	12.6	10.4	15.2	14.5
Hearing	18.0	11.7	24.4	22.7
Speech	21.0	3.4	1.6	3.1
Orthopedic	48.4	74.5	58.8	59.7
Total	100.0	100.0	100.0	100.0
N (1000's)	(2,601)	(3,952)	(32,560)	(39,113)

$$\lambda_{yx} = .00$$

Note that although λ_{yx} is zero, there *is* a difference in percentages compared across rows of this table, and thus an association not detected by a measure concerned with predicting modes. Here again it is important to select measures which are sensitive to desired features of data.

Source: U.S. Dept. H.E.W., 1971:15, Table 3. Data are from household interviews with a national sample of children and youth aged 25 or under from the civilian, noninstitutionalized population by the U.S. Census. Impairments are defined as chronic or permanent defects which cause a decrease or loss of ability to perform various functions.

ing predictions can be made. This is one reason why investigators attempt to preserve more rather than fewer categories of variables which are to be used in statistical analysis. Note that since lambda is suitable for nominal variables, rows and columns may then be reordered without affecting the magnitude of lambda.

λ_{yx} varies in magnitude from 0.0 to +1.0, and it can take on these values for any table regardless of size or marginals. Assuming that, overall, there is some range of scores on the dependent variable, a perfect association is defined as a condition in which all of the cases in each category of the independent variable fall into only one category of the dependent variable (the modal category).* λ_{yx} is zero where the *same* modal prediction is made within all categories of the independent variable as would be made if the overall mode were predicted. Here, quite literally, the extra information about the independent variable is not worth anything in refining predictions of the mode of the dependent variable. Table 7.3 illustrates a situation in which λ_{yx} is zero, but from an examination of percentages one can see that an association exists in a sense other than the meaning involved in predicting modes.

The problem with the data in Table 7.3 is that the distribution of the dependent variable (type of impairment) is decidedly skewed. Six out of every ten impairments are of the orthopedic type. Consequently, the modes within each category of the independent variable are likely, also, to be in the orthopedic category of the dependent variable. In general, lambda is a poor choice as a mea-

*If all scores in the sample are identical, then a perfect prediction could be made with only the overall mode, and λ_{yx} would equal zero, because additional information about an independent variable cannot reduce predictive errors at all. There would be no errors to reduce, in this instance, and thus the independent variable would not be "worth" anything in prediction.

sure of association for data in which there is pronounced skewness in the distribution of the dependent variable.

It should be noted that lambda tells only part of the story of the association between two nominal variables. In addition to knowledge of the *strength* of association, an investigator would want to examine the *nature* of the association. In this case nature would probably best be discussed in terms of the pattern of percentages in a properly percentaged table such as Table 7.3 or Table 7.1.

7.2.2 Goodman and Kruskal's tau-y

Another measure of association suitable for nominal variables uses somewhat different prediction rules than lambda. Like lambda, these taus are asymmetric measures and, again like lambda, they vary from 0.0 for no reduction in error to +1.0 for perfect reduction in error. Goodman and Kruskal's tau-y measures, however, are designed to address the problem of predicting the *distribution* of the dependent variable, Y. Lambda was concerned with predicting an optimal value, namely the *mode* of the dependent variable.

The minimum guessing rule in the case of tau-y is random assignment of cases to categories of the dependent variable so that the marginal distribution of cases is unchanged. Looking back at Table 7.2, we can see this would mean that we would randomly assign 45,501 (1000's) cases to the "married" category, 2,004 to the "separated" category, etc. Such a random allocation of the 64,372 units would, of course, involve some error, and the expected amount of error for this random assignment to categories can be computed for each category of the dependent variable and summed to yield the error expected under this minimum guessing rule. The procedures are as follows:

In Table 7.2, 45,501 are in the "married" category out of 64,372, leaving the difference, 18,871, *not* in the "married" category. We would expect the proportion 18,871/64,372 of the 45,501 cases in the "married" category to be misclassified if 45,501 cases were assigned to this category at random from among the total number of cases.* This results in 13,338.9 expected errors.

$$\frac{18,871}{64,372} (45,501) = .293 (45,501) = 13,338.9 \text{ expected errors}$$

To this are added errors expected to result from random assignment of cases to each other category and, in each case these errors are computed in the same way:

$$\begin{array}{c} \text{Expected Category} \\ \text{Error with Random} \\ \text{Assignment} \end{array} = \begin{array}{c} \text{Proportion } \textit{Not} \text{ In} \\ \text{A Given Category} \end{array} \times \begin{array}{c} \text{That Category's} \\ \text{Frequency} \end{array}$$

---

*The reasoning here is that by chance a certain proportion of cases will be misclassified (*e.g.* a married person might be assigned to the separated category), and that this proportion for any category is simply the proportion of cases *not* in that category to cases *in* that category, based on the overall marginal distribution of the dependent variable. Thus, if all cases were in one category, there could be no error if only that category were predicted. Where there is some spread of scores, there is some chance that random assignment will be correct, and some chance that errors will be made. The chance of error will be less for categories which have high frequencies, and greater for categories which have low frequencies in a given distribution.

These errors are then summed over all categories of the dependent variable. Symbolically, this can be expressed as:

$$E_1 = \sum_{i=1}^{k} \left[\frac{N - f_i}{N} (f_i) \right]$$

where f_i is the frequency in the ith category of the dependent variable, and k is the number of categories of the dependent variable.

Errors in predicting marital status from Table 7.2 would be computed as follows:

$$\frac{64{,}372 - 45{,}501}{64{,}372} (45{,}501) = 13{,}338.9$$

$$\frac{64{,}372 - 2{,}004}{64{,}372} (2{,}004) \quad = \quad 1{,}941.6$$

$$\frac{64{,}372 - 3{,}260}{64{,}372} (3{,}260) \quad = \quad 3{,}094.9$$

$$\frac{64{,}372 - 8{,}779}{64{,}372} (8{,}779) \quad = \quad 7{,}581.7$$

$$\frac{64{,}372 - 4{,}828}{64{,}372} (4{,}828) \quad = \quad 4{,}465.9$$

$$E_1 = 30{,}423.0$$

The improved guessing rule for predicting the exact distribution of the dependent variable makes use of information about the distribution of the dependent variable within categories of a nominal independent variable. Computational procedures are identical, except that the process is carried out separately for the distribution of the dependent variable *within* each category of the independent variable; thus the formula shown below directs summation over all of the c categories of the independent variable.

$$E_2 = \sum_{j=1}^{c} \sum_{i=1}^{k} \left[\frac{N_j - n_i}{N_j} (n_i) \right]$$

where n_i is the cell frequency in the ith category of the dependent variable within one of the c categories of the independent variable and N_j is the total (marginal) frequency for that category of the independent variable. These computations are carried out for each category of the independent variable and then summed over all of the c categories.

For Table 7.2, these computations are totalled as follows:

Computation for male-headed households with related children
under 18 = 788.23

Computation for male-headed households with no related children
under 18 = 8,558.08

Computation for female-headed households with related children
under 18 = 2,829.49

Computations for female-headed households with no related children
under 18 = 5,675.88

Total $E_2 = 17,851.68$

Goodman and Kruskal's tau-y is computed as follows:

(7.3)
$$\text{tau}_y = \frac{E_1 - E_2}{E_1}$$

$$= \frac{(30,423.0) - (17,851.68)}{30,423.0} = .41$$

Goodman and Kruskal's tau-y can be interpreted as the proportion reduction in errors of predicting category placement. Here, the prediction rules involve the overall marginal distribution of the dependent variable (or distributions within categories of the independent variable), rather than only the central-tendency measure which was used in computing lambda. Again, there are two tau-y values which could be computed on the same table if the roles of dependent and independent variable were reversed. In general, these two values do not turn out to be the same value.

Tau-y varies from 0.0 to +1.0. It is +1.0 where all cases within a category of the independent variable are in the same category of the dependent variable, and this can occur only if there are at least as many categories of the independent variable as of the dependent variable. This condition of perfect association would also yield a value of +1.0 for lambda.

Unlike lambda, tau-y will not be zero whenever the modal frequencies of the dependent variable within the various categories of the independent variable all fall into the same category as the modal frequency of the marginal distribution of the dependent variable. Thus, for table 7.3 tau-y is not zero; however, it is very low (.01). In general, in such situations the delta based measures of association (section 6.5.1) will produce higher coefficients than will tau-y because they are based upon a different philosophy of measurement of association (see section 6.5.2).*

7.3   MEASURES OF ASSOCIATION FOR ORDINAL VARIABLES

Prediction of scores on ordinal variables is somewhat different from prediction of scores on nominal variables. Since we are interested in the *ranking* of scores on ordinal variables, it is useful to think of *pairs* of observations. It takes at least two scores before the idea of "rank" is meaningful. For a measure of association we are interested in the rank of pairs of cases on *two* ordinal variables, since we are interested in whether or not knowledge of the rank ordering of pairs of cases on one variable is useful in predicting their rank order on the other variable. If the knowledge of ranking of pairs on one variable is of no use in predicting rank order on the other variable, then we would like an ordinal measure of association to equal zero. This situation is equivalent to predicting the rank order of

*For a 2 × 2 table tau-y is equal to phi squared. Consequently, phi squared can be interpreted as a PRE measure (only for 2 × 2 tables, however).

cases randomly by tossing a coin, half the time guessing that the highest-ranked case on one variable is highest on the second variable, and half the time guessing that that highest-ranked case on the first variable is lowest on the second variable.

Two other prediction rules may prove to be useful, however, in the sense of reducing errors in predicting rank order of pairs of observations on a dependent variable. These rules correspond to situations where there is (a) a positive association, or (b) a negative association between the two variables. In the first instance, we would predict **same rank order** on the *second* variable as the pair had on the *first* variable. The other rule is predicting the **opposite rank order** of cases on the second variable as opposed to the first. Thus if Johnson is higher on social class than Jones, we could probably use the *"same rank order"* rule to predict their rank in terms of prestige of occupation, the second variable, since these two ordinal variables are positively related. Likewise, we could reduce predictive errors in predicting the rank order of their scores on "anomie" by using the "opposite rank order" rule, since social class and anomie are negatively related. In order to express how good these rank-order prediction rules may be, we need to consider the distribution of different kinds of pairings of scores on two variables.

7.3.1 Types of Pairs

The total number of possible, unique pairs of cases which can be formed from N cases can be computed as follows:

$$T = \frac{N(N-1)}{2}$$

where N is equal to the total sample size.* With five cases, ten unique pairings are possible; with $N = 10$, 45 unique pairings of cases can be distinguished. Furthermore, if these "T" unique pairs are measured on two ordinal variables, there are only five possible patterns of ranking on these two variables:

(a) **Concordant pairs** (N_s) are pairs which are ranked in the same order on both variables.

(b) **Discordant pairs** (N_d) are pairs which are ranked in the opposite order on both variables.

(c) Pairs *tied on the independent variable* (X) but not tied on the dependent variable (Y). These are symbolized T_x.

(d) Pairs *tied on the dependent variable* (Y) but not tied on the independent variable (X). These are symbolized T_y.

(e) Pairs *tied on both variables,* symbolized, T_{xy}.

These five types of pairs exhaust the possibilities, and the sum of pairs of these types equals T, the total number of possible, unique pairs of cases. Diagram 7.1 illustrates the computational procedures involved in finding the number of these different types of pairs for a given set of data.

---

*The symbol, T, is the total number of unique pairs in a set of N cases. It has nothing to do with a different measure of association, Tschruprow's T, discussed in the last chapter. Unfortunately, traditional symbolism is not altogether consistent, although in context the meanings are generally quite clear.

DIAGRAM 7.1 Ordinal Measures of Association

The Problem:

We want to describe the association between two ordinal variables in this example, mobility-orientation of 10–19-year-old children in 466 households and the
mobility-orientation held for them by their parents (See Furstenberg, 1971:598)*.
Data were collected from a representative sample of households on the Lower
East Side of New York in 1960. The two mobility-orientation scales included
items about educational and occupational values, goals, and attitudes about
achievement.

     The study was concerned with the extent to which attitudes about mobil-
ity are transmitted from parent to child as indicated by the similarity of parent
and child views. A strong association between the child's view and his (her) parents' view would suggest that mobility-orientation is indeed transmitted between
parent and child. The author's hypothesis was that there was little such transmission, and that the association would thus be rather low. We will use these
data to show how to compute the various types of pairs of cases (in this example, a
"case" is a child, measured on his (her) and his (her) parent's mobility-orientation
for him (her)). Several measures of ordinal association may be computed from
these types of pairs and these are discussed in this chapter. In actuality, only
one of these measures of association would be computed; the criteria for deciding
on which one are also discussed in Chapter 7. The author reported tau-*b* as the
appropriate measure in this instance.

(Y) Child's Own Mobility Orientation	(X) Mobility Orientation a Child's Parents Hold for Him (Her)			
	Low	Medium	High	Total
High	d 29	55	68	152
Medium	59	53	48	160
Low	s 71	37	30	138
Total	159	145	146	450

Computing Number of Pairs of Different Kinds

In practice, if a computer were not used, only those types of pairs needed for the
selected measure of association would be computed. Here all five types of pairs
will be computed for illustration.

*Sixteen cases of "don't know" or "no answer" were excluded from the table in Furstenberg's study.

DIAGRAM 7.1 *(Continued)*

Step 1: Examine the table and determine which diagonal is the "positive" diagonal, that is, the one which extends from the "high-high" cell to the "low-low" cell on both variables. In this example, that diagonal is from the lower left to the upper right on the table. Label one end of this diagonal *s* and label one end of the "negative" diagonal *d*. This step assures that N_s and N_d pairings will be computed properly and, thus, that the coefficient's sign will accurately reflect the direction of association in a table.

Step 2: Compute types of pairs.

T = total number of unique pairings of cases

$$T = \frac{N(N-1)}{2}$$

$$T = \frac{450(450-1)}{2} = 101,025$$

N_s = number of "concordant" pairs. This is computed by locating the cell in the "s" corner of the table as indicated in Step 1, above. This is the first "target cell" and its frequency is multiplied by the sum of all cell frequencies above and to the right of the target cell (in this instance, since "s" is in the lower left corner of the table). This is illustrated in the schematic diagram at the left below, where the darkened, single cell is the target cell which is multiplied by the sum of cell frequencies in cells indicated by shading. To this first product are added similar products formed by taking each additional cell in the table (that has cells above and to the right) as successive "target" cells. In the table above, there are four such products which are summed, as follows:

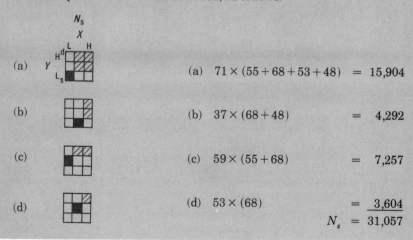

(a) $71 \times (55 + 68 + 53 + 48) = 15,904$

(b) $37 \times (68 + 48) \qquad\qquad = 4,292$

(c) $59 \times (55 + 68) \qquad\qquad = 7,257$

(d) $53 \times (68) \qquad\qquad\qquad = \underline{3,604}$

$\qquad\qquad\qquad\qquad\qquad N_s = 31,057$

DIAGRAM 7.1 *(Continued)*

It is helpful to notice the logic involved in these computations. The 71 cases in the first "target" cell have $55 + 68 + 53 + 48$ cases which are ranked differently and also ranked higher on both variables than the target-cell's 71 cases. The number of pairs which could be created would be equal to 71 times the total of $55 + 68 + 53 + 48$, which is 15,904. Likewise, for each target cell, the total pairings of this concordant kind can be computed for it and the sum of all of these computations equals the total number of unique pairs which are concordant.

N_d = number of "discordant" pairs. This is computed in exactly the same fashion as N_s, *except* that target cells start in the "d" corner and work down the negative diagonal. In this case, target cells are multiplied by the sum of cell frequencies for cells which are below and to its right. These computations yield the following sum:

(e) $29 \times (53 + 48 + 37 + 30)$ $= 4,872$

(f) $55 \times (48 + 30)$ $= 4,290$

(g) $59 \times (37 + 30)$ $= 3,953$

(h) $53 \times (30)$ $= \underline{1,590}$

              $N_d = 14,705$

It is immediately apparent that there are more concordant pairs in these data, and thus that the association is a positive association. This will result in a plus sign on any of the association measures we eventually compute.

T_x = pairs tied on the independent (X) variable but not on the dependent (Y) variable. These pairs are those formed within the same category of the X variable (*i.e.* tied on X) as indicated in the graphic illustration below at the left. Starting with a target cell at the top of a column, this is multiplied by the sum of cell frequencies for cells immediately below the target cell, etc. The computations are given below:

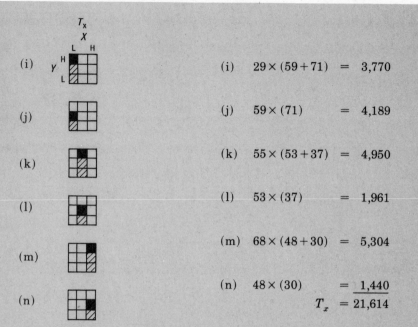

(i)	(i)	$29 \times (59 + 71) = 3{,}770$
(j)	(j)	$59 \times (71) = 4{,}189$
(k)	(k)	$55 \times (53 + 37) = 4{,}950$
(l)	(l)	$53 \times (37) = 1{,}961$
(m)	(m)	$68 \times (48 + 30) = 5{,}304$
(n)	(n)	$48 \times (30) = \underline{1{,}440}$
		$T_x = 21{,}614$

T_y = pairs tied on the Y variable but not on the X variable. These are computed exactly as T_x pairs are, *except* that products are formed within categories of the Y variable, only. In this case, target cells are multiplied only by the sum of cell frequencies to the right, within rows, as illustrated in the diagram at the left. The computations are as follows:

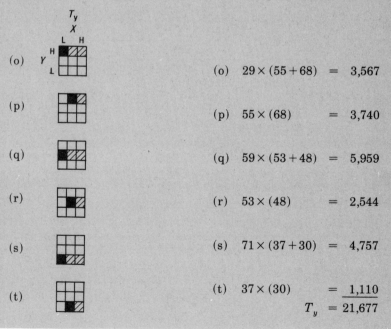

(o)	(o)	$29 \times (55 + 68) = 3{,}567$
(p)	(p)	$55 \times (68) = 3{,}740$
(q)	(q)	$59 \times (53 + 48) = 5{,}959$
(r)	(r)	$53 \times (48) = 2{,}544$
(s)	(s)	$71 \times (37 + 30) = 4{,}757$
(t)	(t)	$37 \times (30) = \underline{1{,}110}$
		$T_y = 21{,}677$

DIAGRAM 7.1 *(Continued)*

T_{xy} = pairs tied on both the X and the Y variables. These consist of the sum of pairs which can be formed out of cases which fall in the same cell (*i.e.* have the identical value on X and also an identical value on Y). These are computed for each cell as follows:

$$\frac{f(f-1)}{2}$$

where f is the cell frequency for a given cell.

These computations on each cell are then summed over all cells to equal T_{xy}. The computations on this table are as follows:

$$29(29-1)/2 = \quad 406$$
$$55(55-1)/2 = \quad 1,485$$
$$68(68-1)/2 = \quad 2,278$$
$$59(59-1)/2 = \quad 1,711$$
$$53(53-1)/2 = \quad 1,378$$
$$48(48-1)/2 = \quad 1,128$$
$$71(71-1)/2 = \quad 2,485$$
$$37(37-1)/2 = \quad 666$$
$$30(30-1)/2 = \quad \underline{435}$$
$$T_{xy} = 11,972$$

As a check, you should notice that the sum of these five kinds of pairs equals the total number of possible, unique pairings:

$$N_s = \quad 31,057$$
$$N_a = \quad 14,705$$
$$T_x = \quad 21,614$$
$$T_y = \quad 21,677$$
$$T_{xy} = \quad \underline{11,972}$$
$$T = 101,025$$

Step 3: Compute the appropriate ordinal measure of association. These measures are as follows:

tau-*a*
tau-*b*
tau-*c*
gamma
Somers' d_{yx}

and they are discussed in the body of Chapter 7.

The difference between the frequency of concordant (N_s) and discordant (N_d) pairs is a measure of which rule is the most accurate predictor of rank order. If there are more N_s pairs than N_d pairs, the *same rank order* rule would be more accurate. If there were a preponderance of N_d pairs in a given set of data, then the *opposite rank order* rule would serve best. In fact, the greater the preponderance of concordant (or discordant) pairs, the better the appropriate rank-prediction rule. If there is no difference between the number of concordant and discordant pairs, then neither rule is better and we would not improve our rank-order predictions over simply using a random guess or flip of a coin.

7.3.2   Measures of Association

For ordinal variables **measures of association** can be created as simple ratios of the various types of pairs distinguished here. In every case, however, the numerator of the ratio turns out to be the same, $N_s - N_d$, the preponderance (if any) of like-ranked or opposite-ranked pairs in a set of data. Each of these measures is a PRE measure, indicating the proportionate reduction in error which could be achieved by using one of the rank-order prediction rules ("same rank" or "opposite rank"), as compared to errors involved in making a random guess about the rank ordering of pairs of cases on some ordinal variable. The different ordinal measures of association we will discuss have different denominators. That is, they differ in terms of the kinds of pairs "at risk," or for which a prediction is attempted. Let us start with the most intuitively meaningful measure of the group, Kendall's tau-*a*.*

7.3.2a Tau-*a*.   *Tau-a* (t_a) is defined as the preponderance of concordant (or discordant) pairs out of all possible unique pairs in the data.

(7.4)

$$t_a = \frac{N_s - N_d}{T}$$

Looking back at the example in Diagram 7.1, the tau-*a* (t_a) coefficient would be:

$$t_a = \frac{31{,}057 - 14{,}705}{101{,}025} = \frac{16352}{101025} = +.16$$

Computing Diagram 1 illustrates how the components of t_a would be computed. This coefficient can vary from -1.0 to 0 to $+1.0$ depending upon whether the association is negative or positive. An association of zero indicates an even split between concordant and discordant pairs (and thus, that neither predictive rule will help in reducing predictive errors over errors one would expect to make just by chance). A t_a of 1.0 would indicate that all of the possible pairs are of one kind (either concordant or discordant depending upon the sign of t_a). This is a symmetric coefficient since no distinction is made between independent and dependent variables in the computation of N_s, N_d or T, and it is appropriate for any sized table or any number of ranks on either of the two ordinal variables. Unfortu-

---

*An interesting discussion of the relationship between some of the ordinal measures of association is contained in Somers (1962:799–811), as well as original work by Kendall, Kruskal and Goodman, and Kruskal, which are cited in Somers' article.

nately, if there are ties, as there usually are, t_a cannot reach the magnitude of 1.0, because the denominator which includes ties will always be greater than either N_s or N_d.

7.3.2b Gamma (G).   One solution to the problem of actually obtaining a coefficient equal to 1.0 when there are ties is to eliminate the ties from the denominator as well. **Gamma** (G) is a frequently used, symmetrical measure for association of two ordinal variables which does just that.* The numerator is the same as that used for t_a and the denominator is simply the sum of pairs which are ranked differently on both variables.

(7.5)
$$G = \frac{N_s - N_d}{N_s + N_d}$$

Gamma (G), like t_a, is a symmetric measure. Unlike t_a, it can always achieve the limiting values of -1.0 or $+1.0$ regardless of the number of ties. In fact, it would be possible to have a G of $+1.0$ (or -1.0) based on only one pair, where all the rest of the pairs are tied on one or both variables. This may be an undesirable property of G, considering it as an overall characterization of the proportionate reduction in errors in predicting rank order of one variable based on a knowledge of ranking on the other variable. Both t_a and G, however, would be appropriate for an $r \times c$ table. It is interesting to note that G in the 2×2 case is the same as Yule's Q, which was introduced in the last chapter, (Section 6.5.2c), except for the traditionally used symbolism. G is a generalized version of Yule's Q for $r \times c$ tables.

Using the illustration from Diagram 7.1, G would be computed as follows:

$$G = \frac{31057 - 14705}{31057 + 14705} = \frac{16352}{45762} = .36$$

G can be interpreted as the proportionate reduction in errors in predicting ranking that would be made by using the "same" (or "opposite") ranking rule rather than randomly predicting rankings among pairs which are ranked differently on both of the two variables in the table.

7.3.2c Somers' d_{yx}.   A somewhat different ordinal measure would be appropriate if one distinguishes between independent and dependent variables. If one were predicting the ranking of cases on a dependent variable (variable Y), using variable X as the independent or predictor variable, he (she) would make a prediction about ranking not only for the pairs that are ranked differently on each variable (*i.e.* the N_s and N_d pairs), but he (she) would also make a prediction for the T_y cases which are different on the predictive variable but tied on the dependent variable. The difference on the independent variable permits a prediction even in the T_y cases, although the prediction may not work out, since there may not be any difference on the dependent variable. The denominator of an association measure, then, should include *all* of the pairs for which a prediction would be risked. This is essentially the definition of this next ordinal measure of association, Somers' d_{yx}.

(7.6)
$$d_{yx} = \frac{N_s - N_d}{N_s + N_d + T_y}$$

*Sometimes gamma is symbolized by the lower-case Greek letter (γ). G is used here.

Notice that the numerator is again the difference between concordant and discordant pairs. Somers' d_{yx} is an asymmetric measure, since it does take account of which variable is the predictor variable and which is the predicted. Ties on the predicted variable are *included*, and ties on the predictor variable are *excluded*. Like G, it can vary from -1.0 indicating a perfect negative association (*i.e.* a situation where all pairs are discordant) to $+1.0$. It can be interpreted as the proportionate reduction in errors in predicting ranking on a dependent variable resulting from using the ("same" or "opposite") predicting rule, rather than chance prediction among pairs which are ranked differently on the independent variable. Here again, as was true of other asymmetric measures such as lambda (λ_{yx} and λ_{xy}), there are two Somers' d's which can be computed on any table, depending upon which variable is treated as the independent variable. In general, too, these values may be (and generally will be) different for any given table.

7.3.2d Tau-b(t_b). An investigator might be interested in a measure of degree of association which is symmetric but, unlike G, takes account of ties on one or the other variable (but not ties on both, T_{xy}). Ties on both variables are considered to be trivial, since they correspond to the number of pairs that could be created out of cases which are identical on both variables and which thus fall within the same cell of a table. In most tables, cell frequencies are greater than one, and it is not the absolute size of any cell frequency, but the *pattern* of frequencies in different cells which is what one means by the association in a table.

Tau-b (t_b) is a kind of average of the two Somers' d's which could be computed on a given set of data. In fact, it can be expressed as the square root of the product of these two d's.

$$t_b = \sqrt{d_{yx}d_{xy}}$$

It is usually computed directly from computations of the number of each type of pairs, and it can also be expressed in this fashion as follows:

(7.7)
$$t_b = \frac{N_s - N_d}{\sqrt{(N_s + N_d + T_y)(N_s + N_d + T_x)}}$$

It can take on values from -1.0 to $+1.0$, depending upon the direction of association, and its magnitude indicates strength of association. t_b cannot achieve a magnitude of 1.0 if a table is not square (*i.e.* if $r \neq c$) since in this case there would have to be more pairs tied on one variable (the one with the fewer categories) than are tied on the other variable.* Unfortunately, the more complicated the denominator the more difficult it becomes to express a clear operational definition in a PRE sense, and this is true of t_b. It is, however, one of the more useful of the ordinal, symmetrical measures, and it is superior to tau-a (t_a) be-

---

*Another measure of association which can achieve 1.0 for tables where the number of rows and columns is not equal is called tau-c. In this formula, m is the smaller of the number of rows or the number of columns.

$$t_c = \frac{2m(N_s - N_d)}{N^2(m-1)}$$

As is true of tau-b, this measure, although normed, is hard to interpret operationally, and tau-c is less frequently used than the other tau measures.

cause it does take account of the non-trivial ties in expressing the relationship between two variables.

For the illustrative data in Diagram 7.1, t_b would be computed as follows:

$$t_b = \frac{31057 - 14705}{\sqrt{(31057 + 14705 + 21614)(31057 + 14705 + 21677)}}$$

$$= \frac{16352}{67407.5} = .24$$

7.3.2e Comparison of Gamma, Somers' d_{yx}, and Tau-b.

All of the preceding measures of association for ordinal variables were simply ratios created out of the number of pairs of the several types we distinguished. In each case the balance of concordant and discordant pairs was evaluated in terms of the number of pairs "at risk," so to speak — that is, the number of pairs which constituted a potential number of errors or correct predictions that might have been made. The essential difference between the measures derives from what pool of differences one would be interested in predicting.

Tau-a takes as its pool all possible pairs; gamma, only untied pairs; Somers' d_{yx}, untied pairs and pairs tied on the dependent variable; and Tau-b, untied pairs and ties on variable y or variable x, but not on both.*

Costner (1965) takes the position that any measure which includes ties of any kind in its pool is not properly a PRE measure because a tie cannot be clearly counted as either a correct or an erroneous prediction and only those two categories are permissible for PRE measures. On the other hand, if one takes the position that the pool of potential errors should include all those for which a prediction is likely to be made, then it is reasonable to include ties. For example, if a pair is untied on the independent variable, a prediction is possible. Although concordance or discordance would be predicted, the possible outcomes would be concordance, discordance, or a tie on the dependent variable. These are, in fact, the possibilities which are included in the pool for Somers' d_{yx}. Since Tau-b is a symmetric measure it makes provision for a tie to occur on either variable Y or variable X.

What consequences do these differences in denominators (pools) have for the resulting measures of association? It is worth spending some time to examine these consequences because they are important both in helping us decide which of the measures of association to use for a particular situation and in helping us interpret what the resulting coefficient means for our data analysis.

Tau-b requires the most restrictive meaning of perfect association. That is, it reaches a value of $+1.00$ or -1.00 when all of the frequencies fall on the diagonal of a square table. If the table is not square, perfect association is not possible. Somers' d_{yx} will reach $+1.00$ or -1.00 either when all of the frequencies in a square table are on the diagonal or in a table which is not square when their progression on the scale of the dependent variable is monotonic. That is, where the frequencies for the dependent variable progress in a stepwise fashion such that a single value of the dependent variable (Y) cannot take more than one

---

*Because of the problem of interpreting Tau-a when ties occur we will not consider it further here.

value on the independent variable (X), but two different values of the independent variable may take the same value on the independent variable. Table 7.4D is an example of a table of frequencies which is characterized by a monotonic progression of the dependent variable (Y). Note that if X is taken as the dependent variable, d_{xy} is not equal to 1.00. That is so because two different values of Y would predict the same value of X. Gamma takes the least restrictive meaning of perfect association of the three measures we are considering. This is so because all sorts of ties are ignored in its computation. Tables 7.4A, 7.4B, 7.4C, and 7.4D illustrate four different patterns of frequencies, all of which result in gammas of +1.00. Notice that Somers' d_{yx} and Tau-b are high for those patterns in which the frequencies all fall close to the diagonal and are low for Table 7.4B where the frequencies are off the diagonal, and incidentally, form an L shaped pattern. Also, take note of the fact that gamma can reach 1.00 even though a table is not square (Tables 7.4B and 7.4D).

In all of the tables of Table 7.4 the value of Tau-b is intermediate to the values of d_{yx} and d_{xy} because, mathematically, the magnitude of Tau-b is equal to the geometric mean of d_{yx} and d_{xy}.

The number of rows and columns in a table has a definite influence on the value of gamma. Collapsing a table (*i.e.* decreasing the numbers of rows and/or columns by combining categories) has the effect of increasing the value of gamma. Examine Tables 7.4E and 7.4F. Both tables contain the same data, but the two middle categories on variables X and Y of Table 7.4E have been combined in Table 7.4F to reduce the table to 3×3. For Table 7.4E gamma equals +.43, but for Table 7.4F it equals +.54. This represents a better than 25 percent increase in predictability — accomplished merely by collapsing the table. However, the table was collapsed at the cost of losing information about the two ordinal scales — that is, the ability to discriminate between four levels of each variable. Somers' d_{yx} and Tau-b also increased, but the increase in those coefficients was slight by comparison (6 percent).

On the other hand, changes in the distributions of marginal frequencies do not have the same affect on gamma as they do on Somers' d_{yx} and Tau-b. Compare Tables 7.4G and 7.4H. The marginal frequencies of Table 7.4G are distributed more evenly across the categories of variables X and Y than are the marginal frequencies of Table 7.4H. In both tables gamma equals +1.00. Somers' d_{yx} is slightly higher in Table 7.4H than in Table 7.4G. The most drastic change occurred in d_{xy}, and Tau-b also changed significantly. The changes that took place in these three measures are attributable to the differences in the numbers of ties in variables Y and X from one table to the other. While the number of ties on Y decreased from 100 to 25 from Table 7.4G to Table 7.4H, the number of ties on X increased from 75 to 225. Naturally, gamma was unaffected by these changes because ties do not enter into the computation of gamma. However, the more ties there are in a table, the less data enter into the computation of gamma. For example, the computation of gamma for Table 7.4G required ignoring data on 175 tied pairs and for Table 7.4H data were ignored for 250 pairs. In extreme cases it is possible that gamma would be computed based upon a minority of the available data. In computing gamma, then, it is usually advisable to pay particular attention to the number of tied pairs in the data. When the proportion of ties is

TABLE 7.4 COMPARISONS OF GAMMA, SOMERS' d_{yx}, AND TAU-b (HYPOTHETICAL DATA)

7.4A

	Variable X		
	Lo		Hi
Hi	0	0	52
Variable Y	0	29	0
Lo	38	43	0

$G = +1.00$

$d_{yx} = +.81$

$d_{xy} = +.85$

$t_b = +.83$

7.4B

	Variable X			
	Lo			Hi
Hi	0	0	0	18
Variable Y	0	0	0	33
Lo	38	14	29	47

$G = +1.00$

$d_{yx} = +.41$

$d_{xy} = +.58$

$t_b = +.49$

7.4C

	Variable X			
	Lo			Hi
Hi	0	0	21	38
	0	61	27	0
Variable Y	14	45	0	0
Lo	32	0	0	0

$G = +1.00$

$d_{yx} = +.84$

$d_{xy} = +.82$

$t_b = +.83$

7.4D

	Variable X			
	Lo			Hi
Hi	0	0	0	61
	0	0	14	0
Variable Y	0	32	0	0
	0	25	0	0
Lo	41	0	0	0

$G = +.100$

$d_{yx} = +1.00$

$d_{xy} = +.92$

$t_b = +.96$

7.4E

	Variable X			
	Lo			Hi
Hi	17	28	39	52
	24	25	21	26
Variable Y	54	46	37	21
Lo	74	28	19	11

$G = +.43$

$d_{yx} = +.33$

$d_{xy} = +.33$

$t_b = +.33$

7.4F

	Variable X		
	Lo		Hi
Hi	17	67	52
Variable Y	78	129	47
Lo	74	47	11

$G = +.54$

$d_{yx} = +.35$

$d_{xy} = +.35$

$t_b = +.35$

7.4G

	Variable X		
	Lo	Hi	
Hi	0	0	15	15
Variable Y	0	20	5	25
Lo	20	0	0	20
Tot.	20	20	20	60

$G = +.100$ Tied Pairs $= 175$

$d_{yx} = +.92$

$d_{xy} = +.94$

$t_b = +.93$

7.4H

	Variable X		
	Lo	Hi	
Hi	0	5	5	10
Variable Y	0	45	0	45
Lo	5	0	0	5
Tot.	5	50	5	60

$G = +1.00$ Tied Pairs $= 250$

$d_{yx} = +.95$

$d_{xy} = +.69$

$t_b = +.81$

large, gamma probably should not be used or, at least, interpretations should be made with caution.

7.3.2f Spearman's rho (r_s). The last ordinal measure we will deal with is interesting, because it takes a different approach to the problem of measuring the direction and strength of association. It is primarily used where

---

BOX 7.2 ORGANIZING

Often it helps to organize statistical measures in terms of some of their main differences—those differences that might lead to the selection of one measure in preference to another for some particular problem. One organizing scheme is suggested here, and the six PRE measures we have discussed thus far in this chapter are entered for illustration. Your own scheme might have more detailed classifications. Selection of an appropriate measure is discussed later in this chapter.

Level of Measurement	Symmetric	Asymmetric
Nominal		Goodman-Kruskal t_y Lambda (λ_{yx})
Ordinal	Kendall's t_a Kendall's t_b Gamma (G)	Somers' d_{yx}

---

rankings of individual cases on two variables are available so that rankings range from 1 to N for each variable. Table 7.5 provides an example of two rankings of the same set of sociology departments. One ranking is in terms of degree productivity and the other is in terms of positions held on the editorial board of the *American Sociological Review*, official journal of professional sociologists in the United States.

TABLE 7.5 RANK OF SELECTED SOCIOLOGY DEPARTMENTS ON PRODUCTIVITY OF DOCTORATES AND REPRESENTATION ON THE EDITORIAL BOARD OF A MAJOR SOCIOLOGY JOURNAL

Sociology Department	(1) Rank on Number of PhDs Produced 1964–8	(2) Rank on No. of Editorial Positions on the ASR for 1948 to 68	Difference Between Ranks D	D^2
Chicago	1	1	0	0
Columbia	2	3	−1	1
Wisconsin	3	4	−1	1
Minnesota	4	5	−1	1
UCLA	5	8	−3	9
Berkeley	6	6.5	−0.5	0.25
Michigan	7	9.5	−2.5	6.25
Ohio State	8.5	6.5	+2	4
Washington (Seattle)	8.5	9.5	−1	1
Harvard	10	2	+8	64
			$\Sigma D = 0$	$\Sigma D^2 = 87.5$

Source: Data from Rossi, 1970, (column 1), and from Yoels, 1971, (column 2). Tied ranks are averaged.

The argument is that the representation of schools on the editorial board of a major journal merely reflects the productivity of these departments. If that is so, then there should be a perfect association — exactly the same ranking — of the departments on both variables, representation and production.

Spearman's rho (r_s) is a measure of association for ordinal variables based on the difference between ranks. If there is no difference, then D will equal zero. Since the sum of the difference between ranks is always zero, as shown in Table 7.5, differences between ranks are squared before summing. In the case of comparisons in ranking of these ten schools, the sum of squared differences between ranks (called ΣD^2) is 87.5. Since this is different from zero we know that the two variables are not identically ranked. But we do not know how to interpret this figure, because we would expect it to vary with the number of individuals ranked in the first place. We could, however, create a ratio of the D^2 obtained and the maximum possible D^2 that could be achieved for a given number of ranked individuals. This maximum D^2 is this: $N(N^2-1)/3$, where N is the number of cases ranked. Then, in order to make it possible for a minus sign to indicate opposite ranking and for the magnitude of 1.0 to be a maximum degree of association, the formula for r_s is written as follows:*

(7.8)
$$\text{Spearman's } r_s = 1 - \frac{6\Sigma D^2}{N(N^2-1)}$$

Rho (r_s) will have a value of +1.0 for a perfect match of ranks, and a value of −1.0 if the ranks are exactly opposite. A r_s of zero indicates no systematic ordering or, rather, no rank pattern between the two variables. In the case of Table 7.5, r_s is as follows:

$$r_s = 1 - \frac{6(87.5)}{10(10^2-1)} = 1 - \frac{525}{990} = 1 - .53 = +.47$$

Intermediate values of r_s can be interpreted in terms of their relative magnitude but r_s does not have a PRE interpretation. However, r_s^2 has a PRE interpretation (for ranks) equivalent to that of r^2 (discussed in section 7.4.2); although there is no prediction equation equivalent to the regression equation (discussed in section 7.4.1) for r_s.

It should be noted that r_s loses its effectiveness as a measure of association as the number of tied ranks increases.

7.4   MEASURES OF ASSOCIATION FOR INTERVAL VARIABLES

As you will recall from the discussion of univariate statistics, the arithmetic mean of an interval-level dependent variable is a useful prediction because the

---

*The formula for Spearman's rho (r_s) is simply a Pearsonian r computed on ranks. Since the ranks for both variables extend from 1 to N, we know that the sum of each variable is $N(N+1)/2$ and the mean of each variable is $(N+1)/2$. The sum of squares becomes $N(N+1)$ $(2N+1)/6$ and the variance in each case is $(N^2-1)/12$. Substitution into the formula for Pearson's r yields the Spearman's r_s formula for the relationship between X ranks on two variables for N cases. The derivation is nicely illustrated in Hammond and Householder (1962:212–214).

mean has the property that the algebraic sum of deviations of actual scores from it is zero. A measure of how badly wrong this prediction is can be derived from these deviations. The variance (or its square root, the standard deviation) is one such measure which expresses the amount of scatter of scores around this mean.

Thus, as a minimum, one could predict the mean of a dependent variable and measure "errors" made in that prediction in terms of the familiar variance (s^2) and this, in fact, constitutes the minimum guessing rule for interval variables shown in Figure 7.1, above. How much better can we do? Is there any way that scores on an independent variable could be used to improve the prediction of a dependent variable? As you might suspect by now, the answer is "yes," although the solution to the problem is new to our line of discussion thus far.

Suppose that we were able to derive a formula which would describe the way the mean of variable Y varied as one moved up the scale of variable X. This would, in effect, be a mathematical description of the *nature* of the relationship between two variables and it would also permit us to "compute" an estimate of an individual's score on the dependent variable from information about his (her) score on the independent variable. With a predicted score (called Y', or Y-prime) and an actually observed score (Y), we could then ask how accurate the prediction equation is. This might take the form of a measure of association (usually called a *correlation coefficient* where the variables are interval-level) which would express the amount by which predictive errors could be reduced, given the prediction equation rather than the overall mean of the dependent variable to use in predicting. This is precisely what we will do to create the next measure of association called **Pearson's Product-Moment Correlation Coefficient, r.** Its square (*i.e.* r^2) will indicate the proportionate reduction in errors resulting from a use of the predictive equation. In order to develop this idea, however, we need to step back and develop a formula for describing the nature of the relationship between two interval variables so as to predict the dependent variable.

7.4.1   Regression Equations

Suppose we start with a small collection of data where two scores are measured on each of six cases.

BOX 7.3   Interval-Level Variables Again

The following section deals with *measures of association* and a description of the *nature of association* for interval-level variables. It is helpful, at this point, to recall some of the concepts from univariate description which were used to summarize interval level variables. In particular, you should feel quite comfortable with the following:

    arithmetic mean   Section 5.3.3
    variance and standard deviation   Section 5.4.4
    scatter diagram   Section 4.4.2

Case	(X) Years of Education	(Y) Income in ($1000)
A	1	2
B	2	4
C	3	6
D	4	8
E	5	10
F	6	12

These scores could be plotted on a graph, as in Figure 7.2a (page 245), where values of the independent variable appear along the X-axis and values of the dependent variable appear along the Y-axis. This is one use of a **scatter diagram** discussed earlier in this book (Chapter 4). Each case is represented by a point on the graph placed above the X-axis and across from the Y-axis at the point which represents that individual's score on each variable. These points can be located by a pair of scores (*e.g.* 1,2 where the first number is the X-score and the second is the Y-score for that point).

7.4.1a In this example, it is clear that the two variables are related in a rather simple and obvious way. In fact, one could predict the Y-score exactly from a knowledge of the X-score merely by doubling the X-score for these cases. This relationship is expressed in the following equation where Y' (called Y-prime) is an estimated or predicted value for that individual on the Y-variable.

$$Y' = 2(X)$$

The predictions fall along a straight line, which is drawn in for Figure 7.2a. The 2 in this equation is called the **slope** of the regression line, and it means that for one-unit increase along the X-axis there will be a predicted increase of two units on the Y-scale. This slope is illustrated in the small triangle shown on the graph, Figure 7.2a. Notice that predictions derived from this equation describe a **straight line,** one of the simplest (*i.e.* easiest to describe) ways in which two variables can be related. One could say that these variables are related in a "linear" fashion; a straight line very nicely describes their relationship. This line is called a linear **regression** line; a regression of the dependent variable, Y, on the independent variable, X.

7.4.1b Let us examine another set of data (top of page 244). Here again we have a very small set of scores for illustrative purposes and, again, we can create a scatter diagram of these five cases as shown in Figure 7.2b. The Y-scores for these five cases can also be accurately predicted by a simple formula.

$$Y' = 1 + 2(X)$$

Given an X-score, we can predict a Y-score for that case by merely doubling the

Case	(X) Years of Education	(Y) Income in ($1000)
A	1	3
B	2	5
C	3	7
D	4	9
E	5	11

X-score and adding a constant, 1. Here too, the equation describes a simple straight line and one would again say that the relationship between these two variables is of a linear nature. The formula, above, describes the specific nature of this relationship for these data. Here again, the 2 expresses the number of units change in Y given a single unit change in X. The constant, 1, indicates the point at which the straight line crosses the Y-axis in this scattergram in Figure 7.2b and it is called the **Y-intercept.**

7.4.1c The general formula for any straight line is expressed as follows:

(7.9)

$$Y' = a_{yx} + b_{yx}(X)$$

where Y' is the predicted or computed value of the dependent variable; a_{yx} is the constant or Y-intercept for this equation; and b_{yx} is the slope coefficient (also referred to as the **b-coefficient**). The first subscript of the a_{yx} and b_{yx} values is the dependent variable, and the second subscript refers to the independent variable. It is immediately clear that these coefficients distinguish between the independent and dependent variables and we might, in fact, expect a different equation if values of X were to be predicted from values of Y. As in the case of asymmetric measures of association, there are two regression equations which could be computed to describe the nature of the relationship between the pair of variables in a scattergram. More of this later.

Notice that the a_{yx} and b_{yx} coefficients in the last two examples are positive values. This need not be the case for all sets of data. In fact, where the *b-coefficient* is negative, that indicates that the variables are negatively related, since an increase of one unit in the X-value would signal a negative change in the predicted Y-score.

Now, not all associations between variables are described very well by a straight line, although many come relatively close to this nature of association. A straight line represents in many respects the simplest relationship we could express, and it is usually the one expressed in theoretical statements about how sociological variables are related. As a matter of fact, some relationships between variables are better described by curved rather than straight lines, but these raise the problem of discovering, by some curve-fitting method, what the proper formula for the curve might be, as well as what the particular form of curve might mean, theoretically. Although the problems are not particularly different, we will limit our attention to the simple straight-line relationship and the development of the linear regression equation.

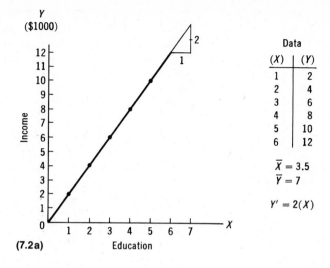

(7.2a)

| Data | |
(X)	(Y)
1	2
2	4
3	6
4	8
5	10
6	12

$\bar{X} = 3.5$
$\bar{Y} = 7$

$Y' = 2(X)$

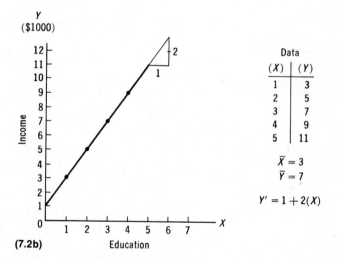

(7.2b)

| Data | |
(X)	(Y)
1	3
2	5
3	7
4	9
5	11

$\bar{X} = 3$
$\bar{Y} = 7$

$Y' = 1 + 2(X)$

FIGURE 7.2 SCATTER DIAGRAMS

Figure 7.3 illustrates the more typical scatter diagram. In this instance, individual cases are represented by the placement of the small numbers on the scatter diagram. The number itself indicates how many cases fall at that same point. The relationship is between two different measures of social class. One measure was developed by Hollingshead using occupation and education; the other social class score was developed by Duncan from income and education for

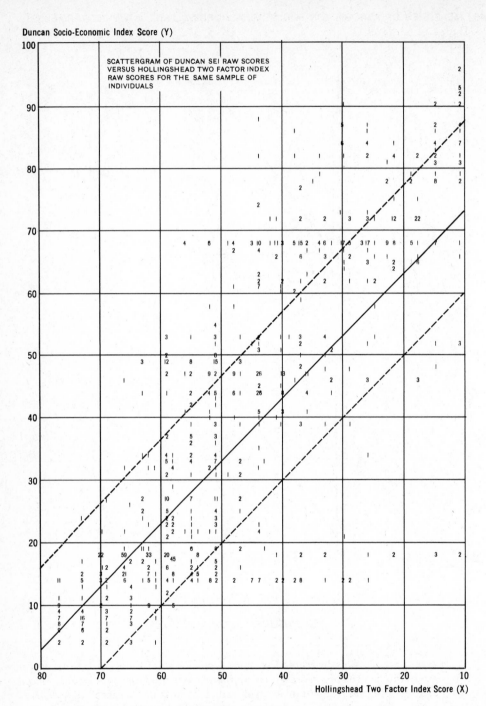

Source: Haug and Sussman, 1971. Used by permission.

FIGURE 7.3 SCATTER DIAGRAM SHOWING THE RELATIONSHIP BETWEEN HOLLINGHEAD'S TWO-FACTOR INDEX AND DUNCAN'S SEI AS MEASURES OF SOCIAL CLASS STANDING different occupation groups. (See Haug and Sussman, 1971.) One would expect that the two measures would be relatively closely related, since they are supposed to measure essentially the same phenomenon. In actual data, however, there is a considerable amount of scatter even though, overall, the data cluster fairly well around a linear regression line. The problem is to place the linear regression line in such a manner that it fits the data as well as possible.

7.4.1d "Best Fitting" Line. The criterion of "fit" of a regression line is still how well the dependent variable can be predicted by the equation the line represents. Let us take another set of data, where a small number of data points are scattered somewhat, and try to discover how one might derive a best-fitting straight line.

The set of five scores shown in Figure 7.4 could be predicted in terms of several different equations. Let us take three to show the kind of error that the "best-fitting" line would help eliminate. First, we could simply predict the mean of Y for every case. This is the regression line shown in Figure 7.4a. Secondly, we could use the prediction: $Y' = 2.5 + .83X$, which is an arbitrarily picked equation (for illustrative purposes) which is *not* the "best-fitting," but it is better than the overall mean of Y as a prediction. Any number of other inadequate equations could be chosen rather than this one, of course. Thirdly, we could use the "best-fitting" regression equation for these data, which happens to be: $Y' = 1.1 + 1.3X$.

7.4.1e Notice in Figure 7.4a that the same prediction will be made for each case, and the inaccuracy of this prediction will be described by subtracting the *predicted* Y-score from the *actual* Y-score, squaring the difference, and dividing this sum by N, the number of cases. You will recall that this is simply the formula for the variance of a set of scores. In Figure 7.4a this variance can be symbolized by s_y^2 to indicate clearly that it is the variance of the Y-variable.

$$s_y^2 = \frac{26.0}{5} = 5.2$$

Using the second prediction equation shown in Figure 7.4b, a different predicted value of Y is shown for each different value of X. Again, we can describe the accuracy of prediction by subtracting predicted from actual Y-score, squaring the difference, summing and dividing by N.* In this case, since the predicted value of Y depended upon the value of X, we will symbolize the variance by $s_{y \cdot x}^2$, which is called the *variance of the estimate;* the square root of this

$$s_{y \cdot x}^2 = \frac{\Sigma (Y - Y')^2}{N}$$

value is called the **standard error of estimate.** Notice that the subscript indicates not only the dependent variable, Y, but the dot followed by the X indicates that different values of X are reflected in predictions upon which this error figure is based. The dotted lines in Figure 7.4a, 7.4b, and 7.4c represent the error of predicted value (on the regression line) and the actual score. These lines are

---

*At this point your attention should again be called to the fact that a better estimate (unbiased estimate) of the population variance includes $N-1$ in the denominator of the variance computed from sample data. The same principle applies to $s_{y \cdot x}$, which would have $N-2$ in the denominator.

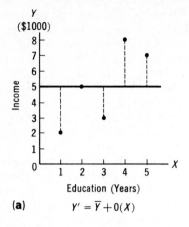

Data

(X)	(Y)
1	2
5	7
3	3
2	5
4	8

$\bar{X} = 3$
$\bar{Y} = 5$

(a) $Y' = \bar{Y} + 0(X)$

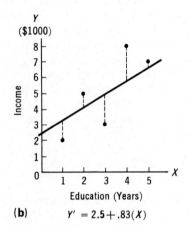

(b) $Y' = 2.5 + .83(X)$

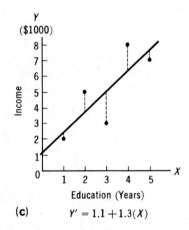

(c) $Y' = 1.1 + 1.3(X)$

	Actual Data		Prediction #1 $Y' = \bar{Y} + 0(X)$			Prediction #2 $Y' = 2.5 + .83(X)$			Prediction #3 $Y' = 1.1 + 1.3(X)$		
	(X)	(Y)	Y'	(Y−Y')	(Y−Y')²	Y'	(Y−Y')	(Y−Y')²	Y'	(Y−Y')	(Y−Y')²
	1	2	5	−3	9	3.33	−1.33	1.77	2.4	− .4	.16
	5	7	5	+2	4	6.66	+ .34	.12	7.6	− .6	.36
	3	3	5	−2	4	5.00	−2.00	4.00	5.0	−2.0	4.00
	2	5	5	0	0	4.17	+ .83	.69	3.7	+1.3	1.69
	4	8	5	+3	9	5.83	+2.17	4.71	6.3	−1.7	2.89
Sum	15	25	25	0	26.0	25.00	.0	11.29	25.0	.0	9.10
Avg	3	5	5	0	5.2	5.0	0	2.26	5.0	0	1.82

Measurement of Prediction Errors

FIGURE 7.4 FINDING THE "BEST" STRAIGHT LINE TO REPRESENT DATA getting shorter as the prediction gets better, and this improvement in prediction is also illustrated by the decrease in the variance figures we have created.

7.4.1f Least squares criterion. Finally, using the third prediction equation shown in Figure 7.4c, we notice that the dotted lines representing inaccuracy in prediction are generally shorter than those for Figures 7.4a and 7.4b. This is reflected in a smaller variance of the estimate, which is again simply the sum of squared differences between actual and predicted scores averaged over the number of cases. This third prediction equation $(Y' = 1.1 + 1.3(X))$ produces the smallest variance. It happens to be the best prediction, and the regression line in Figure 7.4c is called the "best fitting regression line of Y on X." It is *best* in the sense that the sum of squared deviations of scores around this line is the smallest it could be for any straight line and thus, this is called the **least-squares** regression line. It is important to notice that the deviations are all figured in terms of Y-scores, since that is the variable we are interested in predicting. Figure 7.5 shows the "least-squares" regression line for the situation where we are interested in predicting X-scores instead, and you will notice that a different equation is needed to minimize errors in predicting X from a knowledge of Y. It is generally true that two different regression equations can be computed for a given bivariate distribution, depending on which variable is to be predicted.

Computation of the coefficients for the least squares regression line is illustrated in Diagram 7.2. The numerator for b_{yx} is a value that expresses how well the two variables go together, and it is formed from the sum of the product of deviations of each score from its mean. The denominator is the sum of squares for the independent variable.

(7.10)
$$b_{yx} = \frac{\Sigma(X-\overline{X})\,(Y-\overline{Y})}{\Sigma(X-\overline{X})^2}$$

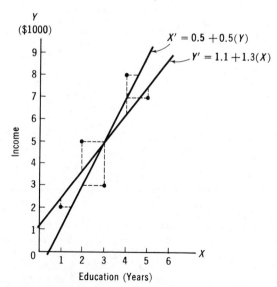

FIGURE 7.5 LEAST SQUARES REGRESSION OF VARIABLE X ON VARIABLE Y TAKEN FROM DATA IN FIGURE 7.4, SUPERIMPOSED ON THE REGRESSION LINE OF Y ON X

A more convenient formula for hand computation, using only the raw scores themselves, is given in Diagram 7.2.

7.4.1g Since the best-fitting regression line always goes through the point where the means of both variables intersect (note that this is true of Figures 7.4c and 7.5), the a_{yx} coefficient can be computed by substituting these mean values in the regression equation and solving for a_{yx}, as follows:

(7.11)
$$a_{yx} = \overline{Y} - b_{yx}(\overline{X})$$

We have now established a simple formula which describes the **nature** of association between two variables and which thus permits us to make use of information about the independent variable to reach a better prediction of the dependent variable. The next problem is to devise a measure of degree of association which will express the proportionate reduction in predictive errors that this formula permits.

7.4.2 Product-Moment Correlation Coefficient (r)

The least-squares regression line computed in Figure 7.4c, permitted us to predict scores on the dependent variable, Y, with somewhat greater accuracy than we could achieve merely by predicting the overall mean of Y. We could say that the regression line helps us "explain" some of the variation in the dependent variable. This left, typically, some of the variation in Y unexplained, and the total of **unexplained** and **explained variation** equals the **total variation** of Y around its mean. This can be shown for Figure 7.4c, as follows:

Actual Scores X Y	Predicted Y-Scores Y'	Unexplained Variation $(Y-Y')^2$	Explained Variation $(Y'-\overline{Y})^2$	Total Variation $(Y-\overline{Y})^2$
1 2	2.4	.16	6.76	9
5 7	7.6	.36	6.76	4
3 3	5.0	4.00	.0	4
2 5	3.7	1.69	1.69	0
4 8	6.3	2.89	1.69	9
15 25	25.0	9.10	16.90	26

$$\overline{Y} = 5$$
$$26 = 9.10 + 16.90$$
$$\Sigma(Y-\overline{Y})^2 = \Sigma(Y-Y')^2 \quad + \Sigma(Y'-\overline{Y})^2$$
$$\frac{\text{Total}}{\text{Variation}} = \frac{\text{Unexplained}}{\text{Variation}} + \frac{\text{Explained}}{\text{Variation}}$$

The object, of course, is to explain as much of the variation as possible. The Pearson product-moment correlation coefficient expresses how well the linear regression line explains the variation in the dependent variable as follows:

$$r = \sqrt{\frac{\text{Explained Variation}}{\text{Total Variation}}}$$

$$= \sqrt{\frac{\Sigma(Y'-\overline{Y})^2}{\Sigma(Y-\overline{Y})^2}} \quad \text{which also equals} \quad \sqrt{\frac{\Sigma(X'-\overline{X})^2}{\Sigma(X-\overline{X})^2}}$$

DIAGRAM 7.2 Regression and Correlation Coefficients

To illustrate the computation of regression coefficients and the Pearsonian correlation coefficient, we will use the illustrative data from Figure 7.4. These are raw data, and $N = 5$.

(X) Education In Years	(Y) Income in $1000's	X^2	Y^2	$X \cdot Y$
1	2	1	4	2
5	7	25	49	35
3	3	9	9	9
2	5	4	25	10
4	8	16	64	32
$\Sigma X = 15$	$\Sigma Y = 25$	$\Sigma X^2 = 55$	$\Sigma Y^2 = 151$	$\Sigma XY = 88$
$\overline{X} = 3$	$\overline{Y} = 5$			

The Problem is to compute coefficients a_{yx} and b_{yx} in the following regression equation: $Y' = a_{yx} + b_{yx}(X)$, and compute the correlation coefficient r. There are a number of different approaches, including using a computer program (which is probably most common and useful), using definitional formulae given in this chapter, or the formula which uses data converted to z-score form prior to computations. Among the hand-computation procedures, the raw-data formulae used below seem most useful.

Step 1: Examine a scatter diagram of the data to determine whether or not a linear regression line is appropriate. Often computer programs provide scatter plots, and computer-driven mechanical plotters also are useful. If a correlation coefficient is low, one reason may be that the linear model does not fit the data very well. (See Figure 7.4.)

Step 2: Find the following sums from the raw data: ΣX, ΣY, ΣXY, ΣX^2, ΣY^2. If data are in a grouped frequency distribution form, then X will refer to category midpoints which have to be multiplied by category frequencies to form sums indicated by: ΣfX, ΣfY, ΣfXY, ΣfX^2, ΣfY^2.

DIAGRAM 7.2 *(Continued)*

Computing b_{yx}, the slope coefficient

The slope coefficient, or *b*-coefficient, of the regression equation expresses the number of units change expected in the dependent variable, *Y*, given a single unit increase along the scale of the independent variable. Expressed differently, it is the ratio of the co-variation between *X* and *Y*, over the variation of the independent variable. Variation is expressed in terms of variances, thus the *b*-coefficient can be expressed thus:

$$b_{yx} = \frac{\text{Covariance of } X \text{ and } Y}{\text{Variance of } X}$$

The variance of *X* may be computed by this familiar formula:

$$s_x^2 = \frac{\Sigma X^2 - (\Sigma X)^2/N}{N}$$

The covariance of *X* and *Y* may be computed by the following formula where *XY* is used instead of *XX* or X^2 and $(\Sigma X)(\Sigma Y)$ is used in place of $(\Sigma X)(\Sigma X)$ or $(\Sigma X)^2$. *N* is, of course, the number of cases (or pairs of scores).

$$s_{xy}^2 = \frac{\Sigma(XY) - (\Sigma X)(\Sigma Y)/N}{N}$$

Thus, b_{yx} could be computed from raw data by this process:

D 1

$$b_{yx} = \left[\frac{\Sigma(XY) - (\Sigma X)(\Sigma Y)/N}{N} \right] \Big/ \left[\frac{\Sigma X^2 - (\Sigma X)^2/N}{N} \right]$$

$$b_{yx} = \frac{N(\Sigma XY) - (\Sigma X)(\Sigma Y)}{N(\Sigma X^2) - (\Sigma X)^2}$$

Computing for this example, the process would be as follows:

$$b_{yx} = \frac{5(88) - (15)(25)}{5(55) - (15)^2} = \frac{440 - 375}{275 - 225} = \frac{65}{50} = +1.3$$

DIAGRAM 7.2 *(Continued)*

Computing a_{yx}, *the Y-intercept constant*

Since a regression line goes through the point which is at the intersection of the mean of X and the mean of Y, we could substitute these mean values into the regression equation and solve for a_{yx}, as follows:

D 2
$$a_{yx} = \overline{Y} - b_{yx}(\overline{X})$$

or,

$$a_{yx} = \frac{\Sigma Y - b_{yx}(\Sigma X)}{N} = \frac{25 - 1.3(15)}{5} = \frac{25 - 19.5}{5} = 1.1$$

Thus:

$$Y' = 1.1 + 1.3(X)$$

Computing the Pearsonian Correlation Coefficient, r

The correlation coefficient is defined by the following formula as the ratio of the covariance of X and Y over the product of the standard deviations of X and of Y.

D 3
$$r = \left(\Sigma XY - \frac{(\Sigma X)\,(\Sigma Y)}{N} \right) \Bigg/ \sqrt{\frac{\Sigma X^2 - (\Sigma X)^2/N}{N} \cdot \frac{\Sigma Y^2 - (\Sigma Y)^2/N}{N}}$$

Simplified, this can be expressed as:

$$r = \frac{N\Sigma XY - (\Sigma X)\,(\Sigma Y)}{\sqrt{[N\Sigma X^2 - (\Sigma X)^2\,]\,[\,N\Sigma Y^2 - (\Sigma Y)^2\,]}}$$

Computing for this example:

$$r = \frac{5(88) - (15)\,(25)}{\sqrt{[\,5(55) - (15)^2\,]\,[\,5(151) - (25)^2\,]}} = \frac{440 - 375}{\sqrt{(275 - 225)\,(755 - 625)}}$$

$$= \frac{65}{\sqrt{6500}} = +.81$$

Pearson's r can be interpreted in a PRE sense by squaring its value, thus, $r^2 = .65$, meaning that 65 percent of the variation is "explained" by the linear relationship between these two variables.

The square of the Pearsonian correlation coefficient (called the Coefficient of Determination) could also be expressed in terms of variances:

(7.12)
$$r^2 = \frac{s_y^2 - s_{y \cdot x}^2}{s_y^2} \;=\; 1 - \frac{s_{y \cdot x}^2}{s_y^2}$$

The numerator in these instances represents the explained variation using the informed guessing rule. The denominator represents the overall variation.*

The Pearsonian correlation, r, is a PRE measure. If squared (r^2), it expresses the proportionate reduction in error in predicting scores for the dependent variable, given the best-fitting linear regression equation rather than the overall mean to use in prediction. Since the regression of Y on X and the regression of X on Y both have the same amount of scatter around their respective

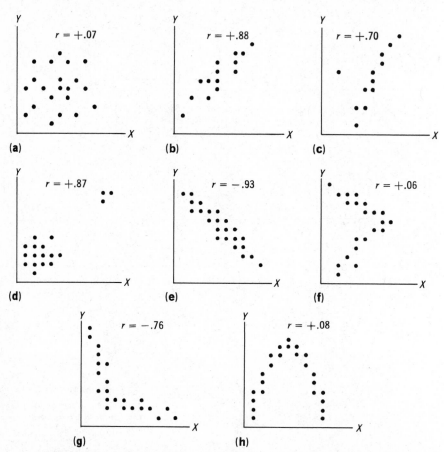

FIGURE 7.6   EXAMPLES OF SCATTERGRAMS AND CORRELATION COEFFICIENTS

---

*The ratio shown in the second formula above is called the coefficient of nondetermination, or K-squared, since it expresses the proportion of total variation which is unexplained by the regression equation.

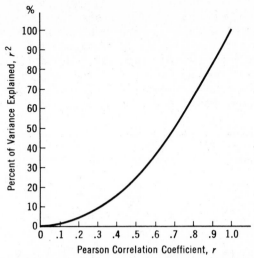

FIGURE 7.7 PERCENT OF VARIATION EXPLAINED BY CORRELATIONS OF DIFFERENT SIZES

regression lines, the same correlation coefficient will result from either predictive equation. Thus r is a *symmetric* measure of degree of correlation. Expressed differently, r^2 indicates the proportion of the variation in one variable which is explained by its linear association with the other variable.

Usually, the correlation coefficient is expressed as r, the square root of r^2, although its interpretation is most useful in its r^2 form as indicated above. The actual computational formula of r is given in Diagram 7.2.

Pearson's r is a symmetric measure of correlation between two interval variables. Its values vary from -1.0 to 0 to $+1.0$, indicating direction and strength of association. Figure 7.6 shows some of the values of r and their related scatter diagrams. Figure 7.7 shows the relationship between r and r^2, that is, the proportion of variation explained for different values of r. Notice in Figure 7.6 that a Pearson r may be quite small, not only where there is a random scattering of cases around the mean of the dependent variable, but also where there is an obvious association, yet the best-fitting straight line is no better than simply the mean of the dependent variable. Where the data are of a nature that is not too close to linear, an alternative, either Eta (symbolized E_{yx}, to be discussed in Section 7.4.4), or a curvilinear regression equation would be more appropriate.*

7.4.3 Another Way to Express Correlation and Regression

In order to get a better grasp of the meaning of correlation and regression it might be helpful to look at them from a somewhat different perspective. Suppose we are interested in determining whether there is a relationship between the

---

*In some cases it is possible to find an equation which expresses in mathematical terms the nature of curvilinear associations. If such an equation can be found, its "fit" to the data can be described by the correlation coefficient, r, by simply substituting the predicted value from the curvilinear equation for the predicted value from the simple linear equation in formulas for Pearson's r, above.

number of years of school completed by the head of household and the monthly rental value of the family residence. Below are ten pairs of scores presumably collected from ten families (the data are fictitious). In the first column are listed the years of education completed by the head of household and in the second column the monthly rental value (in dollars) of the family residence. We have computed the means and standard deviations of the two distributions and they are reported at the bottom of the table.

Years of Education X ·	Monthly Rental Value of Residence (dollars) Y	$X - \overline{X}$ x	$Y - \overline{Y}$ y	xy
12	400	−1.5	27.5	−41.25
10	325	−3.5	−47.5	166.25
16	450	2.5	77.5	193.75
8	350	−5.5	−22.5	123.75
14	200	0.5	−172.5	−86.25
20	525	6.5	152.5	991.25
11	375	−2.5	2.5	−6.25
15	250	1.5	−122.5	−183.75
16	675	2.5	302.5	756.25
13	175	−0.5	−197.5	98.75
$\overline{X} = 13.5$	$\overline{Y} = 372.5$			+2012.5
$s_x = 3.29$	$s_y = 144.24$			

If there is a correlation between these two variables, then it should be reflected in the deviations of the scores for each from their respective means. If the correlation is positive, then scores on variable X (years of education) which deviate in a positive direction from their mean should be paired with scores on variable Y (monthly rental value) which also deviate in a positive direction from their mean; furthermore, X scores which deviate in a negative direction should be paired with Y scores that deviate in a negative direction. Thus, if families with better educated heads tend to live in residences with higher monthly rental values, then for each pair of scores, the head of household's years of education should be higher than the mean for that variable and the monthly rental value of the family's residence should be above the mean on that variable. And, if the head's years of education is lower than the mean, then the monthly rental value should also be below the mean. In the third column of the table are entered deviations of head of household's years of education from the mean years of education $[x = (X - \overline{X})]$ and in the fourth column, the deviations of the monthly rental value of the family's residence from the mean of that variable $[y = (Y - \overline{Y})]$.

Column five of the table contains the products of the multiplication of each X deviation by its corresponding Y deviation (called cross-products). These cross-products are summed algebraically at the bottom of the table. If the sign of the sum of these cross-products is positive, then there is a positive correlation between the variables; if the sign is negative, then there is a negative correlation. Furthermore, the magnitude of the sum should tell us something about the strength of the relationship. The greater the magnitude, the stronger the correlation (whether positive or negative).

If the sum of the cross-products is divided by N (the number of cases in the sample), the resulting figure is the *covariance*. The covariance is literately the average magnitude of the cross-products of the X and Y deviations. The covariance for the data we are considering is +201.25. Since the sign is positive, we know that there is a positive correlation between the two variables. But is 201.25 large in magnitude or not? It is quite a bit larger than the mean value for variable X, but it is smaller than the mean value for variable Y. The problem is that the possible upper limit of the covariance depends upon the units of measurement of the variables being correlated. In this case we are correlating years of education with dollars — two very different kinds of units. By comparing the means of the two distributions we can see that the magnitudes of the dollar values are considerably greater than those of years of education. Furthermore, from the standard deviations we can see that monthly rental values of residences are much more variable than are years of education of heads of households. We find ourselves in a situation of trying to relate two variables which are very different from each other both in central tendency and in variability. It is like trying to compare apples and oranges.

There is a way out of this dilemma, however. We can convert the values on both variables to standard scores (z scores); and, thus convert them into comparable units. When the values on both variables are expressed as standard scores they become comparable because, as you may recall from Chapter 5 (Section 5.4.7), the mean of a set of standard scores is always zero and the standard deviation is always 1. The values for the X variable can be converted into standard score form by using the formula:

$$z_x = \frac{X - \overline{X}}{s_x}$$

$$= \frac{X - 13.5}{3.29}$$

For the value of the Y variable a similar formula is used. That is

$$z_y = \frac{Y - \overline{Y}}{s_y}$$

$$= \frac{Y - 372.5}{144.24}$$

The table below presents the same 10 pairs of scores as were presented in the previous table; but, in this case, the deviations of scores from their means are expressed as standard scores rather than simply deviation scores. Column 3 consists of standard scores for the ten values on the X variable and column 4 consists of those for the Y variable. Column 5 consists of cross-products of the standard scores for the X and Y variables. If the sum of the cross-products of these standard scores is divided by N, the result is a sort of standardized covariance.

Years of Education X	Monthly Rental Value of Residence (Dollars) Y	z_x	z_y	$(z_x)(z_y)$
12	400	−0.46	0.19	−0.09
10	325	−1.06	−0.33	0.35
16	450	0.76	0.54	0.41
8	350	−1.67	−0.16	0.27
14	200	0.15	−1.20	−0.18
20	525	1.98	1.06	2.10
11	375	−0.76	0.02	−0.02
15	250	0.46	−0.85	−0.39
16	675	0.76	2.10	1.60
13	175	−0.15	−1.37	0.21
				+4.26

This standardized covariance is, in fact, Pearson's Product Moment Coefficient of Correlation. That is,

$$r = \frac{\Sigma z_x z_y}{N}$$

(7.13)

In the case of the data under consideration, the r which results from using this formula is +4.26/10 = +.43.

where the +4.26 is the sum of the cross-products of the standard scores of the X and Y variables.

Since z_x can be expressed as x/s_x and z_y as y/s_y, we can substitute those expressions into formula 7.13 as follows:

$$r = \frac{\Sigma \left(\frac{x}{s_x}\right)\left(\frac{y}{s_y}\right)}{N}$$

Then the two standard deviations s_x and s_y may be put in the denominator with N without changing the value of the formula, as follows:

(7.14)

$$r = \frac{\Sigma\, xy}{N s_x s_y}$$

The resulting formula is one which is the basis for the computing formula for r which is given in Diagram 7.2.

Formula 7.13 makes it clear that the computation of r involves the transformation of the pair of scores from their original measurement units (such as years and dollars) to standard score units. You will recall from univariate statistics that the sum of squares of z-scores equals the total number of cases, N. If a case has a score which is in the same relative position on both the X and Y variables (and thus the same sized z-score on each variable), the correlation coefficient will equal 1.0, since the numerator will then sum to N. To the extent that the standing of individuals is different on the two variables (as reflected in different z-scores), the numerator of the equation above will not reach N and the

---

BOX 7.4 STANDARD SCORES

If you are a bit rusty on z-scores, refer back to Section 5.4.7. Briefly, a standard score is the number of standard deviation units above or below the mean that a score falls. The mean of a distribution of z-scores is therefore, zero, and the standard deviation of a distribution of z-scores is one.

---

correlation will be less than 1.0. If one set of the z-scores tends to be negative, while the other tends to be positive, then the correlation coefficient is negative. The point is that here again a simple ratio expresses one feature of the relationship between two variables, a kind of ratio that was used (with somewhat different components, to be sure) for the arithmetic mean, the variance, ordinal measures of association, and now an interval-level measure of association.

The standard-score form of the simple regression equation is also interesting to note.

$$z'_y = r(z_x)$$

Here the estimated value of the z-score on variable Y can be computed from a knowledge of the z-score on variable X, if an adjustment is made. The adjustment corresponds to multiplication of the z_x score by a constant which (for one independent variable) is the Pearson correlation coefficient, r. This is called a **standardized regression equation.**

One final expression for the correlation coefficient will help show similarities between this measure and some of the other symmetric measures. The Pearson correlation coefficient can also be expressed as a kind of average* of two asymmetric regression coefficients, b_{yx} and b_{xy}, as follows:

(7.15)
$$r = \sqrt{(b_{yx})(b_{xy})}$$

You will recall that this way of forming a symmetric correlation coefficient out of two asymmetric measures was also possible for tau-b, which was expressed in terms of Somers' d as follows:

$$t_b = \sqrt{(d_{yx})(d_{xy})}$$

7.4.4 Interpretations and Assumptions

There are several ways to interpret the results of correlation or regression analysis. Each of these interpretations is based upon a different set of assumptions about the data being analyzed. In order to make intelligent use of these interpretations it is necessary to be familiar with the assumptions being made. Accordingly, we will consider these several interpretations and those conditions under which they are justified.

---

*This average is an example of the geometric mean discussed in Chapter 5.

1. The correlation coefficient, r, may be interpreted as a measure of the rate at which the dependent variable (Y), expressed as a standard score, changes with a change of one unit of the independent variable (X), expressed in standard score form. This is so because, as was pointed out in section 7.4.3, r is the standardized form of the regression coefficient (b)*.

This interpretation of r is based upon the assumption that the relationship between X and Y is linear. If X and Y jointly form a bivariate normal distribution, then a linear relationship between the two variables is assured. However, a linear relationship between the variables does not assure that their joint distribution is bivariate normal. It is useful to plot the scores of a bivariate distribution on a scatter diagram to see whether the assumption of linearity is a reasonable one.

The assumption of linearity is a crucial one for this interpretation of r because the standardized regression coefficient is a constant which is multiplied by the z score for X no matter what value X happens to take. If the correlation between X and Y were curvilinear rather than linear, the slope represented by the regression coefficient would assume different values for different segments of the curve. In the case of a parabola, for example, the slope of the curve is positive for half of the length of the curve and negative for the other half.

Note that when the scores of the independent and dependent variables are not expressed in standard score form it is the regression coefficient, b_{yx}, rather than r which measures rate of change in Y associated with a unit change in X.

2. When r takes on a value of either +1.00 or −1.00 the association of variables Y and X can be described by a mathematical function which allows us to transform values on the X variable to values on the Y variable without error. For example, the correlation between measurements on a Fahrenheit scale of temperature and measurements on a Celsius scale is +1.00. By a linear transformation formula the temperature on a Fahrenheit scale can be expressed as a temperature on the Celsius scale without error. Thus, 68 degrees Fahrenheit is equivalent to 20 degrees Celsius.

When r takes on a value of zero it will usually be the case that variables Y and X are independent. A zero correlation assures independence of the variables, however, only when the joint distribution of Y and X is bivariate normal. If Y and X do not form a bivariate normal distribution, it is possible for the correlation coefficient to be zero even though the two variables are not independent.

Intermediate values of r (intermediate between zero and 1.00) indicate that there is a relationship between Y and X, but the strength of that relationship is not intuitively meaningful.

3. The Coefficient of Determination, r^2, has a much more appealing interpretation than r for the intermediate values of r. The r^2 can be interpreted as the *proportion of the variance of the dependent variable (Y) which can be accounted for by its linear relationship to the independent variable (X)*. Or put a little differently, r^2 is the ratio of the variance of Y explained to the total variance of Y. This ratio (the variance *explained*) plus the ratio of the *unexplained* variance of Y over the total variance of $Y,(1 - r^2)$, is equal to 1. The r^2 can be

---

*This relationship between r and b holds only for the case where one independent variable is being correlated with one dependent variable—the bivariate correlation situation.

given this sort of interpretation because it is based upon a ratio of variances as is $1 - r^2$ and variances are additive whereas standard deviations are not.

This interpretation of r^2 is based upon the assumption that X and Y are linearly related. The assumption is necessary because r^2 implies that the portion of the variance of Y accounted for by X is the same no matter what value X takes. Thus, the same regression equation is assumed to be equally effective in predicting Y for any value in the range of values the variable X may take.

4. The Standard Error of Estimate is a measure of the accuracy with which the regression equation predicts the dependent variable from known values of the independent variable. It measures the magnitude of prediction error. There are three assumptions underlying the use of the Standard Error of Estimate: (1) it is assumed that there is a linear relationship between X and Y, (2) it is assumed that the array of Y scores associated with each value of the X variable forms a normal distribution, and (3) it is assumed that the variances of these distributions (arrays) of Y scores are all equal. This latter assumption is known as the assumption of homoscedasticity. These assumptions are necessary for utilizing the Standard Error of Estimate because its value remains constant in measuring error no matter what point along the regression line a prediction is made. Furthermore, assumption two makes it possible to interpret the Standard Error of Estimate in terms of areas under the normal curve. Thus, it can be said that approximately 68 percent of the predictions of an actual value of Y will be within plus or minus one Standard Error of Estimate of that value.

Where there is interest in generalizing conclusions from samples to more general populations, other assumptions are involved (see Chapter 16)*.

7.4.5 The Correlation Ratio, Eta (E_{yx}), a Measure of Association for Non-Linear Relationships

If a straight line is not a very useful description of the nature of the association between two interval variables, and if there is no other formula for a curved line available, one might think of falling back on a procedure of making separate predictions of the mean of the dependent variable within categories of the independent variable. This situation is illustrated in Table 7.6.

Here the dependent variable is the number of wage earners in a family and the independent variable is family income, divided up into a series of categories. The overall mean number of wage earners for these data is 1.7 and this would constitute the minimum prediction rule we could use. On the other hand,

---

*If the data being correlated constitute a random sample of some population and our intention is to generalize to the population (see Chapter 16), then it is assumed that variables X and Y are both normally distributed random variables. If this assumption is tenable, then inferences made from the sample data allow us to make statements about the independence or dependence of variables X and Y in the population. As was mentioned above, a zero correlation between X and Y assures their independence from each other only when X and Y jointly form a bivariate normal distribution.

When inferences are made to population data using the regression equation, it is assumed that the dependent variable is a random variable, but values of the independent variable are assumed to be constants. Thus, specific values of the independent variable may be entered into the regression equation to predict a value of the random dependent variable with a range of prediction error measured by the Standard Error of Estimate.

we could improve our prediction by guessing the category mean for those cases falling in a given income category.

The errors of prediction are calculated as the squared difference between prediction and actual score on the Y-variable. Since there are, again, two predictions, the overall mean and the mean within categories, we have two error figures to contrast.* Eta squared (E^2_{yx}) is simply a ratio of these quantities, as defined below:

$$E^2_{yx} = 1 - \frac{\text{variance within categories of } Y}{\text{overall variance of } Y}$$

(7.16)
$$E^2_{yx} = 1 - \frac{\Sigma(Y - \overline{Y}_i)^2}{\Sigma(Y - \overline{Y})^2}$$

TABLE 7.6 Number of Earners in Families in the U.S. in 1970 by Family Income

			(X) Family Income			
Under 1500	1500– 2999	3000– 4999	5000– 7999	8000– 11999	12000– 49999	50000 and up
1	0	0	0	0	1	1
1	0	1	1	1	1	
	2	1	1	1	1	
		1	1	1	2	
		2	1	1	2	
			2	1	2	
			2	1	2	
			2	2	2	
			3	2	2	
				2	2	
				2	2	
				3	2	
				3	2	
				4	2	
					2	
					3	
					3	
					7	
$N_i =$ 2	3	5	9	14	18	1
$\overline{Y}_i =$ 1	0.7	1	1.4	1.7	2.2	1

$N = 52$
$\overline{Y} = 1.7$
N_i and $\overline{Y}_i$ refer to category size and mean.
N and $\overline{Y}$ refer to the overall size and grand mean.

*The following formula for Eta (E_{yx}) is a more useful computational version.

$$E^2_{yx} = 1 - \frac{\Sigma Y^2 - \Sigma n_k \overline{Y}_k^2}{\Sigma Y^2 - N\overline{Y}^2}$$

where $n_k \overline{Y}_k^2$ is the product of the number of cases times the squared mean of the k th sub-category, and these products are then summed over all k sub-categories. ΣY^2 is simply the overall sum of squared scores and $\overline{Y}$ is the overall mean of Y.

TABLE 7.6 *(Continued)*

(Y) Number of Earners Per Family	Predictions Overall Mean	Category Mean	$(Y-\overline{Y})$	$(Y-\overline{Y})^2$	$(Y-\overline{Y}_i)$	$(Y-\overline{Y}_i)^2$
1	1.7	1	− .7	.49	0	0
1	1.7	1	− .7	.49	0	0
0	1.7	.7	−1.7	2.89	− .7	.49
0	1.7	.7	−1.7	2.89	− .7	.49
2	1.7	.7	+ .3	.09	+1.3	1.69
0	1.7	1	−1.7	2.89	−1.0	1.00
1	1.7	1	− .7	.49	0	0
1	1.7	1	− .7	.49	0	0
1	1.7	1	− .7	.49	0	0
2	1.7	1	+ .3	.09	+1.0	1.00
0	1.7	1.4	−1.7	2.89	−1.4	1.96
1	1.7	1.4	− .7	.49	− .4	.16
1	1.7	1.4	− .7	.49	− .4	.16
1	1.7	1.4	− .7	.49	− .4	.16
1	1.7	1.4	− .7	.49	− .4	.16
2	1.7	1.4	+ .3	.09	+ .6	.36
2	1.7	1.4	+ .3	.09	+ .6	.36
2	1.7	1.4	+ .3	.09	+ .6	.36
3	1.7	1.4	+1.3	1.69	+1.6	2.56
0	1.7	1.7	−1.7	2.89	−1.7	2.89
1	1.7	1.7	− .7	.49	− .7	.49
1	1.7	1.7	− .7	.49	− .7	.49
1	1.7	1.7	− .7	.49	− .7	.49
1	1.7	1.7	− .7	.49	− .7	.49
1	1.7	1.7	− .7	.49	− .7	.49
1	1.7	1.7	− .7	.49	− .7	.49
1	1.7	1.7	− .7	.49	− .7	.49
2	1.7	1.7	+ .3	.09	+ .3	.09
2	1.7	1.7	+ .3	.09	+ .3	.09
2	1.7	1.7	+ .3	.09	+ .3	.09
2	1.7	1.7	+ .3	.09	+ .3	.09
3	1.7	1.7	+1.3	1.69	+1.3	1.69
3	1.7	1.7	+1.3	1.69	+1.3	1.69
4	1.7	1.7	+2.3	5.29	+2.3	5.29
1	1.7	2.2	− .7	.49	−1.2	1.44
1	1.7	2.2	− .7	.49	−1.2	1.44
1	1.7	2.2	− .7	.49	−1.2	1.44
1	1.7	2.2	− .7	.49	−1.2	1.44
2	1.7	2.2	+ .3	.09	− .2	.04
2	1.7	2.2	+ .3	.09	− .2	.04
2	1.7	2.2	+ .3	.09	− .2	.04

TABLE 7.6 *(Continued)*

(Y) Number of Earners Per Family	Predictions Overall Mean	Category Mean	$(Y-\overline{Y})$	$(Y-\overline{Y})^2$	$(Y-\overline{Y}_j)$	$(Y-\overline{Y}_j)^2$
2	1.7	2.2	+ .3	.09	− .2	.04
2	1.7	2.2	+ .3	.09	− .2	.04
2	1.7	2.2	+ .3	.09	− .2	.04
2	1.7	2.2	+ .3	.09	− .2	.04
2	1.7	2.2	+ .3	.09	− .2	.04
2	1.7	2.2	+ .3	.09	− .2	.04
2	1.7	2.2	+ .3	.09	− .2	.04
2	1.7	2.2	+ .3	.09	− .2	.04
3	1.7	2.2	+1.3	1.69	+ .8	.64
3	1.7	2.2	+1.3	1.69	+ .8	.64
7	1.7	2.2	+5.3	28.09	+4.8	23.04
1	1.7	1.0	− .7	.49	0	0
				$\Sigma(Y-\overline{Y})^2 = 68.37$		$\Sigma(Y-\overline{Y}_j)^2 = 56.78$

Source: Data are fictitious in this example, but they follow the distribution of number of earners and family income for families in the U.S. in 1970. Each figure is approximately one million families. See U.S. Bureau of the Census, 1971, p-60, No. 80: Tables 1 and 19. The actual average number of earners per family for the above income categories from lowest to highest are: .76, .69, 1.05, 1.44, 1.73, 2.19, and 1.97. The overall average number of earners per family in the U.S. in 1970 was 1.68.

For these data, substituting figures from the second half of Figure 7.5, we get these computations:

$$E_{yx}^2 = 1 - \frac{56.78}{68.37} = 1 - .83 = +.17$$

$$E_{yx} = \sqrt{.17} = +.41$$

This is essentially the same type of measure as *r*, except that the mean within categories of the independent variable, rather than a regression equation, is used to form the predicted *Y*-score. Unlike *r*, however, there are two Etas depending upon which variable is predicted from which. E_{yx} is asymmetric. E_{yx}^2 indicates the proportionate reduction in error in predicting the *Y*-scores if category means rather than the grand mean are predicted. E_{yx} varies in magnitude from 0 (in which case the grand mean is as useful in prediction as category means) up to +1.0, which indicates that category means permit exact prediction. The independent variable may be nominal, ordinal, or interval just as long as the dependent variable is interval-level. Thus sign of the association does not have meaning, or, in other words, E_{yx} will always be a positive value. Since *r* and E_{yx} have essentially the same form, and differ only in the source of the refined prediction, they can be compared directly. If, for a set of data where both measures are appropriate, E_{yx} is *larger* than Pearson's *r*, then one can infer that category means do not fall along a simple straight line, and thus, to some degree, the *nature of association is curved* or different from a straight line. If they are identical, then it is clear that subcategory means fall exactly along a least-squares regression line, such as we have developed here. (E_{yx} will never be smaller in magnitude than r.) This provides a rather handy basis for making inferences about the nature of an association.

7.5 THE CORRELATION MATRIX

Because of the interest in comparison, it is not at all unusual for investigators to compute a number of similar association measures, showing the relationship between all possible pairs of a set of items. An example of such a correlation matrix is shown in Table 7.7, where Pearsonian correlation coefficients are used. A matrix consisting of gammas (G's) or any other appropriate measure of association could be used, depending upon the variables being correlated.

In Table 7.7, Marsh and Stafford (1967) present a correlation matrix showing the correlation between answers to each pair of eleven questions given male professionals with an M.A. or Ph.D. degree. This was a part of a larger 1962 survey of the U.S. Some respondents were from academic settings and some were from private industry. Agreement with certain of the eleven items was thought to reflect a "professional" orientation and agreement with some of the other items was thought to reflect an "acquisitive" attitude. Examination of the correlation matrix, below, helped them see to what extent these two groups of items— "professional" and "acquisitive"—appeared in their data. Their basic hypothesis in this part of their study was that professionals in academic settings would be more likely to agree with "professional" items and disagree with "acquisitive" items. They thus expected to show that professional and intellectual values of work would serve as "compensation" for the lower incomes academics generally receive as contrasted to professionals in industry. An examination of the correlation matrix in Table 7.7 helped them identify items which could be used to measure these two attitudes.

TABLE 7.7 MATRIX OF CORRELATION COEFFICIENTS BETWEEN ELEVEN MEASURES OF ATTITUDE TOWARD WORK

(Items) Attitudes	(Items)	1	3	10	12	13	11	9	4	5	7	8
1 Opportunity to be original and creative			.36	.29	.42	.49	.14	.25	.31	.24	.18	.27
3 Relative independence in doing my work				.31	.36	.36	.11	.26	.23	.25	.17	.21
10 Freedom from pressures to conform in my personal life					.35	.31	.26	.37	.17	.34	.23	.33
12 Freedom to select areas of research						.47	.12	.19	.18	.24	.20	.21
13 Opportunity to work with ideas							.29	.31	.29	.24	.24	.26
11 Opportunity to work with people								.43	.39	.35	.44	.27
9 Pleasant people to work with									.29	.45	.36	.39
4 A chance to exercise leadership										.26	.34	.31
5 A nice community or area in which to live											.45	.41
7 Social standing and prestige in my community												.36
8 A chance to earn enough money to live comfortably												

correlation of items 4 and 11.

Source: Marsh and Stafford, 1967. Used by permission. The study is based on an April, 1962 self-administered questionnaire given to some 51,505 professional, technical, and related workers in the U.S. This investigation selected 13 professional fields and limited the analysis to males who had M.A. or Ph.D. degrees. The correlations above are Pearsonian r's between responses to the different items on attitude toward work.

Table 7.7 is organized so that each row refers to a different item (the numbers at the left are item numbers in the original questionnaire). Columns also refer to the same items, with each item (numbered as they are at the left of the table) being in a different column. The number at the intersection of a given row and column is a correlation coefficient showing the correlation between the items indicated by row and column headings. The items in Table 7.7 appear to fall into two clusters, one being a "professional" cluster including items about independence, creativity, and working with ideas, and the other being a cluster dealing with "acquisitive" items, such as pleasant work, community settings, and prestige.

For convenience, Marsh organized the attitude items into these two clusters and drew lines around the coefficients which refer to the intercorrelation of items within each cluster. The expectation is that items within a cluster would be more highly related than they would be with items in the other cluster. This appears to be the pattern in Table 7.7. Patterns in matrices such as this will be discussed later in this text. For now, the important point is to call your attention to the *comparative* use of measures of association. Notice that the Pearsonian correlation coefficient, r, which was used in this instance, is a symmetric measure. Thus, only half of the matrix in Table 7.7 needs to be given, because the other half would be identical (*i.e.* the correlation between item 3 and item 1 is the same as the correlation of item 1 and item 3).

7.6   SELECTING A USEFUL MEASURE OF ASSOCIATION

There have been a number of criteria discussed which distinguish between the different measures of association and, in general, these are the differences which lead to the selection of one measure rather than another for a particular problem. In a given problem, certain features of these distinctions may take on more importance for the comparisons that are to be made and, thus, these criteria would override other kinds of differences.

7.6.1   Symmetry and Asymmetry

One of the important kinds of differences between coefficients is the way independent and dependent variables are handled. In problems where explanation and prediction of a dependent variable are of particular interest, such as that shown in the following arrow diagram, an asymmetric measure, if one is

available, would be most useful. If we are interested simply in the way variables covary or relate to each other, which might be the case when we examine the

---

BOX 7.5 THE THEME

In Chapters 6 and 7 we have attempted (a) to show how the various measures of association are relatively straightforward solutions to the problem of describing relationships between variables, and (b) to highlight some of the differences between them which may correspond closely to the substantive problem being investigated.

At this point, you should avoid looking for the easy or oversimplified ways to choose an appropriate measure. It might seem easy to recommend only one or two measures out of the variety we have discussed, but the decision on appropriate measures is not that mechanical. The selection depends upon the substantive problem being investigated as well as some "technical" characteristics of each measure. Part of the skill in creating and using statistical information lies in an understanding of the relationship between problem and tools.

---

relationship between several indicators of the same concept, then a symmetric measure provides this information.

Interrelation of social class measures.

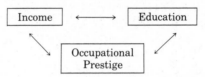

7.6.2 Level of Measurement

The meaning defined into variables also provides an important basis for selecting among the various measures. If only the categorical aspect of a variable is meaningful, then measures suitable for nominal variables would be most appropriate. Diagram 7.3 organizes the measures we have discussed, in this and the last chapter, in terms of level of measurement. As in the case of univariate descriptive statistics, one might find occasion to use lower-level measures on data defined at a higher level of measurement, but this would result in the loss of information contained in the data — a result which is usually undesirable.

In some cases, level of measurement does not make much difference. In a 2×2 table, for example, Pearson's r, tau-b, and the phi coefficient are numerically identical. Spearman's rho is a Pearsonian correlation (a measure designed for interval variables) computed on ranks which may be assigned in the process of measuring an ordinal variable. Generally speaking, however, level of measurement *is* an important consideration in selecting and certainly in interpreting association measures.

DIAGRAM 7.3 Organization of Measures of Degree and Nature of an Association

| Level of Measurement | MEASURES OF DEGREE OF ASSOCIATION | | NATURE OF ASSOCIATION |
	Symmetric	Asymmetric	
Nominal	C $\lvert Q \rvert$ T $\lvert \phi \rvert$ V	ϵ λ_{yx} t_y	percentaged table
Ordinal	Q G ϕ r_s t_a t_b	d_{yx}	percentaged table
Interval-Ratio	r	E_{yx}	$Y' = a_{yx} + b_{yx}(X)$ or some other, specified form.

7.6.3   Nature versus Strength of Association

Increasingly, sociology is addressing itself to the description of the amount of change a unit change in a given independent variable will likely produce in a dependent variable. "Powerful" variables in this sense are those which produce great effect, and this, of course, has both theoretical and practical importance. If this is the interest, then **regression coefficients** become more interesting than correlation coefficients, or **percentaged tables** more interesting than the various measures of association. Association measures express how well the variables go together, in the sense of scatter around a regression line, or errors in predicting a mode or rank order. A "clean" prediction might be possible in this sense, whenever a predictive variable is relatively so "powerless" that changes in it may not reflect or cause much change in the dependent variable. These two interests go together, of course, but in any given problem one may be more important than the other, and this may influence the selection of measures. Regression equation coefficients and the correlation coefficients are particularly useful if the main focus is on the potency of a variable.* On the other hand, if the nature of association is not relatively linear, then one either would have to find a curvilinear regression equation that fits the nature of the data or shift to some measure which is not tied to a specific nature of association.

7.6.4   Interpretation

Finally, measures of association differ in terms of the features of the association toward which they are most sensitive. Some measures (λ_{yx}, r, E_{yx} for example) are oriented toward predicting an optimal or central value of a dependent variable. Others (*e.g.*, t_y) are oriented toward predicting the distribution of a dependent variable. Some measures (*e.g.*, r, G) contrast observed data with a specific model of perfect association or independence. Some of these differences are organized in Figure 7.8.

Many of the measures discussed have PRE interpretations, which makes them more convenient to use in research. Selection among these measures depends upon what one wants to treat "at risk." For asymmetric measures, such as Somers' d_{yx}, the at-risk pairs are pairs which are distinguishable on the independent variable. For gamma (G), they are pairs which are ranked differently on both X and Y variables. Other measures of association do not have PRE interpretations (*e.g.*, C, t_c) but can be interpreted in terms of magnitude between limiting values (often zero and 1.00).

Finally, some of the measures of association we have discussed are most suited to tables which have specific types of layout. Yule's Q and the Phi (ϕ) coefficient from Chapter 6, are oriented toward 2×2 tables, unlike most of the other measures. The contingency coefficient, C, is influenced by the number of rows and columns in a table. Tau-b (t_b) cannot achieve 1.0 in a table which is not square, and measures such as tau-b (t_b), Spearman's rho, and tau-a (t_a), for example, are especially sensitive to ties in data.

---

*For other models expressing the impact of changes in one variable upon another, see Coleman (1964), Chapters 6 and 7.

Measure	Level of Measrmt.	Table Size	Range	Definition of a Perfect Assn.	Formula	Interpretation
ϵ epsilon	Nominal	$r \times c$	0 up to some x $x \leq 100\%$	More restrictive	Usually, difference of 2 extreme corner pcts. in properly percentaged table	A rough degree of association measure for a table. With care, one can compare across tables of the same size and similar marginals. In a 2×2 table, an epsilon of zero means no association. Asymmetric.
Yule's Q	Ordinal (Nominal if sign dropped)	2×2	-1 to $+1$ (0 to $+1$)	Less restrictive	$Q = \dfrac{ad - bc}{ad + bc}$ $\begin{array}{c c}& \text{Lo Hi} \\ \text{Hi} & \boxed{\begin{array}{c c} b & a \\ d & c \end{array}} \\ \text{Lo} & \end{array}$	Symmetric measure identical to gamma. Probability of like (opposite) ranking on two variables among cases ranked differently on both variables.
C	Nominal	$r \times c$	0 up to some x $x \leq 1.00$		$C = \sqrt{\dfrac{\phi^2}{1 + \phi^2}}$	Interpret in terms of magnitude.
ϕ Phi	Ordinal (Nominal if sign dropped)	2×2	-1 to $+1$ (0 to $+1$)	More Restrictive	$\phi = \dfrac{ad - bc}{\sqrt{(a+b)(c+d)(a+c)(b+d)}}$	Degree of diagonal concentration. A symmetric measure.
λ_{yx} Lambda	Nominal	$r \times c$	0 to $+1$		$\lambda_{yx} = \dfrac{\Sigma m_y - M_y}{N - M_y}$	An asymmetric measure indicating the percent improvement in predictability of the dependent variable with information about the independent variable classification.
t_y	Nominal	$r \times c$	0 to $+1$	Y all one category for any given X	$t_y = \dfrac{E_1 - E_2}{E_1}$	PRE in predicting category placement. Asymmetric.

Measure	Level of Measrmt.	Table Size	Range	Definition of a Perfect Assn.	Formula	Interpretation
G Gamma	Ordinal	$r \times c$	−1 to +1	Less restrictive.	$G = \dfrac{N_s - N_d}{N_s + N_d}$	A symmetric measure indicating the relative preponderance of like (unlike) ranked pairs among pairs ranked differently on both variables.
t_a Tau-a	Ordinal (No Ties)	$r \times c$	−1 to +1		$t_a = \dfrac{N_s - N_d}{T}$	Symmetric. PRE among all pairs.
t_b Tau-b	Ordinal	$r \times c$	−1 to +1		$t_b = \dfrac{N_s - N_d}{\sqrt{(N_s + N_d + T_y)(N_s + N_d + T_x)}}$	Symmetric. For square tables.
d_{yx} Somers'	Ordinal	$r \times c$	−1 to +1		$d_{yx} = \dfrac{N_s - N_d}{N_s + N_d + T_y}$	Asymmetric. PRE among predicted pairs.
E_{yx} Eta	Interval Dependent Variable (Other may be nominal)	$r \times c$	0 to +1	Line of category means.	$E_{yx}^2 = 1 - \dfrac{\sum Y^2 - \sum n_k \bar{Y}_k^2}{\sum Y^2 - N\bar{Y}^2}$	PRE Asymmetric.
r_{yx}	Both interval	$r \times c$	−1 to +1	linear	$r = \sqrt{1 - \dfrac{S_{y \cdot x}^2}{S_y^2}} = \dfrac{\sum z_x z_y}{N}$	PRE. r^2 is the proportion of variation in one variable explained by linear assn. with the other. Symmetric.
r_s Spearman's rho	Ordinal	Two sets of N ranks 2 × c	−1.0 to +1	Identical ranks on two vars.	$r_s = 1 - \dfrac{6\sum D^2}{N(N^2 - 1)}$	Symmetric. Sensitive to large differences in ranks. D is difference between the two ranks for a given subject.

FIGURE 7.8 CHARACTERISTICS OF SELECTED MEASURES OF ASSOCIATION

Another aspect of interpretability of a coefficient is its familiarity to an audience. At this point in sociology, coefficients such as Pearson's r, G, t_b, lambda (λ_{yx}), percentage difference (epsilon $-\epsilon$), and Somers' d_{yx} are relatively frequently seen in print. If an appropriate measure for one's data happens to be among these, the more widely known and used measure would probably be chosen.

As you might suspect, there is no flat rule which predetermines which coefficient *must* be used. Usually the choice is clear-cut — we would prefer a PRE measure, appropriate for the level of measurement of variables in the study, asymmetric if the independent/dependent role of variables is important, and sufficiently widely known so that it can be used to communicate information about the data. But often the choice is a matter of balance and judgment between coefficients which offer somewhat different strengths. The reasoning process in selecting a coefficient is closely tied to the logic and purpose of a research problem. Probably the only flat rule is that it is *not* appropriate to compute all possible measures and select one measure for the reason that it is numerically, the largest!

7.7 SOME CAUTIONS ABOUT THE INTERPRETATION OF ASSOCIATION MEASURES

An important purpose of correlation coefficients is, of course, to aid in comparison. Generally we want to know the proportionate reduction in error which can be achieved in some defined respect by using one variable or another. Usually (or hopefully) the contrasts are suggested by some prior reasoning. We expect social class rather than toenail length to have something to do with the distribution of anomie, and a comparison of correlation coefficients bears this expectation out.

7.7.1 Causation and Association

There is sometimes a temptation to attribute more meaning to a measure of association than it really contains. This is particularly likely if the reasoning behind the investigation comes from an interest in explaining the causes of some phenomenon. If knowledge of social class reduces the errors of predicting alienation by quite a large amount, the temptation is to say that social class influences or causes alienation, simply on the basis of the correlation coefficient. Upon reflection, this is not appropriate at all. The correlation may well be the same between doorknob size and education of head of household and we would hardly want to conclude on this basis that knob size has anything particularly to do with education. Even asymmetric measures only state how measured variables relate in the data at hand.

The investigator's theory, however, may lead him to predict that there will be a certain kind of association in the data because of some hypothesized causal link between variables. The data may yield correlations which turn out to be close to predictions, and because of this he (she) may feel that his (her) faith in his (her) theory is well founded, that one of the variables does cause the other to change as his (her) theory stated, and that this is what the static correlation coefficients mean. The cause-effect argument is clearly *in his (her) theory,* however, and *not* something that is included in the correlation coefficient itself. *Correlation is not causation.* More evidence than merely a correlation coefficient is needed to begin to have confidence in a conclusion that one variable influences another under certain conditions.*

7.7.2 Ecological Fallacy

There are some interpretative problems which arise essentially because an investigator is not clear about the kind of subject or case he (she) is examining. This may be an important source of difficulty in sociology, since the field deals with many different levels of units, from dyads to individual people, to individual roles or self-concepts, to small groups or societies or groups within groups. Correlations between variables measured on *groups* are *not the same* as correlations measured on *individuals,* in spite of the fact that those individuals may form the groups. The error of inferring how two variables are related among *individuals* by examining the correlation of similar variables measured on *groups* is called the **ecological fallacy.**†

A classic example of this problem is illustrated in an article by W. S. Robinson (1950) in which he shows the relationship between "foreign-born" and "illiteracy." The correlation between percent foreign-born and percent illiterate was computed to be $-.62$, with these two variables measured on each of several regions of the United States. The correlation used geographic regions as the case. This means that the higher the percent of foreign-born in a region, the lower the percentage of "illiterate" we would expect. Does this mean that foreign-born *people* are less likely to be illiterate? No! The correlation between foreign-born or not, and illiterate or not, for individuals in the United States was $+.12$, a low, positive association. In fact, there is little reason why there should be the same correlation between these variables at the group and individual levels. To start with, the variables are different. In one case they are percentage concentration of foreign-born or illiterates, and in the other case they refer to whether or not an individual is or is not foreign-born or illiterate. Secondly, it is quite possible that all of the foreign-born people are illiterate, or none of the foreign-born people are illiterate, in any given area with a given percentage of foreign-born and illiterate. The moral is the same one raised early in this volume: we must be clear about the meaning of statistical observations before statistical results can be meaningfully interpreted.

---

*Although one would expect a correlation between a causal variable and its effect, there are several other types of information which need to be examined before a casual interpretation of data begins to become useful. One needs to know time-order (that the cause occurred before the effect), and that the association could not be "explained away" by other factors. The next chapter deals with some of these questions.
†Some of the problems in inference between different levels of phenomena are nicely discussed in a research methods book by Matilda White Riley, (1973: see especially unit 12).

7.7.3 Built-in Correlations*

One final set of alternative interpretations of measures of association should be mentioned before we close this chapter. The possibilities to be mentioned here flow fairly directly from the meaning of specific measures of association we have discussed, but they might be pulled together as a set of possible interpretations of association. A measure of association may be low, not because two variables are not related in some way, but because they are not related in the way to which a given measure of association is sensitive. This was pointed out for Pearson's r, where strong curvilinear associations may show up as a low correlation because r is measuring fit to a simple straight line. It is usually a good idea to examine the nature of an association via a scattergram or table.

An association could be "built in," if the original observations are not independent. For example, if a study of income and education of people were designed so that husbands and wives were both measured, then it is likely that some association between education and income would be observed, simply because these variables are probably related for husbands and wives. We would be able to handle the problem in this instance by separating out husbands and wives or by treating them as a pair. Built-in correlations could occur in other ways too — for example, in correlations between a score that is derived from answers to all items and a score that is derived from answers to a sub-set of the same items.

It is possible to have a correlation affected by a few very extreme cases (as shown for Pearson's r in Figure 7.6d), where, except for these extremes, there may be little or no association (or a very strong association) in the data. Correlations based on small numbers of cases are especially vulnerable to these kinds of aberrations. It is possible too, that an association may appear very weak (or strong) within a certain restricted range of the variables, but be quite different overall, or within other restricted ranges of the same variables. Unless we are particularly interested in some restricted range of values of a variable, it is good practice to be sure that the full range of possible scores is considered. Grouping variables (or truncating) could also have the effect of distorting a measure of association. This was illustrated in Table 6.1, but it applies to Chapter 7 as well.

The forms of the distributions of the variables being correlated can have profound effects upon the magnitude which a correlation coefficient may take. For example, the maximum values which may be attained by Pearson's r depend upon the relative forms of the distributions of the independent and dependent variables. Pearson's r only has a potential range from -1.00 to $+1.00$ when both variables X and Y are symmetrical and have the same form. They don't necessarily have to be normally distributed; however, whatever form the symmetry of their distributions assumes, it must be the same for both. If the distributions of X and Y are both skewed in the same direction and both have the same degree of skewness, then r can reach a value of $+1.00$ but it cannot reach a value of -1.00. If the distributions of X and Y are skewed in opposite directions and both have the same degree of skewness, then r can reach a value of -1.00 but it cannot reach a value of $+1.00$ (Edwards, 1976:54–56).

Finally, the reliability of the measurements used for X and Y has an in-

---

*Chapter 6 of Edwards (1976) is a particularly good discussion of factors affecting correlation coefficients.

fluence on the magnitude a correlation will reach. In general, unreliable measurements will depress the magnitude of a correlation coefficient. Consequently, if a correlation is weak, it may be because of poor measurement instruments rather than a lack of relationship between the variables being studied.

7.8   SUMMARY

In this chapter we have developed a series of measures of degree of association for nominal, ordinal, and interval-level variables. In each case our concern was with developing a measure which could be interpreted as an indication of the proportionate reduction in predictive errors to be made by using information about an independent variable, as opposed to using only information about the dependent variable.

We could predict modes, either overall or within categories, for an independent variable. We could predict category placement of cases. We could predict rank orderings, either by chance, or by using a rule based on the direction of association. We could develop an equation to compute a predicted score and compare that prediction with the overall mean of the dependent variable. In each case the association measure amounted to a version of a ratio that expresses the reduction in errors made possible by improved information, as against the total of possible errors that could be accurately predicted.

These measures do not exhaust the range of possible coefficients, and the student may well be able to devise a new coefficient himself (herself). The measures do have utility in sociology, however, because they permit comparisons between tables which would otherwise be difficult to make. By virtue of proper norming and proper selection of measures of association, an investigator is able to raise questions about how well his (her) theory permits him (her) to explain and predict some phenomenon of interest to him (her).

In addition to the theme of the ratio as a means by which many different kinds of valid comparisons are possible, there has been another theme in this chapter. It points back to the original ideas an investigator has in mind in examining data. Many of the differences in the array of coefficients we have presented are the result of the different kinds of information a collection of data may contain, which may be relevant to some problems and not to others. The difference between asymmetric and symmetric measures, for example, reflects the difference in interest in how the two variables go together, versus how well one variable permits us to predict or explain variation in a dependent variable. We would select between measures according to the nature of the association and, in fact, according to what we want to treat as a perfect association. Clear formulation of a research problem will usually permit a clear selection of the most appropriate coefficient.

Up to this point we have dealt with univariate distributions one by one, and then we compared central tendency and variation, for example, between univariate distributions. Then we packaged a series of these univariate distributions into a single table and developed an overall series of measures on the table itself as a whole. We are about to do this same type of thing again. We have discussed single bivariate distributions. Now we will begin to examine sets of bivariate distributions—one for each special condition or value of a further set of control variables. Then we will again ask whether there is some way to put all of these separate tables together into some overall comparison. At each step, we selectively ignore the more irrelevant information contained in data, and we focus an investigative light on those features which contain the comparative information which gives us the answers to the questions we ask.

CONCEPTS TO KNOW AND UNDERSTAND

PRE measures
regression equation
  intercept
  slope
  b-coefficient
  standardized regression equation
  scatter diagram
linear relationship
curvilinear relationship
nature of association
least squares criterion
standard error of estimate
explained variation
unexplained variation
total variation
lambda
asymmetric, symmetric relationships
concordant, discordant pairs

Kendall's tau-*a*
Kendall's tau-*b*
Somers' *d*
Goodman and Kruskal's tau-*y*
Gamma
Pearson's product-moment
  correlation *r*
Eta
Spearman's rho
correlation matrix
selecting appropriate measures
cautions
ecological fallacy
causation vs association
built-in correlations
homoscedasticity
coefficient of determination

PROBLEMS AND QUESTIONS

1. Compute Goodman and Kruskal's tau-*y* for Table 7.2, using "type of household" as the dependent variable and "marital status" as the independent variable. Discuss differences between this computation and the tau-*y* computed in the text and the two lambda's given in the text. What kinds of different information do each of these coefficients provide? Under what conditions would they be quite different from each other?

2. Develop an analytic organization of measures of association. As a start you might use the guides given in the text and consider the points mentioned in the section on selecting measures of association.

3. Find an interesting pair of scores measured on about 70 cases and compute correlation and regression coefficients. Create a scatter diagram, draw in the regression lines, and interpret the statistics you have computed.

4. Using tables A and B from problem 5 of Chapter 6, compute Gamma, Tau-*b*, and Somers' d_{yx}. Compare the three measures. To what do you attribute the differences among them? How was each influenced by the skewed marginal distributions of the tables? Why were they influenced to different extents? Do

these measures of association alter in any way the interpretation of the tables which you based upon percentages of Chapter 6?

5. The table below is from a longitudinal study of white males. The men in this sample were asked when they expected to retire. Later during the same year, after they had actually left the labor force, the reason for their leaving was determined. Using the appropriate measure of association determine whether there is a relationship between their expected retirement age and the reason for leaving the labor force. What is your conclusion?

| | Expected Retirement Age | |
	Less Than 65	65 and Over
Reason for Leaving Job		
Involuntary	10	22
Health	19	51
Retirement	41	38
Other voluntary reason	4	18

Source: *The Pre-Retirement Years Volume 4,* Table 5.13, p. 182.

6. Use an alternate measure of association to measure the relationship between the variables in problem 5. How does this measure compare with the first one you used? How do you explain the difference?

7. Compute Spearman's rho (r_s) and Pearson's Product-Moment r for the following data. How do the two measures compare? To what can the difference be attributed?

TOTAL HOSPITAL BEDS AND TOTAL ELDERLY POPULATION IN NEBRASKA, 1970

Planning and Service Areas	Population 65 and Over X	Hospital Beds Y
A	38,675	3,658
B	16,737	1,069
C	10,695	177
D	10,998	310
E	13,482	381
F	9,957	389
G	17,421	643
H	7,481	381
I	16,014	623
J	6,351	382
K	6,925	275
L	11,902	664
M	4,763	170
N	5,832	198
O	6,323	266

Source: *Social Statistics for the Elderly: State Level System Users Manual,* Table 11.

8. The data below from 52 Standard Metropolitan Statistical Areas are percentages of occupied dwelling units with 1.01 or more persons per room and murder and manslaughter rates (per 100,000 population). Compute r, b_{yx} and a_{yx} for these data. Construct a scatter diagram for the data and plot the regression line. How strong is the correlation? How may it be interpreted?

DENSITY OF DWELLING UNITS AND MURDER AND MANSLAUGHTER RATES FOR STANDARD METROPOLITAN STATISTICAL AREAS WITH 200,000 POPULATION OR MORE IN 1970

Occupied Units with 1.01 or More Persons per Room Percentages	Murder & Nonnegligent Manslaughter Rates (per 1000,000)
5.2	8.3
4.3	3.2
11.5	9.7
4.0	3.8
6.1	3.7
6.1	7.3
8.1	1.4
8.3	20.8
11.0	16.5
9.7	11.1
11.3	11.4
7.1	16.8
12.7	13.9
10.1	12.9
5.1	2.0
9.6	17.7
5.8	5.6
5.8	3.3
5.3	6.0
6.2	5.3
7.7	5.0
11.7	15.8
9.0	18.0
8.9	16.2
8.2	15.9
9.3	7.8
5.5	17.2
6.6	5.6
9.2	19.5
11.4	21.4
6.0	8.0
17.1	13.5
8.5	14.5
6.9	3.0
6.1	13.6
5.4	7.4
6.1	4.4
7.8	20.2
7.5	3.5

(continued)

DENSITY OF DWELLING UNITS AND MURDER AND MANSLAUGHTER RATES FOR STANDARD METROPOLITAN STATISTICAL AREAS WITH 200,000 POPULATION OR MORE IN 1970

Occupied Unites with 1.01 or More Persons per Room Percentages	Murder & Nonnegligent Manslaughter Rates (per 1000,000)
18.6	5.5
6.0	3.0
5.9	3.0
8.8	8.2
11.3	15.1
8.7	13.6
7.1	15.6
6.3	4.8
11.1	12.5
12.1	20.7
6.2	3.8
8.1	9.6
9.4	16.3

Source: *Statistical Abstract of the United States 1975*, p. 886ff.

GENERAL REFERENCES

Blalock, Hubert M., Jr., *Social Statistics,* Revised Edition (New York, McGraw-Hill), 1979.

Coleman, James S., *Introduction to Mathematical Sociology,* (New York, The Free Press of Glencoe), 1964.
See especially Chapters 6 and 7.

Costner, Herbert L., (ed.), *Sociological Methodology: 1971,* (San Francisco, Jossey-Bass), 1971.
See especially the chapter (10) by Robert K. Leik and Walter R. Gove, "Integrated Approach to Measuring Association," which presents a more detailed treatment of the subject.

Freeman, Linton C., *Elementary Applied Statistics for Students in Behavioral Science* (New York, John Wiley and Sons), 1965
See especially Section C.

Mueller, John H., Karl F. Schuessler, and Herbert L. Costner, *Statistical Reasoning In Sociology,* Third Edition (Boston, Houghton Mifflin), 1977.
See especially Chapters 9, 10, and 11.

OTHER REFERENCES

Coleman, James S., 1964; see General References.

Costner, Herbert L. "Criteria for Measures of Association," *American Sociological Review,* 30 (June, 1965) p 341–353.

Edwards, Allen L., *An Introduction to Linear Regression and Correlation* (San Francisco: W. H. Freeman and Company), 1976.

Furstenberg, Frank F., Jr., "The Transmission of Mobility Orientation in the Family," *Social Forces* 49 (June, 1971) p 598.

Hammond, Kenneth R., and James E. Householder, *Introduction to the Statistical Method,* (New York, Alfred A. Knopf), 1962, p 212–214.

Haug, Marie R., and Marvin B. Sussman, "The Indiscriminate State of Social Class Measurement," *Social Forces,* 49 (June, 1971) p 559.

Hunter, Albert, "The Ecology of Chicago: Persistence and Change, 1930–1960," *American Journal of Sociology,* 77 (November, 1971) p 425–444.

Kim, Jae-On, "Predictive Measures of Ordinal Association," *American Journal of Sociology,* 76 (March, 1971) p 891–907.

Marsh, John F., Jr. and Frank P. Stafford, "The Effects of Values on Pecuniary Behavior: The Case of Academicians," *American Sociological Review,* 32 (October, 1967) p 743.

Riley, Matilda White, *Sociological Research: A Case Approach,* vol. 1 (New York, Harcourt, Brace and World), 1963. See especially the commentary in unit 12.

Robinson, W. S., "Ecological Correlation and Behavior of Individuals," *American Sociological Review,* 15 (1950) p 351–357.

Rossi, Alice, "Status of Women in Graduate Departments of Sociology, 1968–1969," *American Sociologist,* 5 (February, 1970) p 4.

Social Statistics for the Elderly: State Level System Users Manual, U.S. Bureau of the Census (Washington, D.C.), 1975.

Somers, Robert H., "A New Asymmetric Measure of Association for Ordinal Variables," *American Sociological Review,* 27 (December, 1962), p 799–811. See also original work by Kendall, Kruskal, and Goodman and Kruskal, cited in Somers' article.

Statistical Abstract of the United States 1975, U.S. Bureau of the Census (Washington, D.C.), 1975.

The Pre-Retirement Years, Volume 4, Manpower R&D Monograph 15, U.S. Department of Labor (Washington, D.C.), 1975.

U.S. Bureau of the Census, "Household Income in 1970" and "Selected Social and Economic Characteristics of Households," *Current Population Reports,* P-60, No. 79, (Washington, D.C.), 1971, p 17, Table 6.

————"Income in 1970 of Families and Persons in the United States," *Current Population Reports,* P-60, No. 80, (Washington, D.C.), 1971, Tables 1 and 19.

U.S. Department of Health, Education and Welfare, Public Health Service, "Children and Youth: Selected Health Characteristics," Series 10, Number 62, (Washington, D.C.), 1971, p 15, Table 3.

Yoels, William C., "Destiny or Dynasty: Doctoral Origins and Appointment Patterns of Editors of the *American Sociological Review,* 1948–1968," *American Sociologist,* 6 (May, 1971) p 135.

Descriptive Statistics: Three or More Variables

Elaborating the Relationship between Two Variables

8

8.1 EXAMINING RELATIONSHIPS BETWEEN VARIABLES

Up to this point we have been concerned with the relationship between two variables, often an independent and a dependent variable. The last chapter dealt with a variety of measures of the *nature* and *strength* of their relationship, and these measures were developed so that comparisons could be made either between bivariate associations, in studies involving different sub-groups, or between associations in studies involving different variables. In this chapter we will begin to introduce more systematic differences between tables whose coefficients we want to compare. In fact, we will begin to define sub-populations for which the basic relationship between the same independent and dependent variables can be examined from sub-population to sub-population. Careful comparison will permit us to draw conclusions about the effects of other variables on the *relationship* between the original pair of variables. Let us begin with an example.

8.1.1 Morale and Interaction—An Illustration of Elaboration

What happens to a person's involvement in society as he (she) ages? Does he (she) continue to be engaged pretty much the way he (she) always has been in the round of social activities, perhaps substituting more leisurely for more strenuous pursuits; or does a person typically disengage from society in the normal sequence of his (her) aging? One theory holds that people disengage from society as they enter old age, and that this is a mutually satisfactory process both for the individual and for society. Such decreases in social interaction would have a positive or at least a minimal impact upon his (her) level of morale or satisfaction. An alternative viewpoint is that individuals attempt to continue to be engaged as they were in middle age, although the form of social engagement may shift. In this view, declines in social interaction have the effect of decreasing the level of morale and satisfaction of an individual. Both lines of reasoning make predictions for the relationship between morale and degree of social interaction for people moving toward the end of life, and data supporting each viewpoint have been presented.

Mark Messer addressed himself to this problem and tried to examine the relationship between morale and interaction in somewhat greater detail (Messer, 1967). He reasoned on the basis of other research that the age concentration of the person's environment is a crucial factor which also has to be considered (i.e., the extent to which older people live only with other older people or with people of a mix of different ages). There is reason to believe, for example, that social interaction rates are higher when a person is in an age-homogeneous environment. This may be because people prefer to interact with age peers. It has been argued that age grouping (especially in the socialization process of young people) helps support transitions from one role to another by insulating an individual from some of his (her) other roles, which may be inconsistent with the new role, until the problems and strains of his (her) role-transition are settled. If this is the case, Messer argues, high morale among those in an age-concentrated environment should be a function of the normative system that this concentration permits rather than a function of the social interaction of an individual. In a mixed-age environment, on the other hand, morale should depend upon a higher rate of social interaction.

To examine his ideas, Messer contrasted 88 older people living in the age-concentrated public housing project for the elderly in Chicago as against 155 elderly people living in a mixed-age public housing project. In each setting, however, he was interested in the relationship between morale and the degree of social interaction, and he expected a stronger positive association in the mixed-age setting. Table 8.1 presents his results.

Interestingly, it does appear that age concentration of the environment has an effect upon the relationship between morale and interaction. In fact, the percentaged tables show a very low and negative relationship between morale and interaction in the age-concentrated environment (Table 8.1b), but the two variables are positively associated in the mixed-age environment. His hypothesis about the effect of age-concentration seems to be confirmed.* The test of his ideas involved an examination of a relationship between two variables within the categories of a third variable. This leads to an important modification and possible linking of the two initial theories.

This kind of examination is called **elaboration** because a basic relationship of interest is examined under a variety of different conditions. In this instance, the introduction of the third variable into the analysis helped *specify*

---

*Messer's hypothesis was that social interaction is causally prior to morale, and he treats his data in these terms. It is possible, however, that morale may have some influences upon future interaction. The importance of theoretical ordering of variables for the way one conducts research will be emphasized later in this chapter in Section 8.2.3.

TABLE 8.1 Percent Distribution of Morale for Elderly in Public Housing

(a) PERCENT DISTRIBUTION OF MORALE BY SOCIAL INTERACTION

| (Y) | Interaction (X) | | |
Morale	Low	High	Total
High	14.3	28.2	21.4
Medium	58.0	50.0	53.9
Low	27.7	21.8	24.7
	100.0	100.0	100.0
	(119)	(124)	(243)

$$d_{yx} = .15$$

(b) PERCENT DISTRIBUTION OF MORALE BY SOCIAL INTERACTION AND AGE OF ENVIRONMENT

(T-1) AGE-CONCENTRATED ENVIRONMENT				(T-2) MIXED-AGE ENVIRONMENT			
(Y)	Interaction (X)			(Y)	Interaction (X)		
Morale	Low	High	Total	Morale	Low	High	Total
High	26.7	25.9	26.1	High	10.1	30.3	18.7
Medium	60.0	50.0	53.4	Medium	57.3	50.0	54.2
Low	13.3	24.1	20.5	Low	32.6	19.7	27.1
	100.0	100.0	100.0		100.0	100.0	100.0
	(30)	(58)	(88)		(89)	(66)	(155)
	$d_{yx} = -.09$				$d_{yx} = +.25$		

Source: Messer, 1967. Used by permission of the publisher.

conditions under which the relationship would be strong or weak. This kind of examination is not at all uncommon in sociology, and the purpose of this chapter is to show some of these patterns of relationship among three or more variables, to look at some of the ways they might be summarized, and to investigate how the different patterns might be interpreted. Let us start with some of the basic terminology.

8.1.2 Total Tables and Conditional Tables

The overall association between two variables is called a *total association,* or a **zero-order association.** Table 8.1a is an example of such a total association. Total tables include, as *N,* all of the available cases to be examined in terms of the pair of variables. As in the example above, however, analysis of an overall bivariate relationship is usually aided if the cases are divided up into sub-groups corresponding to categories of an additional variable. Tables showing the association between two variables within categories of other variables are called **conditional tables,** or conditional associations.* Table 8.1b above shows two conditional

---

*In some texts the "conditional" table is called a "partial" table and the process is called "partialling." We will reserve the term "partial" for the summary coefficients used later in this chapter, which express the relationship between X and Y variables where the effect of a control variable (or variables) has been statistically removed.

BOX 8.1 BASIC IDEAS

Before you get into this chapter very far you should check yourself on the following ideas.

— association of two variables (Section 6.4)
— independent, dependent, and control variables (Section 2.1.4)
— measures of association (Sections 7.2 – 7.7)
— the idea of setting up important contrasts (Sections 3.1 and 3.2)

This chapter has two main parts. The first part deals with comparisons of contrasting associations, that is, associations in several bivariate tables. The second main part concerns ways to make a statistical summary over a set of bivariate tables. In between, topics on experimental design and theory are introduced. These topics are central to an understanding of where and why we would want to use elaboration, standardization, or partial correlation techniques in examining the relationship between three or more variables.

---

tables. The additional variable that is introduced into the analysis as a basic split in the group of cases is called a "control" variable or "test" variable. We are interested in examining the basic relationship between two variables (X and Y) within categories of control variable(s) – (T). Each conditional table includes only cases which have the same value (or range of values within a category) on that control variable, and there is a different conditional table for each category of the control variable(s). This procedure "controls" or "takes out" variation in that additional variable, permitting a clear focus on the relationship between the original independent and dependent variables within any given table.

The *terminology* goes as follows:

zero-order   *No* control variables (*i.e.* no "test" factors). The basic relationship between two variables is examined *overall*.

first-order conditional tables   *One* control variable. The basic $X - Y$ relationship is examined within each category of the control variable.

second-order conditional tables   *Two* control variables. The basic $X - Y$ relationship is examined within categories created by all possible combinations of the categories of the *two* control variables.

third-order conditional tables   *Three* control variables. The basic $X - Y$ relationship is examined within categories created by all possible combinations of the categories of the *three* control variables.

etc.

TABLE 8.2 OTHER TABLES MADE FROM THE FIRST ORDER CONDITIONAL TABLES SHOWN IN TABLE 8.1, ABOVE.

(a) FREQUENCY DISTRIBUTION OF MORALE BY INTERACTION AND AGE ENVIRONMENT. Taken from Table 8.1, above.

AGE-CONCENTRATED ENVIRONMENT

(Y) Morale	Interaction–(X) Low	High	Total
High	8	15	23
Medium	18	29	47
Low	4	14	18
Totals	30	58	88

$d_{yx} = -.09$

MIXED-AGE ENVIRONMENT

(Y) Morale	Interaction–(X) Low	High	Total
High	9	20	29
Medium	51	33	84
Low	29	13	42
Totals	89	66	155

$d_{yx} = +.25$

(b) FREQUENCY DISTRIBUTION OF MORALE BY INTERACTION.

(Y) Morale	Interaction–(X) Low	High	Total
High	17	35	52
Medium	69	62	131
Low	33	27	60
Total	119	124	243

$d_{yx} = +.15$

(c) FREQUENCY DISTRIBUTION OF MORALE BY AGE ENVIRONMENT.

(Y) Morale	Age-Environment–(T) Concentrated	Mixed	Total
High	23	29	52
Medium	47	84	131
Low	18	42	60
Total	88	155	243

$d_{yt} = +.12$

A second-order conditional table would be one in which, for example, morale and interaction (the basic relationship of interest in Table 8.1) were examined in each category of age-concentration of environment and in public *vs.* private-owned categories of housing. The basic table would then be examined separately for these four categories: (a) age-concentrated public housing, (b) age-concentrated private housing, (c) mixed-age public and (d) mixed-age private housing.* Thus there would be only four tables to compare, since there are only two categories of each control variable.

In general, a statistical analysis of the relationship between two variables

---

*The four conditional tables in this example could be increased if the control variables had more categories than two each. If one has three categories and the other has four, then there would be $3 \times 4 = 12$ conditional tables to examine, and all 12 tables would be second-order conditional tables because they

TABLE 8.2 *(Continued)*

(d) FREQUENCY DISTRIBUTION OF INTERACTION BY AGE ENVIRONMENT.

(X) Interaction	Age Environment–(T)		
	Concentrated	Mixed	Total
High	58	66	124
Low	30	89	119
Total	88	155	243

$$d_{xt}=+.23$$

(e) FREQUENCY DISTRIBUTION OF MORALE BY AGE ENVIRONMENT CONTROLLING FOR INTERACTION.

(Y) Morale	Interaction–(X)			
	Low Age Environment (T)		High Age Environment (T)	
	Concentrated	Mixed	Concentrated	Mixed
High	8	9	15	20
Medium	18	51	29	33
Low	4	29	14	13
	30	89	58	66
	$d_{yt}=+.28$		$d_{yt}=-.07$	

(f) FREQUENCY DISTRIBUTION OF INTERACTION BY AGE ENVIRONMENT CONTROLLING FOR MORALE.

(X) Interaction	Morale–(Y)					
	Low Age Environment (T)		Medium Age Environment (T)		High Age Environment (T)	
	Concentrated	Mixed	Concentrated	Mixed	Concentrated	Mixed
High	14	13	29	33	15	20
Low	4	29	18	51	8	9
	18	42	47	84	23	29
	$d_{xt}=+.46$		$d_{xt}=+.22$		$d_{xt}=-.04$	

Source: Data from Table 8.1 (Messer, 1967).

proceeds by introducing control variables one at a time, starting from the original zero-order table and proceeding to first-order conditional tables and then to second-order conditional tables, etc. All first-order tables are examined, then second-order, etc.

resulted from a combination of two control variables. In general, the number of conditional tables equals the product of the number of categories of all control variables. Obviously, the higher the order of the conditional tables, the more exacting and refined the analysis, and also the more cases one needs to include in a study to "fill up" the conditional tables. It is a general practice not to compute percentages where the base of the percentage is less than, say, 15 or 20, because such percentages would ordinarily be too unreliable (a topic for consideration in inferential statistics). Where many control variables are used, alternative ways to handle the problem which require fewer cases are usually used. Some of these will be discussed later in this chapter and in the next.

As Table 8.2 illustrates, a number of different total and conditional associations can be created from the basic information given in higher order conditional tables in Table 8.1b (but one cannot create higher order tables out of lower order tables). For example, from the first order conditionals shown in Table 8.1b and reproduced in Table 8.2 as a frequency distribution, we could create three total associations: (a) one between morale and the degree of social interaction, (b) another between age concentration of environment and morale, and (c) a third between age concentration of environment and degree of social interaction. Notice that these total tables are created out of the marginals of the first order conditional tables or, in the case of Table 8.2b, by adding together the two conditional tables cell by cell.

Table 8.2c is created by simply taking the row marginals for the age-concentrated and for the age-mixed environments in Table 8.2a and using these as the frequencies in columns of Table 8.2c. All 243 cases are classified in this total table. Table 8.2d is made up from the column marginals of Table 8.2a and again, all 243 cases appear in this total table.

Table 8.2e is another rearrangement of Table 8.2a. In this case, the control variable is social interaction and the basic relationship under examination is morale and age-concentration. In Table 8.2f, the basic relationship of interaction and age homogeneity of environment is examined within categories of the morale variable. Although 8.2e and 8.2f could be created from these same data, they would not be of central interest to Messer because they do not permit him to focus readily upon the relationship of morale and social interaction, the two key variables in Messer's study. These examples do begin to illustrate, however, the variety of ways a set of data may be examined and, hopefully, the importance of being clear about what it is that we are interested in studying.

8.1.3 The Role of Theory

What other variables may help explain the relationship between morale and interaction? Probably income, sex, cultural background, age, experience in other kinds of interaction settings, whether one is married or not, etc. The list could be rather long. Just which variables to pick is a serious problem and largely unsolvable without a heavy reliance upon prior theoretical work and results of research aimed at explanation of phenomena in a certain area of interest. Clearly, "toe-nail length" would be ruled out immediately as a general explanation of the relationship of morale and interaction. The role of some of the other variables mentioned above is not as clear-cut, and careful theoretical work would point the way toward finding significant variables. In Messer's research he was able to draw upon three kinds of previous theory, the disengagement theory, the activity theory, and ideas which relate age homogeneity to satisfaction and morale. An understanding of these ideas played a critical role in specifying important variables to control and the way results should be interpreted. Without this kind of starting point, an investigator would hardly be able to pick a relevant independent variable, let alone select those which are relevant from an infinity of variables which could be measured and used as control conditions. Even in very small studies, the possible combinations of variables becomes very large indeed, and as purely a practical matter, theoretical guidance in selection of relationships and control conditions becomes important.*

Research which is able to contrast two theories purporting to explain the same phenomenon is particularly fruitful for building scientific knowledge. Usually, however, research leads to an extension of a theory, to an addition of some condition under which the theory is found to lead to somewhat different predictions, or to the refinement of measurements and of "scope conditions" of a theory (*i.e.* the class of phenomena to which the theory pertains). Much research is of an exploratory nature (as sharply opposed to "sloppy" or unplanned investigation), especially when explicit theories have not been developed.† But whether theories are explicit or not, investigators make use of theoretical knowledge in selecting variables and planning their analysis. The more clearly set out these ideas are, the more fruitful the analysis.

The **theoretical order** of variables is important in interpreting the outcome of the process of elaboration. An independent variable, if it is thought of theoretically as a causal variable, comes before the dependent variable (*i.e.*, it changes first) in any time sequence. The "test factor" or control variable, however, may have its effect at different points in time relative to the independent and dependent variables. If a test factor has its effect *before* both independent *and* dependent variables, it is called an **antecedent** test factor. Here is an example:

SEX	SOCIAL INTEGRATION	SUICIDE
(T) ⟶	(X) ⟶	(Y)
Antecedent Variable	Independent Variable	Dependent Variable

If it has its effect *after* the independent *and* dependent variables, it is called a **consequent** test factor, as in this example:

NO. OF DELINQUENT FRIENDS	NO. OF DELINQUENT ACTS SELF-REPORTED	NUMBER OF TIMES ARRESTED
(X) ⟶	(Y) ⟶	(T)
Independent Variable	Dependent Variable	Consequent Variable

If the test factor has its effect *after* the independent variable but before the dependent variable, it is called an **intervening variable** or intervening test factor as below:

NO. OF DELINQUENT FRIENDS	PARENTAL SUPERVISION	SELF-REPORTED NO. OF DELINQUENT ACTS
(X) ⟶	(T) ⟶	(Y)
Independent Variable	Intervening Variable	Dependent Variable

The basic dependent variable of interest will be referred to as variable-Y in this chapter. Variable-X will be the main independent variable of interest, and vari-

---

*For example, in a modest study with only 5 variables, each having 5 categories, there are 10 zero-order tables, 150 first-order conditional tables, 1500 second-order tables, etc. Needless to say, typical research involves many more variables than five, and the problem quickly becomes one of avoiding irrelevant tables and selecting relevant ones.

†There are a number of excellent discussions on theory, on theory construction, and on the role of research. See especially Reynolds, 1971, Blalock, 1969, and Stinchcombe, 1968.

able-T will be used for the control or "test" variable. Where it is important to specify the category of the test variable, we will use subscripts such as: T_1, T_2, etc. The relationship between Y and X within category T_1, then, may be symbolized as: $YX{:}T_1$.

Drawing an arrow diagram of the relationships between variables included in a study is often a useful device for clarifying both the theory and the kinds of statistical analysis needed to answer a research question. For example, in a study of the relationship of organizational involvement and political participation, William Erbe (1964) surveyed the available research literature which suggested that two other variables were relevant. The theoretical relationships between variables used to guide his study were drawn out as follows:

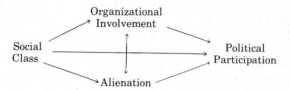

The above model immediately indicates the ordering of the variables, shows which are intervening, and suggests the apparent necessity of taking account of both social class and alienation in examining the "involvement-political participation" relationship.

As we shall see, the theoretical ordering of variables also has a key role to play in the interpretation of results. Virtually identical statistical results may be interpreted as evidence that an hypothesized causal link is spurious, that there is evidence of a causal sequence, or evidence of independent influence, depending upon the ordering of variables.

In many cases, of course, the time order of variables is established simply by the logic of the measurement or the type of variable. If there is any causal relationship between ethnic group and attitudes, the group membership probably is the causal factor. On the other hand, many variables are not clearly time-ordered, especially when the investigator takes measurements of all variables at the same point in time. Where this confusion of time order is likely to occur, an investigator would normally attempt to gather specific information about time order for the effects of the variables he (she) wants to study.

Time order could be established by gathering data at two or more points in time. In voting studies, for example, an investigator may create a "panel," or sample of individuals, who are questioned each month for the two or three months before an election. Measures of attitudes, for example, at "time-point one" could be related to attitudes at "time-point two" and the time order would be clear. This is called a **panel study** design. Other **longitudinal** designs are used, for example, in the study of families or aging individuals over time. Sometimes it is sufficient to ask individuals to report on past events or to use records to establish the time order of variables.*

The centrality of theory as a guide to analysis (and research as a guide to

---

*Additional information on panel studies and ways they may be statistically examined may be found in Riley, 1963: v. I, 559, in Zeisel, 1957: Ch. 10, and in Lazarsfeld and Rosenberg, 1955:231.

development of theory) will become more evident as we proceed through this examination of statistical elaboration.*

8.1.4 Analysis of Conditional and Total Associations

The statistical analysis of *each* conditional table does not differ in any important respect from the analysis one might do on a total association between two variables. There is the full range of measures of association discussed in Chapter 7 from which to choose, and selection depends upon the character of the variables and the problem at hand. In addition to strength and direction of association, one should examine the *nature* of each association. The only difference in dealing with conditional tables is that comparison between conditional tables takes on a more systematic character than it might otherwise. Tables shown above could be examined in terms of, say, Somers' d_{yx} and percentages. In Table 8.1b the association in the two conditional tables is $d_{yx} = -.09$ in the age-concentrated environment and $d_{yx} = +.25$ in the mixed-age environment. The stronger association between these two variables exists in only one of the two conditions — namely the age-heterogeneous condition — an outcome concerning the relationship between morale and interaction which Messer's logic had predicted.

The relationship in the conditional tables, however, is only part of the story. The more complete story would include the overall relationship in the three total associations (Tables 8.2b, 8.2c, and 8.2d) as a further basis for interpreting these data. Notice, for example, that the original zero-order association between interaction and morale (Table 8.2b) is positive but relatively weak ($d_{yx} = +.15$). From a look at the conditional tables (Table 2a), it is clear that part of the reason for this overall weakness is that a weak negative association in one conditional (*i.e.* the age-concentrated group) is thrown together with a stronger positive association in the other conditional table (*i.e.* the mixed-age group), and they tend to cancel each other out (in Table 8.2b) to some extent.

Notice, too, that the control variable, T, is positively related to each of the main variables, morale and interaction. The Somers' d for the table of age environment and morale (Table 8.2c) is $d_{yx} = +.12$, and the association of age environment and interaction (Table 8.2d) is $d_{yx} = +.23$. Part of the association observed between morale and interaction is due to the correlation of each of these two variables with the test factor. These outcomes are organized, below.

$\underline{d_{yx} = +.15}$ *Original total association* between morale and interaction is a result of these:

               Conditional Associations:

$d_{yx} = -.09$ a) The association of morale and interaction in the age-concentrated environment

$d_{yx} = +.25$ b) The association of morale and interaction in the mixed-age environment

               Total Associations With the Test Factor:

$d_{yx} = +.12$ c) The association of morale with the test factor, age environment

$d_{yx} = +.23$ d) The association of interaction with the test factor, age environment

*A classic statement on the bearing of theory on research and vice versa is made by Merton, 1957:pt. I.

8.1.4a The Lazarsfeld Accounting Formula. The relationship between a total association, a set of conditionals, and another set of total associations has been pointed out in the sociological literature by Paul Lazarsfeld, and it is sometimes summarized in what is called the **Lazarsfeld accounting formula.** This equation states that a total association between X and Y can be accounted for by the conditional associations of X and Y within categories of T (test variables), plus total associations of X and Y separately with the test variables. For the three-variable situation where the variables are all dichotomies and where the degree of association is expressed in terms of delta (Δ),* the Lazarsfeld accounting formula is as follows:

$$\Delta\, XY = \Delta\, XY\!:\!T_1 + \Delta\, XY\!:\!T_2 + \frac{N_1 + N_2}{(N_1)(N_2)} (\,\Delta\, XT)(\,\Delta\, YT)$$

If the morale variable in Table 8.2 were grouped into only two categories (by putting the "high" and "medium" categories together, for example) and the expected cell frequencies were computed as discussed in Chapter 6, the resulting Δ values would be as follows:

$$+3.62 = -2.13 + 4.88 + \frac{243}{(88)(155)}(3.73)(13.09)$$

$$3.62 = -2.13 + 4.88 + .87$$

Using Δ on a fourfold table, this situation works out nicely: the total association at the left equals the sum of terms expressed in the equation at the right. In general, however, the weighting factor is not resolved, either for other measures of association or for tables larger than 2×2 tables.† The usefulness of Lazarsfeld's formula holds, however; associations between two variables can be accounted for by a set of conditional and total associations.

An examination of the accounting formula quickly yields a rich set of possible outcomes.‡ An association may or may not exist in the total table which is cross-classifying an independent and a dependent variable. If no association exists, then the right-hand side of the Lazarsfeld formula would have to "balance out" to zero. This might happen if the conditional tables yielded a negative and

---

*Recall from Chapter 6 that delta (Δ) is simply the difference between observed and expected cell frequencies. In a 2×2 table the size of Δ is the same, regardless of which cells are contrasted (although the sign may be different). It is traditional (consistency is necessary for the Lazarsfeld equation to work out) to compute Δ on a 2×2 table by subtracting the expected frequency from the observed frequency for only *one* cell, the high-X, high-Y cell in a table. Delta is not a normed measure of association, so it is generally not used except to illustrate how Lazarsfeld's formula works out numerically.
†A slightly different approach using Yule's Q is described in Davis, 1971:188.
‡In addition to the outcomes which are discussed in the chapter, there are other outcomes which should be noted because they are important, although of a different type than those discussed here. First, it is possible that the observed association between two variables is simply due to *chance*. This would be a possible outcome where the data are gathered by sampling procedures from some population. There is some chance that the sample will misrepresent the population from which it is drawn and show an association between two variables when, in fact, in the population there is none. Secondly, an observed association may be due to the research process itself. If observations are related rather than independent, the observed association could be accounted for largely by whatever research steps led to this relatedness. Generally speaking, an investigator tries to design his research so that observations are, in fact, independent observations.

positive association which, added together in a total table, cancelled each other out. Looking at the right-hand side of the formula, an association may be shown in the total tables, in the conditional tables, or in both. Furthermore, the conditional tables may show the same strength and direction of association, or they may differ among themselves, with some showing a stronger association than other conditional tables, or some showing associations with a different direction or nature than others.

8.1.4b Odds Ratios and Hierarchical Analysis. A system for the analysis of cross-classifications of three or more nominal or ordinal level variables has been developed by Goodman (1965, 1969, 1972).* The complete system is involved and comprehensive and we will not treat it in detail here. We will, however, consider that part which deals with hierarchical analysis. This type of analysis is very useful for sorting out the relationships between dependent, independent, and one or more test variables. A key concept in the system is the concept of an *odds ratio*. An odds ratio is the ratio between the frequencies of two categories of a variable. For example, consider the following frequency table consisting of fictitious data for a dependent (Y) and an independent (X) variable.

		Variable X	
	Low	High	Totals
Variable Y			
High	33	47	80
Low	52	23	75
Totals	85	70	155

Overall, of the 155 cases for which data are available, 80 scored high on variable Y and 75 scored low. Consequently, the odds ratio for variable Y is 80:75 or 80/75 = 1.07. In this case, the frequency for the high scores on Y was divided by the frequency for the low scores. It would have been equally proper to divide the low score frequency by the high (*i.e.* 75/80 = 0.94). To be consistent, whenever the variable is ordinal, we will always divide the frequency for the *higher score* category *by* that for the *lower*. If the variable is nominal, of course, the choice of which category frequency to put in the nominator and which in the denominator is a matter of personal preference. Once the choice is made, however, it should be irrevocable.

An overall odds ratio can also be computed for the X variable as follows: 70:85 or 70/85 = 0.82. Using Davis' (1974) system of notation we will symbolize the odds ratio for Y $\begin{pmatrix} Hi \\ Y \\ Lo \end{pmatrix}$ and the odds ratio for X $\begin{pmatrix} Hi \\ X \\ Lo \end{pmatrix}$. Thus,

$$\begin{pmatrix} Hi \\ Y \\ Lo \end{pmatrix} = 1.07 \quad \text{and} \quad \begin{pmatrix} Hi \\ X \\ Lo \end{pmatrix} = 0.82$$

---

*Davis (1974) has done an excellent job of translating the work of Goodman into terms which do not require mathematical sophistication.

When the frequencies of the two categories which constitute an odds ratio are equal the odds ratio will equal 1.00. As the frequency in the numerator exceeds that in the denominator, the odds ratio will exceed 1.00 and, vice versa. The minimum odds ratio is zero, but there is no defined upper limit to odds ratios. An odds ratio such as 50/0 is undefined since it is not permissible to divide by zero. If such an odds ratio came up in the analysis of a set of data, an adjustment would need to be made.*

Because $\begin{pmatrix} Hi \\ Y \\ Lo \end{pmatrix}$ and $\begin{pmatrix} Hi \\ X \\ Lo \end{pmatrix}$ are odds ratios based upon distributions of frequencies for single variables they are referred to as *first order* odds ratios. If an odds ratio for variable Y is computed taking a particular value of X into consideration, then it is referred to as a first order *conditional* odds ratio of Y. For example, in the table there are two columns of values for variable X, one column in which the value of X is high and another in which the value of X is low. There are two frequencies for variable Y in each of these columns for variable X. When X is high the frequency of high values on variable Y is 47 and the frequency of low values is 23. The odds ratio for these frequencies of Y is $47/23 = 2.04$. This odds ratio is a first order conditional odds ratio and is symbolized as follows:

$$\begin{pmatrix} Hi \\ Y \\ Lo \end{pmatrix} | X = Hi = 2.04\dagger$$

There is a corresponding conditional first order ratio for Y when the value of X is low. It is as follows:

$$\begin{pmatrix} Hi \\ Y \\ Lo \end{pmatrix} | X = Lo = 33/52 = 0.63$$

If they are of interest to us, we can also compute two first order conditional odds ratios for X, one for high values of Y and another for low values.‡ They would be as follows:

$$\begin{pmatrix} Hi \\ X \\ Lo \end{pmatrix} | Y = Hi = 47/33 = 1.42$$

$$\begin{pmatrix} Hi \\ X \\ Lo \end{pmatrix} | Y = Lo = 23/52 = 0.44$$

When an odds ratio is computed from two first order conditional odds ratios, the resulting odds ratio is known as a *relative odds ratio*. For example, we

---

*Davis (1972) suggests adding 1 to each cell of a table in order to circumvent this problem.
†The vertical line in the expression means, "given that." So that part of the expression after the vertical line reads, "given that the value of X is high."
‡Whether we would be interested in these two conditional odds ratios would depend upon the nature of the problem we were studying and the specific comparisons which were relevant to that problem.

could compute a relative odds ratio from the two first order conditional odds ratios for Y as follows:

$$\frac{\left(\begin{array}{c} Hi \\ Y \\ Lo \end{array} \mid X=Hi\right)}{\left(\begin{array}{c} Hi \\ Y \\ Lo \end{array} \mid X=Lo\right)} = \frac{2.04}{0.63} = 3.24$$

This relative odds ratio is also called a *second order odds ratio* and is symbolized as $\left(\begin{array}{cc} Hi & Hi \\ Y & X \\ Lo & Lo \end{array}\right)$.

A relative odds ratio (or second order odds ratio) such as the one just computed gives an indication of the amount of association which exists between two variables. If the second order odds ratio is less than 1.00, the association between the two variables is a negative one. If the odds ratio is greater than 1.00, the association is a positive one. An odds ratio of 1.00 indicates that there is no association between the two variables—they are independent. Notice that the second order odds ratio computed above is 3.24. This ratio indicates that variables X and Y are positively associated. A high value on X is associated with a high value on Y and a low value on X, with a low value on Y.

It should also be noted that second order odds ratios (as well as higher order ones) are symmetrical. That is, the second order odds ratio computed from the two first order conditional odds ratios for X will have the same value as the one just computed from the two first order conditional odds ratios for Y (within errors of rounding). Thus,

$$\frac{\left(\begin{array}{c} Hi \\ X \\ Lo \end{array} \mid Y=Hi\right)}{\left(\begin{array}{c} Hi \\ X \\ Lo \end{array} \mid Y=Lo\right)} = \frac{1.42}{0.44} = 3.23$$

If we were working with data that included a third, test variable (T), it would be possible to extend the notions we have been discussing and compute second order conditional odds ratios such as the following:

$$\left(\begin{array}{cc} Hi & Hi \\ Y & X \\ Lo & Lo \end{array} \mid T=Hi\right) \qquad \text{and}$$

$$\left(\begin{array}{cc} Hi & Hi \\ Y & X \\ Lo & Lo \end{array} \mid T=Lo\right)$$

Furthermore, from these two second order conditional odds ratios a *third order odds ratio* could be computed as follows:

$$\begin{pmatrix} Hi\,Hi\,Hi \\ Y\ \ X\ \ T \\ Lo\,Lo\,Lo \end{pmatrix} = \cfrac{\begin{pmatrix} Hi\,Hi \\ Y\ \ X\ \mid T=Hi \\ Lo\,Lo \end{pmatrix}}{\begin{pmatrix} Hi\,Hi \\ Y\ \ X\ \mid T=Lo \\ Lo\,Lo \end{pmatrix}}$$

Given the availability of more variables, fourth order, fifth order, and higher order odds ratios could be computed.

Third order and higher order odds ratios impart information about interactions between the variables being analyzed. A third order odds ratio which equals 1.00 indicates that there is no interaction among the three variables being analyzed. Consequently, all of the two variable associations will be equal in magnitude for the different categories of the third variable. Thus, the association of X and Y would be the same regardless of whether T were high or low. Also, the association of Y and T would be the same regardless of whether X were high or low. And finally, the association of X and T would be the same regardless of whether Y were high or low. It would not mean, however, that all of the two variable associations, YX, YT, and XT, would have the same strength or that any of them would necessarily depart from independence.

A third order odds ratio that departs from 1.00 indicates that each of the three possible two variable associations will differ in magnitude according to the value of the third variable. For example, the magnitudes of the relationships between variables Y and X will be different depending upon the value of variable T. Thus, the relationship between Y and X might be positive when the value of T is high, but negative when the value of T is low; or the relationship between Y and X might be negative when the value of T is high and positive when it is low. It is also possible, however, for the relationship between Y and X to be positive whether T is high or low, but the magnitude of the relationship would be greater for one condition of T than for the other. Finally, the relationship between Y and X could be negative regardless of the value of T, but the magnitude of the relationship would vary depending upon whether T was high or low.

Since third order odds ratios like second order odds ratios are symmetrical, the relationship between variables Y and T would vary according to the value of X and the relationship between variables X and T would vary according to the value of Y.

An odds ratio might have more than one test or control variable. Furthermore, this is true for odds ratios at any level. Consequently, a first order odds ratio could be conditional upon the values of two or more test variables, as could a second order or a third order ratio. A second order odds ratio conditional upon the values of two test variables would be symbolized as follows:

$$\begin{pmatrix} Hi\,Hi \\ Y\ \ X\ \mid T=Hi \text{ and } Z=Hi \\ Lo\,Lo \end{pmatrix}$$

There would be four such second order conditional odds ratios with two test variables with dichotomous categories. The one presented above, one in which the value of T would be high and the value of Z would be low, one in which the value of T would be low and the value of Z would be high, and one in which the values of both T and Z would be low. Of course, if there were more than two categories for variables T and Z, there would be more possible odds ratios.

Comparison of these conditional odds ratios makes it possible to sort out the relationships between the dependent variable, the independent variable, and the test variables. Analysis of cross-classifications of nominal or ordinal variables through the use of odds ratios constitutes a hierarchical type of analysis because of the different levels or orders that odds ratios may take and because of the interpretations that can be made about the relationships between two or more variables from an examination of these various levels.

To demonstrate the usefulness of this type of hierarchical analysis we have computed various odds ratios for Messer's data as they were presented in frequency form in Table 8.2. First, we computed first order odds ratios for the dependent variable, morale; the independent variable, interaction; and the test variable, age environment. Since there are three categories of morale represented in the data, it is possible to compute three first order odds ratios for morale. Taking the marginal totals for morale (variable Y) in Table 8.2b in pairs, we can compute the following three odds ratios:

$$\left(\begin{array}{c} Hi \\ Y \\ Med \end{array}\right) = 52/131 = 0.40$$

$$\left(\begin{array}{c} Med \\ Y \\ Lo \end{array}\right) = 131/60 = 2.18$$

$$\left(\begin{array}{c} Hi \\ Y \\ Lo \end{array}\right) = 52/60 = 0.87$$

Notice that two of the three odds ratios for morale are below 1.00 while the third is well above. These values result largely from the piling up of frequencies in the medium morale category. In the case of the high versus low odds ratio the ratio is below 1.00 because the high morale cases are slightly outnumbered by the low morale cases. The discrepancy is not large, however.

The first order odds ratio for social interaction (variable X) is as follows:

$$\left(\begin{array}{c} Hi \\ X \\ Lo \end{array}\right) = 124/119 = 1.04$$

There is only one of these since social interaction is a dichotomous variable. This one happens to be very close to 1.00.

The first order odds ratio for age environment (variable T) is as follows:

$$\left(\begin{array}{c} C \\ T \\ M \end{array}\right) = 88/155 = 0.57$$

where C represents the age-concentrated environment and M represents the mixed-age environment.

In this case, those in the mixed-age environment outnumber those in the age-concentrated environment by a sizeable margin, thus the odds ratio is well below 1.00.

From Table 8.2b we also computed first order odds ratios for variable Y, conditional upon the values of variable X. Since variable Y takes three values, there are six ratios. They are as follows:

$$\begin{pmatrix} Hi \\ Y \\ Med \end{pmatrix} X = Hi = 35/62 = 0.56$$

$$\begin{pmatrix} Hi \\ Y \\ Med \end{pmatrix} X = Lo = 17/69 = 0.25$$

$$\begin{pmatrix} Med \\ Y \\ Lo \end{pmatrix} X = Hi = 62/27 = 2.30$$

$$\begin{pmatrix} Med \\ Y \\ Lo \end{pmatrix} X = Lo = 69/33 = 2.09$$

$$\begin{pmatrix} Hi \\ Y \\ Lo \end{pmatrix} X = Hi = 35/27 = 1.30$$

$$\begin{pmatrix} Hi \\ Y \\ Lo \end{pmatrix} X = Lo = 17/33 = 0.52$$

Note that the values of variable Y are compared in pairs again: high values against medium values, medium values against low values, and high values against low values. These, by the way, are not three independent comparisons. Rather, they are overlapping odds ratios because the relationship between the high and the medium values and the relationship between the medium and low values bear directly on the relationship between the high and low values.

While these first order conditional odds ratios are used to compute second order odds ratios, in and of themselves, they impart information about the relationships which exist between the dependent and independent variables. For example, comparing the first two odds ratios in the list we find that, while medium morale is more frequent than high morale when social interaction is high (as indicated by the odds ratio of 0.56), it is even more frequent when social interaction is low (as indicated by the odds ratio of 0.25). To take one more example, let's compare the fifth and sixth odds ratios in the list. The fifth one indicates that high morale is more frequent than low morale when social interaction is high (the odds ratio is 1.30). The last odds ratio indicates that low morale is more frequent than high when social interaction is low (the odds ratio is 0.52).

As mentioned earlier, second order odds ratios may be computed from pairs of these first order conditional odds ratios. The three second order odds ratios that result from comparisons of the high-low pairs on variable X of the preceding odds ratios are as follows:

$$\begin{pmatrix} Hi & Hi \\ Y & X \\ Med & Lo \end{pmatrix} = 0.56/0.25 = 2.24$$

$$\begin{pmatrix} Med & Hi \\ Y & X \\ Lo & Lo \end{pmatrix} = 2.30/2.09 = 1.10$$

$$\begin{pmatrix} Hi & Hi \\ Y & X \\ Lo & Lo \end{pmatrix} = 1.30/0.52 = 2.50$$

Two of these second order odds ratios deviate considerably from 1.00 while the third is fairly close to 1.00. The first and third of these (2.24 and 2.50) indicate that there is a positive relationship between the dependent and independent variables. That is, higher values of Y are associated with higher values of X and lower values of Y, with lower values of X. As was pointed out previously, medium morale is more common than high morale in any case; however, values of high morale are more commonly associated with high social interaction than they are with low.

Since Messer's data included three variables, a dependent variable, an independent variable, and a test or control variable, it is instructive to compute second order conditional odds ratios and third order odds ratios.

To compute second order conditional odds ratios for the three variables it is necessary to compute first order odds ratios for variable Y with conditions on the other two variables (X and T) specified. These odds ratios have been computed and are listed below:

$$\begin{pmatrix} Hi \\ Y & |X=Hi \text{ and } T=C \\ Med \end{pmatrix} = 15/29 = 0.52$$

$$\begin{pmatrix} Hi \\ Y & |X=Lo \text{ and } T=C \\ Med \end{pmatrix} = 8/18 = 0.44$$

$$\begin{pmatrix} Hi \\ Y & |X=Hi \text{ and } T=M \\ Med \end{pmatrix} = 20/33 = 0.61$$

$$\begin{pmatrix} Hi \\ Y & |X=Lo \text{ and } T=M \\ Med \end{pmatrix} = 9/51 = 0.18$$

$$\begin{pmatrix} Med \\ Y & |X=Hi \text{ and } T=C \\ Lo \end{pmatrix} = 29/14 = 2.07$$

$$\begin{pmatrix} Med \\ Y & |X=Lo \text{ and } T=C \\ Lo \end{pmatrix} = 18/4 = 4.50$$

$$\begin{pmatrix} Med \\ Y & |X=Hi \text{ and } T=M \\ Lo \end{pmatrix} = 33/13 = 2.54$$

$$\begin{pmatrix} Med \\ Y & |X=Lo \text{ and } T=M \\ Lo \end{pmatrix} = 51/29 = 1.76$$

$$\begin{pmatrix} Hi \\ Y & |X=Hi \text{ and } T=C \\ Lo \end{pmatrix} = 15/14 = 1.07$$

$$\begin{pmatrix} Hi \\ Y & |X=Lo \text{ and } T=C \\ Lo \end{pmatrix} = 8/4 \quad = 2.00$$

$$\begin{pmatrix} Hi \\ Y & |X=Hi \text{ and } T=M \\ Lo \end{pmatrix} = 20/13 = 1.54$$

$$\begin{pmatrix} Hi \\ Y & |X=Lo \text{ and } T=M \\ Lo \end{pmatrix} = 9/29 \quad = 0.31$$

By pairing these first order odds ratios conditional on both X and T it is possible to come up with second order conditional odds ratios in which the conditional variable is the test variable. From the twelve odds ratios above we can compute the following six second order conditional odds ratios:

$$\begin{pmatrix} Hi & Hi \\ Y & X & |T=C \\ Med & Lo \end{pmatrix} = 0.52/0.44 = 1.16$$

$$\begin{pmatrix} Hi & Hi \\ Y & X & |T=M \\ Med & Lo \end{pmatrix} = 0.61/0.18 = 3.44$$

$$\begin{pmatrix} Med & Hi \\ Y & X & |T=C \\ Lo & Lo \end{pmatrix} = 2.07/4.50 = 0.46$$

$$\begin{pmatrix} Med & Hi \\ Y & X & |T=M \\ Lo & Lo \end{pmatrix} = 2.54/1.76 = 1.44$$

$$\begin{pmatrix} Hi & Hi \\ Y & X & |T=C \\ Lo & Lo \end{pmatrix} = 1.07/2.00 = 0.54$$

$$\begin{pmatrix} Hi & Hi \\ Y & X & |T=M \\ Lo & Lo \end{pmatrix} = 1.54/0.31 = 4.97$$

To go the final step in the process, we can pair these six second order conditional odds ratios and arrive at three third order odds ratios as follows:

$$\begin{pmatrix} Hi & Hi\ C \\ Y & X\ T \\ Med & Lo\ M \end{pmatrix} = 1.16/3.44 = 0.34$$

$$\begin{pmatrix} Med & Hi\ C \\ Y & X\ T \\ Lo & Lo\ M \end{pmatrix} = 0.46/1.44 = 0.32$$

$$\begin{pmatrix} Hi & Hi\ C \\ Y & X\ T \\ Lo & Lo\ M \end{pmatrix} = 0.54/4.97 = 0.11$$

As you may recall, third order odds ratios inform us about the absence or presence of interaction among three variables. Since these three third order ratios all deviate from 1.00, interactions are indicated. The nature of these interactions can be determined by looking back at the second order conditional odds ratios which were used to compute the third order ones. The first such ratio of 1.16 tells us that in age-concentrated communities there is little relationship between high versus medium morale and high versus low social interaction. However, the second ratio of 3.44 indicates that in mixed-age communities these two morale levels are related to level of social interaction. Furthermore, the relationship is a positive one. That is, high levels of morale tend to be associated with high levels of social interaction.

The two second order odds ratios which pair medium and low levels of morale with high and low levels of social interaction are also informative. For age-concentrated communities, while both high and low social interaction are associated with medium rather than with low morale, medium morale is more common in low social interaction situations than it is in high ones. On the other hand, in mixed-age communities medium morale is more common in high social interaction situations than it is in low.

Finally, when high and low morale are compared with high and low social interaction in age-concentrated and age-mixed communities the results differ. In age-concentrated communities high morale is most likely to be found in those situations where social interaction is low; while in age-mixed communities high morale is much more common in high social interaction situations than it is in low.

We have computed 35 different odds ratios from Messer's data and there are many others we could have computed. All of those that we did compute were not actually necessary for the analysis we performed. Some were computed just to show how they are computed. In general, only those odds ratios that are relevant for a specific research problem are computed. Nor is it necessary to hand-compute those that are relevant. Computer programs are available to do the job and they are especially useful when there are more than three variables involved.

Proper conceptualization of a research problem will usually point to those odds ratios which are of particular interest. A model can be developed by specifying, in terms of odds ratios, which variables should be related and which should not be. Such a model would specify which of all possible odds ratios would be expected to depart from 1.00. All of those not specified would be assumed to be 1.00. As a matter of fact, models can be delineated in which all odds ratios above a certain specified order can be set to 1.00. Goodman has provided techniques for specifying such models and then for testing them to see whether they fit empirical data.

8.2   PATTERNS OF ELABORATION

The interpretation of patterns of outcome revealed by the accounting formula and by hierarchical analysis will be illustrated and discussed in the sections that follow.

8.2.1   Specification

Under what conditions does a relationship hold up? Stated differently, under what situations can a given variable (independent variable) explain why a dependent variable is distributed as it seems to be? In Messer's study, discussed above, we used some ideas from theory to suggest a variable, age-concentration of the environment of a person, which we then introduced into the analysis as a control or test factor. The result was that under one condition of the control variable, the relationship remained strong and, in fact, increased but under the other condition the relationship between morale and interaction changed (dropped) markedly. This is a pattern of *specification*. There is a **statistical interaction** between the control variable and the independent variable, and this specifies the level of the dependent variable. The outcome is different for different categories of the test variable and this has helped specify (at least one of the) conditions under which the relationship is maintained.

What other conditions govern the relationship of morale and interaction? We can think of several likely ones just on the basis of our general experience— physical health, whether the move to public housing was voluntary or not, general orientation to life in the past, etc. Does the relationship hold up for younger people? For people in other cultures? For both men and women? These questions suggest a line of further investigation similar to that conducted by Messer. We could select further test factors and examine them for *differences* between the conditional tables.

8.2.2   The Causal Hypothesis

Several astute observers have noticed a curious relationship between dollar loss from fires and whether or not firemen happen to be at the fire. The association, as shown below, is positive, rather strong, and seems to raise a number of embarrassing questions about the activities of those fire departments upon which these data are based. As the hypothesis goes, firemen, rather than reducing the loss from fire, are actually increasing that loss. Graphically, the hypothesis can be shown thus:

$$(X)\longrightarrow(Y)$$

Number of	Dollar Loss
Firemen at	from the Fire
the Fire	

Model 1

In these data 143 fires are examined in terms of the number of firemen present and the ultimate fire loss in dollars.

	(X)		
Dollar Loss	Number of Firemen		
from the Fire	None	One+	Total
Over $500	24	61	85
$500 or less	42	16	58
Total	66	77	143

$$Q=+.74$$

Do these data prove that firemen cause fire loss?

While the data are sufficient to raise the suspicions of those who see them, they are not, for a number of reasons, sufficient to settle the issue. First, there are the usual questions about bias in collecting the data, sampling procedures, adequacy of the measurement procedures, and computational accuracy. Secondly, there is the issue of time-ordering of the variables. Do we have evidence that the firemen arrived in droves *before* the fire loss became severe? Thirdly, the idea of *causal* influence includes more than one association. Can the dependent variable be altered by manipulating the independent variable? Is this possible under all conditions? Finally, we would expect that the relationship shown above, if it is indeed a causal one, would be maintained even if other *antecedent* variables are controlled — even within the categories of other variables.* Is this the case here?

8.2.2a Testing for Spuriousness.   We might suspect that if we could somehow control for the initial size or threat of the fire, then we could explain away the apparent support for the causal hypothesis. If, in controlling for this test variable, the association between loss and number of firemen is *maintained*, then our causal interpretation would be still *not dis*confirmed, and we would continue to hold our initial causal model. On the other hand, we might expect that firemen are rarely called for small fires, and almost always called for larger ones, so that we would observe an overall association simply because firemen

---

*The notion of "cause" is a theoretical notion about the way some set of variables are related. A causal relationship is said to exist if: (a) the independent variable has its effect before, in time, the dependent variable, (b) there is an association between the independent variable and the dependent variable, and (c) if this relationship is maintained even when antecedent test variables are controlled. It is often difficult to argue convincingly that all potential test factors have been introduced, and sometimes the time-order of variables is at issue, particularly in survey research. Generally, too, there are conditions under which the relationship holds and conditions where it does not. What seems well established today may well be modified or explained away tomorrow, but this is the state of any explanatory theory in any field.

and loss are related to the initial threat from the fire (its initial size). We thus have a *competing model* to the causal one above. This model is also a causal model, but it relates each of the original variables to an antecedent variable, and not to each other, as below:

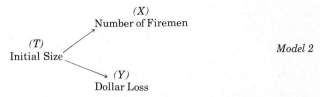

(X)
Number of Firemen

(T)
Initial Size

Model 2

(Y)
Dollar Loss

Under this model, we would expect to be able to show that by controlling for the test factor, the association of X and Y would drop from its overall association of $Q = +.74$ down to no association, $Q = 0.0$. If this happens, we will say that our initial causal interpretation about the link between number of firemen and dollar loss is a **spurious interpretation,** and that Model 2 is a better explanation of the data than Model 1. Let us look at the data.

Table 8.3 shows the relationship of density of firemen and dollar loss separately for the smaller and the larger fires. This is clearly an antecedent variable, because it is measured as the state of the fire prior to any call for firemen and prior to any final determination of loss. The association in the conditional tables does drop to zero. Model 2 works better than Model 1 for these data.

What happened to the association that once was so strong? Lazarsfeld's accounting formula helps explain. If there is an initial association between X and Y and if the association does not exist in any of the conditional tables, then it must exist in the other total associations of each of the other variables and the test variable. This is, in fact, the case, as shown in Table 8.3. Computing delta for these data, the Lazarsfeld equation is as follows:

$$15.2 = 0.0 + 0.0 + \frac{143}{(55)(88)} (27.7)(18.6)$$

Notice that the two conditional tables, added together, do add up to the total association originally observed, and the other total tables can be created from the marginals on the pair of conditional tables in Table 8.3.

This pattern of outcome, when controlling for an antecedent variable, is called "spurious" because the initial causal interpretation of the X-Y relationship is not borne out in these data, and the conditional associations drop to zero. This is another common type of analysis—the search for test factors which might show an hypothesized causal relationship to be spurious. This is the second pattern which would lead an investigator to control for other variables in the process of examining the relationship between two variables.

How much does the association have to drop in the conditional associations before we would conclude that the causal interpretation is spurious? This is hard to say. Any general drop would mean that the test factor helps account for some of the observed, zero-order association between the independent and dependent variables. The more the association drops in the conditional associations, the more the association could be attributed to the test factor. Usually the

TABLE 8.3 Conditional and Total Tables for the Fireman Example*

(a) FIRST-ORDER CONDITIONAL TABLES

(T-1)
SMALL SIZED FIRE
(X)

(Y) *Dollar Loss*	Number of Firemen		Total
	None	*One +*	
Over $500.	4	1	5
$500. or Less	40	10	50
Total	44	11	55

$Q = 0.0$

(T-2)
LARGE SIZED FIRE
(X)

(Y) *Dollar Loss*	Number of Firemen		*Total*
	None	*One +*	
Over $500.	20	60	80
$500. or Less	2	6	8
Total	22	66	88

$Q = 0.0$

(b) TOTAL TABLES

(T)

(Y) *Dollar Loss*	Fire Size		*Total*
	Small	*Large*	
Over $500.	5	80	85
$500. or Less	50	8	58
Total	55	88	143

$Q = +.98$

(T)

(X) *Number of* *Firemen*	Fire Size		*Total*
	Small	*Large*	
One or more	11	66	77
None	44	22	66
Total	55	88	143

$Q = +.85$

*Data are fictitious.

results are somewhat mixed, and an investigator has to be satisfied with the conclusion that there are several contributing factors to an adequate explanation of the distribution of a dependent variable.

When have sufficient control variables been introduced to assure us that a causal interpretation is the best interpretation? There is never a guarantee that this process is at an end. Many of the advances in the theory of an area come about by introducing a further test factor which alters a favorite causal interpretation. With a reasonably well thought out theory, we can be more confident that all *relevant* test factors have been examined, but this is no ultimate proof or guarantee that controlling on some other variable will not show that a long-held causal hypothesis is, indeed, spurious.

8.2.3   A Causal Sequence of Influence

Why do people plan to vote the way they do? How can we explain their political behavior? This question was raised by Lazarsfeld and others in their research report, *The People's Choice.** They hypothesize that a person's placement in the social class structure of his(her) society explains voting intentions, but that this happens because higher social class individuals are more interested in politics, and this leads them to be more likely to plan to vote. It is a causal sequence from social class standing (as measured here by education level) to interest in politics to vote intentions.

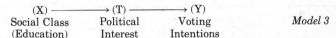

$$(X) \longrightarrow (T) \longrightarrow (Y)$$

Social Class	Political	Voting	*Model 3*
(Education)	Interest	Intentions	

An alternative explanation might be that social class influences political interest but that it also has an independent effect on voting intentions. This could be shown as follows:

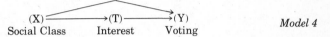

$$(X) \longrightarrow (T) \longrightarrow (Y)$$

Social Class	Interest	Voting	*Model 4*

A further competing model might be that social class and political interest have independent effects on voting behavior.

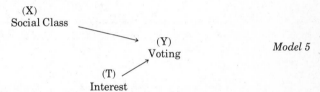

(X)
Social Class

(Y)
Voting   *Model 5*

(T)
Interest

Is it reasonable to think of these three variables as forming a causal sequence, or are the other two models more appropriate?

If there is a causal sequence of the kind in Model 3, we might be able to test it by controlling for the intervening variable, "interest," to see if the relationship between social class and voting drops to zero in the conditional tables. This would happen because, if Model 3 is correct, removing the variation in the data in the "interest" variable would also have the effect of removing variation in social class, and thus would eliminate the association of social class and voting. On the other hand, if the association persists at some level, then Model 4 or 5 would more accurately represent the separate and direct contribution of social class to voting. The data are presented in Table 8.4.

Notice that our expectation tends to be supported in the data. What association there was in the data in the original association of social class and voting plans is reduced by half or more in the conditional associations where political interest is controlled. Our conclusion, then, would be that these variables tend to form a causal sequence. Notice, too, that there was not much variation

*This example is cited in an excellent book on partialling and other procedures for handling survey data, by Rosenberg, (1968:59), and comes from Lazarsfeld *et al.* (1948:47).

TABLE 8.4a PERCENTAGE DISTRIBUTION OF VOTE PLANS BY EDUCATION

	(X)		
	Some High School	No High School	
(Y) Vote Plans			Total
Will not vote	8	14	11
Will vote	92	86	89
Totals	100	100	100
	(1613)	(1199)	(2812)

$$d_{yx} = .06$$

TABLE 8.4b PERCENTAGE DISTRIBUTION OF VOTE PLANS BY EDUCATION CONTROLLING FOR POLITICAL INTEREST

	(T) Political Interest					
	Great		Moderate		Low	
(Y) Vote Plans	Some High School	No High School	Some High School	No High School	Some High School	No High School
Will not vote	1	2	7	10	44	41
Will vote	99	98	93	90	56	59
Totals	100	100	100	100	100	100
	(495)	(285)	(986)	(669)	(132)	(245)
	$d_{yx} = .01$		$d_{yx} = .03$		$d_{yx} = -.03$	

Source: Rosenberg, 1968: Tables 3.3 and 3.4. Copyright © by Basic Books, Inc., Publishers, New York. Used by permission.

in voting plans in the first place and, furthermore, the independent variable as it is measured here does not help explain much of the variation that there is. This would suggest that there may be some separate influence of social class on voting plans, independent of political interest. Model 4 may more accurately reflect the weak streams of influence in this case. The coefficients (.01, .03 and −.03) represent the independent influence of social class (education as measured here) on voting plans when political interest is controlled. Let us examine the association of political interest, controlling for the antecedent variable, social class, to see how much independent influence political interest has for voting plans. These coefficients of association can be computed by using the data in Table 8.4. They are as follows:

$d_{yt} = .15$ The association of political interest (variable T) and voting plans (variable Y) for individuals with *some high school*.

$d_{yt} = .22$ The association of political interest and voting plans for individuals with *no high school*.

Our conclusion would be that these coefficients of association are much stronger than those for the independent effects of social class (education). The X-Y stream in Model 4 would be considerably weaker than the T-Y stream. Finally, looking at the conditionals for social class and interest, controlling for the *consequent* variable, voting intentions, we can see to what extent Model 5 would more accurately represent the data. If the conditional associations drop to zero in this case, we would say that Model 5 would be a better fit. The data are as follows (again computed from information contained in Table 8.4, rearranged):

$d_{tx} = .12$ The association of political interest (here called variable T) and social class (variable X), for those with *no plans* to vote.

$d_{tx} = .14$ The association of political interest and social class among those *planning to vote*.

Comparing the coefficients for the three sets of conditional tables, then, the X-T and T-Y links appear to be much stronger than the X-Y link when the third variable is controlled. Since the association between X and Y drops when T is controlled, the conclusion is that the data fit Model 3 best. There is no reason to give up the idea of causal sequence; at the same time, the direct influence of social class on voting plans is considerably less than its indirect influence through political interest. Our hypothesis is confirmed.

8.2.3a   Independent Causation.   The point to be emphasized is that we have contrasted our expectations for three competing models of the relationship between these variables (Models 3, 4, and 5). By controlling on an intervening variable, we are able to conclude that a causal sequence interpretation is *not disproven* if the association in the partials drops to zero. To the extent that it does not drop to zero, there is evidence of some *independent* effect of the independent variable on the dependent variable, irrespective of the test factor which intervenes. By controlling on the other two variables we are able to examine the relative strengths of association. By controlling on X, for example, we can see whether T is related to Y and how strongly, independently of X. By controlling on Y we can see to what extent X and T are related independently of Y. If the partial associations went to zero in this latter test, we would have concluded that the two variables X and T independently influence the dependent variable Y and are related only by virtue of this common influence. Model 5 would have represented the outcome of **independent causation** in this case.

8.2.4   Suppressor Variables

In all of the cases discussed thus far, we have started with some total association between the independent and dependent variables, and the task was to introduce control variables which would help explain the original association further. Is it sufficient to elaborate only those relationships which are not zero to start with? The answer is *no:* It is possible to have a third variable act to *suppress* the observed relationship between two variables and, in fact, to mask it completely as a **suppressor variable.** This sometimes happens with variables such as sex or race when a relationship in one conditional association is equally strong but opposite in direction to the association in another conditional. The result is that one relationship could cancel the other out, so that the overall as-

DIAGRAM 8.1. Some Strategies of Elaboration

1. Be sure to keep a clear focus on which variable it is that is the *dependent* variable.
2. Contrast explanatory models, if possible, using theory to develop alternatives, and guide the selection of control and independent variables. A useful procedure is to draw out expected relationships in the form of an arrow diagram.
3. Test factors used should be those related to the dependent variable. Take account of the ordering of variables in selecting control variables and interpreting outcomes of analysis.
4. If you are expecting or looking for statistical interaction, examine the conditional tables themselves.
5. Generally, start with a zero-order association and work toward more controls, examining first-order associations, then second order, etc.
6. If you are looking for a general increase or decrease (or no change) in the conditional tables, consider using odds ratios, standardization, or the use of an appropriate measure of partial correlation. (See Section 8.3.)
7. In interpreting outcomes, remember where the data came from, the source of measurements, number of cases, reasonableness of contrasts, quality of the data, whether or not multiple measurements of variables were made, and independence of observations.

sociation may be zero. Without some sort of expectation (*i.e.*, a theory) it is, of course, difficult to separate a situation where there is a zero-order association between toe-nail length and number of languages spoken, when one would not expect any association to appear no matter what controls were introduced, as against a situation where there is zero association between sexual permissive attitudes and social class, which is definitely not expected given previous research and theorizing. This latter case will serve as an illustration of a suppressor effect.

Reiss hypothesized, on the basis of prior research and theory, that there

TABLE 8.5a PERCENTAGE DISTRIBUTION OF PERMISSIVENESS ATTITUDES BY SOCIAL CLASS FOR THE STUDENT SAMPLE

Sexual Permissiveness Attitudes	Social Class			Total
	Low	Medium	High	
High	49	46	50	49
Low	51	54	50	51
	100	100	100	100
	(383)	(189)	(225)	(797)

Gamma = .01

TABLE 8.5b PERCENTAGE DISTRIBUTION OF PERMISSIVENESS ATTITUDES BY SOCIAL CLASS AND CHURCH ATTENDANCE FOR THE STUDENT SAMPLE

HIGH CHURCH ATTENDANCE

Sexual Permissiveness Attitudes	Social Class			Total
	Low	Medium	High	
High	42	26	23	34
Low	58	74	77	66
	100	100	100	100
	(262)	(98)	(102)	(462)

Gamma = −.35

LOW CHURCH ATTENDANCE

Sexual Permissiveness Attitudes	Social Class			Total
	Low	Medium	High	
High	64	67	72	68
Low	36	33	28	32
	100	100	100	100
	(113)	(89)	(119)	(321)

Gamma = +.14

Source: Data from Reiss, 1967: Tables 4.1 and 4.2. Copyright © 1967 by Holt, Rinehart and Winston, Inc. Reprinted by permission of Holt, Rinehart and Winston.

would be a negative relationship between social class and permissiveness attitudes, with attitudes toward sexual permissiveness measured by a series of scales (Reiss, 1967: 59 and 61). For the purposes of this study, he dichotomized permissiveness, although the puzzling findings were checked by including more categories of permissiveness attitudes and several different measures of social class. The data were gathered on about 800 students from five schools in the East. Table 8.5 presents these data.

Using gamma (G) as a measure of association, he found that there was no association between permissiveness attitudes and social class. This finding occurred in each of the five schools, with a dozen different social class measures and with a variety of other controls as well. Prior research by Kinsey indicated that religion was an important explanation of variation in sexual relations, so frequency of church attendance was introduced as a control variable, as shown in Table 8.5b. Among the high-church-attenders, the social class-permissiveness association was −.35 (gamma) and it was +.14 among the low-church-attenders. The higher the social class, the lower the level of permissiveness among high-church attenders, and the opposite was true of low-church-attenders. Further investigation indicated that the same kind of outcome occurred whenever the student group was divided into categories of generally "conservative" and generally "liberal" on a number of other variables such as political preference, beliefs about integration of schools, civil rights activity, etc. This and related analyses eventually led Reiss to a general proposition, namely: "The stronger the amount of general liberality in a group, the greater the likelihood that social forces will maintain high levels of sexual permissiveness" (Reiss, 1967:73). Social class standing is one of those forces.

Had he stopped his analysis upon finding no association between social class and permissiveness attitudes, he would have missed an important series of findings about variables which masked the relationship.

8.2.5   Social System Analysis

For many sociological problems an investigator is interested in two levels (or more) of analysis, the *individual* level and the *group* level. He(she) may argue, for example, that differences between groups stem from the different kinds of individuals who compose the group (called structural analysis), or he(she) may feel that individuals behave in certain ways not only because of their own characteristics but because of the character of the group itself of which they may be members (called contextual analysis). In addition to these two types of analysis are one-level analyses using groups as the unit and one-level analyses using individuals as the unit.*

In a social system analysis involving three variables and two levels, each individual case is characterized by three measured variables. Usually two of these variables, the independent and dependent variables, are individual-level

---

*In this chapter we will illustrate some of the possibilities that involve conditional tables and an examination for interaction between group and individual level variables, one of the patterns of relationship between three variables. An excellent discussion of social system analysis with some examples from the field of sociology is found in Riley (General References, 1963:V.I, 800, Unit 12).

characteristics and the third variable is a group-level variable which is used to characterize the kind of group context the individual is in. We could consider creating one table cross-classifying the independent and dependent variable for all individuals in a given *kind* of group. One such table would be created for each of the different kinds of groups included in the study, and systematic comparisons would be made within and between groups.

8.2.5a. A **structural analysis** might involve comparison of a specific cell (*e.g.* percent psychotic among foreign-born) across *groups* which differ in some interesting respect (*e.g.* degree of social disorganization of the area), to see what effects the differences between groups might have when the structure of the group is held constant (*e.g.* only foreign-born are compared, or only native-born). Riley points out another kind of structural analysis, used by Durkheim, in which group differences are controlled by examining the association of two variables (*e.g.* religion and suicide) within separate groups (*e.g.* within different types of countries).

8.2.5b. A **contextual analysis** combines information from all three variables, examining the way the relationship between two variables (*e.g.* two individual-level variables, or perhaps an individual and a context variable) may change systematically, for *individuals*, across groups which are set up to differ in a systematic way on a group-level variable. Here the analysis will focus on individuals as the unit being investigated, and it will measure individuals on variables which reflect individual properties as well as some aspect of the kind of group context within which an individual is located. Let us turn to an example at this point.*

In a study of job satisfaction, previous research has supported the notion that older individuals are more likely to be satisfied than are younger workers. It has also been argued that the "age" of the company or department of a company has an influence on job satisfaction (and on the creativity and productivity of the organization), but the data are conflicting and generally show little difference by group age. Is it possible that job satisfaction, for example, is influenced not only by individual age of the worker, but also by the organizational age of the group context within which the worker is situated?

One of the authors examined this question using data on some 235 individuals who worked in 35 branches or departments of a large research and development organization. Each individual was asked his(her) own age and a series of questions designed to measure his(her) degree of job satisfaction. One of these job satisfaction questions was whether or not the individual felt his(her) department was a highly supportive work environment. In addition, the departments themselves were "measured" on organizational age, and this was done by computing the arithmetic mean age of individuals in that department. Each individual, then, was characterized by his(her) own age, the age of the department he(she) worked in, and his(her) job satisfaction.

Both individuals and departments were classified on age into "old," "middle age," and "young" in this analysis. There were, of course, a number of different departments in each age bracket, but rather than examine each organiza-

---

*An excellent example of contextual analysis which makes use of some of the procedures to be discussed in the next chapter is given in McDill *et al.,* 1967. At issue is whether or not the socioeconomic context of the school influences individual achievement. Data came from 20 public high schools.

tion separately as we might do in one kind of structural analysis, we put the data together in three summary tables, one for each kind of group (*i.e.* old, middle-aged, and young groups). These data are presented in Table 8.6, below.

Overall, 48.1 percent of the workers felt their work context to be "highly supportive." Older individuals were more likely to agree that their context was highly supportive, as is indicated in the following distribution. The pattern by group age is reversed in that the percentage is lower for the older-age groups than for younger-aged groups.

PERCENTAGE OF WORKERS FEELING THEIR BRANCH PROVIDES A HIGHLY SUPPORTIVE WORK ATMOSPHERE, BY GROUP AND BY INDIVIDUAL AGE

By Group Age		N(100%)		By Individual Age		N(100%)
31–35yrs	52.2%	(86)	Young	22–32yrs	42.9%	(63)
36–39yrs	46.1%	(89)	Middle	33–39yrs	45.6%	(90)
40–55yrs	45.0%	(60)	Older	40–64yrs	54.9%	(82)

TABLE 8.6 PERCENTAGE FEELING THEIR DEPARTMENT IS A HIGHLY SUPPORTIVE WORK ATMOSPHERE, BY INDIVIDUAL AGE AND ORGANIZATIONAL AGE

YOUNGER DEPARTMENTS

Feelings About Support of Work Context	Individual Age			Total
	Young	Middle	Older	
Highly Supportive	51.4	48.6	64.3	52.2
Not Highly Supp.	48.6	51.4	35.7	47.8
	100.0	100.0	100.0	100.0
	(35)	(37)	(14)	(86)

MIDDLE-AGED DEPARTMENTS

Feelings About Support of Work Context	Individual Age			Total
	Young	Middle	Older	
Highly Supportive	30.4	45.9	58.6	46.1
Not Highly Supp.	69.6	54.1	41.4	53.9
	100.0	100.0	100.0	100.0
	(23)	(37)	(29)	(89)

OLDER DEPARTMENTS

Feelings About Support of Work Context	Individual Age			Total
	Young	Middle	Older	
Highly Supportive	(40.0)	37.5	48.7	45.0
Not Highly Supp.	(60.0)	62.5	51.3	55.0
	100.0	100.0	100.0	100.0
	(5)	(16)	(39)	(60)

Source: Data are from 235 individuals in 35 branches of a research and development organization, collected in 1966, by Robert Biller, who kindly made these data available for the author's study of individual and organizational age. Branch heads and managers have been eliminated in the above data.

Data in the three summary tables (Table 8.6) show an interesting added pattern. Young or middle-aged individuals in middle-aged or older group contexts are less likely to feel their department is a highly supportive work context. Compare these figures to, for example, older individuals in young groups, older individuals in middle-aged groups, or the young or old individuals in their own-aged groups. The differences in percentage feeling their department work context is supportive in Table 8.6 show a pattern of *statistical interaction* between individual and group age. Not only are there differences by individual age or by organizational age, separately, but there are much more impressive differences when specific combinations of these two age measures are examined. This effect can be shown clearly in the graph in Figure 8.1.

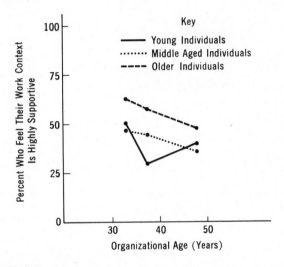

Source: Data from Table 8.6 above.

FIGURE 8.1 PERCENTAGE DISTRIBUTION OF A JOB SATISFACTION MEASURE BY AGE OF INDIVIDUALS AND THEIR ORGANIZATION CONTEXT FOR WORKERS IN DEPARTMENTS OF A RESEARCH AND DEVELOPMENT ORGANIZATION

In these graphs, each of the dots indicates one of the combinations of individual age and group context age. There are nine such combinations here and, thus, nine dots. Each dot is placed at a point which corresponds to (a) the percentage of individuals who feel that the work context is supportive, and (b) the average age of the department within which they are located. The dots representing young individuals are connected and, separately, those dots for the middle-aged individuals and for the older individuals. There are three lines on the graph, one for each category of individual age and the lines reflect changes in the dependent variable, percentage feeling the context is supportive, across the differently aged departmental contexts. These lines converge or cross each other, which means that there is statistical interaction; the level of the dependent variable depends not only on the addition of a group effect to an individual age effect but depends on the specific combination of those two variables.

Figure 8.2 illustrates some of the other outcome possibilities one might find in two-level analysis of this sort. In Figure 8.2a, for example, the two lines

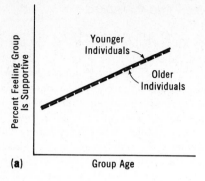

(a)   Group Age

(a) *Group Effect Only.* Here there is a difference in the dependent variable for contexts of different kinds, but there is no difference within any of the contexts in terms of the other individual characteristic. If the lines were together as they are here, but parallel with the X-axis, there would be no group or individual difference in the dependent variable.

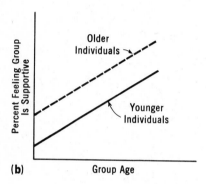

(b)   Group Age

(b) *Separate Individual and Group Effects.* In this graph, the lines are parallel, sloped, and different. This indicates that there is a change in the dependent variable for different kinds of contexts and that the effect is the same on each kind of individual within these contexts. There is also an individual effect since the two lines are separated. Within each kind of context, there is an individual difference on the dependent variable. Note that the dotted line might be above or below the solid line. That is, the correlation of the group variable may differ in sign from the correlation of the individual-level variable.

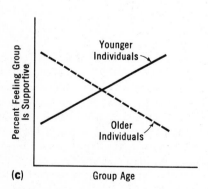

(c)   Group Age

(c) *Interaction Effects.* Here the lines are different, both sloped, and they are not parallel. This indicates that the dependent variable depends upon the specific combination of individual and group level independent variables. It is not a simple case of adding together separate group and individual effects as in situation (b), above. Here, as the context examined is older, older individuals tend to decline on the dependent variable, but younger individuals tend to increase.

FIGURE 8.2 SOME POSSIBLE PATTERNS OF GROUP AND INDIVIDUAL EFFECTS [For a Discussion of Some of These Patterns, See General References, Riley (1963), and also Davis *et al.* (1961).]

BOX 8.2 STATISTICAL INTERACTION

If two variables, *A* and *B*, each explain some of the variation in a dependent variable, their combined explanatory power may be simply a sum of their separate effects. That is, they each contribute a part to their combined effect on a dependent variable. They have an *additive effect*.

Contrasted with this additive effect are *interaction effects* in which specific combinations of categories of two or more independent variables (or independent and control variables) explain more of the variation in a dependent variable than would be expected from a simple additive combination of their separate effects. The combination of certain categories of individual age and group age, for example, seems to result in differences in satisfaction beyond what would be expected from a sum of their separate effects.

Statistical interaction is an important kind of outcome of the relationship of two or more independent variables to a dependent variable, not only because it helps determine the kind of analysis to use, but also because our theories often lead us to expect interesting and unusual effects of particular combinations of values of two or more independent variables.

---

which represent the different categories of one of the independent variables are identical. This means that there is no individual-level effect. The fact that the lines have a slope indicates, however, that there is a group-level effect on the dependent variable. Here the dependent variable within a given context for a certain kind of people is measured in terms of percentages, although it could be measured in terms of medians or means, etc., depending upon the kind of measurement or feature of interest.

In Figure 8.2b, there is an added effect due to individual-level differences in the independent variable. Notice that the individual differences are the same within the different categories of group context, and thus both the group and the individual level variables make separate, additive contributions to the explanation (or prediction) of the dependent variable. To say it somewhat differently, the individual-level difference occurs regardless of the particular group-level context one is talking about (and vice versa). In Figure 8.2c, however, the level of the individual-age effect on the dependent variable depends upon the group context in important ways. Older individuals will have a higher level of the dependent variable than younger individuals only if the group context happens to be younger. The reverse is true in this example, where group contexts are older and there is a type of middle-age group context for which there are no individual-level differences. This is an example of statistical interaction, since the relationship between two variables (independent and dependent) depends critically upon the value of the other independent variable. In this case, since one of the variables is group-level and the other is individual-level, the analysis permits us to draw conclusions about group and about individual effects. Interaction could also occur, of course, between variables both at the same level of analysis (*e.g.* both group-level or both individual-level), and this was the case in the example of specification discussed earlier in this chapter.

8.2.6 Recapitulating Patterns of Elaboration

Some of the more frequent and interesting patterns of outcome are of the type referred to as *specification* or *statistical interaction* (Section 8.2.1), a condition in which the conditional tables differ substantially from each other. Patterns where there is an original association but the conditional associations are near zero are interpreted as evidence that a causal hypothesis is *spurious* (8.2.2a), that a *causal sequence* hypothesis is supported (8.2.3), or that there is *independent causation* (8.2.3a), depending upon the antecedent, intervening, or consequent theoretical ordering of a test factor. Finally, there is the pattern in which a *suppressor* variable (8.2.4) operates to mask out an original association, and this is discovered only after examining other conditional and total tables. The *Lazarsfeld accounting formula* (8.1.4a) is useful in organizing patterns of outcome when total and conditional associations are examined. In general, the procedure used is one of (a) examining lower-order tables first and higher-order tables later in an analytic sequence, and (b) controlling on other relevant variables in the examination of an *X-Y* relationship of interest. What is controlled and how depends critically upon the ideas brought to the research by the investigator. *Hierarchical analysis* (8.1.4b), which utilizes odds ratios, is useful for sorting out all of the patterns of statistical interaction to be found in total and conditional tables.

8.3 THE AVERAGE PARTIAL TABLE

The lines of elaboration we have been discussing direct the investigator to examine (a) any difference there may be in the conditional tables, or (b) the overall change in degree of association in the conditionals. In situation (a), the only analysis procedure is elaboration (or some form of dummy-variable regression analysis, as discussed toward the end of the next chapter), and elaboration involves the examination of the different conditional tables. In situation (b), however, where we are merely interested in whether the association in the conditionals remained the same, decreased, or increased, there are some other types of statistical summary which could be carried out to simplify the analysis. Two of these procedures will be discussed in this section. One is *standardization* and the other is *partial association*. In both instances the result is a kind of "average" of the outcome in a set of conditionals. In the first instance an average partial table is the result, and in the second it is an average of the measures of association that might be computed on separate conditionals.

In either case, these further descriptions are used in the same way they were used earlier. One would compare the original association of two variables with the association after a test factor has been "partialled out" or controlled. If, for example, the test factor were an antecedant variable, and if the partial association dropped to zero when the test factor was controlled, the conclusion would be as it was when individual conditional tables were examined—any causal hypothesis would then be a *spurious* interpretation of the relationship between the main independent and dependent variables. Let us discuss the idea of standardization first and then turn to partial association coefficients.

BOX 8.3 CONDITIONAL TABLES AND PARTIAL ASSOCIATION

One of the organizing themes in this book has been the search for ever more powerful summary statistics. This search led to the step from a univariate distribution to summary index numbers such as the mean and standard deviation. Again, at the point where several univariate distributions were being contrasted, we stepped up to an overall treatment of these separate distributions as columns in a bivariate table, and this led to the use of measures of strength and nature of association. In the first part of this chapter we began by looking at a basic table relating two interesting variables within categories of other control variables. Now, in this section, we are again looking for a more powerful summary procedure which will express in one table or in one number what we would otherwise get from a potentially large number of conditional tables.

Although terminology is not altogether consistent in statistics, we will here observe a distinction between the examination of *conditional* tables and their associated *conditional association* measures (the last section) as against *partial association* measures, which summarize a set of conditional relationships and express the part of the total association which exists after the effects of control variables have been statistically removed. It should be noted, however, that some texts refer to the examination of conditional tables as "partialling."

8.3.1 Standardization

Standardization is a procedure which statistically removes fluctuations in the data due to some control variable, thus permitting an investigator to examine and compare tables without this contaminating effect. Two procedures will be discussed, both of which are called "direct" standardization. The first procedure applies to figures, such as rates, which we may want to compare, and the second applies to whole tables. We will use the first to explain the idea of standardization. Computational procedures for the second procedure are presented in Diagram 8.2.

8.3.1a STANDARDIZING RATES

Suppose that we are interested in studying the rate at which women of different educational backgrounds produce children. Examining Census reports, we discover that among those women in the child-bearing ages (*i.e.* 15–49 years) there are 302 children under 5 years old per 1000 women who had 8 years of elementary school education, and that the rate increases to 393 per 1000 for women with 4 years of high school, and it drops again to 383 per 1000 for those with 4 or more years of college (U.S. Bur. Census, 1970). There appears to be a difference in number of children under 5 by education.

At this point we might be somewhat suspicious of our comparison because, after all, women in their 20's are more likely to have young children than women in their 40's. If there is a higher percentage of young women in the high-school-educated group, we would expect a higher rate. Some of the difference in rates may be due to differences in age distribution of elementary, high-school, and college trained women. The problem, then, is to create three rates which somehow take out the effects of any differences there may be in the distribution of age.

To do this, we will choose some standard age distribution as our **standard population**. The choice is arbitrary but the rates will be more interesting if the standard population is realistic. For this purpose we will use the actual overall age distribution of women in the U.S. in 1969. This distribution is as follows:

"STANDARD POPULATION" NUMBER OF WOMEN BY AGE BETWEEN AGES 15 AND 49, U.S., MARCH, 1969 DATA

Age of Woman	Number (in 1000's)
15 to 24	17,109
25 to 29	6,608
30 to 39	11,388
40 to 49	12,472
Total	47,577

Source: U.S. Bur. Census, 1970.

Now we need detailed "children-under-five" rate figures for women of different ages and for the three different education categories. These rates are given below.

NUMBER OF CHILDREN UNDER 5 YEARS OLD PER 1000 WOMEN, BY AGE OF WOMAN AND NUMBER OF YEARS OF SCHOOL COMPLETED, U.S., 1969

Age of Woman	Schooling Completed Elementary	High school	College
15 to 24	251/1000	413/1000	110/1000
25 to 29	916/1000	879/1000	673/1000
30 to 39	456/1000	407/1000	572/1000
40 to 49	96/1000	76/1000	88/1000

Source: U.S. Bur. Census, 1970.

The standardization procedure involves multiplying each of the *rates by the number of women in the standard population who are in that age bracket.* This produces an "expected" number of children under 5 if there were as many women in that age group as there are in the standard population. The expected number of children under 5 can then be summed separately for each of the levels of schooling, divided by the number of women in the standard population, and expressed as a rate per 1000 women. Thus 17,109 (1000's) women aged 15 to 24 would be *expected* to have 4,294,359 children if they all had completed only 8 years of elementary school (*i.e.* multiply the rate—251 from the table above for those in the 15–24 age bracket with elementary schooling—by the 17,109 women in that age group in the standard population). On the other hand, one would expect 7,066,017 children under 5 years of age ($413 \times 17,109 = 7,066,017$) if all of the women of this age in the standard population had a high-school education. Finally, one would expect fewer children under 5 if this same group had a college background ($110 \times 17,109 = 1,881,990$). These computations are shown below:

(a) Computation Using Rates for Women with Elementary School Education.

Age of Woman	Rate per 1000		Standard Population (1000's)		Expected Number
15–24	251	×	17,109	=	4,294,359
25–29	916	×	6,608	=	6,052,928
30–39	456	×	11,388	=	5,192,928
40–49	96	×	12,472	=	1,197,312
Totals			47,577		16,737,527

$$\text{Standardized Rate} = (1000)\frac{16,737,527}{47,577,000} = 352$$

(b) Computation Using Rates for Women with a High-School Education.

Age of Woman	Rate per 1000		Standard Population (1000's)		Expected Number
15–24	413	×	17,109	=	7,066,017
25–29	879	×	6,608	=	5,808,432
30–39	407	×	11,388	=	4,634,916
40–49	76	×	12,472	=	947,872
Totals			47,577		18,457,237

$$\text{Standardized Rate} = (1000)\frac{18,457,237}{47,577,000} = 388$$

(c) Computation Using Rates for Women with College Education.

Age of Woman	Rate per 1000		Standard Population (1000's)		Expected Number
15–24	110	×	17,109	=	1,881,990
25–29	673	×	6,608	=	4,447,184
30–39	572	×	11,388	=	6,513,936
40–49	88	×	12,472	=	1,097,536
Total			47,577		13,940,646

$$\text{Standardized Rate} = (1000)\frac{13,940,646}{47,577,000} = 293$$

As a result of applying the different rates to the same standard population, we can compute rates which can be *compared:* 352, 388, and 293 per 1000 women. The following table summarizes the rate of children-under-5-years-old per 1000 women aged 15 to 49 both before standardization on age and after standardization.

RATE OF CHILDREN UNDER 5 YEARS OLD PER 1000 WOMEN AGED 15 TO 49, U.S., 1969, BEFORE AND AFTER STANDARDIZATION ON AGE

Education Completed	Not Standardized	Standardized On Age
Elementary	302	352
High school	393	388
College	383	293

The age structure of the population upon which the rates are computed is the *same* for each education group, so the overall differences between age-standardized rates must be due to the different rates which apply to differently educated women. These are called age-standardized rates because the effect of age structure of the population is removed. There are a number of uses of this standardization procedure in demography in comparisons of death rates or birth rates or disease rates for populations of different makeup.

8.3.1b STANDARDIZED TABLES

The second application of standardization is called "test factor standardization" (*cf.* Rosenberg, 1963), and it too statistically removes the effect of control variables so that the relationship between independent and dependent variables can be examined without this source of contamination. This permits an investigator to compare an original total association of two variables and the same association where the effects of some test factor have been statistically removed. This would permit us to make some kind of judgment about what is happening to the strength of association in the conditionals in general. Test factor standardization is illustrated in Diagram 8.2.

To explain the logic behind standardization, consider the two conditionals from Table 8.3, above. They have been percentaged and reproduced in Diagram 8.2.

In these conditional tables there is no association between X and Y, although, when cell frequencies are added together, they produce a rather striking association ($\phi = .44$) in the total table. Looking at the differences between the total and conditional tables, we can see that the effect of the test factor has been to "rearrange" cases within each conditional table, and thus to rearrange the distribution of column and row totals, so that each cell frequency in a given conditional contributes a *differently weighted amount* to the total table. In the T-1 partial, for example, the "4" in the upper left cell contributes 17 percent of the 24 in that cell of the total table, while the "1" contributes only 2 percent to its combined-cell total, and the cell with 40 cases contributes 95 percent of the cases in its cell in the total table. This differential weighting of cells does not reflect the relative number of cases in the distribution in a column of the conditional table and, thus, it does not reflect the pattern in the conditionals.

To handle this differential weighting, we can take two steps. *First*, per-

DIAGRAM 8.2 Test Factor Standardization

Conditional Tables:

		T-1					T-2	
		X					X	
FREQUENCY		4	1	5		20	60	80
	Y	40	10	50	Y	2	6	8
		44	11	55		22	66	88
PERCENTAGE		9%	9%	9%		91%	91%	91%
		91	91	91		9	9	9
		100	100	100		100	100	100

		Total		
		X		
		24	61	85
	Y	42	16	58
		66	77	143
		36%	79%	59%
		64	21	41
		100	100	100

Weights:

   T-1 conditional contains 55/143 = .385 of the total number of cases.
   T-2 conditional contains 88/143 = .615 of the total number of cases.

               sum of weights = 1.000

Computation:

T-1	9	9	9	each cell	3.5	3.5	3.5
	91	91	91	→ multiplied →	35.0	35.0	35.0
	100	100	100	by .385	38.5	38.5	38.5
	(44)	(11)	(55)				
T-2	91	91	91	each cell	56.0	56.0	56.0
	9	9	9	→ multiplied →	5.5	5.5	5.5
	100	100	100	by .615	61.5	61.5	61.5
	(22)	(66)	(88)				

summed, cell by cell, to form the standardized table below

	STANDARDIZED PERCENTAGE TABLE				STANDARDIZED FREQUENCY TABLE		
						X	
	59.5	59.5	59.5		39	46	85
	40.5	40.5	40.5	Y	27	31	58
	100%	100%	100%		66	77	143
	(66)	(77)	(143)				

Source: Data from Table 8.3.

DIAGRAM 8.2 *(Continued)*

Summary of Steps In Creating A Standardized Table.

1. Create conditional tables, controlling on desired test factors. The table may be of any size $(r \times c)$.*
2. Properly percentage all conditional tables.
3. Select a weight to apply to all cells of a given conditional table. These weights might be the proportion of cases which are in a given conditional table, as in the example above. Weights should sum to 1.00 over all tables.
4. Multiply each cell percentage by the weighting factor for each conditional table.
5. Sum these weighted percentages across all conditional tables, cell by cell. The resulting table is the standardized percentage table.
6. Using the marginal frequencies from the total table, convert the standardized table percentages back into frequency form.
7. Compute measures of association on the standardized frequency table for comparison with measures computed on the total table.
8. Compare total and standardized tables (and association measures) to assess the effect of the test factor/s.

*Note that the conditional tables must have some non-zero frequency in each category of the independent variable (*i.e.* the base of percentages) so that percentages in each category of the independent variable in the standardized table will total to 100 percent.

centage each conditional table so that the effect of different column (or row) totals is removed. If Y is the dependent variable, percentaging would be in the direction of the independent variable. In effect, this gives proper "rates" of appearance of the different values of Y within the separate categories of the independent and control variables. Then, all of the cells in a conditional table (*i.e.* all cells in one category of the control variable) are weighted equally (*i.e.* multiplied by the same weight) and the weighted percentages can then be summed over all conditional tables to create a standardized percentage table.

$$\Sigma(w_i P_i) = \text{standardized percentage for a cell}$$

where w is the weight and P is the cell percentage, summed over all of the conditional tables.

There is no mathematical reason for picking any particular set of weights. The important point is that whatever they are, they should be *uniform* within any one conditional table. There are some subsidiary reasons for selecting certain weights, however. If the sum of weights used in the different partial tables is equal to 1.00, then the resulting standardized table will be a percentaged table with column (row) totals equal to 100 percent. If weights total to more than 1.00 then column (row) totals in the standardized table will be larger than 100 and would have to be re-percentaged, following the usual procedures but being careful to percentage in the same direction that the conditionals were percentaged. It is also useful to select a weighting factor for a given conditional table that is equal to the proportion of all of the cases which fall in that category of the control variable. This, in effect, gives more weight to the conditional table which has the greater number of cases (and may, therefore, be more stable) and gives less weight to conditionals with very few cases (which may therefore be less stable). This is the procedure described in Diagram 8.2, although again, the important point is that weights within a partial not be changed from cell to cell. Computations for a 2×2 table are shown in Diagram 8.2, although test factor standardization is a technique which applies equally well to any sized table and any number of categories of control variables.

The standardized percentage table may be converted back to frequencies, using the total table marginals as the base of the corresponding 100-percents. Appropriate measures of strength of association could be computed on the standardized frequency table in the usual way, and the original total association could be compared with the association computed on the standardized table.

Standardization was used as an analysis technique by McAllister in a study of residential mobility among blacks and whites in the U.S. (McAllister *et al.*, 1971). The literature, they note, suggests that blacks move more often than whites and that blacks' mobility is more likely to be local. Using data from a national survey, the investigators examined moving behavior between 1966 and 1969 for blacks and whites. The data are given in Table 8.7. Gamma (G) equals $-.20$ meaning that whites were more likely to have stayed than is true of blacks. The comparison, as they point out, is not an appropriate one, however, because blacks are more likely to be renters, and renters of any ethnic status tend to move more frequently than owners. The standardized table in Table 8.7 controls for the effects of the owner-renter variable. You will notice that the association

TABLE 8.7 MOBILITY DIFFERENCES BETWEEN RACES: STANDARDIZED AND TOTAL

Total Association

Mobility Behavior	Ethnicity		Total
	Black	White	
Stayed	48.7	58.6	56.9
Moved	51.3	41.4	43.1
	100.0	100.0	100.0
	(263)	(1226)	(1489)

gamma $(G) = -.20$

Total Association Standardized on Owner/Renters

Mobility Behavior	Ethnicity		Total
	Black	White	
Stayed	58.4	55.9	56.3
Moved	41.6	44.1	43.7
	100.0	100.0	100.0
	(263)	(1226)	(1489)

gamma $(G) = +.11$

Source: McAllister *et al.*, 1971. Used by permission.

drops to a G of +.11, indicating that there are minor ethnic differences in moving behavior and, if anything, blacks tend to be the stayers.

Test factor standardization is useful whenever we are interested in what happens in a set of conditional tables in general. It does not permit us to examine *differences* between conditional tables (*i.e.* interaction effects).

8.3.2   Measures of Partial Association and Correlation

The third approach to an analysis of three or more variables to be discussed in this chapter, is the computation of some type of "average" over the measures of association for each of the conditional tables. These coefficients are called **coefficients of partial association** and, like the standardized table, they are compared with the original total association to determine what happened in general to the strength of association when one or more test factors were introduced. Like the standardized table, partial association coefficients do not provide information about any pattern of *differences* between separate conditional tables over which they are computed. Two procedures for computing partial coefficients will be introduced here, one for ordinal data and one for interval data.

8.3.2a   ORDINAL PARTIAL ASSOCIATION COEFFICIENTS

A relatively direct approach to the creation of ordinal partial association coefficients involves combining computations based on each conditional table into a single, "average" coefficient. For those ordinal measures of association which make use of counts of pairs (*i.e.* gamma, Somers' d_{yx}, etc.), the combination is accomplished by simply adding together the count of a certain type of pair computed on each of the conditional tables. Thus a total of concordant pairs (N_s) could be arrived at by adding up the N_s computations on each of the conditional tables. A similar summing could be made for other types of pairs, such as the discordant pairs (N_d), pairs tied on X but not on Y (T_x), pairs tied on Y but not on X (T_y), and pairs tied on both X and Y (T_{xy}). The appropriate totals would then be substituted into the formula for the ordinal coefficient one wants. For example, ΣN_s could be used instead of N_s, ΣN_d could be used instead of N_d, and so on.

The formula for *partial gamma* (G_p) then becomes this:

(8.1)
$$G_p = \frac{\Sigma N_s - \Sigma N_d}{\Sigma N_s + \Sigma N_d}$$

where G_p refers to "partial gamma" and the N_s and N_d sums are taken over the N_s and N_d components computed on each conditional table separately. This formulation of partial coefficients for ordinal measures would be appropriate for Somers' d_{yx} and other ordinal measures as well.

The creation of a partial G can be illustrated with data from an article by Ransford on "Skin Color, Life Chances, and Anti-White Attitudes" (1970), in which some 312 black males were interviewed shortly after a mid-60's race-riot in the Watts area of Los Angeles. His hypothesis was that skin color itself has an influence on the structure of opportunity, even when such variables as educational experience are taken into account. Table 8.8 presents the total and conditional associations of occupation and skin color, controlling for three categories of formal education.

The original, zero-order G was $-.26$, indicating that the lighter the skin color, the higher the occupational standing. The N_s and N_d counts are indicated in Table 8.8. N_s and N_d are also shown in Table 8.8 for each of the three conditional tables, where education is controlled. What is the effect of education controls? The partial G can be computed to answer this question by summing N_s and N_d counts over the three tables as follows:

$$\Sigma N_s = 412 + 754 + 833 = 1999$$
$$\Sigma N_d = 714 + 526 + 1695 = 2935$$

The partial G, then, is:

$$G_p = \frac{1999 - 2935}{1999 + 2935} = \frac{-936}{4934} = -.19$$

Comparing the original G of $-.26$ with the G_p of $-.19$ indicates that indeed the association does drop when education is taken into account. It is also clear, however, that the association does not drop to zero, thus partially supporting the author's initial expectation.

Again it should be pointed out that the partial G, like partial correlation coefficients, answers only the question of what happens in the conditional tables in general. Notice in Table 8.8 that there are interesting differences in the strength of association between the G coefficients computed on the separate tables. The G_p indicates the proportionate reduction in error in predicting rank on

TABLE 8.8 PERCENTAGE DISTRIBUTION OF OCCUPATIONAL LEVEL BY SKIN COLOR AND EDUCATION FOR BLACK MALES INTERVIEWED IN LOS ANGELES IN THE MID-1960's

| | TOTAL | | | Less than H.S. | | | EDUCATION H.S. Grad. | | | Some College | | |
| | Skin Color | | | Skin Color | | | Skin Color | | | Skin Color | | |
Occupation	Lt	Md	Dk	Lt	Md	Dk	Lt	Md	Dk	Lt	Md	Dk
	%	%	%	%	%	%	%	%	%	%	%	%
White Coll.	52	40	29	15	05	13	32	17	9	83	70	58
Blue Collar	39	48	50	69	70	45	50	67	76	17	26	36
Unemployed	9	12	21	15	24	42	18	15	14	0	4	6
	100	100	100%	100	100	100%	100	100	100%	100	100	100%
	(64)	(159)	(85)	(13)	(37)	(31)	(22)	(46)	(21)	(29)	(76)	(33)

$$N_s = 6640 \qquad\qquad N_s = 412 \qquad\qquad N_s = 754 \qquad\qquad N_s = 833$$
$$N_d = 11239 \qquad\qquad N_d = 714 \qquad\qquad N_d = 526 \qquad\qquad N_d = 1695$$
$$G = -.26 \qquad\qquad G = -.27 \qquad\qquad G = -.18 \qquad\qquad G = -.34$$

Source: Based on rearranged data from Ransford, 1970:171, Table 1.

one variable from rank on the other variable after the effects of education have been taken out.*

8.3.2b PARTIAL CORRELATION COEFFICIENT FOR INTERVAL VARIABLES

A frequently used partial correlation coefficient for interval-level variables is $r_{yx \cdot z}$ which is computed from Pearson's r (discussed in Chapter 7). Although r is symmetric, subscripts are used to indicate which variables the correlation is between (the first two subscripts in front of the dot), and the variable(s) used as controls (those which are listed after the dot).

The partial correlation coefficient which statistically controls the effects of one control variable is called a **first-order partial correlation,** and it is computed from zero-order correlation coefficients as follows:

(8.2)
$$r_{yx \cdot t} = \frac{r_{yx} - (r_{yt})(r_{xt})}{\sqrt{(1 - r_{xt}^2)(1 - r_{yt}^2)}}$$

Higher-order partial correlation coefficients can be computed in a similar manner, using the next lower partial correlation coefficient in the general formula above. Thus, for two control variables, the second-order partial becomes:

(8.3)
$$r_{12 \cdot 34} = \frac{r_{12 \cdot 3} - (r_{14 \cdot 3})(r_{24 \cdot 3})}{\sqrt{(1 - r_{14 \cdot 3}^2)(1 - r_{24 \cdot 3}^2)}}$$

Higher-order partials may be formed in similar ways.

Like the total correlation, r, the partial correlation varies from -1.00 to

---

*A more precise statement of the meaning of gamma (G) is found in Section 7.3.2b. It is the proportionate reduction in error resulting, in this case, from using the "opposite-rank order" rule to predict the rank of pairs on one variable from a knowledge of the rank order on the other variable, rather than making a random guess of rank order (the minimum guessing rule), among pairs which are ranked differently on both variables (*i.e.* the $N_s + N_d$ pairs), after the effects of education have been taken into account. It is relevant to note that education was taken into account to the extent that three categories were used. Somewhat different results may occur if more or fewer categories of the control (and other variables, for that matter) are used.

+1.00. Its square expresses the proportion of the variation in Y (or X) explained by its linear association with the other variable, X (or Y), after the linear effects of the control variables have been taken into account (statistically removed).*

An example of the use of the partial correlation coefficient is provided by Lightfield (1971), who studied factors influencing the recognition of sociologists by their peers. His data are based upon responses of 200 university sociologists, and as independent variables he used (a) status of the department from which the rated person received his PhD., (b) the quality of his publications, and (c) his quantity of publications. The dependent variable was a peer-recognition score. In order to guide his analysis he presented an arrow diagram shown in Figure 8.3.

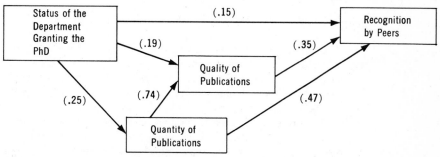

Source: Based on data from Lightfield, 1971.

FIGURE 8.3 DIAGRAM OF THE EXPECTED RELATIONSHIPS BETWEEN THREE INDEPENDENT VARIABLES AND PEER RECOGNITION OF SOCIOLOGISTS (Partial Correlation Coefficients Are Entered on the Diagram)

The numbers in brackets on the diagram are partial correlation coefficients, controlling on prior variables which influence the relationship between a given independent and dependent variable. Thus the .35 on the line between quality of publications and recognition by peers is the partial correlation of these two variables, with quantity of publications and status of the PhD. department controlled. The .47 between quantity and recognition is the first-order partial correlation controlling for department status, etc. From this set of partial correlation coefficients Lightfield is able to examine the strength of association of publication variables, without the "contamination" of the general linear effects of departmental status. The relationship between recognition and either of the publication variables is positive and relatively strong even when departmental status is taken into account.

These partial correlation coefficients can also be expressed in terms of the

*Another way to express the meaning of the partial correlation coefficient is in terms of a correlation between two sets of residuals. Think of two linear regression equations, $Y' = a_{yt} + b_{yt}T$ and $X' = a_{xt} + b_{xt}T$, where T is the control variable whose effects are to be partialled out, statistically, and X and Y are the main independent and dependent variables whose relationship is of interest. If T is related to X and to Y then the two regression equations will be able to improve the prediction of Y and X somewhat, but there will likely be some residual in both Y and X yet to be explained. That is, for many scores, there will be both a $Y - Y'$ and an $X - X'$ difference. The partial correlation coefficient is the Pearsonian correlation, r, between these two sets of residuals which remain after the linear effect of T has been statistically removed. After T has explained all it can in terms of a simple linear regression, the remaining correlation between X and Y is the partial correlation coefficient.

proportion of variance in one variable explained by its linear association with the other variable, by simply squaring the coefficient. Here, quality contributes about 12 percent to the explanation of recognition and quantity contributes 22 percent.

It should be recalled that the correlation coefficient r indicates how well a straight line explains the relationship between two variables. One of the reasons for a low correlation (or low partial correlation) may be that the relationships are not very close to linear, either in general, or within categories of test variables.

8.4　STATISTICAL CONSIDERATIONS IN RESEARCH DESIGN

Statistical procedures for controlling on test factors are closely related to more general issues of research design. Courses and texts in research methods deal with these issues, and while we will not pursue these topics here, we would like to reemphasize their importance. Experimental designs approach the problem of handling other relevant variables somewhat differently, often through the use of random assignment of cases to control and experimental groups and through manipulation of an independent variable. Some of the advantages and possibilities of this approach are discussed in a classic article by the sociologist Samuel Stouffer (1962:Ch. 15).

The explanation of a dependent variable usually leads an investigator to consider not only other measured independent variables, but also the representativeness of the dependent variable (avoiding, for example, stratification on a dependent variable), independence of observations, the amount of variation on dependent and independent variables, and a number of other possible explanations of outcomes (such as chance or history, etc.), which are nicely described by Campbell and Stanley (1963).

Since the dependent variable generally is critical to a study, investigators often spend extra effort on its measurement to assure themselves that they are measuring the kind of phenomenon they wish to measure. Sometimes this care takes the form of including several separate measures of the dependent variable so that parallel analyses can be conducted using the different measures. In fact, there are recommended styles of analysis which specifically call for research designed to include different measurements and different methods to get at the same research problem (Campbell and Fiske, 1969).

There is a very close relationship between research design and statistical analysis. This book is designed to introduce you to some of the possibilities and requirements of statistics for sociological research.

8.5　SUMMARY

In this chapter we have been primarily concerned with the problem of simultaneously handling more than two variables. Our general concern is with conditions under which a relationship between an independent and dependent variable holds or varies. The problem, ideas, or theory of an investigator become particularly important in selecting relationships to examine and control variables to introduce into an analysis.

Three general approaches to the analysis of relationships between three or more variables were introduced. The first approach was *elaboration* of a relationship of interest (summarized in Section 3.2.6). This involved creation of conditional tables showing the relationship of key independent and dependent variables within categories of control variables. Lazarsfeld's accounting formula and hierarchial analysis helped summarize possible patterns of elaboration involving both the idea of statistical interaction or differences in strength and direction of association among conditional tables, and the idea of a general decrease, constancy, or increase in the strength of association in conditional tables.

When an investigator is interested in the general effect of introducing a control or test variable upon an original relationship, two other approaches may prove especially useful. The first of these is standardization. Test factor standardization, for example, is a procedure for creating a single table showing the relationship between an independent and dependent variable with the effects of control variables statistically removed. This permits a direct comparison of the original total association and a standardized table. Standardization provides for a more compact analysis of the general effect of introducing a test factor, but it does not permit an examination of differences in conditional tables. Likewise, the third approach, partial correlation, permits an investigator to contrast two summary numbers. One number is an original coefficient of association and the other is a kind of average of the associations in conditional tables or, to put it differently, a measure of association between two variables with the effects of control variables statistically removed. Again, as in the case of standardization, partial association permits us to examine what happens in general when control variables are introduced into the analysis of a relationship.

The next chapter directs attention toward another way to put together information from a series of independent variables in order to improve predictions of a dependent variable and to examine theoretically proposed models more fully. Our movement to this point has meant a selective dropping of detail to achieve more comprehensive and pertinent descriptions of interesting segments of the social world. The pursuits described in the next chapter follow this general progression as well.

CONCEPTS TO KNOW AND UNDERSTAND

conditional tables
control variables, test factors
zero-, first-, second-order associations
role of theory
theoretical ordering of variables
antecedent variables
intervening variables
hierarchical analysis

odds ratio
consequent variables
Lazarsfeld accounting formula
specification
causal model
testing for spuriousness
causal sequence
suppressor variables

independent causation
statistical interaction
structural analysis
contextual analysis
social system analysis
standardization

standardizing rates
standard population
standardized tables
partial correlation coefficients
partial association coefficients

QUESTIONS AND PROBLEMS

1. Using data from Table 8.8, create a standardized table and compare it with the original association of occupation and skin color. Discuss the advantages and disadvantages of using a test factor standardization rather than a partial association measure to test the investigator's ideas. It may help to look up the article.

2. Select a theoretical model which has been drawn out as an arrow diagram (such as that in the middle of Section 8.1.3 or in the data sets of a statistics workbook). Explain how one might use conditional tables, standardization, or partial correlation coefficients to test the model. What relationships would the model lead us to expect of data?

3. Does the high-school environment have an effect on student performance? Select some "environment" variable, such as social class of the high school, an individual characteristic such as IQ, and student performance, and then draw out a graph such as those shown in Figure 8.2 to illustrate (a) only an individual-level effect on performance, (b) only a group-level effect on performance, and (c) both group and individual effects. Carefully label each graph and describe what the graphs show. It might help to read further about this type of analysis in Riley (1963) and the example of contextual analysis discussed by McDill *et al*, (1967).

4. The data from Table 8.6 are presented below in frequency form. Analyze these data by using the techniques of partial association discussed in this chapter. Do the results agree with those reported in section 8.2.5b? Now compute the relevant odds ratios for these data. Do these results agree with those reported in section 8.2.5b? How do they compare with the analysis you did using techniques of partial association?

FREQUENCY TABLES FOR EMPLOYEES FEELING THEIR DEPARTMENT IS A HIGHLY
SUPPORTIVE WORK ATMOSPHERE BY INDIVIDUAL AGE AND ORGANIZATIONAL AGE

| Table A | Individual Age | | | |
	Young	Middle	Older	Total
Feelings About Support of Work Contexts				
Highly Supportive	27	41	45	113
Not Highly Supp.	36	49	37	122
Total	63	90	82	235

FREQUENCY TABLES FOR EMPLOYEES FEELING THEIR DEPARTMENT IS A HIGHLY
SUPPORTIVE WORK ATMOSPHERE BY INDIVIDUAL AGE AND ORGANIZATIONAL AGE

Table B

YOUNGER DEPARTMENTS

| | Individual Age | | | |
	Young	Middle	Older	Total
Feelings About Support of Work Contexts				
Highly Supportive	18	18	9	45
Not Highly Supp.	17	19	5	41
Total	35	37	14	86

MIDDLE-AGED DEPARTMENTS

| | Individual Age | | | |
	Young	Middle	Older	Total
Feelings About Support of Work Contexts				
Highly Supportive	7	17	17	41
Not Highly Supp.	16	20	12	48
Total	23	37	29	89

OLDER DEPARTMENTS

| | Individual Age | | | |
	Young	Middle	Older	Total
Feelings About Support of Work Contexts				
Highly Supportive	2	6	19	27
Not Highly Supp.	3	10	20	33
Total	5	16	39	60

---

5. The following tables are from a study of retired English men. They compare
the type of work engaged in before retirement (blue-collar versus white-collar)
with a measure of adjustment to retirement for different levels of self-perceived
health status. Analyze these data using the techniques discussed in this chapter. Discuss the results of your analysis and explain why you used the specific
techniques which you did.

| | Health Status: Poor | |
| | Type of Work | |
	Blue-Collar	White-Collar
Level of Adjustment to Retirement		
Poor	8	2
Fair	9	2
Good	6	6

Health Status: Fair *continued*

Health Status: Fair

| | Type of Work | |
Level of Adjustment to Retirement	Blue-Collar	White-Collar
Poor	6	2
Fair	13	3
Good	28	8

Health Status: Good

| | Type of Work | |
Level of Adjustment to Retirement	Blue-Collar	White-Collar
Poor	9	1
Fair	8	3
Good	54	59

TOTAL TABLE

| | Type of Work | |
Level of Adjustment to Retirement	Blue-Collar	White-Collar
Poor	23	5
Fair	30	8
Good	88	73

GENERAL REFERENCES

Blalock, Hubert M., Jr., *An Introduction to Social Research,* (Englewood Cliffs, N.J., Prentice-Hall), 1970.
Chapter 4 on "Explanation and Theory" is especially relevant to this chapter.

Davis, James A., *Elementary Survey Analysis,* (Englewood Cliffs, N.J., Prentice Hall), 1971.

Davis, James A., Joe L. Spaeth, and Carolyn Huson, "A Technique for Analyzing The Effects of Group Composition," *American Sociological Review* 26, (April, 1961), p 215–225.

Hyman, Herbert, *Survey Design and Analysis,* (Glencoe, Ill., The Free Press), 1955.

Riley, Matilda White, *Sociological Research: A Case Approach* volume 1, (New York, Harcourt, Brace and World), 1963.
See especially the commentary after Unit 12, pages 700–738, which discusses social system analysis, structural and contextual analysis.

Rosenberg, Morris, *The Logic of Survey Analysis,* (New York, Basic Books), 1968.

Stouffer, Samuel A., *Social Research To Test Ideas,* (New York, The Free Press of Glencoe), 1962.
See especially the 15th chapter, "Some Observations on Study Design."

OTHER REFERENCES

Campbell, Donald T., and Donald W. Fiske, "Convergent and Discriminant Validation by the Multitrait-Multimethod Matrix," *Psychological Bulletin* 56, (March, 1969), p 81–105.

Campbell, Donald T., and Julian C. Stanley, *Experimental and Quasi-Experimental Designs For Research,* (Chicago, Rand McNally), 1963.

Davis, James A., "The Goodman Log Linear System for Assessing Effects in Multivariate Tables" (Chicago, National Opinion Research Center), 1972. Litho.

———, "Hierarchical Models for Significance Tests in Multivariate Contingency Tables: An Exegesis of Goodman's Recent Papers," in Herbert L. Costner, editor, *Sociological Methodology 1973–1974* (San Francisco, Jossey Bass), 1974, p 189–231.

Erbe, William, "Social Involvement and Political Activity: A Replication and Elaboration," *American Sociological Review*, 29 (April, 1964), p 198–215.

Goodman, Leo A., "On the Multivariate Analysis of Three Dichotomous Variables," *The American Journal of Sociology*, 71 (November, 1965), p 290–301.

———, "How to Ransack Social Mobility Tables and Other Kinds of Cross-Classification Tables," *The American Journal of Sociology*, 75 (July, 1969), p 1–40.

———, "A Modified Multiple Regression Approach to the Analysis of Dichotomous Variables," *American Sociological Review*, 37 (February, 1972), p 28–46.

Lazarsfeld, Paul F., Bernard Berelson and Hazel Gaudet, *The People's Choice* (New York, Columbia University Press), 1948.

Lazarsfeld, Paul F., and Morris Rosenberg (editors), *The Language of Social Research* (Glencoe, Ill., The Free Press of Glencoe), 1955, p 231.

Lightfield, E. Timothy, "Output and Recognition of Sociologists," *American Sociologist* 6, (May, 1971), p 128–133.

McAllister, Ronald, Edward Kaiser, and Edgar Butler, "Residential Mobility of Blacks and Whites: A National Longitudinal Survey," *American Journal of Sociology,* 77 (November, 1971), p 445–456.

McDill, Edward L., Edmund D. Meyers, and Leo C. Rigsby, "Institutional Effects on the Academic Behavior of High School Students," *Sociology of Education* 40, (Summer, 1967), p 181–199.

Merton, Robert K., *Social Theory and Social Structure,* 1968 Enlarged Edition (New York, The Free Press), 1968, Part I.

Messer, Mark, "The Possibility of an Age-Concentrated Environment Becoming a Normative System," *Gerontologist* 7 (December, 1967), p 247–251.

Ransford, H. Edward, "Skin Color, Life Chances, and Anti-White Attitudes," *Social Problems* 18, (Fall, 1970), p 164–179.

Reiss, Ira L., *The Social Context of Premarital Sexual Permissiveness,* (New York, Holt, Rinehart and Winston), 1967, p 59 and 61.

Reynolds, Paul Davidson, *A Primer in Theory Construction* (Indianapolis, Ind., Bobbs-Merrill Company), 1971.

Rosenberg, Morris, "Test Factor Standardization As A Method of Interpretation," *Social Forces* 41, (October, 1962), p 53–61.

Stinchcombe, Arthur L., *Construction of Social Theories,* (New York, Harcourt, Brace and World), 1968.

U.S. Bureau of the Census, "Population Characteristics," *Current Population Reports,* Series P-20, No. 205, Washington, D.C., 1970.

Zeisel, Hans, *Say It With Figures,* Revised Fourth Edition, (New York, Harper and Row), 1957, Chapter 10.

9 Path Analysis and Multiple Regression

There are several analysis techniques used in sociology which fit under the general label of multiple regression analysis. They are appropriate where an investigator is interested in predicting or analyzing scores on one dependent variable by combining the predictive power of several independent variables by means of an equation which is called a **multiple regression equation.** How well this equation is able to predict scores on the dependent variable is indicated by the **multiple correlation coefficient,** R. In many ways this procedure is simply an extension of correlation and regression procedures we have discussed where only one independent variable is used to predict one dependent variable.

This chapter is an introduction to some of the multiple regression techniques you are likely to encounter in reading reports of sociological research. Topics we will discuss include multiple correlation and multiple regression, path analysis, the use of dummy variables, and stepwise multiple regression procedures. These are all quite easily understood at an interpretative-theoretical level, even though various computational alternatives and their derivation are details we will reserve for a separate course in regression analysis.*

*Clerical mathematics needed to compute some of the coefficients will generally be left to existing computer programs, as done by most investigators. Your instructor will explain how to use your computer and the programs that are available for multiple regression. These are generally quite easy to use in problems you may be assigned.

It is well to use some caution in accepting the output of "canned" computer programs which are not known to have been checked out by someone with sufficient statistical and mathematical insight to enable him/her to detect possible errors. Some checking procedures are provided in this chapter, but our focus will be on the theory necessary for an interpretative understanding of these techniques. With an adequate computer program you will be able to make use of the procedures yourself. An interesting commentary on the accuracy of some existing multiple regression programs is provided in Wampler (1970).

It is quite worthwhile and interesting to work through a more mathematical treatment of these topics in other books and courses. This would be especially helpful in interpreting results of more unusual applications of these techniques. See, for example, Draper and Smith (1966), in General References.

9.1 MULTIPLE REGRESSION AND MULTIPLE CORRELATION

9.1.1 Some Assumptions of Multiple Regression

Multiple regression techniques are among the more interesting and useful because they help handle the kind of complexity that begins to reflect theoretical notions sociologists have about the social world they are trying to explain. Even so, a number of simplifications are involved in multiple regression analysis, and these may mean that the regression model as described here simply is not appropriate or useful for many research problems. Figure 9.1 indicates other "multivariate" procedures which take into account more dependent variables, dichotomous dependent variables, underlying "factors" and the like.*

One basic assumption of multiple regression analysis is that variables are

Number of Independent Variables To Be Used At One Time	Number of Dependent Variables To Be Used At One Time	
	One	Two or More
None	(A) Univariate descriptive statistics.	(B) Same as cell (A). Where several items measure the same variable, one might use: Index Forming– Scaling– Factor Analysis– Smallest Space Analysis–
One	(C) Bivariate statistical description: Tables– Measures of association– Correlation regression–	(D) Same as cells (B), (C), (E), or some combination. Treating dependent variables one at a time and/or in scales.
Two or More	(E) Partial tables– Standardization– Partial correlation coefficients– Multiple regression and correlation– Discriminate function analysis–	(F) Same as cells (B), (E), in combinations treating dependent variables as scales or individually. Canonical correlation.

FIGURE 9.1 SOME ALTERNATIVE PROCEDURES FOR STATISTICAL DESCRIPTION

*Note that it is not always possible to divide variables neatly into "independent" and "dependent" categories.

related to the dependent variable in a simple *linear* fashion, and it is usually a good procedure to construct scatter diagrams to check this assumption. Sometimes simple transformations (such as a logarithmic transformation) can be used to uncover a linear relationship. Another assumption is that effects of variables can be *added* together to form a prediction of the dependent variable. "Statistical interaction" cannot be handled unless it is "scored" and included as a separate variable in the regression equation. Further, all of the variables included in the multiple regression equation should be interpretable as *interval-level* measures. Finally, there is an assumption of zero correlation between independent variables, so that the effects of each variable on the dependent variable, with others taken into account, can be reliably computed.

Some other assumptions are involved in specific applications of multiple regression (see Figure 9.2) and these will be discussed when introduced later in this chapter.*

Basic assumptions.

1. Independent variables are related in a linear fashion with the dependent variable and among themselves.
2. Effects of independent variables can be added together to yield a prediction of the dependent variable.
3. Independent variables are not correlated (or minimally correlated).**
4. All variables are interval-level variables.

Added assumptions if one is interested in running statistical tests of hypotheses about a population from random sample data.

5. The dependent variable is normally distributed within categories of independent variables, singly and in combination.
6. The variance in the dependent variable is equal across categories of the independent variables.

Added assumptions if multiple regression analysis is to be applied to testing causal models of the relationship between variables.

7. The theoretical ordering of independent and dependent variables should be known and such that independent variables change first and dependent variables later.
8. The set of independent and dependent variables should be inclusive of all (major) variables influencing the dependent variables. That is, it should be a closed system.
9. Measures should have high (demonstrated) reliability and validity.
10. Disturbance terms (error terms) should be uncorrelated with each other or with independent variables directly connected to the same variable.

**For a discussion of correlated independent variables, see Blalock, (1970).

FIGURE 9.2 THE ASSUMPTIONS OF MULTIPLE REGRESSION ANALYSIS

*Assumptions should be taken seriously and consequences of departures from one assumption upon the importance of other assumptions is the subject of debate and inquiry in the field. The interval-level measurement assumption, for example, can be handled in some cases for even nominal-level variables by using what are called "dummy variables." The "no-intercorrelation among independent variables" assumption is usually interpreted to mean "low intercorrelation" (low multicollinearity), but checking and caution are needed where there are small departures. It may also be the case that the theoretical idea of variables related in a simple, linear fashion is intrinsically interesting as a model (or first model) of some aspect of the social world. In this case, the results of a regression analysis would be interesting, even though all of the regression assumptions may not fit the world particularly well.

BOX 9.1 Reviewing Correlation and the Nature of Association

At this point you may want to check back on the idea of *correlation* (Chapter 7) and particularly the *nature* of association and possibilities for linear and curvilinear association (end of Section 7.4.4). Sometimes some function of the scores (like logarithms) rather than the scores themselves will have a simple linear relationship with another variable and thus satisfy a presumption of the Pearson correlation coefficient and linear regression analysis. The transformation of scores to produce this effect is something that requires experience and usually involves double-checking through the use of scatter diagrams both before and after transformations.

9.1.2 One Independent Variable: A Review

In Chapter 7, "simple" (one independent and one dependent variable) linear regression equations were expressed in the following form where the b-value represents the slope or amount of change in the dependent variable for each unit change in the independent variable and a_{yx} represented the Y-intercept where the regression line crossed the Y-axis.

(9.1)
$$Y' = a_{yx} + b_{yx}X$$

If the scores for both the Y and X variables were expressed not as raw measures but as standard scores (z-scores), the regression equation could be expressed as follows.† Note that we are changing the notation slightly so that all variables including the dependent variable are referred to by a different subscript.**

(9.2)
$$z_1' = b_1^* + b_{12}^* z_2$$

The b_1^* (b-star or beta-weight) coefficient in (9.2) corresponds to a_{yx} in equation (9.1) and in fact the two equations are identical, term for term, except for the notation and the use of z-scores rather than raw Y and X scores in (9.2).

Since the mean of a distribution of z-scores is zero, and since the regression "line" passes through $\overline{X}_1$, $\overline{X}_2$, then the value of b_1^* will always be zero, and the regression equation in (9.2) can be simplified to this:

---

†Recall that z-scores are simply the difference between a score and the mean of its distribution expressed in terms of the number of standard deviation units it is from the mean. The formula for z-scores is this: $z = (X_i - \overline{X})/s$. It is discussed in Section 5.4.7.

**Note on symbolism. Dependent variables are referred to in a number of different ways, for example, as: Y or X_0 or X_1 etc. To indicate that this is a predicted value, some texts use a prime (X') and some use a "hat" ($\hat{X}$). We will use subscripts starting at 1, generally, to refer to variables and explain in context whether each variable is to be treated as an independent or dependent variable. A predicted dependent variable will be indicated by a prime, X_1'. Weights in the regression equation will be written in two ways, b for weights in an equation which uses raw scores, and b^* (b-star) for weights in an equation where standard scores (z-scores) are used. Appropriate subscripts will also be added: the first indicating the dependent variable; the second indicating the independent variable (and other variables in the equation will be listed following a dot in the subscript). Some texts also introduce Greek letters for the a and b coefficients where the population value (rather than sample value) is used. To simplify symbolism at this point, sample notation is used. Later on (in inferential statistics), where the distinction between population and sample values is of concern, further symbolism may be introduced and explained.

(9.3)
$$z_1' = b_{12}^* z_2$$

It can be shown that in the case where there is only one independent variable, $b_{12}^* = r_{12}$. (Of course $r_{12} = r_{21}$ since r is a symmetric measure of association.) Thus the regression equation in Formula (9.3) could be expressed (with one independent variable) as follows:

(9.4)
$$z_1' = r_{12} z_2$$

The correlation coefficient is simply the slope coefficient in a two-variable situation where standard scores are used. The correlation coefficient can be seen as a correlation between the score on the dependent variable *predicted* by using the regression equation and the *actual* score of the dependent variable. The closer the prediction comes to the actual score, the higher the degree of linear association and the closer r comes to plus or minus one.

The coefficient of determination (the square of the correlation coefficient), r_{12}^2, indicates the proportion of variation in one variable (for example the dependent variable) which is explained by its linear association with the other variable expressed in the right-hand side of the linear regression equation. Unexplained variation is expressed as $1 - r_{12}^2$ and is referred to as the **coefficient of non-determination.** In the two-variable situation, these interpretations also apply to b_{12}^*, b_{12}^{*2}, and $1 - b_{12}^{*2}$.

9.1.3   Two or More Independent Variables

When more than one independent variable is used to explain variation in a single dependent variable, the basic regression equation in (9.3) can be extended as in (9.5), below, to include terms which combine a beta-weight (b^*) coefficient and the z-score of k independent variables. Thus we have the **multiple regression equation:**

(9.5)
$$z_1' = b_{12 \cdot 34...k}^* z_2 + b_{13 \cdot 24...k}^* z_3 + b_{14 \cdot 23...k}^* z_4 + b_{1k \cdot 234...}^* z_k$$

Here the standard scores for each independent variable are weighted according to the contribution that variable makes to the overall predicted sum, z_1'. Each b^* value represents the relative amount of contribution of that variable, after contributions of the other variables included in the regression equation are taken into account. In that sense, the b^* values are like a partial correlation coefficient in "holding constant" or "correcting" for the contribution of other included variables. Often this controlling effect is indicated in the subscripts for the b^* coefficients, as illustrated in the three-variable regression equation, below. The dependent and independent variables to which the b^* value refers are written in front of the dot in the subscript, and the other variables included in the regression equation are listed after the dot. Those after the dot are "held constant," *i.e.* their contribution is partialled out. As usual, the first subscript is the dependent variable and the second is the independent variable for that particular b^* coefficient. Using two independent variables, Formula (9.5) becomes this:

(9.6)
$$z_1' = b_{12 \cdot 3}^* z_2 + b_{13 \cdot 2}^* z_3$$

Beta-weights (b^*) are computed in a way that minimizes the sum of squared deviations between predicted and actual scores on the dependent variable. This is called the **least-squares criterion.**

(9.7)
$$\Sigma(z_1 - z_1')^2 = \text{a minimum}$$

Using only two independent variables and substituting the right-hand side of equation (9.6) for z_1' , the quantity to be minimized is this:

(9.8)
$$\Sigma(z_1 - b_{12\cdot3}^* z_2 - b_{13\cdot2}^* z_3)^2 = \text{a minimum}$$

This results in a maximum linear correlation between predicted and actual scores on the dependent variable.

To carry out the computations implied by the least squares criterion, a set of normal equations (*not* related to the idea of a normal curve in any way), predicting correlations between each variable and the dependent variable, are created and solved for unknown b^*s. For the three-variable problem, these normal equations would be as follows:

(9.9)
$$b_{12\cdot3}^* + r_{23}b_{13\cdot2}^* = r_{12}$$
$$r_{23}b_{12\cdot3}^* + b_{13\cdot2}^* = r_{13}$$

Note that solution for the b^*s is possible because there are as many equations as unknowns. The intercorrelation of variables can be computed directly from the data in the study being used in the multiple regression analysis. Thus, in a three-variable regression equation with two independent variables, the two normal equations in (9.9) could be solved for the two betas, $b_{12\cdot3}^*$ and $b_{13\cdot2}^*$.

Hand computation of the b^* coefficients is illustrated in such books as Walker and Lev (General References, 1953; Chapter 13, discusses the Doolittle method of solution). Usually b^* coefficients are computed by means of computer programs for multiple regression, where raw scores on all variables are transformed into standard scores as a first (often optional) step in the computer program. The beta coefficients can also be computed from a correlation matrix of the interrelationship between all pairs of variables to be included in the regression equation, but this procedure becomes tedious for more than one or two independent variables. Diagram 9.1 illustrates these procedures.

The b^* coefficients provide an investigator with a basis for comparing the relative contribution of one variable to a prediction of the dependent variable with the contribution of other variables in the equation. If $b_{12\cdot3}^*$ were larger than $b_{13\cdot2}^*$, then one would be able to conclude that a given amount of change in z_2 would result in more change in the dependent variable than the same amount of change in the z-score of variable z_3. z_2 is a more potent influence than z_3. If $b_{12\cdot3}^*$ is twice the size of $b_{13\cdot2}^*$, then a given change in z_2 has twice the effect of the same change in z_3 . Variables with b^* coefficients which are very nearly zero have very little separate influence on the dependent variable, at least in the linear, additive way described by the multiple regression equation. We might, on this basis, argue that such variables could well be eliminated from the prediction equation, and the beta coefficients could be re-computed.

At this point we are only talking about the way changes in some variables are related to changes in other variables, and not about influence in a "cause-

DIAGRAM 9.1 Standardized Regression Coefficients

The problem: Compute the standardized regression weights (b^*) for the following 3-variable, multiple regression equation.**

(D1)
$$z_i' = b_{ij \cdot k}^* z_j + b_{ik \cdot j}^* z_k$$

Computation from correlations among the variables:

(D2)
$$b_{ij \cdot k}^* = \frac{r_{ij} - r_{ik} r_{jk}}{1 - r_{jk}^2}$$

where the subscript i refers to the dependent variable and j is the independent variable for which the regression-weight is being computed. Variable k is the third variable involved in the regression equation in (D1). Note that two such regression weights must be computed, the one given in (D2) and this:

$$b_{ik \cdot j}^* = \frac{r_{ik} - r_{ij} r_{jk}}{1 - r_{jk}^2}$$

Computation from unstandardized regression coefficients:

(D3)
$$b_{ij \cdot k}^* = b_{ij \cdot k} \frac{s_j}{s_i}$$

where i refers to the dependent variable and j to the independent variable for which the b^* coefficient is being computed. k is the third variable in the equation. s_j is the standard deviation on the independent variable. s_i is the standard deviation on the dependent variable. b_{ij} is the *unstandarized* or *raw* score form of the regression coefficient.

In addition to (D3), one would have to compute this:

$$b_{ik \cdot j}^* = b_{ik \cdot j} \frac{s_k}{s_i}$$

Computation using standard computer programs:

Many universities have existing computer programs which will compute regression coefficients from information on each variable for each case. To compute standardized regression coefficients, select the control-card option which converts the raw score data into standard score form prior to computations. The transformation and regression computations generally occur in the same computer

---

**Note that these formulas can be extended easily to the situation where there are more independent variables, simply by adding subscripts to specify the specific betas of interest and adding terms to the standardized regression equation, one for each additional variable.

DIAGRAM 9.1 *(Continued)*

run. The computational technique used is likely to be one that makes use of matrix algebra. Although this topic is not discussed in this volume, it would provide one clear basis for expressing the relationship between correlations of the input data and standardized regression coefficients. If you are interested in this way of solving the problem of standardized coefficients you should see (in General References) the following: Walker and Lev (1953), Chapter 13; Draper and Smith (1966), Chapter 6.8; or Cooley and Lohnes (1962), Chapter 3.

Example

Duncan (1961:124) presents data on a socioeconomic index for occupation groups (a social status measure), income level and educational level for the same occupational groups. The correlation matrix is as follows:

	X_1	X_2	X_3
X_1 Duncan"s SEI		.84	.85
X_2 Income Level			.72
X_3 Education Level			

Compute coefficients for this:

$$z_1' = b_{12 \cdot 3}^* z_2 + b_{13 \cdot 2}^* z_3$$

from formula (D2) above:

$$b_{12 \cdot 3}^* = \frac{r_{12} - r_{13} r_{23}}{1 - r_{23}^2}$$

$$= \frac{.84 - (.85)(.72)}{1 - (.72)^2}$$

$$= \frac{.84 - .61}{1 - .52} = \frac{.23}{.48} = .48$$

and

$$b_{13 \cdot 2}^* = \frac{r_{13} - r_{12} r_{23}}{1 - r_{23}^2} = \frac{.85 - (.84)(.72)}{1 - (.72)^2} = \frac{.85 - .60}{1 - .52} = \frac{.24}{.48} = .50$$

The equation would thus be written as follows, predicting the standard score of variable 1, Duncan's Socioeconomic Index from variable 2, income, and variable 3, education.

$$z_1' = .48z_2 + .50z_3$$

Income makes about the same contribution as education to the prediction of Duncan's SEI score.

effect" sense, although clearly, given other kinds of information, such as the theoretical ordering of the influence of variables, we might use evidence from a multiple regression equation in evaluating a theoretical cause-effect argument.

9.1.4 Regression Weights and Partial Correlation Coefficients

The standardized beta weight (b^*) and the partial correlation coefficient share some characteristics. They both reflect the effect of an independent variable on a dependent variable when the effects of other included independent variables are taken into account, statistically. The dependent variable scores to be predicted are adjusted by subtracting out predictions that would be made by linear regression equations that omit a given independent variable. Thus the b^* for a given independent variable is computed so that it best predicts (in the least squares sense) the adjusted scores for the dependent variable. The partial correlation coefficient expresses the relationship between such predicted scores and the adjusted scores on the dependent variable.

The two coefficients provide different information, since the b^* indicates the *amount of change in the dependent* variable which is associated with a unit change on the independent variable (when other independent variables are taken into account). As such it is an asymmetric measure. The partial correlation coefficient is a symmetric measure which indicates the closeness of relationship, overall, between a dependent and independent variable when scores of the dependent variable have been adjusted to take out the effect of variation in other independent variables implied by their linear relationship with the dependent variable. The partial correlation coefficient provides a measure of the accuracy of prediction and the beta coefficient provides a measure of the contribution of a variable to the prediction. By squaring the partial correlation coefficient, one can measure the proportion of the variation in the dependent variable that is explained by the direct contribution of an independent variable when effects of other included variables are taken into account. Both coefficients provide useful and somewhat different information.

9.1.5 Multiple Correlation

Multiple correlation ($R_{1 \cdot 23}$), like the simple product-moment correlation coefficient r_{yx}, is simply the correlation between the actual scores on the dependent variable, and the scores on the dependent variable predicted by use of the multiple regression equation. The beta weights (b^*), in fact, are computed so that this correlation is as high as possible for a given set of data. The multiple correlation coefficient is symbolized by the capital letter, R, and it varies on a scale from 0 to +1.00. The smaller the coefficient the poorer the correlation, and the larger the coefficient, the stronger the correlation.

Like the coefficient r, the multiple correlation coefficient can be interpreted more usefully by squaring it. R^2 has an interpretation quite akin to r^2, namely, the proportion of the variation in the dependent variable which is explained by the regression equation.

$$R^2 = \frac{\text{Explained variation in } X_1}{\text{Total variation in } X_1}$$

(9.10)

$$R^2 = \frac{\Sigma (X_1' - \overline{X}_1)^2}{\Sigma (X_1 - \overline{X}_1)^2}$$

R^2 is called the coefficient of multiple determination. $1 - R^2$ is the proportion of variation in the dependent variable left unexplained by the multiple regression equation.

Diagram 9.2 explains the computation of the multiple correlation coefficient. If all of the intercorrelations of independent variables were zero, then the square of the multiple correlation coefficient would simply be the sum of squared correlations between each independent variable and the dependent variable as in Formula (9.11), below.

(9.11)

$$R^2_{1 \cdot 23} = r^2_{12} + r^2_{13}$$

If, as is usually the case, some independent variables are related to each other, the "overlap" in contribution to the explanation of the dependent variable would have to be taken into account and eliminated from (9.11) in order to arrive at R^2. This is done simply by adjusting each r by multiplying it times the related beta weight, b^*, as in (9.12).

(9.12)

$$R^2_{1 \cdot 23} = r_{12} b^*_{12 \cdot 3} + r_{13} b^*_{13 \cdot 2}$$

Both (9.11) and (9.12) could be extended to include more than two independent variables by simply adding additional terms.

It is apparent from (9.11) that R will be zero if all the correlations between dependent and independent variables are zero, and R cannot be less than the highest r relating any independent and dependent variable. R exceeds the highest r relating any independent and dependent variable by the largest amount when independent variables are independent of each other (*i.e.* have zero intercorrelations), since each variable contributes an added, separate amount to the prediction of the dependent variable.**

9.1.5a Corrected R. The multiple correlation coefficient computed from a sample tends to overestimate the coefficient for the population (the parameter). The extent to which the sample coefficient will be inflated depends upon the size of the sample and the number of independent variables in the regression equation. This bias in the direction of a higher R for sample data is due to the fact that the multiple regression equation is "tailored" to the sample data to produce the highest R possible. This operation takes advantage of any chance difference between the sample and population distribution of scores to find a higher R by "tailoring" the b^*s to the data. Since a small sample is likely to have a

---

**Your attention should be called to another variation on the multiple correlation coefficient. It is a "multiple partial correlation coefficient," and it indicates the multiple correlation between a set of independent variables and a dependent variable when other independent variables are statistically controlled. Thus it is useful in examining the explanatory power of one set of independent variables on a dependent variable, controlling for effects of another set of independent variables. The multiple partial coefficient is introduced by Blalock, (1972), in General References, pages 458–459.

DIAGRAM 9.2   The Multiple Correlation Coefficient

The Problem:

Compute a measure of the proportion of the variation in a dependent variable which is explained by the linear combination of a set of independent variables, R^2 (the coefficient of multiple determination).

Computation of R^2 from Correlation Coefficients and Standardized Regression Betas:

(D1)
$$R^2_{1 \cdot 23} = r_{12} b^*_{12 \cdot 3} + r_{13} b^*_{13 \cdot 2}$$

where variable 1 is the dependent variable and 2 and 3 are independent variables. Correlation coefficients are zero-order associations between the dependent and each independent variable.

Note that this can be extended for any number of independent variables by adding comparable terms to the right-hand side of (D1) and listing all independent variables in the subscript of R.

Computation of R^2 from Zero-Order and Partial Correlation Coefficients; Illustrated for Three Variables:

(D2)
$$R^2_{1 \cdot 23} = r^2_{12} + (1 - r^2_{12}) r^2_{13 \cdot 2}$$

where variable 1 is the dependent variable.

Note that r^2_{12} is the proportion of variation in the dependent variable explained by the linear association with variable 2. $1 - r^2_{12}$ is the unexplained portion. Of this unexplained portion, variable 3 can explain an added amount symbolized by the partial correlation coefficient $r^2_{13 \cdot 2}$.

Computation of R^2 from Zero-Order Correlation Coefficients; Illustrated for the Three-variable Case:

(D3)
$$R^2_{1 \cdot 23} = \frac{r^2_{12} + r^2_{13} - 2r_{12} r_{13} r_{23}}{1 - r^2_{23}}$$

where variable 1 is the dependent variable.

Note, if the intercorrelation of the independent variables is zero, equation (D3) reduces to this:

$$R^2_{1 \cdot 23} = r^2_{12} + r^2_{13}$$

and this can be generalized to any number of independent variables.

Corrected R^2:

(D4)
$$R^2_c = 1 - \left(\frac{N-1}{N-k}\right)(1 - R^2)$$

where N is the sample size; k is the number of independent variables included in the multiple regression equation; R^2 is the uncorrected coefficient; R^2_c is the corrected coefficient.

DIAGRAM 9.2 *(Continued)*

Example:

In the data from Duncan where occupational social status scores (Duncan's SEI) are the dependent variable and income and education are independent variables, the following partial correlation and zero-order correlation coefficients are provided.

$$r_{12} = .84 \qquad r_{12 \cdot 3} = .61$$
$$r_{13} = .85 \qquad r_{13 \cdot 2} = .65$$
$$r_{23} = .72$$

From (D3), above:

$$R^2_{1 \cdot 23} = \frac{r_{12}^2 + r_{13}^2 - 2r_{12}r_{13}r_{23}}{1 - r_{23}^2}$$

$$= \frac{.84^2 + .85^2 - 2(.84)(.85)(.72)}{1 - (.72)^2}$$

$$= \frac{.706 + .723 - 2(.514)}{1.00 - .518}$$

$$R^2_{1 \cdot 23} = \frac{.400}{.482} = .83 = \begin{array}{l}\text{Coefficient of}\\ \text{Multiple Determination}\end{array}$$

$$R_{1 \cdot 23} = \sqrt{.83} = .91 = \begin{array}{l}\text{Multiple Correlation}\\ \text{Coefficient}\end{array}$$

Note that variable 2 explains $.84^2 = .706$ of the variation in variable 1 and leaves $(1 - .706)$ or .294 unexplained.

Of the .294 left unexplained by variable 2, variable 3 is able to explain an additional 42 percent (*i.e.* $.65^2$), or .423 of .294 is .124. Thus the two variables can explain a total of $.706 + .124$ or .83 of the variation in variable 1 by their linear association with variable 1.

Eighty-three percent of the variation in variable 1 is a rather high amount to explain. Perhaps by adding another carefully chosen variable, one could explain even more of the variation in variable 1.

Because of bias involved in taking advantage of sampling variability to find the highest possible R, a corrected R could be computed from Formula (D4). Duncan's sample size was 45 occupations. There were two independent variables.

$$R^2_{1 \cdot 23} = 1 - \frac{45 - 1}{45 - 2}(1 - .83) = 1 - (1.02)(.17) = 1 - .173 = .827 \text{ or } .83$$

DIAGRAM 9.2 *(Continued)*

Here, the sample size compared to the number of independent variables was sufficiently large to result in a negligible correction.

Alternative computation of R^2:

Using Formula (D1), the same R^2 could be obtained since the b^* values are available from computations in Diagram 9.1, as follows:

$$b^*_{12 \cdot 3} = .48 \qquad r_{12} = .84$$
$$b^*_{13 \cdot 2} = .50 \qquad r_{13} = .85$$

from (D1)

$$R^2_{1 \cdot 23} = r_{12} b^*_{12 \cdot 3} + r_{13} b^*_{13 \cdot 2}$$
$$R^2_{1 \cdot 23} = (.84)(.48) + (.85)(.50) = .403 + .425 = .828 \text{ or } .83$$
$$R_{1 \cdot 23} = .91$$

larger sampling error than a larger sample, an R computed from a small sample is more likely to deviate from the parameter it is meant to estimate. Furthermore, as the number of independent variables in the regression equation is increased, the job of tailoring the R to the sample data will be more efficient, thus exaggerating the bias. This bias can be corrected by using the following formula:

(9.13)
$$R_c^2 = 1 - \left(\frac{N-1}{N-k}\right)(1-R^2)$$

where R^2 is the uncorrected coefficient; R_c^2 is the corrected coefficient; N is the sample size; k is the number of independent variables included in the multiple regression equation.

Generally, the investigator is interested in finding sets of independent variables which will provide the best prediction of the dependent variable, the highest value of R. If he(she) preselects independent variables by looking for independent variables which are most highly correlated with the dependent variable in his(her) sample data, and then introduces these into a multiple regression equation, the inflation of R is even greater than that eliminated by the correction in Formula (9.13) above. An R, re-computed on a larger sample or on the population would be less than that found by preselecting independent variables. For this reason, an investigator may want to replicate his(her) study on other new data and use relatively large samples as the basis of computing multiple regression coefficients and R.

Notice in Formula (9.13) that when the sample size, N, and the number of independent variables, k, are equal, R^2 and R will always be equal to 1. This is because b^*s would be calculated to perfectly fit each individual score, and there would be no difference between predicted and observed scores. The numerator in Formula (9.10) would be equal to the denominator simply because of this, and R^2 would equal 1. To compute the multiple regression and multiple correlation coefficients, then, the number of cases must exceed the number of variables.

9.1.6 An Illustration

Koslin and a group of investigators at Princeton University studied a group of 29 eleven to thirteen year old boys at a boys camp in Canada (Koslin et al., 1968). In this study they developed a regression equation to predict the standing of a boy in a group, using several measures of the expectations group members had for each other's performance on several tasks. They argue that individuals within groups develop ways of evaluating the performance of their associates, and that expected performance is related to status in the group. Higher-status individuals are expected to perform better than lower-status individuals on tasks of central concern to the group, regardless of actual performance. If this is the case, then one should be able to predict status within a group by knowing whether an individual's performance is "over-rated" or "under-rated" when compared with his(her) actual performance.

Four different tests were devised to measure over-under rating: (1) a rifle task, (2) a canoe task, (3) a sociometric "preference" test and (4) a height perception test. The rifle test, for example, involved having each boy in turn shoot four shots at a target while other group-members watched. The target disappeared immediately after a shot was fired and the boy who shot as well as others in the group was asked to record how close to the "bull's eye" the shot hit. The score on this task for each group member was the average "over" or "under" estimate of his shooting accuracy (compared with where the actual shot hit) as made by group members who watched him shoot. The canoe task involved group estimates of the time a boy took to go a given distance. The sociometric questionnaire asked for teammate preferences, and the height test involved comparisons of an individual's height in relation to other individuals in the group. In each case, the test was conducted in an ambiguous situation (*i.e.*, participant members had no watches to measure time; people were too far apart to allow actual comparisons of height, etc.). The dependent variable, group status, was determined by the investigators through careful observation of the boys over the course of the experiment. Data from the study are as follows:

TABLE 9.1 CORRELATION MATRIX FOR THE PREDICTOR AND CRITERION VARIABLES

	(2) Canoe test	(3) Socio- metric test	(4) Height Guess test	(5) Observed Social Status
1. Rifle test	.52	.58	.48	.67
2. Canoe test		.61	.65	.67
3. Sociometric			.66	.64
4. Height guessing				.51

The regression equation computed by Koslin using standardized beta weights was as follows:

$$z'_5 = .37z_1 + .37z_2 + .25z_3 - .07z_4 \qquad \text{(9.14)}$$

where z_1 is the rifle test; z_2 is the canoe test; z_3 is the sociometric test; z_4 is the height guessing test; z_5 is the investigator's rating of individual social status in the group on the basis of careful observational data.

It is immediately evident that the rifle and canoe tests are equally good predictors of observed status, the sociometric test is somewhat poorer but still a positive contributor to the prediction of observational status independently of the other three variables. Only height-guessing makes a very minor contribution to the prediction, and that contribution is a negative one. Thus their expectations were supported; over-under ratings on relevant tasks could be used to predict observed status and over-under ratings on irrelevant tasks contributed little in addition. Notice that this conclusion is not immediately apparent in Table 9.1, where zero-order correlation coefficients are all relatively high. In fact,

TABLE 9.2 MULTIPLE CORRELATION USING OBSERVED SOCIAL STATUS AS THE CRITERION (DEPENDENT) VARIABLE

Predictor	$b*$	Partial r
1. Rifle test	.37	.42
2. Canoe test	.37	.39
3. Sociometric	.25	.26
4. Height guess	−.07	−.08

Multiple R = .79

Source: Koslin, 1968. Used by permission.

from the $b*$ coefficients, one could say that a unit change in the rifle and canoe tests results in the same amount of change in observed status; but the sociometric test contributes only two-thirds as much as either of these variables (.25/.37 = .676). The impact of height is only 19 percent of that of the rifle or the canoe tests, and only 28 percent of the impact of sociometric choice.

Partial correlation coefficients provide somewhat different but consistent information. The partial correlation coefficient .42 in Table 9.2 is the correlation between the rifle test scores and observed status when effects of the canoe test, sociometric choice, and height-guessing were taken into account. The square of these coefficients provides a way of judging the proportion of the variation in the dependent variable accounted for by a given independent variable after other effects are taken into account. Thus the rifle test explained 18%, the canoe test 15%, the sociometric test 7%, and the height guessing test about one-half of one percent.

Overall, the multiple regression equation yielded a multiple correlation between predicted and observed social status scores of .79. Squaring this coefficient, it becomes evident that 62% of the variation in the dependent variable is accounted for by the multiple regression equation and 38% (the difference between 62% and 100%) is left unexplained. Since there were four independent variables and a sample size of only 29, a better estimate of a population multiple correlation coefficient would be R = .76, a corrected value computed from Formula (9.13).

9.2 PATH ANALYSIS: AN APPLICATION OF MULTIPLE REGRESSION TO PROBLEMS OF THEORY

9.2.1 Path Analysis

The use of standardized multiple regression equations in examining theoretical models is called **path analysis.** The objective is to compare a model of the direct and indirect relationships that are presumed to hold between several variables to observed data in a study, in order to examine the fit of the model to the data.

If the fit is close, the model is retained and used or further tested. If the fit is not close, a new model may be devised, or, more likely, the old one will be modified to better fit the data and then be subjected to further tests on new data.

This is precisely the interest we have if we want to build theoretical explanations of social phenomena. Where the underlying assumptions of path analysis are reasonably met, it provides a very pertinent way to relate theory and data when many variables are to be handled simultaneously. Correlation analysis alone, although helpful, does not provide as useful a measure of the impact of variables directly and indirectly on others.

In addition to the basic assumptions of multiple regression, path analysis includes two others (see Figure 9.2). First, the variables included in a model must be known to fall in some specific theoretical order in their effect on individuals; the independent variables must fall in causal order *before* the dependent variable. This is important and not always obvious.** Secondly, the model must be treated as a closed system in the sense that all important variables are explicitly included in the model. To some extent one can assess the extent to which this is true in the process of conducting the path analysis itself. Finally, for the purposes of this presentation, it is assumed that the influence of one variable on another is a one-way influence, or, in other words, the model is *recursive*; there is no "feedback."

Path analysis models are generally illustrated, as in Figure 9.3, by means of one-headed arrows connecting some or all of the variables included in the model. Variables are distributed from left to right, depending upon their theoretical-ordering, with the first independent variables at the extreme left. Intercorrelations (zero-order) between variables not influenced by other variables in the model are given as numbers on curved, two-headed arrows. These are cor-

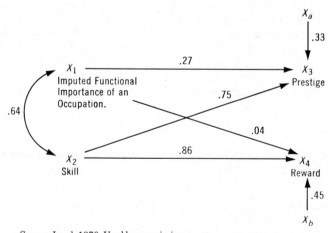

Source: Land, 1970. Used by permission.

FIGURE 9.3 Davis-Moore Model of Social Stratification

**At this point you may wish to refer back to discussions in Chapter 8 on the ordering of variables and the use of diagrams to make explicit the relationships between variables implied by a theory.

relations between *exogenous* variables, those influenced only by variables prior to and outside the model. Exogenous variables provide a link between a model under consideration and outside variables.

Numbers entered on the direct paths reflect the amount of direct contribution of a given variable on another variable when effects of other related variables are taken into account. Path coefficients are symbolized by the letter p, with subscripts indicating the two variables it connects. It turns out that path coefficients are identical to the b^* coefficients in the standardized multiple regression equation discussed earlier, where the regression equation reflects the structure of the model one is testing. These are computed as described earlier, from scores ultimately measured on a group of subjects. Thus the path model and the path coefficients provide a picture of the bit of the social world an investigator is interested in explaining, together with the coefficients describing the impact of independent variables. The impact is in terms of the amount of change in the dependent variable which is associated with a unit change in a given independent variable, over and above the contribution of other variables to the dependent variable.

The basic theorem of path analysis, and the way by which a specific structural model is reflected in the computation of **path coefficients** is as follows:

(9.15)
$$r_{ij} = \sum_k (p_{ik} r_{kj})$$

where: k includes each of the variables connected directly to the dependent variable i and prior to i in theoretical ordering shown in a path diagram. Let us develop these ideas in terms of a particular model and some actual data.

9.2.2 An Example of Path Analysis in Sociology

The Davis-Moore theory of social stratification argues that some type of social stratification will be found in every society because some types of duties must be performed for the society to continue as a functioning unit, and by some means individuals must be motivated to accept and perform these duties (see Land, 1970). The stratification system is one means of assuring the placement and motivation of individuals in the social structure. Rewards used in the process include income, leisure, and prestige. These are distributed among occupations, for example, in terms of (a) the importance of that occupation in the society, and (b) the scarcity of appropriately trained individuals needed for the occupation. Figure 9.3 describes this model, except that income and leisure are combined as "other rewards," distinguished from "prestige." Note that two other "variables" (X_a and X_b) are included to reflect variables external to the model and measurement errors that may influence the dependent variables.

Using X rather than z to stand for the *standard score* of a given variable (since this is the convention here), one could express the graphic model in terms of a "Path Model" as follows:

(9.16a)
$$X_3 = p_{31} X_1 + p_{32} X_2 + p_{3a} X_a$$

(9.16b)
$$X_4 = p_{41} X_1 + p_{42} X_2 + p_{4b} X_b$$

This is a "fully" recursive model, since all of the possible one-way arrows are drawn between the four explicit variables included in the model. These structural equations correspond to the multiple regression equations in (9.6), above, except for the addition of the residual variables X_a and X_b.† These equations can be rewritten in terms of path coefficients and zero-order correlation coefficients in a fashion parallel to that in Formula (9.9), remembering that path coefficients are standardized b*s. In fact, in this example, one could use a computer program for a multiple regression equation with standardized data using X_1 and X_2 to predict X_3, and X_1 and X_2 to predict X_4.** From the information provided all of the path coefficients would be readily available.

Using path coefficients and correlation coefficients, one could write a series of equations which would predict the correlation of independent and dependent variables. The basic theorem of path analysis (9.15), provides the model for constructing these path estimating equations.

(9.17a)
$$r_{31} = p_{31}r_{11} + p_{32}r_{21} \text{ (or, since } r_{11} = 1.0) \ p_{31} + p_{32}r_{21}$$

(9.17b)
$$r_{32} = p_{31}r_{12} + p_{32}r_{22} \text{ (or, since } r_{22} = 1.0) \ p_{31}r_{12} + p_{32}$$

(9.17c)
$$r_{41} = p_{41}r_{11} + p_{42}r_{21} \text{ (or, since } r_{11} = 1.0) \ p_{41} + p_{42}r_{21}$$

(9.17d)
$$r_{42} = p_{41}r_{12} + p_{42}r_{22} \text{ (or, since } r_{22} = 1.0) \ p_{41}r_{12} + p_{42}$$

Taking (9.17a) as an example and referring to the path diagram in Figure 9.3, notice that the correlation between variables 3 and 1 can be written in terms of the path between variables 3 and 1, *times* the correlation of variable 1 with itself (which is taken as 1.00, and so r_{11} and r_{22} could be dropped from these equations), *plus* the path from 3 to 2 and the correlation of variable 1 and 2. Here, k in the path theorem, Formula (9.15), has taken on two values, 1 and 2, reflecting the fact that variables 1 and 2 are directly connected and prior to variable 3. Four such equations are possible to predict each of the four correlations between independent and dependent variables. Since these four equations have only four unknown path coefficients, they can be solved for the path coefficients, and there are computer programs which carry out the calculations. All we need at this point is data from a study which measure variables included in the model. Land (1970) examines this model by using data gathered from 185 junior and senior high school students in Nevada and Massachusetts in 1962, where they made judgments of the functional importance, skill, prestige and rewards of 24 occupations (Lopreato and Lewis, 1963).

The zero-order intercorrelation (r) of variables is as follows:

Variables	(2)	Variables (3)	(4)
(1) Imputed functional importance of an occupation	.64	.75	.59
(2) Skill	—	.92	.89
(3) Prestige	—	—	.87
(4) Reward	—	—	—

†The X_a and X_b terms are called residual variables or, sometimes, the "random shock" or "error" terms.
**A helpful discussion of path analysis and a program for computing path coefficients is provided by Nygreen (1971), listed in General References.

From these data and the path estimating equations in (9.17), path coefficients can be computed for the model. A multiple correlation coefficient could be obtained for the regression equation predicting each dependent variable as illustrated in Diagram 9.2. Both path coefficients and the multiple correlation coefficients are generally provided by standard computer regression programs.

With computed coefficients shown in Figure 9.3, one can turn to the crux of the problem, the evaluation of the "fit" of the model to the data. "Goodness of fit" of the model can be examined in a number of ways. Land (1970) cites three general approaches.

(a) One could examine the amount of *variation* in dependent variables which is *explained* by variables linked as specified in the model.
(b) One could examine the *size of path coefficients* to see whether they are large enough to warrant the inclusion of a variable or path in the model.
(c) One could evaluate the ability of the model to *predict correlation* coefficients which were not used in computation of the path coefficients themselves.

In each of these cases, an investigator usually contrasts the usefulness of his(her) model in these three respects with alternative models, and this is the heart of explanatory progress in any science.

Taking the "goodness of fit" criteria one at a time, and applying them to the example, we see the following. The two multiple correlation coefficients, squared, provide an estimate of the proportion of variation in the dependent variable explained by the linear combination of specific independent variables. In this case:

$$\begin{array}{ccc} & Explained & Unexplained \\ R_{3 \cdot 12} = .94 & R_{3 \cdot 12}^2 = .89 & 1 - R_{3 \cdot 12}^2 = .11 \\ R_{4 \cdot 12} = .89 & R_{4 \cdot 12}^2 = .80 & 1 - R_{4 \cdot 12}^2 = .20 \end{array}$$

Explaining eighty and eighty-nine percent of the variance in a dependent variable is quite high, although clearly not perfect. The model leaves unexplained only 11% of the variation in variable 3, and 20% in variable 4. In this respect the model is relatively satisfactory. The "unexplained" variation is due to variables or measurement error not included in the model and, for the sake of completeness, the square roots of these $(1 - R^2)$ values are ascribed to the residual variables a and b, in Figure 9.3.

(9.18)
$$p_{3a} = \sqrt{1 - R_{3 \cdot 12}^2} \qquad p_{3a} = .33$$
$$p_{4b} = \sqrt{1 - R_{4 \cdot 12}^2} \qquad p_{4b} = .45$$

If these "residual" paths become large, then an investigator would begin to search for an alternative model which included other independent variables and/or he(she) would examine his(her) measurement process for measurement error.

In terms of the second criterion of goodness of fit, the model shows some weaknesses. Most of the path coefficients are moderately high except the path from variable 1 to 4. One might consider eliminating this path from the model for this reason and recompute path coefficients. The indirect effect of functional importance on rewards can be found by multiplying path coefficients (or correlation coefficients on curved lines in the model) times each other, along each route connecting the two variables of interest, and then by summing over all connecting paths. In this case $(.64)(.86) = .54$, which represents the indirect effect of variable 1 on variable 4. This is considerably stronger than the direct effect of 1 on 4.

Thirdly, one could examine the "fit" between observed correlation coefficients not previously used in formulas for calculating path coefficients and predictions of correlation coefficients which would be derived from the model. If the fit is good, the model is supported; if not, some modification is perhaps needed. In this instance, the correlation between variables 3 and 4 was not used to estimate path coefficients in equations shown in (9.17).

Since a correlation equals the sum of products of coefficients along all connecting paths, the model would lead to the following prediction of r_{34}

(9.19)
$$r'_{34} = p_{41}p_{31} + p_{41}r_{12}p_{32} + p_{42}r_{12}p_{31} + p_{42}p_{32}$$

or
$$
\begin{aligned}
(.04)(.27) &= .011 \\
(.04)(.64)(.75) &= .019 \\
(.86)(.64)(.27) &= .149 \\
(.86)(.75) &= \underline{.645} \\
r'_{34} &= .824
\end{aligned}
$$

This corresponds quite closely with the observed correlation of prestige and reward of .87.**

9.2.3 Testing an Alternative Model

Land goes on to contrast the goodness of fit of the Davis-Moore model with that of the Parsonian model, postulating an added variable, *values,* an unmeasured, underlying variable which is thought to account for the other four variables: skill, importance, prestige, and reward. This model and the path estimation equations are presented in Figure 9.4. Land concludes that neither the model in Figure 9.3 nor the model in Figure 9.4 can be rejected on the basis of the goodness of fit criteria, but that the Parsonian model somewhat more consistently fits these data. Other theoretical work suggests that the Davis-Moore theory of social stratification would be a better representation in social systems where there is *high* interdependence of work activities, and the Parsonian model would be better under conditions of *low* interdependence – theoretical ideas that suggest greater refinement and the possibility for further productive research.

Path analytic models may be extended to include more variables, as is illustrated in Figure 9.5, in which Sewell et al. (1970) revised and extended earlier work aimed at explaining the occupational attainment of Wisconsin high school seniors. The data upon which path coefficients in this model are based are from 4388 high school seniors first interviewed in 1957 and then re-interviewed in 1964.

**What is "quite closely" and what is not depends to a large extent upon the investigator's experience and judgment. Differences of .05 or .10 are likely to be considered small unless a meaningful alternative model is able to make closer predictions.

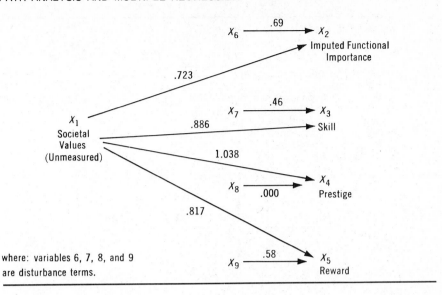

where: variables 6, 7, 8, and 9
are disturbance terms.

Path Model

$$X_2 = p_{21}X_1 + p_{26}X_6 \qquad X_4 = p_{41}X_1 + p_{48}X_8$$
$$X_3 = p_{31}X_1 + p_{37}X_7 \qquad X_5 = p_{51}X_1 + p_{59}X_9$$

Path Estimation Equations

$$r_{23} = p_{21}p_{31} \qquad\qquad p_{26} = \sqrt{1 - p_{21}^2}$$
$$r_{24} = p_{21}p_{41} \qquad\qquad p_{37} = \sqrt{1 - p_{31}^2}$$
$$r_{25} = p_{21}p_{51} \qquad\qquad p_{48} = \sqrt{1 - p_{41}^2}$$
$$r_{34} = p_{31}p_{41} \qquad\qquad p_{59} = \sqrt{1 - p_{51}^2}$$
$$r_{35} = p_{31}p_{51} \text{**}$$
$$r_{45} = p_{41}p_{51} \text{**}$$

Correlation Matrix (From Lopreato-Lewis matrix above, following Formula 9.17)

	(3)	(4)	(5)
2. Imputed Functional Importance	.64	.75	.59
3. Skill		.92	.89
4. Prestige			.87
5. Reward			

Goodness of Fit of Model in Figure 9.4

1. Predictability of dependent variables.
   Unexplained Variation in:
   Functional Importance   48%
   Skill   21%
   Rewards   25%
   Prestige   00%
2. Path coefficients large. The unmeasured, hypothetical variable, "values," is virtually identical to prestige.
3. Predicted $r_{35} = p_{31}p_{51} = (.886)(.817) = .72$ (Actual $r_{35} = .89$)
              $r_{45} = p_{41}p_{51} = (1.038)(.817) = .85$ (Actual $r_{45} = .87$)

**These equations, not needed to compute path coefficients, are used to check model's fit.
Source: Land, 1970. Used by permission.

FIGURE 9.4 PARSONIAN MODEL OF SOCIAL STRATIFICATION

Intercorrelation of Variables (Pearsonian r):

	Variables						
	(2)	(3)	(4)	(5)	(6)	(7)	(8)
(1) 1964 Occupational Attainment	.618	.483	.463	.438	.384	.331	.363
(2) Education		.632	.696	.609	.535	.417	.486
(3) Level of Occupational Aspiration			.771	.565	.470	.366	.445
(4) Level of Educational Aspiration				.611	.459	.380	.418
(5) Influence of Significant Others					.473	.359	.438
(6) Academic Performance						.194	.589
(7) Socioeconomic Status							.288
(8) Mental Ability							

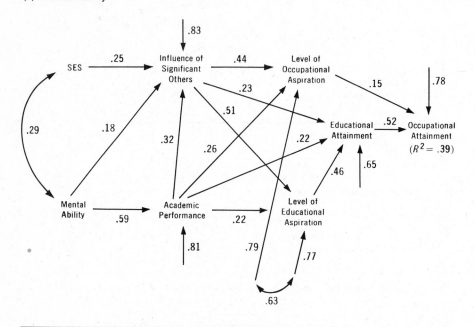

Source: Sewell et al., 1970. Used by permission.

FIGURE 9.5 THE SEWELL MODEL OF OCCUPATIONAL ATTAINMENT

9.3 MULTIPLE REGRESSION USING UNSTANDARDIZED SCORES

Up to this point, we have been converting raw scores on each variable to z-scores before computing coefficients for a multiple regression equation. This transformation sets the mean of each variable equal to zero and the standard deviation equal to unity. Because of this transformation, we could legitimately compare beta weights (b^*) or path coefficients (p) to determine the relative contribution of each variable to the explanation of variation in a dependent variable. In this sense we could determine which of several independent variables is "most important."

Often, however, one is more interested in *predicting* the value of a dependent variable rather than analyzing a theoretic model or comparing the importance of independent variables. Using z-scores as data in a multiple regression equation yields a prediction of the z-score of the dependent variable, rather than a predicted raw score expressed in units such as miles or feet, pounds or ounces, number of births, or number of friendship choices, etc. The raw scores of variables used as independent variables in the prediction equation also have units of measurement, and it may be useful not to convert them to standard scores in making predictions. This means that the beta coefficients (b^*) would have to be altered somewhat to take account of the particular units of measurement used in the regression equation.

This type of situation would occur most frequently in an applied condition in which the multiple regression equation is used to make some actual prediction of the value of a dependent variable. Such predictions might then be used to guide actions such as turning on air-conditioners, adding more of some ingredient, hiring more truck drivers, building more schools, deciding on closer supervision of new parolees, or admitting a college applicant. Regression weights (b) for predicting height in inches, using age in years and weight in pounds, would be quite different from what they would be if the prediction were to be made in feet rather than inches, or if the weights were to be expressed in ounces rather than pounds.

Currently, much of the research in sociology is oriented toward testing general ideas about which variables are most important to include in a theory, rather than toward predicting actual score outcomes, although there is an indication that this situation may be changing. Voting studies, for example, predict the actual percentage vote for a given candidate. Population studies predict the actual birth rate or the size of a population, and those involved in controlling crime predict the likelihood that new parolees will commit the same crimes again (recidivate).

The following equation is written in a form appropriate for raw data. Notice that a constant term, $a_{1 \cdot 234}$, has been added to adjust for the location of scores on the dependent variable. This corresponds to the Y-intercept term in a simple regression equation discussed earlier in this book, $Y' = a_{yx} + b_{yx}X$. The regression coefficients are now written without an asterisk ($*$) to indicate that we are not using standard score input data, but rather the raw scores themselves in their original measurement units.

(9.20)
$$X'_1 = a_{1 \cdot 234} + b_{12 \cdot 34}X_2 + b_{13 \cdot 24}X_3 + b_{14 \cdot 23}X_4$$

Again, the regression coefficients are *partial* regression weights, in the sense that they reflect the additional contribution of a variable beyond the contributions of other variables in the specific equation.

In order to change from a standardized beta weight (b^*) to an unstandardized one (b), the relative amount of variation in the dependent variable and independent variables must be taken into account. This is accomplished by multiplying the standardized beta by the *ratio* of the standard deviation of the dependent variable to the standard deviation of the specific independent variable to which the standardized beta weight refers. This is shown and illustrated in Diagram 9.3.

DIAGRAM 9.3 Converting to Unstandardized Regression Coefficients in a Four-Variable Equation†

The Problem

To convert from a standardized regression equation, as follows:

$$z_i' = b_{ij \cdot kl}^* z_j + b_{ik \cdot jl}^* z_k + b_{il \cdot jk}^* z_l$$

to an unstandardized regression equation as below:

$$X_i' = a_{i \cdot jkl} + b_{ij \cdot kl} X_j + b_{ik \cdot jl} X_k + b_{il \cdot jk} X_l$$

Converting Standardized into Unstandardized Beta Weights

(D1)
$$b_{ij \cdot kl} = b_{ij \cdot kl}^* \frac{s_i}{s_j}$$

where subscript i refers to the dependent variable, and j to the independent variable for which the beta is being computed. Variables k and l are two other independent variables in the multiple regression equation for four variables illustrated here.

In this problem, two other beta-weights would have to be computed as follows:

$$b_{ik \cdot jl} = b_{ik \cdot jl}^* \frac{s_i}{s_k}$$

$$b_{il \cdot jk} = b_{il \cdot jk}^* \frac{s_i}{s_l}$$

Computing the Regression Constant, $a_{i \cdot jkl}$

(D2)
$$a_{i \cdot jkl} = \overline{X}_i - b_{ij \cdot kl} \overline{X}_j - b_{ik \cdot jl} \overline{X}_k - b_{il \cdot jk} \overline{X}_l$$

where i refers to the dependent variable, and other subscripts refer to the three independent variables in this example.

COMPUTING BETA WEIGHTS FOR A FOUR-VARIABLE REGRESSION EQUATION FROM BETA WEIGHTS FOR A THREE-VARIABLE REGRESSION EQUATION

(D3)
$$b_{ij \cdot kl} = \frac{b_{ij \cdot k} - b_{il \cdot k} b_{lj \cdot k}}{1 - b_{jl \cdot k} b_{lj \cdot k}}$$

Example

Hal H. Winsborough, in a study discussed by Duncan (1955:8), uses three variables which, in combination, serve to define population density in the 74 community areas of Chicago in 1940. In the study, the logarithms of each variable were taken, because this transformation resulted in the components being additive, an assumption of the regression model. The variables were these:

†Note that these formulae can be extended easily to the two or to the more-than-three variable cases by adding (or deleting) subscripts to specify the specific regression weights of interest and adding (or deleting) terms (one for each independent variable) from the regression equation.

DIAGRAM 9.3 *(Continued)*

X_1 = Density (*i.e.* population divided by land area).
X_2 = Household Density (*i.e.* population divided by the number of dwelling units in the area).
X_3 = Dwelling Unit Density (*i.e.* number of dwelling units divided by the number of structures such as apartments, hotels, single family dwellings, etc.)
X_4 = Structure Density (*i.e.* the number of structures divided by the land area).

Correlation Matrix

	(1)	(2)	(3)	(4)
(1)		−.419	.636	.923
(2)			−.625	−.315
(3)				.305

The standardized regression equation relating these variables was this:

$$z_1 = .132z_2 + .468z_3 + .821z_4$$

Standard deviations in this study were:

for variable $X_1 = .491$
for variable $X_2 = .065$
for variable $X_3 = .230$
for variable $X_4 = .403$

from formula (D1)

$$b_{12 \cdot 34} = b_{12 \cdot 34}^* \frac{s_1}{s_2}$$

$$= .132 \frac{.491}{.065} = .132(7.55) = .997$$

similarly,

$$b_{13 \cdot 24} = .468 \frac{.491}{.230} = .468(2.13) = .997$$

$$b_{14 \cdot 23} = .821 \frac{.491}{.403} = .821(1.22) = 1.00$$

The constant, $a_{1 \cdot 234}$ could be computed from (D2), above, if means of variables were available, simply by substituting these values below.

$$a_{1 \cdot 234} = \overline{X}_1 - b_{12 \cdot 34}\overline{X}_2 - b_{13 \cdot 24}\overline{X}_3 - b_{14 \cdot 23}\overline{X}_4$$
$$a_{1 \cdot 234} = \overline{X}_1 - .997\,(\overline{X}_2) - .997\,(\overline{X}_3) - 1.00\,(\overline{X}_4)$$

$$b_{12} = b_{12}^* \frac{s_1}{s_2}$$

where s_1 is the standard deviation of the dependent variable, and s_2 is that for the independent variable.

In general, the conversion can be expressed as follows for a four-variable problem:

$$b_{ij \cdot kl} = b_{ij \cdot kl}^* \frac{s_i}{s_j}$$

where the subscript i refers to the dependent variable, and subscript j to the specific independent variable in a regression equation with variables k and l.

The intercept coefficient can be computed as follows, again for the four-variable problem. These procedures can be generalized for more variables simply by, including comparable terms for each of the variables.

(9.21) $$a_{i \cdot jkl} = \overline{X}_i - b_{ij \cdot kt}\overline{X}_j - b_{ik \cdot jt}\overline{X}_k - b_{il \cdot jk}\overline{X}_l$$

Computation of these coefficients would usually be left to a checked-out multiple regression computer program, where raw scores are specified as input. It is helpful and highly desirable, however, to be able to double check some of the computations and make simple conversions by hand, as illustrated in Diagram 9.3.

9.3.1   An Example of Multiple Regression Using Raw Scores

A very nice illustration of raw-score multiple regression procedures comes from the social stratification literature, where these procedures were applied to the problem of predicting an occupational prestige score which could then be used in further research involving a social class variable.

You have probably read about the NORC-North-Hatt scale of occupational prestige, since it is widely used in research, and it is frequently quoted in introductory books in social sciences. Cecil North and Paul K. Hatt directed a study by the National Opinion Research Center in March, 1947 in which some 2920 respondents were interviewed. Each respondent was asked to give his (her) own "personal opinion of the general standing" of some 88 different occupations.

Responses on a five-point scale (excellent, good, average, somewhat below average, and poor) were summarized to create scores which could vary from 20 to 100 (actually ranging from 33 for shoe shiner to 96 for U.S. Supreme Court Justice), and these scores on 88 occupations are referred to as the North-Hatt scale.

Unfortunately for later investigators desiring to use these prestige scores for occupations in their own research, the 88 occupations did not cover the range of occupations that would come up in their data, so they were faced with the problem of guessing an approximate score, or of predicting missing scores by using some prediction formula based on other available information. Otis Dudley Duncan helped solve the problem by multiple regression techniques, using income and education as predictor variables (Duncan, 1961). Forty-five of the 88 North-Hatt occupations matched sufficiently well with one of the occupational categories of the inclusive Census classification of all jobs into 425 categories

(U.S. Bur. Census, 1960). For each of these categories, information was available on (a) the proportion of men in the occupation with 1949 incomes of $3,500 or more, and (b) the proportion of men in the occupation with four years of high school or more formal education. These variables were then used to predict the percentage of respondents who judged an occupation to be of "excellent" or "good" standing in the North-Hatt data.** The regression equation developed on these 45 occupations is as follows:

$$X_1 = -6.0 + .59X_2 + .55X_3$$

where X_1 is the occupation score; X_2 is income; X_3 is education.

Using these weights in the multiple regression equation, Duncan could create social class scores for the rest of the 425 Census occupational categories for which income and education data were available in the Census publications. Truck drivers, for example, are in an occupation in which 21 percent of the males in 1950 had 1949 incomes of $3,500. or more, and 15 percent had graduated from high school or had further education (both standardized for age).

Using the regression equation above, one could predict that the percentage of people who would rate that occupation as "excellent" or "good" would be 15 percent; this then becomes Duncan's Socioeconomic Index Score. The original NORC prestige rating was actually 13 percent, so the prediction missed the mark by only two percentage points. The coefficient of multiple determination, R^2, using the data on the 45 occupations, was .83; that is, for the 45 occupations, 83 percent of the variance in occupational prestige was accounted for, statistically, by this particular linear combination of indicators of income and education for these occupations.

Prior to using the multiple regression procedure, scatter plots of X_1 by X_2, X_2 by X_3, and X_1 by X_3 were inspected, and the relationships were found to be essentially linear. One could compute the regression weights and R^2 from the following information using formulas discussed earlier in this chapter.

$$r_{12} = .84 \qquad r_{12 \cdot 3} = .61$$
$$r_{13} = .85 \qquad r_{13 \cdot 2} = .65$$
$$r_{23} = .72$$

It is interesting to note that a correlation of .99 was found between prestige scores for the 90 occupations in the 1947 North-Hatt study and a 1963 replication reported by Hodge and others (1964). Regression analysis was also used in this study to examine shifts over the 16-year period. One of the findings was that there had been an upward shift of scores for blue-collar occupations as compared to professionals and white-collar workers.

9.4 DUMMY VARIABLE MULTIPLE REGRESSION

Sometimes it happens that an independent variable we would like to use in a multiple regression analysis is a nominal-level variable, rather than an inter-

**Because, overall, income and education are related in a curvilinear fashion with age—the young and old are low on both—these two variables were standardized on age, and the age-adjusted proportions were used in the multiple regression equation expressed here. Standardization is discussed in Chapter 8.

val-level as the model assumes. It is possible to include these variables in the analysis by creating what is called a "dummy variable." These are dichotomous variables indicating the presence (scored 1) or absence (scored 0) of a certain characteristic for each individual respondent.

If five categories for marital status were useful as an independent variable, one could represent the same information by four dummy variables as follows:

Marital Status	Dummy Variables
1 Currently married	1 Currently married (1 = Yes) (0 = No)
2 Never married	2 Never married (1 = Yes) (0 = No)
3 Widowed	3 Widowed (1 = Yes) (0 = No)
4 Separated	4 Separated (1 = Yes) (0 = No)
5 Divorced	

Using this scheme, if a person were widowed, rather than have a score of 3 on the marital status variable, he(she) would have four scores, one on each of the four dummy variables: 0,0,1,0. That is, he(she) is 0 on currently married, 0 on never married, 1 on widowed, and 0 on separated dummy variables. The "0,0,0,0" pattern would indicate a person who is *none* of the four explicitly mentioned marital statuses—so he(she) would have to be "Divorced," the only other coded possibility. In general, it is conventional to create one dummy variable fewer than there are categories of the nominal-level variable we are interested in including.[†] These dummy variables are all included in the usual multiple regression analysis. If an individual has a score of zero on one of the dummy variables, then that regression weight and term in the regression becomes zero for him(her). Otherwise, the regression weight represents the value of that particular marital status —that is, the amount that is to be added in the regression equation to represent the value of being of some particular marital status. Dummy variables can be used as independent variables, whether one is expressing scores in standard-score form or raw-score form, and they can be used in path analysis problems as well. Dummy variables provide a useful way to "score" specific combinations of values of variables in order to handle expected statistical interaction in multiple regression equations where otherwise the model would assume an additive relationship among variables.

Duncan used dummy variables in predicting the percentage of respondents rating occupations as "excellent" or "good" as part of the study discussed in the previous section (Blau and Duncan, 1967). Instead of the education variable he used the dichotomy "white-collar" versus "manual workers." Since there were only two categories in this variable, only one dummy variable is needed, with $Z = 1$ if the occupation is white-collar and $Z = 0$ if the occupation is a manual occupation. Again using the data on 45 occupations from the North-Hatt study for which Census data on the independent variables was available, Duncan derived the following regression equation:

$$X_1' = 3.9 + .79X_2 + 19.8Z$$

$$R_{1 \cdot 2Z}^2 = .76$$

[†]For an explanation of dummy variables, see Suits (1957), in the General References.

This means that 19.8 points are added if the occupation is a type where $Z = 1$, that is, "white-collar." If $Z = 0$, then the 19.8 points would not be added, since 19.8 times 0 is 0.

The correlation between social class scores for occupations using the first formula and the dummy-variable formula above is .96. That is, education and the "white-collar/manual worker" variables produced very nearly the same ordering of scores.

When the dummy variable in a multiple regression problem is the dependent variable rather than an independent variable the analysis becomes what is known as a *discriminant function analysis*. That is, one in which the values of a number of continuous variables are used to predict categories of a nominal variable. Consideration of discriminant function analysis will not be included here since it is usually treated in specialized courses in multivariate analysis. If you wish to pursue the subject further on your own, see Van de Geer (1971).

9.5 STEPWISE REGRESSION PROCEDURES

If an investigator is interested in searching his(her) data for the best set of predictor variables, he(she) may utilize a procedure which is called stepwise multiple regression. Usually carried out with the help of a computer, the stepwise regression procedure examines a larger number of potential predictors, starting with a single independent variable which is the best predictor of the dependent variable. Then a further variable is added, and this added variable is one which explains as much of the remaining variation in the dependent variable as possible. Then the "next best" variable is added, and so forth, each time adding a term to the multiple regression equation. The purpose is to find a small set of independent variables which produce the highest R^2 possible. An investigator could drop variables out of the multiple regression equation whenever their addition would produce little increase in the coefficient of multiple determination, R^2.

It sometimes occurs that the most highly predictive set of variables is not the one "discovered" by starting with the "most predictive" variable in the procedure outlined above. Some computer programs solve this situation by examining essentially *all* of the possible combinations of different numbers of independent variables chosen from the original data set.** This procedure would permit an investigator to pick the "best" set of predictor variables, in the sense of explaining the maximum amount of variation in the dependent variable with a minimum number of linearly related independent variables.

Regardless of which of the approaches is used, stepwise regression analysis is a search procedure which capitalizes on any unusual predictive power in the data at hand. The results are, so to speak, "tailored" to the existing data, and because of this the regression equation may not be applicable to data beyond that included in the study. It is a helpful search procedure, but as in the case of results from any search, we would want to replicate the study to see if variables included are still the best predictors beyond the data on which they were derived.

**Some of these computational alternatives are presented in the early part of chapter 6 in Draper and Smith (1966) in General References.

9.5.1 An Example of Stepwise Multiple Regression

One example of the use of stepwise multiple regression procedures is in a study of factors predictive of the number of demonstrations among 104 selected, state-supported, non-technical, non-specialized colleges and universities during 1964 and 1965 (Scott and El-Assal, 1969). Representatives of 69 of the 104 schools responded with a variety of information including size of school. It was hypothesized that "multiversities" (those more complex in terms of number of degrees granted, size of departments offering various degrees, the ratio of dormitory to non-dormitory students, professor-student ratio, etc.) ferment more student unrest than other types of schools.

A stepwise multiple regression analysis among size, complexity (the multiversity measure), quality, and community size was conducted, with the dependent variable being number of student demonstrations. The table below shows the multiple correlation coefficient and the coefficient of multiple determination, R^2, at each step as new variables were added to the multiple regression equation. It is evident that in these data, school size is the best predictor of incidence of student protest demonstrations. Other variables added little additional explanatory power to the regression equation.

STEPWISE MULTIPLE REGRESSION ANALYSIS, PREDICTING NUMBER OF STUDENT DEMONSTRATIONS AT SELECTED, STATE, NON-TECHNICAL, NON-SPECIALIZED COLLEGES AND UNIVERSITIES DURING 1964–5

Independent Variables	R	R^2
School Size	.580	.336
School Size & Complexity	.591	.349
School Size, Complexity & School Quality	.593	.352
School Size, Complexity, School Quality & Community Size	.594	.353

Source: Scott and El-Assal, 1969.

The authors note that the multiple regression model is not a very good fit, since some of the variables are not "interval level of measurement," and, in fact, statistical interaction may be expected, rather than additive relationships between variables. In spite of this, they argue, the multiple regression analysis helped pinpoint school size as a central variable in explaining the incidence of demonstrations.

9.6 CURVILINEAR ASSOCIATION IN MULTIVARIATE ANALYSIS

The basic assumption underlying the discussion of linear regression in this chapter is that variables are related in a simple straight-line fashion. This, of course, does not exhaust the possibilities and, in fact, a fair number of relationships in the field of sociology are either known to be or suspected to be curvilinear in nature. The relationship between income and age is an example. Where relationships do not come very close to the linear expectation, the multiple regression equation will not be a very good fit, will not explain very much of the variation in a dependent variable and, at worst, may be misleading to an investigator who is interested in the relative importance of variables for theory-construction purposes.

Figuring out exactly the most appropriate form of relationship for a given set of variables is beyond this introductory book, although two kinds of comments may help to provide a basis for further inquiry. First, it is an excellent practice to examine the *nature* of relationship between all pairs of variables in an analysis using either (a) a scatter plot of pairs of variables, or (b) a careful comparison of the Pearsonian Product-Moment Correlation coefficient r with eta (E), as discussed in Chapter 7. Computer programs are generally available which will print out scatter plots (or r's and E's). If the relationship is not linear between some of the variables, we might attempt a transformation of the data, such as the logarithmic transformation used by Winsborough (see Diagram 9.3). This may straighten the relationships and permit us to use the multiple regression procedures we have discussed. It may also be possible to standardize or "adjust" the raw data, to take account of curved patterns in the data, before the regression analysis is attempted. Duncan, for example, adjusted income and education scores for age differences prior to predicting social status scores. His regression equation included only education and income scores after these scores had the effects of age "standardized out," so to speak.

Secondly, we might attempt to deduce or compute an equation which describes the non-linear relationship between variables. Some computer programs are available which search data and fit a more complex equation.

9.7   SUMMARY

In this chapter we discussed the general topic of multiple regression and multiple correlation, in which interval-level variables are related in a linear, additive fashion. The problem is to explain as much as possible of the variation in a single dependent variable by the linear association of two or more independent variables. R^2, the coefficient of multiple determination, provides a measure of the proportion of the variation in the dependent variable which has been explained by the multiple regression equation.

One use of the multiple regression equation is based on the use of standard scores (z-scores) as the input data. Beta-weights (b^*), which result, can be used to compare the relative importance of independent variables used in the regression equation. With the aid of some notion of how variables are linked theoretically, we could use the procedure called "path analysis" to examine the fit between actual data and our theoretical model.

A second use of multiple regression analysis involves an interest in predicting the actual value, rather than standard score, of the dependent variable. Computation procedures for finding non-standardized regression weights (b) were discussed.

Finally, we discussed some of the procedures encountered when multiple regression analysis is used. *Dummy variables* help handle categorical independent variables or anticipated interaction patterns. *Stepwise multiple regression* is a procedure for selectively adding more variables to a regression equation in a search for a maximally predictive set of independent variables. While the multiple regression equations we have discussed assume a linear relationship between variables, it is possible to deal with curvilinear relationships by removing them, either through some type of transformation of the data prior to regression analysis, or through the use of more complex regression equations. Computer programs become exceedingly useful in handling the clerical complexity of multiple regression mathematics.

The multiple correlation coefficient, R, like the bivariate measure, r, expresses how well a regression equation permits one to predict scores on a dependent variable. Like r^2, the square of the multiple correlation coefficient, R^2, can be interpreted as the proportion of the variation in the dependent variable, which is explained by the use of the regression equation rather than merely the mean of the dependent variable.

In the beginning of this volume we talked about statistics and parameters. A **statistic** (defined in Chapter 1) is a description of a set of data which was defined by the investigator to be a **sample** from some population, and a **parameter** is a description of scores which the investigator had determined to be *all* of the relevant scores, that is, a total **population.**

So far this volume has been concerned only with the idea of *describing*, whether or not the scores were from samples or populations. To be sure, we did couch the description in what we take to be the typical situation of sociologists, namely describing samples, but eventually we are interested in the populations from which the samples are drawn.

For a number of reasons—expense, necessity, convenience—investigators with interests in parameters base their descriptions on carefully drawn samples from those populations rather than on a complete census or enumeration of the population. The question immediately posed, of course, is how to leap from a description of one group (sample) to conclusions about a description of another group (population). The answer, which is partly developed in the field of inferential statistics, is that we know quite a bit about how certain kinds of samples (probability samples) behave. This permits us to develop expectations about the relationship between a description of a sample (a statistic) and a description of a population (a parameter). Furthermore, we are able to make statements about the accuracy of these expectations. We can describe and take account of "chance" variation in the samples we might draw from a population. The study of these chances and their role in supporting the logical leap from statistics to parameters (or vice versa, for that matter) is the subject of the remainder of this volume.

CONCEPTS TO KNOW AND UNDERSTAND

multiple regression equation
assumptions of multiple regression
regression weights and partial correlation coefficients
standardized regression (beta) weight, b^*

unstandardized regression weight, b
multiple correlation
coefficient of multiple determination, R^2
path analysis
corrected R
path coefficient
basic theorem of path analysis
path diagram
stepwise multiple regression
dummy variables
discriminant function

PROBLEMS AND QUESTIONS

1. Find an example of the use of multiple regression for either predictive or analytic purposes and describe its use for the problem the author describes. Are assumptions met? If not, what consequences may there be as far as you can tell? Is multiple regression the most appropriate technique for the problem? What alternatives might there be? Express clearly in your own words what it is that the multiple regression equation, the regression weights, and the multiple correlation coefficients mean.

2. Blau and Duncan (1967:170) present a correlation matrix showing the zero-order relationships between five variables, as follows.

Zero-order Correlations of Five Status Variables				
	(2)	(3)	(4)	(5)
(1) 1962 Occupational Status	.541	.596	.405	.322
(2) First job status		.538	.417	.332
(3) Education			.438	.453
(4) Father's Occupational Status				.516
(5) Father's Education				

Using the 1962 occupational status of the respondent as a dependent variable, create a path diagram showing the linkages between variables you would hypothesize to exist. Compare this with the Blau and Duncan model. Now, using information in this chapter about path analysis, create the equations needed to compute path coefficients for your model from the zero-order correlation matrix, above. If you have access to computational facilties, compute the path coefficients and appropriately label your path diagram. Again, compare these coefficients to those computed by Blau and Duncan. Finally, examine the model to reach a conclusion about how well the model fits the data (Blau and Duncan's data are from a national survey of some 20,700 men aged 20 to 64 years old, gathered in 1962).

3. Explain the difference between standardized beta weights, unstandardized regression weights, partial correlation coefficients, and multiple correlation coefficients.

4. The data in the correlation matrix below are from a study known as **PROJECT TALENT** (Flanagan, et al., 1964; Lohnes, 1966) based on a stratified random sample of senior males from American high schools and a follow-up of them five years after graduation. The correlations in the matrix are from the 14,891 whites who participated in the follow-up. The variables of the study are the following: 1. Occupational Attainment—the Duncan socioeconomic status index of the occupation of the respondent at the time of the follow-up; 2. Household Head's Occupation—the Duncan socioeconomic status index of the head of household in which the boy lived as a 12th grade student; 3. Intelligence—composite score for respondent from several academic aptitude instruments; 4. Grades—self-reported grade point average of respondent for five academic subjects; 5. Occupational Aspiration—Duncan socioeconomic status index of occupation 12th grade boy said he would most like to enter; and 6. Educational Attainment—highest level of education reported as completed by respondent at time of follow-up.

a. Using occupational attainment as the dependent variable, select two or three independent variables and draw a path diagram which predicts how they relate to each other. Do the path analysis and test the model using the three criteria suggested by Land (1970) and listed on p. 355.

b. Compute the multiple R for the dependent variable and the independent variables you have selected. Compute a residual path coefficient from this R.

CORRELATION MATRIX FOR SIX VARIABLES FROM PROJECT TALENT DATA

	1	2	3	4	5	6
1. Occupational Attainment	——	.197	.337	.236	.262	.489
2. Household Head's Occup.		——	.290	.122	.209	.303
3. Intelligence			——	.341	.323	.516
4. Grades				——	.237	.377
5. Occupational Aspiration					——	.337
6. Educational Attainment						——

Source: Porter, 1974.

5. The data in the table below are from Standard Metropolitan Statistical Areas in the United States with populations of 200,000 or more in 1970.

a. Select one of the variables as a dependent variable and construct a path diagram to show the relationship between it and the other three independent variables.

b. Compute the path coefficients and test the model using the three criteria suggested by Land (1970) and listed on p. 355. If the model does not fit well, try an alternate model.

c. Compute the multiple R and use it to compute the residual path coefficient.

d. Compute the standardized regression equation from these data, then convert the standardized beta weights into unstandardized b's. Use the unstandardized regression equation to predict two or three values of the dependent variable. How well does it work? How do you explain the magnitude of the prediction error?

DATA FROM STANDARD METROPOLITAN STATISTICAL AREAS WITH 200,000 POPULATION
OR MORE IN 1970

% Occupied Units with 1.01 or More Persons Per Room	Birth Rate*	Murder & Nonnegligent Manslaughter*	% Families with Incomes Less than $5,000
5.2	13.9	8.3	24.5
4.3	13.0	3.2	23.0
11.5	16.9	9.7	19.4
4.0	12.0	3.8	17.9
6.1	14.6	3.7	33.7
6.1	15.1	7.3	34.8
8.1	13.5	1.4	19.6
8.3	17.1	20.8	24.2
11.0	18.4	16.5	13.6
9.7	17.1	11.1	19.9
11.3	16.2	11.4	17.5
7.1	12.8	16.8	24.4
12.7	17.9	13.9	18.2
10.1	14.5	12.9	15.6
5.1	12.7	2.0	20.0
9.6	15.0	17.7	14.0
5.8	11.9	5.6	28.7
5.8	11.6	3.3	39.8
5.3	12.4	6.0	24.1
6.2	13.9	5.3	18.5
7.7	14.7	5.0	14.5
11.7	18.4	15.8	13.2
9.0	16.1	18.0	18.7
8.9	17.2	16.2	13.2
8.2	15.5	15.9	32.0
9.3	15.1	7.8	21.9
5.5	13.5	17.2	28.2
6.6	20.4	5.6	17.1
9.2	17.6	19.5	16.2
11.4	20.2	21.4	11.3
6.0	15.8	8.0	22.1
17.1	20.7	13.5	14.6
8.5	17.0	14.5	23.0
6.9	14.6	3.0	21.6
6.1	14.8	13.6	26.9
5.4	14.9	7.4	24.8
6.1	15.0	4.4	22.8
7.8	15.4	20.2	32.5
7.5	13.1	3.5	12.6
18.6	23.8	5.5	14.2
6.0	15.6	3.0	15.6
5.9	14.5	3.0	16.7
8.8	14.1	8.2	14.5
11.3	23.2	15.1	9.3
8.7	17.3	13.6	26.2
7.1	11.9	15.6	22.0

DATA FROM STANDARD METROPOLITAN STATISTICAL AREAS WITH 200,000 POPULATION
OR MORE IN 1970

% Occupied Units with 1.01 or More Persons Per Room	Birth Rate*	Murder & Nonnegligent Manslaughter*	% Families with Incomes Less than $5,000
6.3	16.8	4.8	21.8
11.1	16.3	12.5	17.3
12.1	16.8	20.7	23.9
6.2	15.7	3.8	21.9
8.1	14.2	9.6	15.9
9.4	16.3	16.3	12.6

Source: U.S. Bureau of Census, *Statistical Abstract of the United States 1975,* 96th edition, Washington, D.C., 1975.
*Rates are per 100,000 population.

GENERAL REFERENCES

Blalock, Hubert M., Jr., *Social Statistics,* Second Edition, (New York, McGraw-Hill), 1972.
See especially Chapter 19 and Section 20.4.

Cooley, William W., and Paul R. Lohnes, *Multivariate Procedures for the Behavioral Sciences,* (New York, John Wiley and Sons), 1962, Chapter 3.

Costner, Herbert L., editor, *Sociological Methodology, 1971,* (San Francisco, Jossey-Bass), 1971.
See especially Chapter 5 by George Bohrnstedt and T. Michael Carter for a discussion of "Robustness in Regression Analysis"; also Chapter 6 by Morgan Lyons, on "Techniques for Using Ordinal Measures in Regression and Path Analysis."

Draper, Norman R., and Harry Smith, *Applied Regression Analysis,* (New York, John Wiley and Sons), 1966.

Duncan, Otis Dudley, "Path Analysis: Sociological Examples," *American Journal of Sociology* 72, (July, 1966), p 1–16.

Land, Kenneth C., "Principles of Path Analysis," Chapter 1 in Edgar F. Borgatta, editor, *Sociological Methodology, 1969* (San Francisco, Jossey-Bass), 1969.

Nygreen, G. T., "Interactive Path Analysis," *American Sociologist* 6, (February, 1971), p 37–43.

Suits, Daniel, "The Use of Dummy Variables in Regression Equations," *Journal of the American Statistical Association* 52, (1957), p 548–551.

Walker, Helen M., and Joseph Lev, *Statistical Inference,* (New York, Holt, Rinehart and Winston), 1953.
See especially Chapter 13 on multiple regression and Chapter 17 on transformation of scales.

Wonnacott, Thomas H., and Ronald J. Wonnacott, *Introductory Statistics,* (New York, John Wiley and Sons), 1969, Chapters 13 and 14.

OTHER REFERENCES

Blalock, Hubert M., Jr., *An Introduction to Social Research,* (Englewood Cliffs, N.J., Prentice-Hall), 1970, p 73.

Blau, Peter M., and Otis Dudley Duncan, *The American Occupational Structure,* (New York, John Wiley and Sons), 1967, p 125.

Duncan, Otis Dudley, "A Socioeconomic Index for All Occupations," Chapter 6 in Albert J. Reiss, *Occupations and Social Status* (New York, The Free Press of Glencoe), 1961,

Duncan, Otis Dudley (See General References, 1966), p 8.

Flanagan, John C., F. B. Davis, J. T. Dailey, Marion F. Shaycoft, D. B. Orr, I. Goldberg, and C. A. Neyman, *The American High School Student,* U. S. Office of Education, Cooperative Research Project No. 635 (Pittsburgh, Project TALENT Office, University of Pittsburgh), 1964.

Hodge, Robert W., Paul M. Siegel, and Peter H. Rossi, "Occupational Prestige in the United States, 1925–1963," *American Journal of Sociology* 70, (November 1964), p 286–302.

Koslin, Bertram L., Robert N. Haarlow, Marvin Karlins, and Richard Pargamet, "Predicting Group Status from Members' Cognitions," *Sociometry* 31, (March, 1968), p 64–75.

Land, Kenneth C., "Path Models of Functional Theories of Social Stratification as Representations of Cultural Beliefs on Stratification," *Sociological Quarterly,* 11, (Fall, 1970), p 474–484.

Lohnes, Paul R., *Measuring Adolescent Personality,* PROJECT TALENT Five-Year Follow-up Studies, Interim Report I (Pittsburgh, University of Pittsburgh, 1966).

Lopreato, J., and L. S. Lewis, "An Analysis of Variables in the Functional Theory of Stratification," *Sociological Quarterly* 4 (Fall, 1963), p 301–310.

Scott, Joseph W., and Mohamed El-Assal, "Multiversity, University Size, University Quality, and Student Protest: An Empirical Study," *American Sociological Review* 34, (October, 1969), p 702–709.

Sewell, William H., Archibald O. Haller, and George W. Ohlendorf, "The Educational and Early Occupational Status Attainment Process: Replication and Revision," *American Sociological Review* 35, (December, 1970), p 1014–1027.

U.S. Bureau of the Census, *1960 Census of Population, Alphabetical Index of Occupations and Industries* (Revised edition), (Washington, D.C.), 1960.

Van de Geer, John P., *Introduction to Multivariate Analysis for The Social Sciences* (San Francisco, W. H. Freeman and Company), 1971.

Wampler, Roy H., "A Report on the Accuracy of Some Widely Used Least Squares Computer Programs," *Journal of the American Statistical Association* 65, (June, 1970), p 549–565.

V Inferential Statistics: Sampling and Probability

10 *Probability*

10.1 BACKGROUND TO INFERENTIAL STATISTICS

Up to this point, we have been dealing with the problem of *describing* distributions of a single variable or the relationship between two or more variables. As we have seen, there are several aspects of a distribution that one might want to describe and there are many different techniques designed to get at one or more of these features. The chief problem in descriptive statistics is to select those measures which accurately characterize interesting aspects of a distribution one needs for particular analytic purposes.

The next part of this text deals with *inferential statistics,* a body of tools often used to help an investigator make inferences about a population from a description of a sample. This is a necessary leap when investigators only have data from samples to examine, yet their interest is in making generalizations about a whole population from which the sample was drawn. As we shall see, it is usually unnecessary and often less accurate to attempt to gather data from a whole population when a carefully drawn sample is available. It is the leap from conclusions about a sample to inferences about a broader population that will now occupy our attention.

LOOPBACK BOX 10.1 SAMPLES, POPULATIONS

See Chapter 1.1 to refresh yourself on these terms. Samples and populations will be discussed in Chapter 11.

LOOPBACK BOX 10.2 IDEAS TO REMEMBER

The next chapters will draw upon ideas already covered in earlier chapters. In particular, you may want to refresh yourself on:

Univariate Description—how central tendency, variation and form are described (Chapter 5). In particular, you should be familiar with the arithmetic mean (Chapter 5.3.3) and the standard deviation (Chapter 5.4.4) since these descriptions will be used repeatedly in the chapters to follow.

z-scores and the Standard Normal Curve—for reasons explained below, you will find that the normal curve (Chapter 5.5.2) and standard (*z*) scores (Chapter 5.4.7) are frequently used. You will need to use tables of areas under a normal curve (Appendix, Table B) to find the area between the mean and a given *z*-score (Chapter 5.5.2b explains this).

Statistical Descriptions—finally, you will notice that the following chapters are concerned with the inferential leap between a description of a sample and a description of a population. The descriptions involved are exactly those you have already covered—descriptions of the distribution of a single variable (Chapter 5), descriptions of the relationship between two variables (Chapter 7), and descriptions of the relationship between three or more variables (Chapters 8 and 9). In the chapters to follow, we will only make use of certain of these descriptions to illustrate the logic of inferential statistics. Nevertheless, you will want to refer back to earlier sections from time to time to refresh yourself on the characteristics of a particular descriptive statistic. Cross-references later in this book will help you make these connections.

All of the basic tools for handling this leap have been covered in the past chapters, although we will use these tools in a somewhat different way. For example, the leap from sample description to population description is a leap from one description to another. We have already dealt with ways to describe data (whether they are from a sample or a population). Thus, we may have a sample mean and we want to make inferences about the mean of the population from which the sample was drawn. We have a correlation coefficient describing the relationship between two interval variables in a sample and we want to make some statement about how these two variables are likely to be related in the population. The leap involves possibilities of error, of course, and it is the task of inferential statistics to help assess these inferential risks.

There is a second way in which your understanding to this point will be used in the following chapters. A key idea in inferential statistics is just another univariate distribution (called a sampling distribution). Ways you have already learned to describe univariate distributions (*e.g.* central tendency, variation, form) are used to describe these distributions as well. The distributions, however, are distributions of sample outcomes rather than distributions of individual scores. The "case" has changed but the ideas are the same. You will notice that in some instances these ideas will be referred to with different terms (*e.g.* "standard error" rather than "standard deviation") and these ideas are used in a certain line of reasoning that will be discussed later on. In an important sense, then, you have already covered the ideas which will be put together around the problem of inference in the chapters which are to follow. Perhaps the key to understanding inferential statistics is to understand the notion of *sampling distribution* — a familiar old idea (univariate distribution) applied to the inferential problem most researchers have to face.

Recent work suggests that certain groups are superior to single individuals at problem-solving tasks. This superiority, however, is inversely related to how much hierarchical difference there is in the group, because hierarchically differentiated groups may exhibit less risk-taking behavior, be less efficient and less productive than groups which are relatively undifferentiated (Blau and Scott, 1963; also Bridges, 1968).

A sociologist interested in testing these ideas might decide to examine data on public high schools, because they are organized groups that vary in degree of hierarchical differentiation. He(she) might draw a probability sample of public high schools throughout the country and collect data on them.† His(her) data might include size of student body, size of faculty, number of levels of administration, size of school district, amount of risk-taking evident in seeking solutions to problems or making changes in the organization, percentage of graduates entering college, etc. The investigator would use these data to test his(her) ideas and generalize his(her) findings to all public high schools in the country. In generalizing from sample to population, he(she) would utilize probability theory to gauge the likelihood of error in his(her) generalizations. In this and the following chapters we will come to see how probability theory allows him(her) to do this.

In order to make scientific generalizations, then, the sociologist employs sampling theory and probability theory. He(she) takes a sample of data relevant to theoretical ideas about which he(she) wishes to generalize, studies the characteristics of his(her) sample, and uses probability theory as a basis for extrapolating his(her) findings to the population that the sample represents. This process is induction, which is central to the scientific method.*

Both probability theory and sampling theory are basic to the inductive process. Furthermore, they are so intricately interrelated that it is difficult to divorce one from the other. In order to present the ideas underlying these theories as simply as possible, however, we will treat the two topics separately.

This first chapter on *probability* (Chapter 10) will introduce some of the basic ideas used in dealing with risk, errors, outcomes and chance. For many readers this will be a refresher of ideas covered long ago in classes on the "new math". Building upon these ideas, Chapter 11 discusses ways in which samples may be drawn. This is particularly important where one is interested in drawing conclusions about populations from samples of them. Then Chapter 12 examines what we know about certain outcomes which result from drawing samples in certain ways. It is here that the key concept in inferential statistics — sampling distribution — is discussed and described. Later chapters make use of these notions in the research process of making inferences from descriptions of samples.

†The techniques of probability sampling will be discussed in Chapter 11.
*There are some exceedingly interesting and important issues involved in induction in the social sciences and these are usually discussed in philosophy of science or research methods courses. For a start you might want to explore a general book in this area (see Brown, 1963, in References).

10.2   SOME BASIC NOTIONS ABOUT SETS

It is useful to use some of the concepts of set theory as a basis for understanding probability; so before we get into a discussion of probability *per se*, we will review some relevant ideas about sets.

10.2.1   Definition of a Set

The concept of set is generally treated as a primitive term in mathematics and is not defined rigorously. Intuitively, a **set** may be defined as any specified collection of objects. For example, all of the nonprofit organizations in the United States may be considered a set. Another set may consist of all families with children under the age of eighteen. Any well-defined (*i.e.*, clearly identifiable) collection of objects may be treated as a set.

10.2.2   Definition of an Element

A single object that belongs to a set is called an **element** of the set. Thus, the Society of Colonial Wars might be an element of the set consisting of all nonprofit organizations and the J. L. White family might be an element of the set consisting of all families with children under the age of eighteen.

10.2.3   The Universal Set

It is frequently of use to consider a set that may include other sets but which itself is not included in any other set. Such a set is known as a **universal set.** In sociology the population to which we wish to generalize constitutes a universal set.*

A universal set may be either finite or infinite. The universal sets with which sociologists ultimately concern themselves are infinite sets: that is, sets whose elements cannot be enumerated because there is no limit to their number. It is the nature of science to strive to develop generalizations that predict for infinite sets of data. In point of fact, however, at the present stage of development of sociology, sociologists usually work with finite sets (limited in time and space) because of unresolved methodological problems and because the mathematics involved in working with finite sets is somewhat less complicated than it is for infinite sets.

10.2.4   The Empty Set

A set which contains no elements is known as an **empty set** or a **null set.** This set is usually designated set E (also commonly by $\varnothing$). An example of an empty set

---

*As a matter of fact, the concepts *population* and *universe* are often used interchangeably by sociologists.

would be members of Congress under the age of 25, or members of the American Medical Association who are uneducated.

10.2.5 Subsets

Any set may include a number of other sets. These other sets are referred to as **subsets** of the original set. All of the elements included in a subset are also elements of the set to which the subset belongs. Generally, however, a subset includes fewer elements than are included in the set. If we take as our set all nonprofit organizations, then nonprofit religious organizations would constitute a subset.

Note that if we view a population as a universal set, then a sample of the population can be defined as a subset. The trick is to select the subset, the sample, in such a way that it can be used to generalize to the universal set, the population.

10.2.6 Operations on Sets

There are numerous and complicated operations that may be performed on sets, but for this discussion we will only need to cover two basic operations known as **union** and **intersection.***

10.2.6a THE UNION OF SETS

Given two sets, set A and set B, the union of A and B, denoted by $A \cup B$ (read A union B), is defined as that set of elements that is included in set A or in set B. It is important to note that in mathematical logic the conjunction *or* takes a special meaning. A or B, means A, B, or both A and B.† The union of sets A and B, therefore, includes elements that are only in set A, elements that are only in set B, and elements that are in both sets A and B. Perhaps the clearest way of portraying this situation is through Venn diagrams such as Figure 10.1.‡ The total shaded area of Figure 10.1 represents the union of sets A and B.

As an example of a situation such as is portrayed in Figure 10.1, suppose that set A consists of all college graduates in a city and set B consists of all those in the city who have annual incomes in excess of $30,000. College graduates who do not have incomes in excess of $30,000 constitute the unique elements of set A. Those who have annual incomes in excess of $30,000 but are not college graduates constitute the unique elements of set B. Those who are college graduates and who also have annual incomes in excess of $30,000 belong to both

---

*The algebra of classes or operations on sets is called Boolean Algebra after its founder, the English mathematician George Boole (1815–1864).
†This is sometimes called the "inclusive" rather than the "exclusive" use of the logical connective "or."
‡These diagrams illustrating logical relationships are named after John Venn (1834–1923) an English logician. They are also sometimes referred to as "Euler Diagrams" stemming from their use by Euler in his explanations in *Letters to a German Princess* (1772).

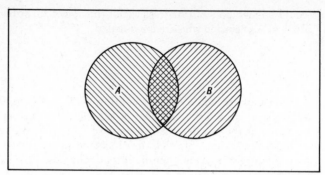

FIGURE 10.1   The Union of A and B, $A \cup B$, Shown by All of the Shaded Area

sets A and B and form an **overlapping set** represented in the diagram by the overlapping area in the center.

Sets that do not overlap (*i.e.*, have no elements in common) are known as **disjoint sets.** The union of two disjoint sets C and D is portrayed in Figure 10.2. The total of the shaded areas represents $C \cup D$.

As an example of this kind of situation, suppose that set C consists of all of the dentists in a city and set D consists of all of the unskilled workers. Then all of the dentists belong to set C, but none of them belongs to set D (although you may occasionally run across a dentist who leads you to wonder). All of the unskilled workers belong to set D, but none belongs to set C. Sets C and D are, therefore, disjoint sets.

Although the examples used to explain the union of sets have been limited to cases where there are only two sets, the operation can be extended to any number of sets. All of the sets involved may be overlapping, all may be disjoint, or the union may include both overlapping and disjoint sets.

10.2.6b   The Intersection of Sets

Given two sets, set A and set B, the **intersection** of A and B, denoted by $A \cap B$ (read A intersect B), is defined as the set of elements common to both

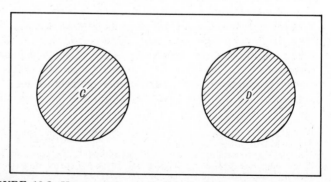

FIGURE 10.2   Union of C and D, $C \cup D$, Shown by the Shaded Areas

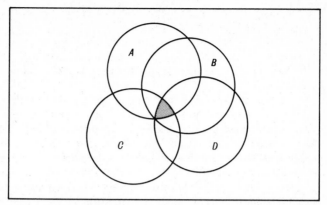

FIGURE 10.3 INTERSECTION OF FOUR SETS, $A \cap B \cap C \cap D$, SHOWN BY THE SHADED AREA

sets. In Figure 10.1 the cross-hatched section in the middle is the intersection of A and B. This intersection consists of those college graduates who have incomes in excess of $30,000.

In Figure 10.2 sets C and D are disjoint, so the intersection of C and $D = 0 = E$ (the empty set).

As was the case with the union of sets, the intersection of sets may be extended to cover any number of sets. Thus, one can meaningfully speak of the intersection of sets $A, B, C,$ and D. In Figure 10.3 the shaded center area represents $A \cap B \cap C \cap D$. As an example of a situation such as this, suppose that set A represents barbers, set B represents union members, set C represents Catholics, and set D represents married persons. Then the intersection of sets A, $B, C,$ and D consists of married union barbers who are Catholics.

Now that we are armed with some of the rudimentary tools of set theory, we will turn to a consideration of probability.

10.3 AN INTRODUCTION TO PROBABILITY

Probability is a concept intuitively familiar to all of us. In our everyday experiences we are frequently concerned with the probability that an event of a certain kind will or will not happen. What is the probability that it will rain today? How likely am I to get stuck in a traffic jam? What are my chances of winning some money at Las Vegas?

The sociologist, too, frequently asks probability questions: What is the probability that children from broken homes will become delinquent? What is the probability that convicted murderers will be successful on parole? What is the probability of successful marriages for couples with very diverse religious backgrounds? What is the probability of survival of organizations that serve a single function?, etc. If the sociologist can establish probability figures as answers to questions such as these, then he(she) can add to his(her) store of scientific generalizations and use his(her) knowledge in predicting future outcomes.

In popular usage the concept of probability takes on many different shades of meaning. However, in most instances of applied statistics, probability is formally defined in a very specific way.

10.3a   Probability as an *A Priori* Concept.   Probability refers to the *theoretical relative frequency of occurrence of a certain kind of event in the long run*. This is an **a priori** concept of probability, theoretically rather than empirically based.*

Let us analyze this definition of probability. Think of an activity or process which may have different *outcomes*. For example, giving birth to a boy or girl baby, drawing a sample of workers with different occupations, tossing a coin that has two sides, conducting an experiment in which one class is shown a film about older people and another class is not but both classes are asked to answer a questionnaire indicating whether they do or do not have favorable attitudes toward older people, or recording the class standing (freshman, sophomore, junior, senior, other) of people walking into a college cafeteria, or noting whether or not a person who has been released from prison has returned to prison within a year, etc.

In each of these examples, we have identified a set of possible *outcomes* of the activity or experiment. The set of all possible outcomes of such an activity or experiment can be taken to define a *universal set* — 100 percent of possible outcomes for that activity or experiment. The outcome of interest to us we can call an "*event*." We may be interested, for example, in the outcome "girl baby" of a birth, the outcome "blue collar" in the sampling of workers, the outcome "heads up" in a toss of a coin, the outcome "favors older people" in the film experiment, the outcome "freshman" in the cafeteria visiting activity, or the outcome "returned" in the prison release problem. We could, of course, be interested in any of the possible outcomes.

Turning next to the phrase, *theoretical relative frequency of occurrence,* imagine repeating the activity or experiment an infinite number of times, each time recording the outcome. The theoretical relative frequency of occurrence is the proportion of "events" (outcomes of interest to us) out of the total of all outcomes of that repeated activity or experiment. The probability is theoretical in the sense that it is derived from some theory of how the activity or experiment should turn out or by what would happen if the process were repeated an infinite number of times. For example, there appear to be biological reasons why there is a slightly higher probability that a birth will result in a boy baby. If only chance were operating in the selection of a sample of workers, one might expect somewhat higher likelihood of drawing a blue collar worker simply because there are usually more such workers in most industrialized societies. If a coin is "fair" we assign the theoretical relative frequency of occurrence of one-half (.50) to the event "heads." That is what we define fairness of coins to mean. If 40 percent of the college population were freshmen, we might expect that chance would result in 40 percent freshmen among those visiting the cafeteria, and so on.†

Another important phrase in the definition of probability is the final one,

---

*There is another concept of probability, known as Baysian probability, which makes use of subjective or intuitive probability estimates (Mosteller *et al.*, 1961).
†With more information we may discover that we can actually make more accurate predictions than that implied in the use of the theoretical (chance) probability used in this situation.

in the long run. This phrase implies that the definition of probability does not apply to a single repetition or a few repetitions of the activity or experiment. It applies to a large number of such instances. In the student body example, we would expect that approximately 40 percent of the persons entering the cafeteria will be freshmen. Does this mean that of the next ten persons entering the cafeteria, four will be freshmen? Not necessarily! Although that is a possibility, ten observations are not enough to constitute an adequate test. Probability refers to the relative frequency of occurrence of an event *in the long run.* The more persons we observe entering the cafeteria, the more closely the proportion of freshmen should approach 0.4, if indeed that theoretical probability applies.

Prediction in terms of probabilities is not prediction of a single outcome; rather, it refers to a large number of outcomes. This principle of prediction is that upon which the insurance business is based. When you take out a life insurance policy you pay a premium based upon the probability that a certain number of people similar to you will die during the year. The insurance company is unable to predict individual deaths except in certain obvious cases – these are the uninsurable people who are already dying. Predictions of death rates for large categories of people are, however, relatively accurate. The insurance company considers you as an element of a set of a large number of people with similar characteristics. For example, your set may be composed of white males between the ages of 20 and 24 (such sets are known as cohorts). Year-by-year studies of death rates for specific sets of people provide the insurance company with estimates of how many people in a particular set are likely to die during a given year. The amount of your premium is determined, in part, by the predicted death rate for your set. If the death rate for your set is high, then the cost of your insurance will be high.

In a study which has become a classic in sociology, the French sociologist Emile Durkheim demonstrated that the suicide rates in certain parts of Europe prior to the turn of the century were consistent from year to year and could be predicted with impressive accuracy (Durkheim, 1897 and 1951). Durkheim could not predict which particular individuals would commit suicide, but he could predict the number of suicides that would occur during a given year in a given area.

10.3b Experiential Probability. Perhaps you have noticed that the concept of probability used in these two examples does not jibe with our original definition of probability. Our original definition was based upon theoretical relative frequencies – an *a priori* notion. Our theoretical notions of how a "fair" coin behaves when tossed provides an obvious example of *a priori,* theoretical probability. The death rates used by the insurance company and the suicide rates used by Durkheim come from empirical observations. The probabilities assigned to these events, therefore, are based upon **experience** rather than theory. In most instances, estimates of theoretical probabilities are derived from experience.

10.3.1 Some Probability Notation

It is useful to apply some of the elementary set ideas to probability and to think of probabilities in terms of areas in a Venn diagram. The set of all possible outcomes

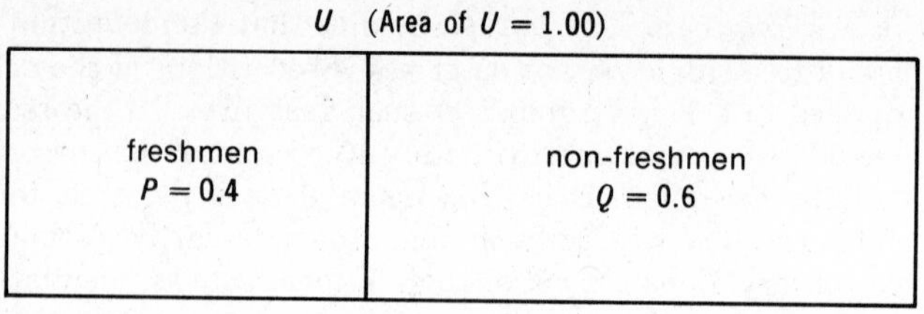

U   (Area of U = 1.00)

FIGURE 10.4   UNIVERSAL SET FOR STUDENT BODY DIVIDED 40%–60% BETWEEN FRESHMEN AND NON-FRESHMEN.

that can occur in an activity or experiment is the *universal set.* In the example of the student body divided into 40 percent freshmen and 60 percent non-freshmen (sophomores, juniors, seniors or others) the universal set is made up of two disjoint subsets, freshmen and non-freshmen. We can conceptualize our universal set as consisting of a large number of elements, 40 percent of which are freshmen and 60 percent of which are non-freshmen. This universal set is portrayed in Figure 10.4.

In the dichotomous case such as this one the convention is to let P stand for the probability of occurrence of an outcome from one of the two subsets and Q for its non-occurrence, that is, the probability of occurrence of an event from the remaining subset. The probabilities are expressed as proportions which total 1.0; therefore, $P + Q = 1.0$. If P equals the probability of occurrence of a freshman, then $P = 0.4$ and $Q = 0.6$, the probability of occurrence of a non-freshman. Note that $0.4 + 0.6 = 1.0$.

The assignment of P to freshmen and Q to non-freshmen is arbitrary. We could just as well have assigned P to non-freshmen and Q to freshmen.

The probability notation that we have introduced here will be used in future chapters whenever we are faced with a dichotomous probability situation.

10.3.2   Some Properties of Probability

There are some basic rules of probability that are of particular interest to us as sociologists because they express basic properties of probability. These include the **general addition rule** (Section 10.3.2a), the **special addition rule** (Section 10.3.2b), the rule of **conditional probability** (Section 10.3.2c), the **general multiplication rule** (Section 10.3.2d), and the **special multiplication rule** (Section 10.3.2f). We will examine each of these, in turn, and see what applications they might have in sociology. In passing, we will take a brief look at stochastic processes, since they flow naturally from a discussion of the general multiplication rule and also since they are becoming increasingly important in sociology.

10.3.2a   THE GENERAL ADDITION RULE

The general addition rule states that for events A and B, the probability that the event A *or* B will occur is equal to the probability that A will occur plus

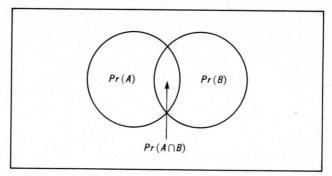

FIGURE 10.5 $Pr(A \cup B)$

the probability that B will occur minus the probability that the event "both A and B" will occur.* Symbolically the rule is stated as follows:

(10.1) $$Pr(A \cup B) = Pr(A) + Pr(B) - Pr(A \cap B)$$

where Pr stands for "probability."

This addition rule gives the combined probability that event A will occur exclusively, event B will occur exclusively, or that both A and B will occur simultaneously. Figure 10.5 is a Venn diagram of this situation.

You may be surprised that we subtracted the probability of the intersection of A and B in the formula. If you compare the formula and the Venn diagram, the reason for the subtraction can be shown. The probability of occurrence of event A, $Pr(A)$, is represented by the whole circle bounding subset A and the probability of occurrence of event B, $Pr(B)$, is represented by the whole circle bounding subset B. Since subsets A and B are overlapping, the area in which they overlap, $Pr(A \cap B)$, has been added *twice* into the sum of $Pr(A)$ and $Pr(B)$; therefore, it is necessary to subtract this area to avoid duplication.

Let us apply this addition rule to our earlier example of college graduates (subset A) and those who have incomes in excess of \$30,000 (subset B).

When we do not have theoretical bases for assigning the probabilities of given outcomes of events, we can find the proportions that the given outcomes represent of all such events in the population of data available to us and use them as the desired probabilities. This experiential assignment of probabilities might be applied to the example as follows: Suppose that 15 percent of the population are college graduates, 10 percent have incomes in excess of \$30,000, and 6 percent are both college graduates and have incomes in excess of \$30,000. Using these data we can substitute into the formula for the general addition rule to find the probability of randomly drawing from the city population either college graduates, or those having incomes in excess of \$30,000, or both, as follows:

$$Pr(A \cup B) = Pr(A) + Pr(B) - Pr(A \cap B) = 0.15 + 0.10 - 0.06 = 0.19$$

The Venn diagram for this situation would be as in Figure 10.6.

*Recall that the *or* includes A, B, or both. See Section 10.2.6a.

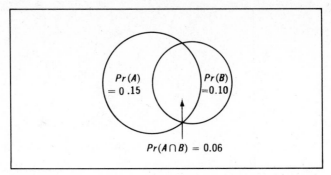

FIGURE 10.6 Pr(College Graduates ∪ \$30,000 + Incomes)

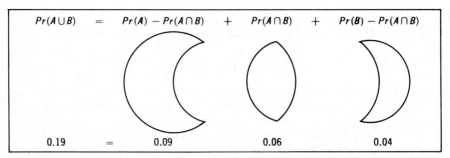

FIGURE 10.7 Pr(College Graduates ∪ \$30,000 + Incomes) with Overlap Eliminated

If we eliminate the overlap in these probabilities, we can represent the probabilities of interest as in Figure 10.7.

Represented here are these elements: (a) the probability of graduates who will not have annual incomes in excess of \$30,000 — *i.e.* the part of circle A that excludes the overlap with circle B, or $Pr(A) - Pr(A \cap B)$; (b) the probability of non-college graduates who will have incomes in excess of \$30,000 — *i.e.* the part of circle B that excludes the overlap with circle A, or $Pr(B) - Pr(A \cap B)$; and (c) the probability of college graduates who will also have incomes in excess of \$30,000 — *i.e.* the overlap between circles A and B, or $Pr(A \cap B)$. These three probabilities added together give us the desired probability of 0.19.

The addition rule can be extended to any number of events. For example, if three events were possible, A, B, and C, the addition rule would be as follows:

(10.2) $$Pr(A \cup B \cup C) = Pr(A) + Pr(B) + Pr(C) - Pr(A \cap B) - Pr(A \cap C)$$
$$- Pr(B \cap C) + Pr(A \cap B \cap C)$$

Suppose that we are interested in the unemployment rates for certain subsets of the civilian labor force, namely, males, teenagers, and nonwhites. Furthermore, assume that the following probabilities are currently accurate:

PROBABILITIES THAT UNEMPLOYED
PERSONS ARE IN GIVEN SUBSETS

0.30 are teenagers (A)
0.20 are nonwhites (B)
0.50 are males (C)
0.07 are nonwhite teenagers $(A \cap B)$
0.16 are teenage males $(A \cap C)$
0.09 are nonwhite males $(B \cap C)$
0.03 are nonwhite teenage males $(A \cap B \cap C)$

Designate teenagers as subset A, nonwhites as subset B, and males as subset C. The probability that randomly sampled unemployed persons will be (1) teenagers or (2) nonwhites or (3) males or (4) teenage males or (5) nonwhite teenagers or (6) nonwhite males or (7) nonwhite teenage males—*i.e.,* $Pr(A \cup B \cup C)$ will be as follows:

$$Pr(A \cup B \cup C) = 0.30 + 0.20 + 0.50 - 0.07 - 0.16 - 0.09 + 0.03 = 0.71$$

10.3.2b THE SPECIAL ADDITION RULE

In the case where the *events* that we are interested in are *mutually exclusive* (represented by disjóint subsets), the formula for the addition rule can be simplified. For example, if A and B are mutually exclusive events (cannot occur together), the probability that either event A or event B will occur is given by the formula:

(10.3)
$$Pr(A \cup B) = Pr(A) + Pr(B)$$

The probability of the intersection of A and B is 0 (an empty set), so it can be dropped from the formula.

Let $Pr(A)$ represent the probability that voters register as Republicans and let $Pr(B)$ represent the probability that they register as Independents. Since it is not possible to be registered simultaneously both as a Republican and as an Independent, these two events are mutually exclusive. If $Pr(A) = 0.38$ and $Pr(B) = 0.15$, then the probability that randomly sampled voters will register either as Republicans or Independents is given by the following:

$$Pr(A \cup B) = 0.38 + 0.15 = 0.53$$

Literally, the special addition rule says that *the probability of occurrence of any of a number of mutually exclusive events is equal to the sum of their separate probabilities.* In the case of three possible events, A, B, and C, the probability would be as follows:

(10.4)
$$Pr(A \cup B \cup C) = Pr(A) + Pr(B) + Pr(C)$$

This special case of the more general addition rule is of particular interest to the sociologist because it fits many of the kinds of problems he(she) is interested in and for which data are available. As we shall see later in Chapter 12, for example, the special addition rule plays an integral part in the generation of a binomial sampling distribution and the binomial sampling distribution is useful for the analysis of small group data.

10.3.2c THE RULE OF CONDITIONAL PROBABILITY

In the case of the addition rule we were primarily interested in the occurrence of a composite event (although the general addition rule did include an adjustment for the possibility of the joint occurrence of events). In the example of college graduates and those with incomes in excess of $30,000, we might be interested in the proper technique for computing the probability that college graduates will have incomes in excess of $30,000.

It has been demonstrated that there is a positive relationship between level of education and annual income. Those persons who have completed higher levels of education tend to have higher annual incomes than persons with less education. Because of this relationship the probability of having an income in excess of $30,000 should be conditional upon level of education. The desired probability can be computed by using the rule of conditional probability. The formula for a conditional probability is as follows:

(10.5)
$$Pr(B \mid A) = \frac{Pr(A \cap B)}{Pr(A)}$$

This means the probability that event B will occur, given the fact that event A has already occurred, is equal to the ratio of the probability of the joint occurrence of A and B, $Pr(A \cap B)$, to the probability of the occurrence of event A, $Pr(A)$.

We are basically interested in a subset of the universal set. The question we wish to answer is: "If event A has happened, what is the probability that event B will happen?" Rather than focusing on the universal set, we look at that part of it representing A and ask ourselves how much of this subset also represents event B.

We found earlier that 15 percent of the population were college graduates, 10 percent had incomes in excess of $30,000, and 6 percent were college graduates with incomes in excess of $30,000. With these data we can compute the probability that randomly sampled persons will have annual incomes in excess of $30,000 given the fact that they are college graduates. Letting A represent college graduation and B having an income in excess of $30,000, and substituting the appropriate data into the conditional probability formula we get:

$$Pr(B \mid A) = \frac{0.06}{0.15} = 0.40$$

Thus, given the fact that persons sampled are college graduates, there is a 40 percent chance that they will have incomes in excess of $30,000 per year. This probability figure differs considerably from the overall figure of 10 percent with incomes in excess of $30,000.

If we know that those we are considering do not have a college education, what can we say about the probability that they will have incomes in excess of $30,000? Substituting the appropriate figures into the formula we get:

$$Pr(B \mid \sim A) = \frac{Pr(\sim A \cap B)}{Pr(\sim A)} = \frac{0.04}{0.85} = 0.047$$

where $Pr(\sim A)$ refers to the probability that persons are not college graduates and $Pr(\sim A \cap B)$ is the probability that they are not college graduates but have annual incomes in excess of $30,000 (see Figure 10.7). Thus, given the fact that the persons we are considering do not have a college education, there is less than a 5 percent chance that they will have annual incomes in excess of $30,000.

10.3.2d THE GENERAL MULTIPLICATION RULE

Elementary algebra makes possible the rearrangement of the formula for conditional probability to yield a formula for computing the probability of the joint occurrence of events. This formula is as follows:

(10.6)
$$Pr(A \cap B) = Pr(B \mid A)\, Pr(A)$$

or its equivalent:

(10.7)
$$Pr(A \cap B) = Pr(A \mid B)\, Pr(B)$$

If we know the probability that event A will happen and the conditional probability that event B will happen given that A has already happened, we can compute the probability that both events A and B will happen. This formula for computing the probability of the joint occurrence of events is known as the **general multiplication rule.**

Let us look at a possible application of this rule in sociology. Assume that 45 percent of the married couples in a city were engaged for less than six months. Further, assume that for those couples who were engaged less than six months, 20 percent of the marriages end in divorce within the first five years of marriage, while for those whose engagements were of six months or more duration the divorce rate is 10 percent during the first five years of marriage.

Given these data, we might ask, if we were to randomly select couples who married five years ago, what is the probability that they were engaged less than six months (call this event A) and that their marriages were subsequently dissolved (call this event B)? The desired probability can be computed as follows:

$$Pr(A \cap B) = Pr(B \mid A)\, Pr(A) = (0.20)(0.45) = 0.09$$

We might also ask what the probability is of finding couples who had been engaged six months or more and whose marriages had been dissolved. Call the event being engaged six months or more $\sim A$, substitute the appropriate probabilities into the altered formula, and the resulting probability is as follows:

$$Pr(\sim A \cap B) = Pr(B \mid \sim A)\, Pr(\sim A) = (0.10)(0.55) = 0.055$$

This joint probability is somewhat lower than the first, demonstrating that short engagement is more likely to be accompanied by divorce than is longer engagement. Figure 10.8 is a tree diagram which includes the two joint probabilities we have just computed and the two remaining possibilities: those whose engagements were short and who are still married and those whose engagements were longer and who are still married. Notice that the four joint probabilities sum to 1.0. This is so because they represent all of the possible combinations of length of engagement and marital status. Notice, also, that each of the four joint probabilities is arrived at by multiplying the probabilities along the branches of the tree leading to it. Thus, the probability 0.495 is arrived at by multiplying 0.55 and 0.90: $(0.55)(0.90) = 0.495$

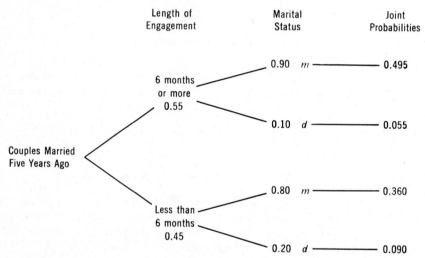

FIGURE 10.8 TREE DIAGRAM OF JOINT PROBABILITIES FOR LENGTH OF ENGAGEMENT AND MARITAL STATUS

The general multiplication rule also can be extended to more than two events. In the case of three events, A, B, and C, the formula for the joint occurrence of the three would be as follows:

(10.8)
$$Pr(A \cap B \cap C) = Pr(C \mid A \cap B) \, Pr(A \cap B)$$
$$= Pr(C \mid A \cap B) \, Pr(B \mid A) \, Pr(A)$$

Either expression may be used to compute the probability.

10.3.2e STOCHASTIC PROCESSES: AN APPLICATION OF THE GENERAL MULTIPLICATION RULE

Whenever we look at a sequence of events in time and compute the probabilities of their joint occurrence, we are dealing with **stochastic processes.** The general multiplication rule is appropriate for the computation of certain simple types of stochastic processes. The example we just considered involving the dissolution of marriages was, in effect, a two-step stochastic process. We will consider another example involving more steps in the process.

The subject of sports is receiving increasing attention from sociologists. This is so because recreation and sports become increasingly important in American society as the work week is reduced, retirement becomes more widespread, and the average person is faced with more free time. Furthermore, sports is an appealing area for study because athletic teams epitomize social behavior in one of its most researchable forms and more complete records are kept of athletic teams and games than, perhaps, any other kinds of social organizations.

One interesting phenomenon for the sociologist to study with respect to athletic teams is what the sports writers have called "the psychological advantage." It is claimed that, in addition to skills and luck, winning teams are supported in their efforts by a momentum which builds to their advantage and to the detriment of their opponents. This momentum is variously attributed to the home field advantage, the advantage of scoring first, the advantage of winning the first game of a series, the advantage of past records, etc. We will see how we might make use of the idea of stochastic process to study "the psychological advantage" in major league baseball.

The one alleged factor in "the psychological advantage" that we can bring into focus for this example is the advantage of winning the first game of a series. We will do this by examining the pattern of wins and losses of eventual World Series winners over the years. Such an analysis offers a straightforward application of the general multiplication rule to the generation of a stochastic process. We went back through major league records and gathered data on the World Series for purposes of the following example. These data allow us to compute the probabilities for various patterns of wins and losses leading to the World Series championship.

There were 62 years between 1905 and 1969, inclusive, when the World Series championship team was required to win four games in the Series. This eliminates for purposes of this analysis the tie games played in 1907, 1912, and 1922 and the Series of 1919, 1920, and 1921 when five wins were required for the championship. Focusing our attention on the eventual winner of each Series for the 62 years under consideration, we find the patterns of wins and losses distributed themselves as is shown in Table 10.1 and the tree diagram depicted in Figure 10.9.

Table 10.1 summarizes the various patterns of wins and losses, records the frequency with which each pattern occurred and gives the probability of occurrence of each pattern. The probability of occurrence of each pattern is arrived at by dividing the frequency of occurrence by 62. An examination of this table reveals that the most common pattern is for the championship team to win four games in a row. This pattern has occurred 19 percent of the time over the 62 years studied. If we wished to predict the patterns of wins and losses for future World Series champions, our best prediction would be that the eventual winner will win four games in a row. This pattern of four wins in a row does seem to be consistent with the notion of a "psychological advantage," although it could be the case that the winning team was indeed superior.

TABLE 10.1 PATTERNS OF WINS AND LOSSES BY WORLD SERIES CHAMPIONS FOR 62 YEARS BETWEEN 1905 AND 1969

Pattern of Wins and Losses	Frequency of Occurrence	Probability of Occurrence
win win win win	12	0.19
win loss win win win	6	.10
win win loss win win	4	.06
win win win loss win	2	.03
loss win win win win	3	.05
win loss win loss win win	1	.02
loss win win win loss win	5	.08
win win loss loss win win	3	.05
win loss win win loss win	1	.02
loss win loss win win win	3	.05
win loss win loss win loss win	2	.03
win loss win win loss loss win	2	.03
loss win loss win loss win win	4	.06
loss win loss win win win win	2	.03
loss win win loss loss win win	1	.02
loss win win loss loss win win	1	.02
win loss win loss loss win win	1	.02
loss loss win win win loss win	2	.03
win win loss loss win loss win	1	.02
loss loss win win win loss win	3	.05
loss loss win loss win win win	1	.02
win loss loss win win loss win	2	.03
Total	62	1.01*

*Due to errors in rounding.

The tree diagram traces the progress of the 62 years of World Series contests, game-by-game. Using this diagram and the general multiplication rule we can compute various probabilities of interest to us. For example, we can examine the evidence in support of the notion that winning the first game of the Series is a factor in the "psychological advantage." We can ask: "What is the probability that eventual World Series winners will win the first game of the Series?" According to the tree diagram, over a period of 62 World Series, the eventual winner has won the first game 37 times and lost it 25 times. The probability, then, of eventual Series winners winning the first game is 37/62 or 0.60. This finding would also seem to support the idea of "psychological advantage."

Notice that of the 37 times the Series champion won the first game it also won the second game 22 times. The probability that Series champions who win the first game will go on to win the second game is 22/37 or 0.59. If we want to know the probability that eventual Series winners will win the first two games of the Series, we can apply the multiplication rule as follows:

$$Pr(W_1 \cap W_2) = Pr(W_1)\,Pr(W_2 \mid W_i) = (37/62)\,(22/37) = 0.35$$

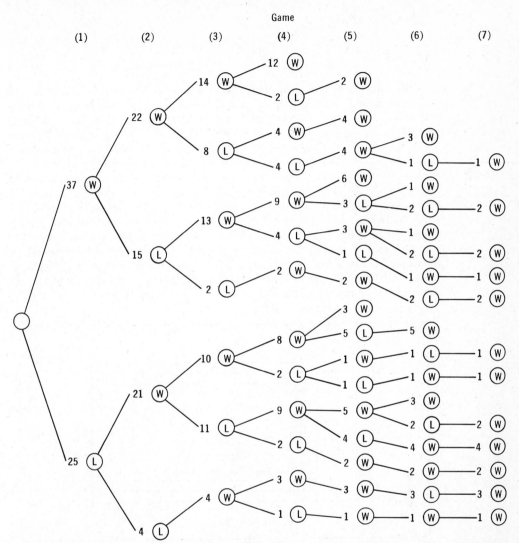

FIGURE 10.9 Patterns of Wins and Losses by World Series Champions for 62 Years, 1905 to 1969

where W_1 refers to winning the first game and W_2 refers to winning the second game.

Notice that we could also compute the probability by taking the number of teams winning the first two games (22) and dividing by the total number of teams we are considering (62): $22/62 = 0.35$. As a matter of fact, this was precisely what we did in computing the probability of winning four games in a row (0.19). Tracing this pattern of four wins in a row in the tree diagram and substituting into the general multiplication formula for a four-event sequence, we get the following:

$$Pr(W_1 \cap W_2 \cap W_3 \cap W_4)$$

$$= Pr(W_1)\, Pr(W_2 \mid W_1)\, Pr(W_3 \mid W_1 \cap W_2)\, Pr(W_4 \mid W_1 \cap W_2 \cap W_3)$$

$$= (37/62)\,(22/37)\,(14/22)\,(12/14) = 12/62 = 0.19$$

We can follow any path along the branches of the tree diagram and compute desired probabilities in a like manner. For instance, assume we are interested in finding the probability that eventual winners will lose the first game and then win the next four [$Pr(L_1 \cap W_2 \cap W_3 \cap W_4 \cap W_5)$]. This probability would be given by $(25/62)(21/25)(10/21)(8/10)(3/8)$ which is equal to 3/62 or 0.05.

In each of these applications, what we are doing in effect is starting with our universal set of 62 and replacing it with successive subsets as we move to each subsequent step (in this case, each game) of the stochastic process. For example, in the pattern of an initial loss followed by four wins, we start with a universal set of 62. The universal set is replaced by a subset of 25 in the second step by virtue of the fact that 25 of the eventual winners lost the first game. In the next step the subset of 25 is replaced by one of 21; then that one is replaced by one of 10; and finally, that one is replaced by one of 8. Of these eight, three won the fifth game. This situation is depicted in Figure 10.10.

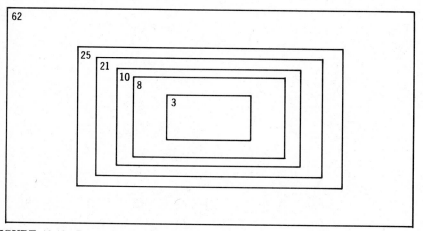

FIGURE 10.10 REDUCTION-OF-POSSIBILITY SPACES FOR THE PATTERN OF ONE LOSS AND FOUR WINS IN WORLD SERIES GAMES

Using the tree diagram in Figure 10.9 and our general multiplication rule, we could compute probabilities for any sequence of World Series wins and losses of interest to us. We could examine any combinations of wins and losses that we thought relevant to the notion of "psychological advantage." If we were actually to pursue the study of "psychological advantage," we would examine data other than those we have considered here. But it is obvious that the kind of analysis we have considered here is very useful for examining certain aspects of the phenomenon.

As sociologists move more toward the analysis of longitudinal data, the concept of the stochastic process will find its way increasingly into the literature of sociology.

10.3.2f THE SPECIAL MULTIPLICATION RULE

The general multiplication rule applies to the probability of the joint occurrence of events regardless of whether the events are related or independent. If two events, A and B, are related, the probability of B is altered by the occurrence or non-occurrence of A. When *two events* are *independent,* however, the occurrence or non-occurrence of A will have no effect on the probability of occurrence of B. Furthermore, the occurrence or non-occurrence of B will have no effect on the probability of occurrence of A. That is, $Pr(B \mid A) = Pr(B)$ and $Pr(A \mid B) = Pr(A)$.

Much of the data analysis engaged in by sociologists is aimed at determining whether variables (or events) are related or independent. In the field of descriptive statistics, considerable effort is devoted to an examination of techniques of analysis designed to determine whether variables are related or independent.*

When a sociologist asks a research question, such as, "Are divorce rates higher among childless couples than among couples who have children?" – he is asking whether the presence or absence of children in marriages alters the probability of divorce occurring. Essentially this asks, "Is the presence or absence of children related to or *independent* of the probability of divorce?" If the events (having or not having children and getting or not getting a divorce) are independent, the divorce rates for childless couples will be comparable to those for couples who do have children. Therefore, knowing that couples do or do not have children will not provide information that is useful for predicting whether they will or will not get a divorce.

It should be noted that lack of independence does not imply cause and effect. It merely implies that the occurrence or non-occurrence of event A is associated with the occurrence or non-occurrence of event B.

Given that events of interest are independent events, the general multiplication rule reduces to the special multiplication rule. This rule states that *the probability of the joint occurrence of independent events is equal to the product of their separate probabilities.* Therefore, if A and B are independent, the probability of their joint occurrence would be as follows:

(10.9)
$$Pr(A \cap B) = Pr(A)\, Pr(B)$$

This formula can be extended to include any number of independent events. The formula for the joint occurrence of the three mutually independent events, $A, B,$ and C, would be as follows:

$$Pr(A \cap B \cap C) = Pr(A)\, Pr(B)\, Pr(C)$$

As an example of the application of the special multiplication rule, assume that we are interested in the probability that a randomly selected set of pregnant women will give birth to twins, will have red hair, and will be married to

*See Chapters 6 through 9.

unemployed men. Presumably these three events are mutually independent. At least there is no apparent reason to expect them to be related. If the probability of giving birth to twins is 0.01, the probability of having red hair is 0.15, and the probability of having an unemployed husband is 0.03, these figures can be substituted into the formula for the special multiplication rule to find the probability of the joint occurrence of the three events.

$$Pr(A \cap B \cap C) = (0.01)\ (0.15)\ (0.03)$$

$$= 0.00004$$

It is obvious that this combination of events is not very likely to occur by chance. As a matter of fact, given the individual probabilities we have used, such a combination would only be expected to occur by chance about 4 times out of 100,000 observations.

It should be noted that the events of concern do not need to take place simultaneously. The special multiplication rule applies equally well to events during a period of time, as long as the independence assumption is tenable.

In making use of the special multiplication rule, the sociologist is often faced with a problem. In the actual research data with which the sociologist works, the independence of variables and observations being investigated must be carefully determined. This is frequently a difficult determination to make.

The difficulty of establishing independence of variables is illustrated nicely by an incident reported in *Time* magazine (April 26, 1968, p. 41). A witness reported that an elderly woman was mugged by a blonde girl with a ponytail who left the scene in a yellow car driven by a bearded Black. An interracial couple fitting the description given by the witness was apprehended and tried for the crime. An expert witness from the mathematics department of a nearby college was called by the prosecution to testify as to the probability of such a set of events occurring by chance. Using the special multiplication rule and what he considered conservative estimates of the probabilities involved (*e.g.*, that the

---

BOX 10.3 RULES TO REMEMBER

You will find that the special addition rule (Section 10.3.2b) and the special multiplication rule (Section 10.3.2f) are the two rules of probability that will be most used in connection with the techniques discussed in the remainder of this volume. They are relatively elementary rules and should not be too hard to grasp. If they are not clear to you at this point, however, take time to review Sections 10.3.2b and 10.3.2f. As you go over the discussion of the special multiplication rule, pay special attention to the concept of **independence** because this will be a key concept in later discussions of hypothesis testing.

probability of a car's being yellow was 0.1, that the probability of a couple in a car being interracial was 0.001, etc.), the expert estimated that there was about 1 chance in 12 million that another couple shared the same characteristics as the defendants. Largely on the basis of this testimony, the defendants were found guilty. However, the California Supreme Court overturned the conviction, holding that (1) there was not complete agreement among the witnesses on the characteristics in question, (2) that the probabilities assigned to the various characteristics were not shown to be even roughly accurate, and (3) that the special multiplication rule did not apply since the characteristics were not independent. On the latter question, the court pointed out that the probability of observing a Black with a beard overlaps the probability that a bearded Black might be part of an interracial couple.

In spite of the difficulties of establishing independence of variables, the special multiplication rule is often useful to the sociologist in his research. Moreover, as we shall see later, the rule plays an important part in the test of certain statistical hypotheses.*

10.4 COMBINATORIAL RULES

In the applications of probability theory it is sometimes necessary to determine the number of sequences in which elements of a set can be arranged. For example, in our discussion of stochastic processes we examined the outcomes of baseball World Series to determine the patterns of wins and losses of the eventual winning team for each series. We discovered that in series that lasted five games, there were four patterns of 4 wins and 1 loss that appeared. Do these four patterns represent all of the possibilities? Is there a general way of determining how many possibilities there are in such a situation? The answers to these questions is provided by one of a number of rules which constitute what is sometimes called combinatorial analysis. We will discuss only two of these combinatorial rules – the *permutation rule* and the *combination rule* – because these are the ones most relevant to our study of statistics.

10.4.1 The Permutation Rule

Permutations are ordered arrangements of elements. For example, assume that we have set up an experiment designed to test the reactions of subjects to different types of social pressures to conform. Furthermore, assume that there are four experimental conditions and ten subjects to be tested. We would wish to know in how many different orders the ten subjects might be assigned to the four experimental conditions, and we might even plan to assign the subjects to the experimental conditions in every possible order. In Table 10.2 we have worked out a number of the different orders available to us. The letters A, B, C, and D repre-

*E.g., see the discussion of chi-square (χ^2) in Chapter 16.

sent the four experimental conditions and the numbers represent subjects 1, 2, 3, and 4. The other subjects would be numbered 5, 6, 7, 8, 9, and 10.

Notice that in Table 10.2 we have used only the first four subjects and we already have 24 different orders in which they can be assigned to the experimental conditions. How many other arrangements are possible?

The permutation rule states that the number of ordered arrangements of M elements taken N at a time is given by the following formula:

(10.10)
$$P(M,N) = \frac{M!}{(M-N)!}$$

where $P(M,N)$ stands for the number of permutations of M elements taken N at a time (*i.e.* M = population size, N = sample size, and $P(M,N)$ is the number of ordered samples of size N drawn from a population of size M).

The exclamation point behind each term stands for the factorial of the number specified by the term. Thus, $M!$ stands for the factorial of M. A factorial is the product of a sequence of consecutive integers down to and including 1.

TABLE 10.2 SOME PERMUTATIONS OF TEN SUBJECTS ASSIGNED TO FOUR EXPERIMENTAL CONDITIONS

| Trial No. | *Experimental Conditions* | | | |
	A	B	C	D
1	1	2	3	4
2	2	3	4	1
3	3	4	1	2
4	4	1	2	3
5	4	3	2	1
6	3	2	1	4
7	2	1	4	3
8	1	4	3	2
9	1	2	4	3
10	2	4	3	1
11	4	3	1	2
12	3	1	2	4
13	1	4	2	3
14	4	2	3	1
15	2	3	1	4
16	3	1	4	2
17	1	3	2	4
18	3	2	4	1
19	2	4	1	3
20	4	1	3	2
21	1	3	4	2
22	3	4	2	1
23	4	2	1	3
24	2	1	3	4
etc.				

Therefore,

$$M! = M(M-1)(M-2)(M-3)(M-4) \ldots (3)(2)(1)$$

For example, if $M = 5$, the factorial of M will be as follows:

$$5! = (5)(4)(3)(2)(1) = 120$$

Take note of the fact that the factorial of 0 is defined as 1; that is, $0! = 1$.

Given the formula for permutations and an understanding of the meaning of a factorial, we can now substitute into Formula (10.10) and find the possible number of ordered arrangements of 10 elements taken 4 at a time:

$$P(10,4) = \frac{10!}{(10-4)!} = \frac{10!}{6!}$$

$$= \frac{(10)(9)(8)(7)(6)(5)(4)(3)(2)(1)}{(6)(5)(4)(3)(2)(1)}$$

By cancelling we get:

$$= (10)(9)(8)(7) = 5{,}040$$

Therefore, there are 5,040 permutations or ordered arrangments of 10 elements taken 4 at a time. It is obvious that it is impractical to try to include all of these permutations in our experimental design and impossible without repeatedly measuring each element. What we would need to do instead is to select some of the permutations to be used and ignore others.

10.4.2 The Combination Rule

While a permutation is an ordered arrangement of elements, a combination is a selection of elements without regard to order. Thus, the 24 permutations in Table 10.2 would all be treated as one combination because they all contain the same elements — individuals 1, 2, 3, and 4 — only the order in which they appear differs.

The combination rule states that the number of combinations of M elements taken N at a time is given by the following formula:

(10.11)

$$\binom{M}{N} = \frac{M!}{(M-N)!\,N!}$$

Note that the expression

$$\binom{M}{N}$$

is not a fraction; it is merely the symbolic name for the combinations formula and is read, "The combination of M elements taken N at a time."

Suppose again that we have 10 subjects and that we wish to arrange them into as many groups of size 4 as are possible for purposes of studying their behavior in a problem-solving situation. How many different ways can they be arranged in groups of 4 without regard for order of arrangement? The answer is given by the following:

$$\binom{10}{4} = \frac{10!}{(10-4)!\,4!} = \frac{10!}{6!\,4!}$$

$$= \frac{(10)(9)(8)(7)(6)(5)(4)(3)(2)(1)}{(6)(5)(4)(3)(2)(1)\quad(4)(3)(2)(1)}$$

By cancelling we

$$= \frac{(10)\,(9)\,(8)\,(7)}{(4)\,(3)\,(2)\,(1)} = \frac{5{,}040}{24} = 210$$

Therefore, there are 210 combinations of 10 elements taken 4 at a time as compared to 5,040 permutations. However, the 210 combinations are still probably too numerous to make use of in arranging groups of 4 to be tested. Again, we probably would select only certain combinations to test and ignore others.

By a slightly different interpretation of the meanings of M and N we can use the combinations formula to answer the original question posed. That is, how many different possible patterns of 4 wins and 1 loss are there in a World Series that lasts five games? We will let $M=5$, the number of games played, and $N=4$, the number of games won out of the 5. Substituting into the combinations formula with these figures we get the following:

$$\binom{5}{4} = \frac{5!}{(5-4)!\,4!} = \frac{5!}{1!\,4!}$$

$$= \frac{(5)(4)(3)(2)(1)}{(1)\quad(4)(3)(2)(1)}$$

By cancelling we get:

$$= \frac{5}{1} = 5$$

Therefore, we conclude that there are five different patterns of 4 wins and 1 loss that are possible in 5 games. The patterns that appeared in Table 10.1 were the following:

win loss win win win
win win loss win win
win win win loss win
loss win win win win

The single pattern that is missing is as follows:

win win win win loss

Of course this pattern is not a possibility in a baseball World Series because the first team to win 4 games wins the Series; therefore, this pattern reduces to a 4 game Series.

It may appear that order of occurrence was considered in this example, although we said earlier that order was not taken into account for combinations. The only distinction made here, however, was between wins and losses. We did not consider the four wins as four distinct events. If we had labeled the four wins as win A, win B, win C, and win D, and made distinctions in the order in which they occurred, then we would have permutations rather than combinations.

10.5   SUMMARY

In this chapter we have presented a brief introduction to some of the concepts of set theory so that we might make use of them to study probability. We have discussed and used two alternative concepts of probability and have presented a number of rules of probability useful to the sociologist. We have tried to present examples of a sociological nature to suggest the kinds of applications that sociologists might have for these rules. We ended the chapter with two useful combinatorial rules.

In the following chapter we will see how probability theory plays an integral part in the sampling process, and in later chapters we will see that probability theory serves as the basis for the whole inductive process.

CONCEPTS TO KNOW AND UNDERSTAND:

set	event
element	experiential probabilities
universal set	stochastic process
empty set	conditional probability
subset	independent events
disjoint sets	mutually exclusive events
overlapping sets	general and special multiplication
union	rules
intersection	general and special addition rules
a priori probabilities	permutation
outcome	combination

QUESTIONS AND PROBLEMS

1. Select some set that a sociologist might be interested in and list the criteria for membership in that set. That is, list those characteristics that distinguish members of the set from non-members.

2. Give examples of finite and infinite universal sets that are of interest to sociologists. Read five article abstracts from a professional journal in sociology (such as those that appear in the data section of *Study Guide to Accompany Descriptive and Inferential Statistics: And Introduction,* 2nd ed., by H. J. Loether and D. G. McTavish, prepared by Sally Goralnik) Determine which universal sets the authors were interested in. Determine whether each of the universal sets is finite or infinite.

3. Consider your college or university to be a universal set. List as many sub-

sets of that universal set as you can. Specify whether each subset is disjoint or intersecting. If a subset is intersecting, specify which of the other subsets intersects with it.

4. Explain what is meant by the statement that a probability is not a prediction of a single event, but of a large number of events. If 75 percent of the students who take a course in elementary statistics earn at least a C for a final grade, can you assume that you have a 75 percent chance of earning at least a C?

5. Give an example of two variables of interest to a sociologist to which the general addition rule would apply (that is, two variables that are not mutually exclusive). Give an example of two variables of interest to a sociologist to which the special addition rule would apply (that is, two variables that are mutually exclusive).

6. Write the formula for the general addition rule for $Pr(A \cup B \cup C \cup D)$.

7. Which of the following pairs of variables are independent of each other, and which of them are conditionally related? For those pairs that are conditionally related, specify which variable of each pair is logically prior to the other.
   a. religious affiliation and income
   b. size of family and income
   c. number of children under 18 and color of mother's eyes
   d. number of years of formal education and number of books read per year
   e. length of women's skirts and unemployment rates
   f. political affiliation and marital status
   g. age and country of birth

8. Think of two examples for which the permutation rule would be applicable and two examples for which the combinations rule would be applicable.

9. Rules of probability are often illustrated with coin tosses or throws of dice. Suppose that you have a "fair" pair of dice; what is the probability of the event:
   a) Getting two 6's?
   b) Getting a 6 and a 2?
   c) Getting either a 4 and a 2 or a 5 and a 1?
   d) Getting an outcome that adds to six?

10. Your job is to set up tennis matches between pairs of people and you have a list of 7 people, below, who have signed up to play. One person in each pair will be given the responsibility of picking up equipment for his or her match.
   Sally
   Joe
   Jim
   Laura
   Sue
   Steve
   Debby

a) How many different pairs can be created, where each person is assigned responsibility for the equipment for each pair?

b) Ignoring who has responsibility for the equipment, how many different pairs can be created?

c) How many pairs can be created where Sally is included on a team? Which rule of probability did you use to arrive at this conclusion?

d) Now suppose that you want to create teams of 4 players from the above group of seven. How many different teams could be created (ignoring order of team selection)?

e) How many different ways could the seven be arranged in order on a list?

GENERAL REFERENCES

Adler, Irving, *Probability and Statistics for Everyman: How to Understand and Use the Laws of Chance* (New York, The New American Library, A Signet Science Library Book), 1963.

This is a well written little book that covers many of the ideas of probability theory in an understandable but solid fashion. He includes many interesting examples and puzzles.

Freund, John E., *Modern Elementary Statistics,* Third Edition (Englewood Cliffs, N.J., Prentice-Hall), 1967.

Part II is a lucid discussion of probability theory and probability distributions. Freund's approach to the subject of probability is more general than we have presented in this book.

Kemeny, John G., J. Laurie Snell, and Gerald L. Thompson, *Introduction to Finite Mathematics* (Englewood Cliffs, N.J., Prentice-Hall), 1956.

Consult this book for a more detailed treatment of set theory. It is also a good source for illustrations of how mathematics is applied to the social sciences.

Mosteller, Frederick, Robert E. K. Rourke, and George B. Thomas, Jr., *Probability with Statistical Applications* (Reading, Mass., Addison-Wesley Publishing Company), 1961.

This book was written as the text for a televised course in probability and statistics. While it is mathematically sound the approach is intuitive.

OTHER REFERENCES

Blau, Peter M., and W. Richard Scott, *Formal Organizations* (San Francisco, Chandler Publishing Company), 1963.

Bridges, Edwin M., *et al.,* "Effects of Hierarchical Differentiation on Group Productivity, Efficiency, and Risk Taking," *Administrative Science Quarterly,* vol. 13 (September, 1968), p 305–319.

Brown, Robert, *Explanation in Social Science* (Chicago, Aldine), 1963.

Durkheim, Emile, *Le Suicide* (Paris, 1897). See also the translation: Emile Durkheim, *Suicide* (Glencoe, Ill., The Free Press) 1951.

Time, April 26, 1968, p 41.

11 *Sampling Principles*

Sociologists are engaged in the business of developing generalizations about social behavior. They attempt to generate explanatory schemes that may predict the orderly operating processes of such social systems as total societies (*e.g.*, the United States), communities (*e.g.*, Solvang, California), organizations (*e.g.*, the American Legion), and groups (*e.g.*, the corner gang). These explanatory schemes are generally tested through the analysis of limited bodies of data, called **samples**. The samples are not inherently interesting in themselves, however. They are only of interest to the extent that they represent larger, more significant bodies of data known as **populations**. In sociology, as in most other sciences, samples are used to represent populations because they are more economical to study and because selection of a sample is often feasible when complete enumeration of a population is not.

11.1 SAMPLE REPRESENTATIVENESS

Both for the process of selecting a representative sample, and for the process of generalizing results to the population sampled, probability theory is the crucial field of operations. In this chapter we will examine how probability theory operates in the selection of a representative sample, and in the following chapters we will examine the part probability theory plays in the generalization process itself.

To take an illustration: It has been found that the level of academic performance of students is related to the degree of racial segregation of the school they attend. For example, the students of predominantly black schools tend to score lower, on the average, on standardized reading tests than the students of predominantly white schools. There are several possible explanations for this phenomenon. One possibility is that the quality of teaching is lower in the black schools than in the white schools. Another is that facilities in the white schools are superior. A third possibility is that the typical educational program is geared to white children and is not relevant and meaningful to black children. A fourth possibility is that the black children are not being offered sufficient motivation to learn to read and achieve. Numerous other factors also suggest themselves as possible explanations.

Suppose we wish to do a study of elementary schools with varying percentages of minority children to see how much their students vary in average reading ability, and to try isolating those factors that best explain any differences we find. The public elementary schools of Los Angeles vary in percentage minority enrollment from a low of 2 per cent to a high of 100 per cent. Furthermore, since Los Angeles is a city with a heterogeneous population largely free of sectional peculiarities, it might be considered a good choice of a population to study.

Since there are 435 public elementary schools in Los Angeles (or were, as of 1971), a study of the whole population of schools would probably be beyond the limits of our resources in staff, dollars, and time. Therefore, we may decide to sample the schools and carry on an intensive analysis of those schools in the sample. We would naturally wish to get a representative sample of schools so that we would have a good spread in percentage of minority enrollment and a realistic distribution of those other variables relevant to differences in average reading ability of students. If the sample is representative, we can feel justified in generalizing the findings from the sample data to the population of all public elementary schools in the city. The basic sampling technique that is indicated for drawing a representative sample from a known population is called a *simple random sample*.

It should be noted that not all simple random samples are representative of the population from which they are drawn. It is the case, however, that simple random samples are more likely to represent accurately characteristics of the population than not. Any differences between characteristics of a sample and the characteristics of the population it represents will be random differences for which we will learn later how to compensate.

11.2   THE SIMPLE RANDOM SAMPLE

A **simple random sample** is one in which the *elements* of the population are listed, and then N of them are randomly selected to be elements of the sample. The random selection is carried out in such a way that (1) every element in the population has an equal probability of being included in the sample, and (2) every possible combination of N elements has an equal probability of constituting the sample. Simple random selection *does not mean haphazard selection;* rather it means a selection process that gives *each element in the population an equal chance of appearing in the sample.*[*]

---

[*]In general, a random sample is one where each element has a calculable probability of being selected. Simple random samples are those where this probability is equal for each population element.

If there are M elements in a population, according to condition (1) above, the probability of each individual element being included in the sample should be $1/M$. Therefore, in drawing a simple random sample of the 435 public elementary schools in Los Angeles, the selection process should be such that each and every school should have a probability of $1/435$ of being included.

Furthermore, the probability of each school being included in the sample should be independent of the probability of any other school being included. This is necessary to meet the second condition for a simple random sample; namely, that every possible combination of N elements has an equal chance of constituting the sample.

11.2.1   Sampling with and without Replacement

Assume that we number the 435 schools in the population serially from 001 through 435, and then we randomly select 100 of these numbers to represent the elements of the sample. (We will explain below how the random selection is accomplished.) If we take as the elements of the sample the first 100 numbers between 001 and 435 that we select, even though some of the numbers may appear two or more times in the sample, we are said to be **sampling with replacement.** If in drawing the 100 numbers we insist that they be 100 unique numbers, then we are said to be **sampling without replacement.**

Both of these sampling procedures meet the first condition of the simple random sample; that is, that every element in the population have an equal probability of being included in the sample. When we sample *with replacement*, for example, the first element drawn in the sample has a probability of $1/435$ because there are 435 numbers to choose from; the second element drawn has a probability of $1/435$ because there are still the same 435 elements to choose from, etc.

When we sample *without replacement* the first element drawn has a probability of $1/435$. The second element drawn is drawn from a pool of the remaining 434 elements (the first element drawn no longer being eligible to be drawn a second time), but its probability of being drawn on the second draw is *conditional* upon its *not* having been drawn on the first draw. The conditional probability of the second element being drawn, then, is $(434/435)(1/434)$, that is, the probability that it was not drawn on the first draw, times the probability of its being drawn on the second draw. Notice that this probability is $(434/435)(1/434) = 1/435$, which is the same probability that the first element had of being drawn on the first draw. Similarly, the third element in the sample will have a probability of $(434/435)(433/434)(1/433) = 1/435$ of being drawn third; and thus each succeeding element, according to the rule of conditional probability, will also have a probability of $1/435$ of being included in the sample.

While the process of sampling *with replacement* also meets the second condition for a simple random sample, the process of sampling *without replacement* does not. Some of the samples which are possible when sampling with replacement are not possible to achieve if a finite population such as this is sampled without replacement. For example, it would be impossible to achieve the 435 samples which include the same element drawn repeatedly 100 times. Nor is it possible to get any of the samples in which an element appears more than once. Thus the second condition for simple random samples, that all possible samples of N elements are equally probable, does not hold for sampling without replacement from finite populations.

The number of possible samples of size $N = 100$ that could be drawn from the population (M) of 435 schools where one samples without replacement (and order of drawing specific elements is not considered), is computed by the following formula:

(11.1)
$$\binom{M}{N} = \frac{M!}{(M-N)!\,N!}$$

which you will undoubtedly recognize as the formula for the number of combinations of M elements taken N at a time.

If we were to draw random samples of size 100 *without replacement* from the population of 435 elementary schools, we would have the following possible number of different samples:

$$\binom{435}{100} = \frac{435!}{(435-100)!\,100!} = \frac{435!}{335!\,100!} = 3{,}236{,}398 \times 10^{94}$$

Technically speaking, sampling with replacement is a necessary condition for a simple random sample, and we will reserve the term **simple random sample** for those random samples drawn *with* replacement. Those samples drawn without replacement we will call **random samples drawn without replacement.** In actual practice sociologists seldom draw simple random samples because of the possibility of drawing the same case more than once. They often have qualms about counting one person's questionnaire responses more than once, for example, or giving the same person more than one questionnaire to fill out. Sampling without replacement is justified on the basis that, if the size of the sample N is relatively small as compared to the size of the population M, it makes very little difference whether the sampling is done with or without replacement.

It is our position that sampling with replacement is technically correct and that it is entirely legitimate to count data from a population element more than once if that element appears in the sample more than once. If there is sound reason for sampling without replacement (*e.g.*, because of the nature of the problem being investigated or because of the quality and quantity of the data available), then that procedure is justifiable and should be followed. However, sampling without replacement should not be preferred merely on the grounds that sample elements should all be unique.

11.2.2 The Basic Nature of the Simple Random Sample

The simple random sample is the basic sampling technique for inferential statistics because it is the sampling technique that was posited when most statistical theories and techniques were originally derived. Furthermore, the simple random sample serves as a model from which other random sampling techniques developed.

If we consider the arithmetic mean and its computation, we can begin to appreciate the fact that the simple random sample served as a basis for the derivation of its formula.

Given a set of scores (X's) we may find the population arithmetic mean (μ, Greek letter mu)* by multiplying each score by the probability of its occurrence—$Pr(X)$ X—and then summing all of the resulting products. The formula for the arithmetic mean, then, may be expressed as follows:

(11.2)
$$\mu = \Sigma \, [\, Pr(X)X \,]$$

In the case of a simple random sample, the probability of occurrence of each score in the population as an element of a sample is assumed to be equal and to be independent of that of every other score; therefore, the probability can be expressed as $1/M$. We can substitute $1/M$ for $Pr(X)$ in the formula for the mean giving us the following:

(11.3)
$$\mu = \Sigma (1/M)(X)$$

Since $1/M$ is a constant and X is a variable, we can apply the principle that *the sum of a constant and a variable is equal to the constant times the summation of the variable.* We thus rearrange the terms as follows:

(11.4)
$$\mu = (1/M)\Sigma X$$

which is equivalent to the familiar

(11.5)
$$\mu = \Sigma X/M$$

Thus, when the formula for the arithmetic mean was derived, the assumption was made that every element in the population has an equal probability of occurring. This assumption stems from and is consistent with the conditions necessary for a simple random sample. The assumption of equal probability of occurrence within the total population is carried over to the formula for computing the sample mean. Thus, all of the elements in a sample are also assumed to be equally probable, and the formula for the sample mean is given as follows:

(11.6)
$$\overline{X} = (1/N)\Sigma X \quad \text{or} \quad \Sigma X/N$$

This same line of reasoning was followed in the derivation of the formulae for most statistical techniques. Therefore, *whenever you use the techniques covered in these volumes to make inferences to populations, you are, in effect, assuming that your data constitute a simple random sample.* Any use of these techniques to generalize from a sample that is *not* a simple random sample must be recognized as a violation of this assumption, and this must be taken into account in interpreting results.

*Note that in order to keep sample and population means separate, the Greek letter mu (μ) is used as the symbol for a population mean and X-bar ($\overline{X}$) is used to symbolize a sample mean. There will be other points at which different symbols are used to distinguish sample and population values.

11.2.3 Random Selection

Crucial to simple random sampling is the process by which elements are selected for inclusion in the sample. If the conditions of simple random sampling are to be met, the selection process must be random.

If we wished to draw a simple random sample of 100 schools from the population of 435, how would we go about doing it? It might occur to you that a good procedure would be to write the names of the 435 schools on slips of paper, put the slips in a container, mix the slips thoroughly, and draw slips from the container. In order to sample with replacement, we would replace each slip drawn after recording the name of the school on it (if we wished to draw a random sample without replacement, we would not replace the slips drawn). Although this procedure seems to be a logical way of drawing a simple random sample, it is not generally satisfactory because it is difficult to mix the slips thoroughly enough to approximate a random process.

When the first draft lottery was conducted in December, 1969, using birth dates encased in capsules and placed in a bowl, a statistician conducted an analysis of the order in which birth dates were selected and came to the conclusion that the process was not random. He found that the birth dates in the last six months of the year were more likely to be drawn early than were birth dates in the first six months (he estimated that there were only 5 chances in 1,000 that the results would turn out as they did if each date had an equal probability of being drawn).

The statistician, John Ware, said:

> I would guess that they probably placed the capsules containing the numbers into the bowl in a chronological order. . . . In the course of shuffling the capsules, the months tended to stay together. The first six months may have been placed in the bowl first and they tended to stay together and be drawn last. . . . The cards were stacked against its being fair. . . . It wasn't the best procedure with our current knowledge of probability (Berman, 1969).

The best approximation to a random process is achieved through the use of a table of random numbers.* Tables of random numbers are generated in such a way that all integers from 0 through 9 occur with about equal frequency. The digits appear on a page in a random fashion; that is, each individual digit has an approximately equal probability of appearing in any position on the page. The digits are usually grouped in columns of two or more digits to make them easier to read, but the groupings are without any other significance. A partial table of random numbers is reproduced as Table 11.1.†

To illustrate the use of this table we will draw a simple random sample of 100 schools from the Los Angeles population of 435 schools. First, we take the list of schools and number them serially from 001 to 435. Second, in order to minimize bias in entering the table of random numbers, we use some reasonably random process of deciding where to start. There are 48 vertical, single-digit columns and 20 horizontal rows of numbers on the page. We might combine the

---

*Computers can be programmed to produce numbers which behave like random digits (called pseudo-random numbers). Other physical processes also can be designed to have "random" appearing outcomes.
†A more extensive table of random numbers appears in the back of this volume as Table A.

TABLE 11.1   RANDOM NUMBERS

```
93 71 61 68 94 32 88 65 97 80 92 05 24 62 15 95 81 90 68 31 39 51 03 59
05 27 69 90 64 94 92 96 26 17 73 10 27 41 22 02 75 86 72 07 17 85 78 34
76 19 35 07 53 39 49 56 62 33 44 42 36 40 98 32 32 57 62 05 26 06 07 39
93 74 08 28 82 53 57 93 12 84 38 25 90 83 82 45 26 92 63 01 19 89 01 88
32 58 08 51 49 36 47 33 31 12 36 91 86 01 97 37 72 75 85 85 13 03 25 52
45 81 95 29 79 61 95 87 71 00 17 26 77 09 43 10 06 16 88 29 35 20 83 33
74 35 66 35 29 72 01 91 82 83 16 70 07 11 47 36 11 13 30 75 86 78 13 86
65 59 27 48 24 54 76 88 65 12 25 96 74 84 39 34 13 28 59 72 04 05 74 02
28 46 17 65 74 11 40 14 90 71 22 67 69 08 81 64 74 49 16 43 59 15 29 26
65 59 08 02 41 32 64 43 44 96 24 04 36 42 03 74 28 38 73 51 97 23 78 67
54 84 65 47 59 65 13 00 48 60 88 61 81 91 61 71 29 92 38 53 27 95 45 89
09 80 86 30 05 14 33 56 46 07 80 90 89 97 57 34 78 03 87 02 67 55 98 66
64 85 87 53 90 88 23 16 81 86 03 11 52 52 75 80 21 56 12 71 92 55 09 97
33 34 40 32 30 75 75 46 10 51 82 16 15 83 38 98 73 74 91 87 07 61 50 27
12 46 70 18 95 37 50 58 71 20 71 45 32 95 04 61 89 75 53 31 22 30 84 20
40 18 90 11 94 75 04 46 60 32 28 46 66 87 95 77 76 22 07 91 55 70 34 48
83 14 65 17 78 55 48 94 97 23 06 94 54 13 74 08 07 55 35 89 72 65 34 46
74 15 52 84 33 85 31 08 00 74 54 49 34 28 07 86 43 93 71 62 81 85 64 60
32 14 17 49 84 71 41 44 72 88 84 36 39 72 43 23 80 77 98 38 78 52 02 92
45 95 74 96 44 03 12 65 11 69 85 50 03 58 65 40 15 03 64 04 71 36 69 94
```

use of a single die and today's report of the volume of sales on the New York Stock Exchange as a means of picking the number of the column and the number of the row at which we will enter the table. For example, we might determine the number of the column by rolling one die to find the first digit of a two-digit number between 01 and 48. If a 1, 2, 3, or 4 comes up we would use that as our first digit. If a 5 comes up, we could treat it as a 0 and use it as our first digit. If a 6 comes up, we could ignore it and roll the die again. To determine the second digit we could look at the next to last digit in today's report of the volume of sales on the New York Stock Exchange. If it happens that that digit is a 9, we could look at the digit immediately preceding it, etc. The appropriate row to start at would be determined through a similar procedure. A roll of the die would determine the first digit in the row number and the last digit in the report of the volume of sales on the New York Stock Exchange would determine the second digit of the row number.

If you examine Table 11.1, you will notice that the digits 486 are underlined. The 4 is in the seventeenth vertical, one-digit column and in the eleventh horizontal row. This is the starting point arrived at by rolling the die and consulting the stock market report. The first roll of the die produced a 1 and the next to the last digit in the report of the volume of sales on the stock exchange was a 7; hence, the appropriate column to enter was the seventeenth. The second roll of the die produced a 1 and the last digit in the report of the volume of sales was also a 1; hence, the appropriate row to enter was the eleventh.

Since we need three-digit numbers (001 through 435), we will consider the 4 the first digit and also include the digits in vertical, single-digit columns 18 and 19. Starting at that point, then, we read down the column (or in some other previously determined direction), reading three digits at a time, until we come to a three-digit number between 001 and 435 inclusive. The first such number happens to be 105. The school numbered 105, therefore, will be the first element of the sample. We continue reading down the column until we find another eligible number (the next one is 007) and the school bearing that number is the second element of the sample. We read down the three-digit column looking for more eligible numbers, then shift to the next columns of three digits (single-digit columns 20, 21, and 22), etc., reading as far along as necessary to draw the sample of 100 schools. Table 11.2 includes the actual list of 100 eligible random three-digit numbers drawn from the table. The schools bearing these numbers would constitute the sample of 100 elements.

If you examine the list of 100 numbers drawn, you will notice that there were nine numbers that were drawn twice. These were 122, 052, 267, 358, 393, 275, 401, 359, and 333. The number 122, for example, was drawn tenth and drawn a second time as the seventy-fourth number. Since we wished to draw a simple random sample, we drew the sample with replacement. The data collected from the nine schools that were selected twice would be counted twice in the analysis and the sample size would be 100 although there are actually only 91 different schools represented in the sample.

If a researcher could not bring herself or himself to sample with replacement and count cases more than once, he(she) could use the table of random numbers to draw a random sample without replacement. To sample without replacement, the only alteration necessary in the procedure described would be to ignore a number whenever it was duplicated and continue selecting numbers from the table until 100 unique numbers had been drawn. If the sample were drawn with-

TABLE 11.2 ONE HUNDRED ELEMENTS DRAWN WITH REPLACEMENT, USING THE TABLE OF RANDOM NUMBERS

105	043	275	164	152
007	115	323	385	237
116	161	377	127	308
092	363	100	079	366
344	083	361	358	359
236	191	341	139	188
017	252	347	119	333
316	295	080	204	386
225	413	238	359	402
122	428	401	197	027
088	358	190	327	420
182	159	257	267	292
071	220	275	192	052
228	431	113	122	325
306	393	328	155	284
052	037	156	281	098
102	389	374	060	333
423	046	393	032	124
259	078	077	208	401
267	432	307	057	321

out replacement, however, the researcher would be obligated to take account of that fact in interpreting the results of his(her) data analysis.

11.2.4   A Preliminary Word about Sampling Error

Sampling error is a term which refers to the variability that could occur from random sample to random sample if repeated random samples of the same size were drawn from the same population. The term will be discussed in detail later on but a preliminary word should be introduced at this point. One of the attractive features of the simple random sample is that there is a straightforward way of measuring *sampling error* and, thus, of gauging the degree of representativeness of the sample. Fortunately, because of the procedures used to draw a random sample, whatever sampling error there is will be random error. In the next chapter (Sections 12.2.2 and 12.4.4 in particular) we will consider, in detail, the measurement of sampling error.

The magnitude of sampling error in a simple random sample is a function of the size of the sample. Interestingly enough, what is crucial is *not* the relative proportion of the population represented by the sample. Rather, as the number of cases in the sample is increased, sampling error decreases without regard to population size. It can generally be assumed that large samples will have small sampling errors. It is desirable, then, to take as large a sample as is feasible. It would not make sense, however, to draw a sample so large that it would be a major part of a population. It would be better to take a complete enumeration of the population and eliminate sampling error. Sampling is, in part, an economy measure, and there is no economy in taking a sample which includes most of the cases in the population.

11.2.5   Feasibility of Simple Random Samples

Although the simple random sample is fundamental to inferential statistics, most of the studies in the sociological literature are not based on data from simple random samples. One exception to this assertion is a study by Alejandro Portes (1969). Portes studied factors involved in the integration of Cuban refugee families who migrated to Milwaukee. His data were based on a simple random sample from the population of names of family heads found in lists provided by the Cuban Association of Wisconsin and other similar organizations. He drew a sample of 48 families from the population of 152 families. In this case, a simple random sample was feasible because the population was small and a list of population elements was available.

In order to draw a simple random sample it is necessary to have a **list of population elements.** If such a list is not available or can not be created, then a simple random sample is not feasible.

If the units of study are social systems, it is more likely that a list of the population elements can be obtained than if the units of study are individuals. Populations whose elements are social systems are generally more stable and permanent than populations whose elements are individuals. For example, we were able to obtain a list of the public elementary schools in Los Angeles; furthermore, we had some confidence that the list was accurate because schools generally exist over a period of many years. It would be a much more difficult task to compile a single list of all of the students enrolled in those 435 schools, and we would have much less confidence in the accuracy of such a list if it were available. Populations involving a large number of individual elements are generally fluid. That is, they are in a state of constant change. People are born, they die, and they move. Thus, publications such as city directories, telephone directories, and mailing lists are seldom accurate to begin with and they "decay" at the rate of about $\frac{1}{3}$ per year.

When sociologists are interested in studying large populations of individuals (e.g., the labor force of the United States) they are not likely to be able to get a list from which to draw a simple random sample. And of course, those populations which are most likely to be significant from a theoretical standpoint are also those for which lists are least likely to be available. If random samples of such populations are desirable, but simple random samples are not feasible, then how may random samples be drawn? The answer is that the sociologist must turn to some sampling design that is random, but is a variation of the basic simple random sample. Such a sampling design is the **cluster sample.**

11.3 THE CLUSTER SAMPLE

A **cluster sample** is usually a multi-stage random sample. It is drawn by taking a series of simple random samples. That is, the population is first divided into a set of clusters, and a simple random sample of the clusters is drawn. Then the clusters sampled at the first stage are each divided again into clusters, and a simple random sample of those clusters is drawn, etc. The final elements sampled may be either clusters or individuals.

The clusters from which the initial random sample is drawn are called **primary sampling units;** the sub-clusters sampled next are called **secondary sampling units,** etc. The final clusters or individuals sampled are called the **ultimate sampling units.**

Assume that we wish to draw a random sample of all of the registered voters in the United States. It might be drawn as follows: First, we could take a simple random sample of the fifty states and the District of Columbia, using a table of random numbers. The states selected would constitute our primary sampling units. Second, from each of the states drawn in the first stage we could draw simple random samples of counties or Congressional Districts. These would be the secondary sampling units. Next, we might divide the sample counties or districts into voting precincts and draw simple random samples of precincts. These would be tertiary (third-stage) sampling units. Finally, we could collect data from all of the registered voters in the sample precincts, or we could draw simple random samples of registered voters from the lists of voters in the sample precincts. The registered voters in the sample precincts would constitute the ultimate sampling units. Table 11.3 outlines such a multi-stage random or cluster sample. Notice that lists of registered voters are required only for the sample precincts drawn and not for all precincts in the country. The cluster sample has the virtues of (1) not requiring a single list of all of the elements of the population, and (2) having the elements in the sample geographically clustered.*

11.3a Homogeneous and Heterogeneous Clustering. The cluster sample is most efficient when the clusters which form the primary sampling units are homogeneous with respect to relevant variables, and the ultimate sampling units are heterogeneous. That is, when between-cluster variability is small relative to within-cluster variability. This cuts the costs of a study because it is possible to take a smaller sample of primary sampling units where geographic dispersion is a problem, and to do complete enumerations or to select relatively large samples of ultimate sampling units which tend to be geographically clustered.

In the example just cited, the states would be quite heterogeneous with respect to political party membership. Ideally, they should be homogeneous; therefore, it would be wise to include a sizeable number of states in the sample in order to assure a more representative sample.† Ideally, the precincts sampled should be quite heterogeneous. Actually, some precincts would be heavily Republican, others heavily Democratic, and still others mixed. Thus, individual precincts would differ in their contributions to the representativeness of the total sample. From the standpoint of sampling efficiency, it would be desirable if all of the precincts that appeared in the final sample were of the mixed type.

Cluster samples generally have a larger sampling error than simple random samples of the equivalent size depending upon the magnitude of between-cluster and within-cluster variation. When the clusters that form the primary sampling units are heterogeneous, and the ultimate sampling units are homogeneous within the final clusters sampled, then the sampling error of a cluster sample is maximized as compared to a simple random sample of the same size. As the clusters that form the primary sampling units become more homogeneous, and the ultimate sampling units within the final clusters sampled become more heterogeneous, the sampling error of the cluster sample becomes smaller and approaches that of a simple random sample (when between-cluster variation is zero).

TABLE 11.3 EXAMPLE OF MULTI-STAGE RANDOM SAMPLE OF REGISTERED VOTERS

Sampling Level	Subset or Element	Example
Primary sampling units	States	Missouri
Secondary sampling units	Counties	Howard County
Tertiary sampling units	Precincts	Precinct #10
Ultimate sampling units	Registered voters	Thomas Anderson

*Clusters need not be geographically identified, but this is common and has the advantage of reducing travel costs involved in personal interviewing.
†In order to increase the homogeneity of the states they could be grouped into regions first and then sampled so that various regions would be represented. This procedure would introduce the notion of stratification to be discussed in the following section of this chapter.

The disadvantage—that sampling error tends to be greater in a cluster sample than in a simple random sample—must be weighed against the advantages accrued from the use of the technique. An important advantage is that the cost per case of collecting data from a cluster sample tends to be less than that for a simple random sample because of lower locating and travel-related costs. Another important advantage of the cluster sample is that it is often feasible when a simple random sample is not. While the simple random sample requires a single list of the elements in the population, a cluster sample does not.

11.3.1   Further Examples of Cluster Sample Designs

We described how we might draw a cluster sample of all of the registered voters in the United States. If we wished to draw a national sample but did not wish to restrict it to registered voters, how might we go about it? We might start as we did in our previous example, sampling states and then counties. Instead of sampling voting precincts, however, we might sample Bureau of the Census enumeration districts and residences within the enumeration districts.*

The residences to be used as the ultimate sampling units could be sampled using detailed maps of enumeration districts. Once the residences had been sampled, interviewers could be sent to them with instructions to interview the head of the household or some other previously determined inhabitant.

A cluster sample reported in the sociological literature was used in a study by Joe L. Spaeth (1968). The sample design was a rather simple two-stage design used in a four-year longitudinal study of June, 1961, college graduates, conducted by the National Opinion Research Center. The first stage was a sample of 135 of the accredited colleges and universities in the United States that grant baccalaureate degrees. The second stage consisted of prospective June, 1961, graduates of the 135 colleges and universities. All 135 schools cooperated in the study and 85 percent of the prospective graduates sampled returned completed questionnaires. The sample consisted of some 41,000 students.

11.3.2   Cluster Samples and Statistical Analysis

As we pointed out earlier in this chapter, statistical techniques generally are based on the assumption that the data to be analyzed constitute a simple random sample of some population of data. Therefore, when the data to be analyzed come from a cluster sample rather than a simple random sample, caution must be exercised in interpreting results. Since sampling error is generally greater in cluster samples than in simple random samples, tests of statistical significance may be misleading. If the actual sampling error is underestimated, a result which appears to be statistically significant may not be significant at all. If ad-

---

*The Bureau of the Census has divided the United States into thousands of enumeration districts (ED's) which are sometimes as large as a county and sometimes as small as part of a residential block. In any case each is small enough to be enumerated by a single field worker. (See Scott, 1968.)

justed formulae are available for use with cluster samples, these should be used in lieu of the formulae that appear in this volume.* For many techniques, however, no such adjusted formulae have been derived. When adjusted formulae are not available and unadjusted formulae are used, it is advisable to make very stringent tests of hypotheses and to be conservative in interpreting results.

11.4 THE STRATIFIED RANDOM SAMPLE

Another variation from the simple random sample used by sociologists is the stratified random sample. The stratified random sample serves different purposes than the cluster sample. It is used when it is desirable to make comparisons among subpopulations of a population, or when it is desirable to reduce either sampling error or the costs of a study.

A **stratified random sample** *is one in which the population is divided into subpopulations* (**strata**) *and then simple random samples are drawn from each of the subpopulations.* When it is desirable to make comparisons among subpopulations, the subpopulations are usually differentiated on the basis of one or more of the independent variables of the study. For example, if a sociologist were going to study comparative family planning practices of Protestants, Catholics, and Jews, religious affiliation could be used to divide the population into three strata—one consisting of Protestants, one of Catholics, and one of Jews (other religions would, of course, be excluded). Then a simple random sample would be drawn from each stratum.

As was mentioned above, stratified random samples may also be used in lieu of simple random samples to reduce sampling error or the costs of a study. This is so because a properly executed stratified random sample (*i.e.* one with small within-strata variation) will have a smaller sampling error than a simple random sample of the same size; or, alternatively, a stratified random sample with fewer cases than a simple random sample, if properly drawn, will have a sampling error of the same magnitude as the larger simple random sample. Stratified sampling is also used as a method of controlling extraneous variation.

When the purpose of drawing a stratified random sample is to reduce sampling error (or sample size), the population is stratified using a variable or variables correlated with the independent and/or dependent variables of the study. For example, if a researcher wished to conduct a study of the relationship between high school students' scholastic performance and their income aspirations ten years hence, he(she) might use a stratified random sample which is stratified on sex and socioeconomic status. Sex would be an appropriate variable on which to stratify because males and females differ to some degree both in their academic performance in high school and in their aspirations. Socioeconomic status would also be appropriate to use as a basis for stratification since it is correlated with at least one of the other two variables (academic performance and aspiration) and probably both (cf. Sewell, 1971).

The result of dividing the population into subpopulations based on sex and

*For a discussion of proper formulae for cluster samples, see Kish (1965:151 ff.).

socioeconomic status is to make of the population a number of subpopulations —
each of which is more homogeneous in composition than is the total population.
If sex and socioeconomic status were used to stratify a population (where socioeconomic status was divided into the categories high, moderate, and low) six
subpopulations would result as follows:

High socioeconomic status males
High socioeconomic status females
Moderate socioeconomic status males
Moderate socioeconomic status females
Low socioeconomic status males
Low socioeconomic status females

Since both sex and socioeconomic status are correlated with the independent and dependent variables of this study, each of the six strata (subpopulations)
should be more homogeneous on academic achievement in high school and/or on
income aspirations than the total population. If such were the case, a composite
sample of 200 drawn by taking six separate simple random samples from the
strata would have a smaller sampling error than a simple random sample of 200
drawn from the total population. This is so because the size of the sampling error
is affected by the degree of homogeneity in the population or subpopulation being
sampled.

A stratified random sample might be preferred to a simple random sample
as a means of reducing sampling error, or it might be preferred as a money sav-
er due to the reduced sample size needed. If the sampling error of a simple random sample of 200 cases was small enough to satisfy the researcher, he(she)
might opt instead to reduce the size of his(her) sample. If, for example, he(she)
could reduce sampling error 10 percent by using a stratified random sample in
lieu of a simple random sample, then a stratified random sample of 180 cases
would have a sampling error equal to a simple random sample of 200 cases.

The crucial factor in using the stratified random sample to reduce sampling error (or the size of the sample) is to select variables on which to stratify
that are correlated with the independent and/or dependent variables. If uncorrelated variables are selected, the resulting strata will be no more homogeneous
than the total population, and the stratification process will have been a wasted
effort.* In the study of the relationship between high school academic performance and income aspirations, if we were to use hair color as a basis for stratifying and were to divide the population into three strata consisting of brunettes,
redheads, and blondes, we would probably be stratifying to no avail. Hair color
is probably not correlated either with high school academic performance or with
income aspirations.

In order to execute a stratified random sample design successfully one
must have more information about the population than is necessary for a simple
random sample. Not only is a list of the population necessary, but it is also necessary to be able to divide the overall list into a number of "subpopulation lists"
differentiated on some relevant variable or variables. As a matter of fact, one of
the strengths of the simple random sample is that when one is ignorant of the

*And "degrees of freedom" will have been lost, another statistical consequence of sampling procedures
which will be discussed later.

relevant variables other than the independent and dependent variables, one can execute a simple random sample with some confidence that the unknown but relevant variables will be sampled in approximately the proportions they are in the total population. At least, any difference between their proportions in the total population and their proportions in the sample will be a random difference, attributable to chance.

11.4a   Proportionate and Disproportionate Stratified Random Samples.   Stratified random samples may be either proportionate or disproportionate. In the **proportionate stratified random sample** *the variables used as a basis for stratifying are represented in the sample in the same proportions as they are in the population.* For example, if the population from which we were drawing the stratified random sample of high school students included all of the high school seniors in the state of Illinois, and if sex and socioeconomic status were distributed throughout that population as in Table 11.4, we would draw our sample so that the variables would be distributed throughout the sample in the same proportions.

Assuming that the population of high school seniors in Illinois numbers 90,000, let us consider how we would go about drawing a proportionate stratified random sample of 450 cases. First, the population would be stratified by sex and socioeconomic status into the six strata listed above. Then we would prepare six separate lists of population elements, one for each stratum. From each of these lists we would draw a simple random sample of 450/90,000 = .005 of the elements, using a table of random numbers. Using a constant fraction (450/90,000 or .005) to sample from each of the six lists would result in an overall sample of 450 cases distributed as in Table 11.5.

Note that the 450 cases in the sample will be distributed by sex and socio-economic status in almost exactly the same proportions as they are in the total population, if our sampling lists are accurate.

Once the proportionate stratified random sample has been drawn and data have been collected, the data can be analyzed in a number of different ways. First, the cases in the six strata could be compared separately on the relationship between their high school achievement and their income aspiration levels; second, socioeconomic status could be held constant by comparing the data for

TABLE 11.4 DISTRIBUTION OF SEX AND SOCIOECONOMIC STATUS IN A POPULATION OF HIGH SCHOOL SENIORS

	Males	*Females*	*Total*
High Socioeconomic Status	10,800 (12%)	11,700 (13%)	22,500 (25%)
Moderate Socioeconomic Status	14,400 (16%)	18,000 (20%)	32,400 (36%)
Low Socioeconomic Status	18,000 (20%)	17,100 (19%)	35,100 (39%)
Total	43,200 (48%)	46,800 (52%)	90,000 (100%)

TABLE 11.5 COMPOSITION OF PROPORTIONATE STRATIFIED RANDOM SAMPLE OF 450/90,000 OF HIGH SCHOOL SENIORS

	Males	*Females*	*Total*
High Socioeconomic Status	54 (12%)	58 (13%)	112 (25%)
Moderate Socioeconomic Status	72 (16%)	90 (20%)	162 (36%)
Low Socioeconomic Status	90 (20%)	86 (19%)	176 (39%)
Total	216 (48%)	234 (52%)	450 (100%)

the 216 males with those for the 234 females; third, sex could be held constant by comparing the data for the 112 high socioeconomic status students, the 162 moderate socioeconomic status students, and the 176 low socioeconomic status students; or, finally, the whole sample of 450 cases could be analyzed as a unit, ignoring differences in sex and socioeconomic status. This can be done without distortion because the sampling fractions are the same for each stratum. Since the two variables used to stratify are correlated with the independent and dependent variables, we would expect each of the six strata to be more homogeneous than the overall population.

In this example the number of cases appearing in each of the six sample strata are similar enough to make comparisons among the strata. However, in the case of a population in which some categories of the stratifying variable have few cases and others many, it might facilitate analysis to take a **disproportionate stratified sample** in order to make the sample strata more equal in size. For example, if we were doing a study in which we first wished to stratify on race, using Caucasoid, Mongoloid, and Negroid categories, and we then wished to compare strata on an equal basis, it would probably be wise to draw a *disproportionate* stratified random sample, because the racial distribution in the United States is lopsided—about 89 percent Caucasoid, 11 percent Negroid, and less than 1 percent Mongoloid. A proportionate sample would probably not include enough of the Mongoloid category to provide an adequate basis for comparisons with the other two categories. To make such comparisons it would be desirable to draw samples with more than a representative number from the Mongoloid category. Thus, one might draw sub-samples of the same size (*e.g.*, 200 cases from each of the three strata).

It should be reemphasized that stratified random samples are generally more difficult to execute than simple random samples because they require more information about the population. Not only is it necessary to have lists of the elements in the population, but information is also needed on the distribution of relevant variables in the population. When the necessary information is available the stratified random sample is a useful one for the sociologist. Marjorie Fiske Lowenthal and her associates have been engaged in a series of studies of adult socialization patterns and adaptations of elderly people in the San Francisco area (*e.g.*, Lowenthal and Haven, 1968). In her research she has utilized stratified random sample techniques. For example, a stratified random sample of 600 persons aged sixty or older was drawn from 18 census tracts in San Francisco, and those in the sample were interviewed three times at approximately annual intervals. This sample was stratified on sex, three age levels, and social living arrangements (alone or with others). The very elderly, males, and persons living alone were disproportionately represented in the sample to make comparisons possible. Dr. Lowenthal, in effect, used stratification as a means of controlling relevant variables.

It is important to note that adjustments may need to be made in the usual statistical formulae when analyzing data from stratified random samples (Kish, 1965: 77ff.). This is particularly important when the strata are sampled disproportionately. The formulae will need to be adjusted to take sizes of strata into account. If sampling is proportionate, the usual formulae may generally be used, but tests of hypotheses will usually underestimate the significance of results, because the sampling error is smaller for stratified random samples than for simple random samples.

11.5   NON-PROBABILITY SAMPLES

In each of the sampling techniques discussed so far, the notion of randomness has been involved, and in each case it has been possible to estimate sampling error. It is crucial that we be able to estimate sampling error, because such estimates serve as bases for the inferential leap from sample data to population data.

There are sampling techniques in use which are non-random (and thus, non-probability) in nature. These techniques are sometimes used because they involve lower data collection costs or because they avoid the problems involved in trying to draw random samples. They should be studiously avoided by sociologists, because there is no legitimate basis for estimating sampling error from them.

We merely wish to mention several of these non-random techniques in passing because they do appear in the literature, and you should be made aware of the fact that they are non-random.

Among the non-random sampling techniques in use are the following:

(A) *Systematic samples.* A systematic sample is one in which cases separated from each other by some set interval (such as every tenth case) are drawn from a list. While the procedure is frequently used to sample from files, it can also be used to sample houses on a street (such as every eighth house), or people waiting in a line. Such samples are obviously not random. If every tenth person in a line was selected, the tenth person, twentieth person, etc. would have a probability of 1.00 of being included in the sample while persons not in those positions would have probabilities of 0.00 of being included.*

---

*If the starting point is randomly chosen, systematic samples can be treated as probability samples and sampling error computed.

(B) *Judgment Samples.* A judgment sample consists of those cases believed by the investigator to be "typical" of the population. What is considered typical may be nothing more than a matter of intuition or guessing. Even if the typical is based upon averages, it is still a matter of judgment rather than randomness.

(C) *Quota Samples.* Perhaps the most popular sampling technique utilized by market researchers and pollsters is a non-random technique known as a quota sample. It is an attractive technique because it is economical, convenient, and a fast way of getting data. A quota sample is drawn by specifying the characteristics desired in the subjects to be interviewed, and then depending upon each of the interviewers to find and interview a quota of persons who possess the desired characteristics. Obviously, the procedures used by interviewers to fill their quotas are non-random.

(D) *Voluntary Samples.* The voluntary sample is composed of subjects who volunteer to participate in a study. It would be naive to assume that people who volunteer to participate in a study do so randomly. Those who volunteer undoubtedly have some good reason for doing so.

As an example of a well-known study that was based on a non-random sampling technique we will look briefly at the "Kinsey Report." Kinsey, Pomeroy, and Martin's study, *Sexual Behavior in the Human Male* (1948), was based on interviews with a voluntary sample of 5,300 white American males. Kinsey and his associates approached the members of 163 groups of various kinds (penal institutions, mental institutions, male prostitutes, junior high school students, speech-clinic groups, rooming-house groups, conscientious objectors, hitch-hikers, N.Y.A. workers, children's homes, etc.), lectured to them, and asked for volunteers. In those cases where "perhaps half or more" of the members of a group volunteered, the researchers attempted to persuade all of the group members to submit to interviews (Kinsey, 1948:95). In some groups 100 per cent of the members were interviewed.

Sampling and statistical problems associated with the Kinsey Report formed the basis for a whole volume of critical essays sponsored by the American Statistical Association (Cochran *et al.,* 1954). Especially vulnerable to attack was the implication that the findings of the study could be generalized beyond the 5,300 men interviewed. It was pointed out, for example, that the responses of those who were in the 100 per cent groups differed from the responses of those who were strictly volunteers. This difference was attributed to the presence of subjects in the 100 per cent groups who had to be pressured into being interviewed and who submitted reluctantly (Cochran *et al.,* 1954: 55ff.).

One critic pointed out that there was evidence that many men volunteered to be interviewed because they had personal sex problems with which they needed help or because they had questions about sex that they wanted answered. Kinsey and his associates attempted to dismiss this as a possible source of bias on the grounds that the problems and questions that motivated these men to volunteer were "the everyday sexual problems of the average individual . . ." (Cochran *et al.,* 1954: 57).

The Kinsey study and the controversy that followed its publication offer a good object lesson about the hazards of non-random sampling techniques such as voluntary samples.

Although non-random sampling techniques are often more economical and convenient than random sampling techniques, these advantages are outweighed by the disadvantage of being unable to estimate sampling error. Therefore, the use of non-random techniques by sociologists should generally be discouraged.

11.6   METHODOLOGICAL PROBLEMS ASSOCIATED WITH INFERENTIAL LEAPS

As a scientist the sociologist is ultimately interested in developing generalizations about social behavior that are applicable to infinite populations, sometimes referred to as **general universes** (the terms *population* and *universe* may be used interchangeably). Because of the methodological problems involved, however, he(she) generally does research on special populations or samples of special populations. In fact, sampling theory, as it is practically applied, focuses on **special populations** or **universes** rather than on general universes. By virtue of the fact that a simple random sample requires a list of population elements, the researcher is generally restricted to working with finite populations. He(she) is faced, then, with the methodological problem of how to make the inferential leap from the special universe with which he(she) must work to the general universe in which he(she) is ultimately interested.

As Sjoberg and Nett (1968:130) have said:

> An essential step in conceptual clarification of the selection procedure is distinguishing between the special, or working, universe and the general one. The *special* (or *working*) *universe* is that specific, concrete system (or subsystem) from which one selects his units of study, notably his respondents. Statisticians refer to such a system as a *universe* or *population*, and usually they are content to work within its narrow boundaries. On the other hand, any theoretically oriented social scientist envisions still another kind of universe. If he studies a particular group or social system, he entertains the notion that his findings will, in part at least, hold for other groups or systems — not just in the United States but in other parts of the world as well. For his ultimate goal is establishing generalizations that extend beyond any time-bound social setting. We therefore define the *general universe* as that abstract universe to which the scientist assumes, however tentatively, that his findings will apply. Put another way, every sample is a subsample of a broader type — mankind being, for purposes of generalization, the ultimate category.

The question of how the theoretical leap is made from the special to the general universe is not a simple one to answer.* As a matter of fact, it is a question that is frequently not faced by sociologists when they do research. In many cases sociologists select some special universe of interest to them, sample from it, and limit their generalizations to it. For example, Sewell and Shah (1968) conducted a study of the relationship between educational aspirations, intelligence, socioeconomic status, and parental encouragement. The data for their study came from a survey of graduating seniors in all public, private, and parochial schools in Wisconsin in 1957. They generalize their findings from the 10,308 seniors in their sample to all high school seniors in Wisconsin in 1957. They do not imply that their findings are generalizable beyond that population.

*See Sjoberg and Nett (1968:129–159) for a discussion of the nuances of this problem.

Other sociologists attempt to justify the leap to a broader universe (not necessarily a general universe) than the one from which they sampled. Barnett and Griffith (1970), for example, conducted a study of the relationship between anomia, achievement values, and attitudes toward population growth. They sampled women 18 years of age or older living in a public housing project in Flagstaff, Arizona. In a footnote (Barnett and Griffith, 1970:48–49) they say,

> It should be pointed out, in justification of the limited population represented in this survey, that a limited population contains fewer extraneous sources of variation which can affect a relationship under study than a broad population; hence, drawing from a limited population all the cases one can afford to draw increases the probability of making correct decisions about hypotheses stating causal relationships in that (limited) population (Herbert Hyman, *Survey Design and Analysis*, New York: Free Press, 1955:81). And even though the findings were technically then applicable only to the limited population, it must nonetheless be kept in mind that "it is more probable that a hypothesis holds true outside the population on which it is confirmed than the contrary of the hypothesis holds true in the new population" (Hans Zetterberg, *On Theory and Verification in Sociology*. Third Edition, Totowa, New Jersey: Bedminster, 1965:128).*

Wallis (1949) offers an alternative rationale for making the inferential leap from sample data to a general universe. He says,

> On the other hand, any batch of data that we do get is a random sample from some population; that is to say, indefinitely many repetitions of the procedure that produced the sample will produce a population. We have two handles to manipulate in generalizing from human samples: one is to define our population as closely as we can and then attempt to approximate randomness in our sampling procedure; the other is to analyze our actual sampling procedure as well as we can and then attempt to describe the actual population to which it relates.

Wallis' statement may be adopted as a basis for making the theoretical leap from conclusions about a special universe to generalizations about a general universe, even when the data we examine are only a sample from a special universe. By studying the characteristics of this sample and making explicit the sampling design used to arrive at the sample, the sociologist can get some clues as to the nature of the general universe.

There are, then, two leaps typically involved in the process of drawing conclusions about general universes. One is the leap from sample to special universe from which the sample was drawn. The other is the leap from special to general universe. The theoretical leap from sample to special universe is the subject of this volume, and the simple random sample is the technique designed to justify the leap. The leap from the special to the general universe is neither directly addressed nor justified by current theories and techniques of statistical inference. This leap is a serious methodological problem which sociologists have to face squarely as they develop general laws of social behavior. (Figure 11.1 diagrams the relationship between sample, special universe, and general universe).

As much as possible the special universe to be sampled should be selected on the basis of theoretical considerations, that is, in terms of its logical connec-

*Reprinted by permission of the Publisher, Sage Publications, Inc.

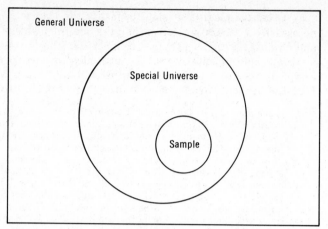

FIGURE 11.1 Relationship between General Universe, Special Universe, and Sample

tions with the general universe to which we ultimately wish to predict. Too often, unfortunately, the selection of a special universe is based instead upon such mundane factors as convenience, manpower resources, time and money.

11.7 SUMMARY

In this chapter we have attempted to show that probability theory is the foundation upon which sampling theory is built. We pointed out that the simple random sample is the basic sampling model for the theory and techniques of inferential statistics. It was explained that other random sampling techniques are variations from the basic simple random sample design.

We explained that the simple random sample requires that sampling be done with replacement and pointed out that the random sample without replacement does not satisfy all of the necessary conditions for a simple random sample. It was noted, however, that sociologists frequently use random samples without replacement because of reluctance to count data more than once (or convenience, or efficiency, or because it approximates a simple random sample in their case). The procedure of selecting sample elements from a table of random numbers was described and, in this connection, the importance of being able to acquire a list of population elements was emphasized. It was pointed out that sociologists often study populations for which a single list of elements is not available. Because simple random samples are not feasible in such situations, the cluster sample may be used in lieu of the simple random sample.

The cluster sample was described as a multi-stage simple random sample, and the procedures involved in drawing a cluster sample were discussed. It was noted that a cluster sample generally has a larger sampling error than a simple random sample, but also generally costs less per case to collect data.

The stratified random sample was described and discussed as another possible alternative to the simple random sample. Its value for making subpopulation comparisons or for reducing sampling error and/or study costs was explained. It was pointed out that the stratified random sample generally has less sampling error than a simple random sample, but that it requires more information and is thus more difficult to execute.

Various non-random sampling techniques were noted in passing, but the reader was advised against using these techniques because they do not allow estimation of sampling error, and estimation of sampling error is crucial to statistical inference.

Finally, we discussed the methodological problem involved in making the inferential leap from sample data and special universes to the general universes in which we are ultimately interested as scientists. We explained that this problem is not faced directly by statistical theory or sampling theory, so it must be attacked at a more abstract and more general methodological level.

We feel obligated to mention the fact that the sampling designs as presented in this chapter were generally oversimplified. Sample designs actually used in research tend to be more complex. It has been our experience that a sampling design must be tailored to a particular study and a particular population. A detailed discussion of the fine points of sample design is beyond the purview of this volume.

In the next chapter we will introduce the concept of the sampling distribution, and we will learn specific techniques for estimating sampling error. At that point we will have available to us the tools necessary to launch into a discussion of parameter estimation and hypothesis testing.

CONCEPTS TO KNOW AND UNDERSTAND

sample
simple random sample
random sample without
    replacement
sampling error
random numbers
population list
cluster sample
homogeneous cluster
heterogeneous cluster

primary sampling units
ultimate sampling units
strata
proportionate stratified random
    sample
disproportionate stratified random
    sample
non-random sample
special universe
general universe

QUESTIONS AND PROBLEMS

1. Make a list of the U.S. states and the District of Columbia (fifty-one entries). Number the entries, and using the table of random numbers at the back of the book draw a simple random sample of 20 states. Draw a random sample without replacement of 20 states. Are there any differences in the two samples? How great is the difference?

2. Describe how you would draw a simple random sample of the student body of your college or university.

3. Make a systematic comparison of the simple random sample, the cluster sample, and the stratified random sample. In what circumstances would a cluster sample be used instead of a simple random sample? In what circumstances would a stratified random sample be used instead of a simple random sample?

4. Explain how you would go about drawing a simple random sample of the names in the telephone book.

5. Read a research article in a sociology journal and analyze the sampling design used. Was it random or non-random? Was the sampling design well planned and well executed? How could the design be improved?

6. Why is a random sample generally preferable to a non-random sample?

7. Explain the relationship between a sample, a special universe, and a general universe. Give an example that illustrates the relationship among the three.

GENERAL REFERENCES

Ackoff, Russell L., *The Design of Social Research* (Chicago: The University of Chicago Press, 1953).
> Chapter 4 of this book, the chapter on sampling, is an excellent short summary of various sampling designs. A series of tables are included which make some important points about various techniques of sampling.

Kish, Leslie, *Survey Sampling* (New York: John Wiley & Sons, Inc., 1965).
> This book offers a good, technical discussion of sample designs and the practical problems facing the researcher in designing a sample. It features illustrations based upon survey samples of human populations.

Sjoberg, Gideon, and Roger Nett, *A Methodology for Social Research* (New York: Harper & Row, Publishers, 1968).
> The authors spend some time discussing the methodological problem of generalizing from special to general universes. They include a consideration of the factors that enter into the selection of a special universe to be sampled.

Stephan, Frederick F., and Philip J. McCarthy, *Sampling Opinions* (New York: John Wiley & Sons, Inc., 1958).
> This book is a good, basic source for those readers who wish to learn sampling procedures for large scale surveys. The second part of the book examines a number of empirical studies and analyzes their sampling designs.

OTHER REFERENCES

Barnett, Larry D., and Jerry J. Griffith, "Anomia and Achievement Values and Attitudes toward Population Growth in the United States," *Pacific Sociological Review*, vol. 13, no. 1 (Winter, 1970), p 47–52.

Berman, Art, "Statistician Brands Draft Lottery Unfair," *Los Angeles Times*, December 3, 1969, part 1, p 13.

Cochran, William G., Frederick Mosteller, and John W. Tukey, with the assistance of W. O. Jenkins, *Statistical Problems of the Kinsey Report on Sexual Behavior in the Human Male* (Washington, D.C.: American Statistical Association, 1954).

Kinsey, Alfred C., Wardell B. Pomeroy, and Clyde E. Martin, *Sexual Behavior in the Human Male* (Philadelphia: W. B. Saunders, 1948).

Lowenthal, Marjorie Fiske, and Clayton Haven, "Interaction and Adaptation: Intimacy as a Critical Variable," *American Sociological Review*, vol. 33, no. 1 (February, 1968), pp. 20–30.

Portes, Alejandro, "Dilemmas of a Golden Exile: Integration of Cuban Refugee Families in Milwaukee," *American Sociological Review,* vol. 34, no. 4 (August, 1969), p 505–518.

Scott, Ann Herbert, *Census U.S.A.* (New York: Seabury Press, 1968), p 128–129.

Sewell, William H., "Inequality of Opportunity for Higher Education," *American Sociological Review,* vol. 36 (October, 1971), p 793–809.

Sewell, William H., and Vimal P. Shah, "Social Class, Parental Encouragement, and Educational Aspirations," *American Journal of Sociology,* vol. 73, no. 5 (March, 1968), p 559–572.

Spaeth, Joe L., "Occupational Prestige Expectations among Male College Graduates," *American Journal of Sociology,* vol. 73, no. 5 (March, 1968), p 548–558.

Wallis, W. A., "Statistics of the Kinsey Report," *The Journal of the American Statistical Association*, vol. 44 (1949), p 468.

12 Sampling Distribution: A Pivotal Concept

Very often a social scientist wants to make a statement about a whole society, a large group or frequently occurring behaviors. Sampling is an essential tool of inquiry. First, and most importantly, study of a carefully drawn sample yields all the information that is needed — study of an entire population would not add more. Secondly, study of a large and unwieldy population may introduce more error by making the task of processing the information too complex and cumbersome or by using up resources that would better be focused on careful data collection and analysis. In some instances the process of inquiry actually requires destruction of sampled items (perhaps in testing water glasses for strength) so that testing of all items would be self-defeating.

Our knowledge about what to expect of samples is based upon drawing samples in a way which permits us to utilize probability theory. As we have seen in the last two chapters, there are several different ways of going about the business of drawing samples from populations. Some of these sampling plans permit us to apply our knowledge of probability and some sampling procedures prevent us from applying probability theory and thus from knowing what to expect of these samples. Inferential statistics makes use of what we know about how samples behave.

Most introductory statistics texts develop the logic of inferential statistics by limiting their discussion to the situation where a simple random sample is drawn with replacement or from a relatively large or infinite, defined population. We shall also follow this practice, leaving discussions of alternative probability sampling procedures to more advanced texts.*

*For a discussion of how samples behave when they are drawn by other, more complicated sampling plans, see Leslie Kish, *Survey Sampling*, N.Y., John Wiley and Sons, 1965.

Our first step will be to illustrate how it is that we can use probability theory to describe what we might expect from drawing simple random samples. Then we will show how you can describe in general how simple random samples are expected to behave. Finally, we will use this knowledge in a logical process of drawing conclusions from sampled data.

12.1   SAMPLING: A CASE APPLICATION

Corporations represent one special way in which individuals form a social relationship. In our society, these relationships have something of the status of 'artificial persons'. They are legally acknowledged through articles of incorporation which explicitly formalize objectives of the relationship and rules by which those involved are to carry on their relationship over time. In 1972, according to the U.S. Census, there were 1,913,000 retail establishments, of which 566,000 or about 30 percent had the legal form of a corporation.

Suppose we decide to draw a sample of retail establishments – a simple random sample from this relatively large population of establishments – in order to study differences between retail establishments which have corporate versus other legal forms of organization. For purposes of illustration, let us draw a sample of size 4 (N=4).

We shall have to find a listing of retail establishments for the U.S. in 1972 in order to be able to draw our small sample by simple random sample means. Presumably, because retail establishments are identified for tax purposes, such a complete listing of establishments would exist. Following simple random sampling procedures, then, we can draw a sample of four establishments (see Chapter 11.2).

Can we describe how we would expect samples of this sort to behave? Because we are using a probability sampling procedure, the answer is yes. For example, one possible sample would be to draw three corporations (C) followed by a retail establishment that is not a corporation (N).

C   C   C   N

We can readily compute the probability of drawing such a sample because we know that:

   a. The probability of drawing a corporation is .30 since there are known (in this instance) to be 30 percent corporations in the population of retail establishments. The probability of drawing a non-corporation would be $1.00 - .30 = .70$, (see Chapter 10.3a).

   b. The outcome of each draw is independent of the outcome of other draws if we follow simple random sampling procedures and thus we can use the special multiplication rule (Chapter 10.3.2f) to compute the probability of this series of four independent events. The probability of drawing such a sample is:

$$(.30)\,(.30)\,(.30)\,(.70) = .0189$$

c. Each sample is mutually exclusive of other possible samples which means, as we shall see, that we can make use of the special addition rule as well (see Chapter 10.3.2b).

We could have drawn many different samples other than the C C C N pattern noted above. Specifically, we could have drawn 2^N (where N is the sample size and 2 is used because there are two outcomes on each draw, being a corporation or not) or 2^4 or 16 different samples (taking order in which each element was drawn into the sample into account). These different samples are listed below.

Our sample of size four could have been any one of the sixteen different sequences of sample draws listed below. The likelihood of any one of these different samples is given at the right in Table 12.1. For example, the probability of drawing a sample with no corporations at all is .2401. This means that, in the long run, if we were to draw random samples of size four from this population, we should expect to get a sample with zero percent corporations about 24 percent of the time. We know this because we can utilize probability rules and we can do that because of the way we carefully drew the samples. The way samples are drawn is thus very important.

For nearly every research purpose, we would have no interest in the order in which cases are drawn into our sample, whether corporations were drawn first or last. We would only be interested in the number (or percentage) of corporations our entire sample contains. While there is only one way to draw a sample with 100 percent corporations and only one way to draw a sample with no corporations, there are several ways to draw a sample with only one corporation. Table 12.1 shows that there are four samples with one corporation, six samples with two corporations and four samples with three corporations out of the four establishments in the sample.

TABLE 12.1 THE SIXTEEN POSSIBLE SAMPLES OF SIZE FOUR

		Sample Draw			Probability of this sample
	1st	2nd	3rd	4th	
1.	C	C	C	C	(.3) (.3) (.3) (.3) = .0081
2.	C	C	C	N	(.3) (.3) (.3) (.7) = .0189
3.	C	C	N	C	(.3) (.3) (.7) (.3) = .0189
4.	C	N	C	C	(.3) (.7) (.3) (.3) = .0189
5.	N	C	C	C	(.7) (.3) (.3) (.3) = .0189
6.	C	C	N	N	(.3) (.3) (.7) (.7) = .0441
7.	C	N	C	N	(.3) (.7) (.3) (.7) = .0441
8.	N	C	C	N	(.7) (.3) (.3) (.7) = .0441
9.	C	N	N	C	(.3) (.7) (.7) (.3) = .0441
10.	N	C	N	C	(.7) (.3) (.7) (.3) = .0441
11.	N	N	C	C	(.7) (.7) (.3) (.3) = .0441
12.	C	N	N	N	(.3) (.7) (.7) (.7) = .1029
13.	N	C	N	N	(.7) (.3) (.7) (.7) = .1029
14.	N	N	C	N	(.7) (.7) (.3) (.7) = .1029
15.	N	N	N	C	(.7) (.7) (.7) (.3) = .1029
16.	N	N	N	N	(.7) (.7) (.7) (.7) = .2401
					Total = 1.0000

In general, we could have computed the number of different ways of getting a particular number of corporations in the sample by making use of the combinations formula (Formula 10.11) discussed in Chapter 10. If we let N equal the total number of choices or the sample size, r equal the number of corporations, and $N - r$ the number of non-corporations, we can substitute into the combinations formula as follows to find out, say, the number of ways one could draw a sample with two corporations and two non-corporations. These are the samples number 6 through 11 listed in Table 12.1.

(12.1)
$$\binom{N}{r} = \frac{N!}{(N-r)!\,r!}$$
$$\binom{4}{2} = \frac{4!}{(4-2)!\,2!} = \frac{4!}{2!\,2!} = \frac{(4)\,(3)\,(2)\,(1)}{(2)\,(1)\,(2)\,(1)} = \frac{24}{4} = 6$$

Using the special addition rule, we can sum the probabilities of drawing a sample with two corporations and two non-corporations for the six possible ways of drawing this kind of sample. Table 12.2 lists these summary results for each of the possible kinds of samples.

By the adoption of a simple notation system and the combined expression of the special multiplication rule and the special addition rule (as represented by the combinations formula) we can come up with a formula for computing the probability of the chance occurrence of any particular pattern of corporations and non-corporations.

(12.2)
$$Pr(N,P,r) = \frac{N!}{(N-r)!\,r!}\, P^r Q^{N-r}$$

where N = the size of the sample, r = the number of corporations the sample contains, $N-r$ = the number of non-corporations in the sample, P is the probability of drawing a corporation from the population, and $Q = 1-P$ = the probability of drawing a non-corporation.

To demonstrate how Formula (12.2) works, we will apply it to the situation of drawing one corporation and three non-corporations:

$$Pr(4,0.3,1) = \frac{4!}{3!\,1!}\,(.3)^1\,(.7)^3$$

$$= \frac{(4)\,(3)\,(2)\,(1)}{(3)\,(2)\,(1)\,(1)}\,(.3)\,(.343)$$

$$= \frac{24}{6}\,(.1029) = (4)\,(.1029) = .4116$$

BOX 12.1 REVIEWING RULES OF PROBABILITY

Your understanding of Formula (12.2) depends upon your grasp of three rules that were discussed in Chapter 10: the *special addition rule of probability* (Section 10.3.2b), the *special multiplication rule of probability* (Section 10.3.2f), and the *combination rule* (Section 10.4.2). If these three rules are still not clear to you, now would be an excellent time to review these sections; then reread the first nine pages of this chapter.

TABLE 12.2   DISTRIBUTION OF POSSIBLE SAMPLE OUTCOMES FOR SAMPLES OF SIZE 4 DRAWN FROM A POPULATION OF RETAIL ESTABLISHMENTS IN WHICH THE PROBABILITY OF RANDOMLY DRAWING A CORPORATION IS .3 AND THE PROBABILITY OF DRAWING A NON-CORPORATION IS .7

Possible Outcomes of Samples of N=4	Computational Form	Probability of this Kind of Sample
4 (100%) corporations	$1 \ (.3)^4 \ (.7)^0$	.0081
3 (75%) corporations	$4 \ (.3)^3 \ (.7)^1$	.0756
2 (50%) corporations	$6 \ (.3)^2 \ (.7)^2$	.2646
1 (25%) corporations	$4 \ (.3)^1 \ (.7)^3$	.4116
0 (0%) corporations	$1 \ (.3)^0 \ (.7)^4$	.2401
		Total = 1.0000

Thus there are four ways to achieve a sample with one corporation and three non-corporations. The probability of any one of these ways is .1029 and the probability of achieving a sample with one corporation from samples of size four is the product, or .4116. This figure is given in Table 12.2 as well. The formula would work as well for figuring the probability of achieving a sample with no corporations (or all corporations), remembering that 0! = 1 and any number raised to the power of 0 is equal to 1.

In Table 12.2 are listed the five kinds of sample outcomes, the computational forms for computing the probability of each of these outcomes, and their associated probabilities. Note that if the five probabilities are added, they sum to 1.0000 (except, perhaps, for rounding error in some instances). These are all the possible kinds of samples.

Table 12.2 shows that there is about eight-tenths of one percent chance of drawing a sample with 4 out of 4 corporations if indeed we are using random sampling procedures and are drawing a sample from a population in which there are 30 percent corporations. On the other hand, under these conditions we might expect to get one corporation in our sample about 41 percent of the time (if we were to draw repeated samples of size 4). This distribution of probabilities is called a **sampling distribution.** It simply describes how often we can expect samples of this kind. This knowledge will be the basis upon which we build a logical argument that helps us make inferential statements about populations on the basis of samples we have drawn from them. For example, if we did draw one sample from this population and found that we drew all corporations, we might begin to: a) doubt whether we carefully applied random sampling procedures, b) doubt whether the population we were actually drawing from really had only 30% corporations, c) decide that this was merely one of the more unusual but expectable chance samples we might get. Our doubts stem from a knowledge of how rare this sample outcome is under the circumstances. On the other hand, we would not be at all surprised to draw a sample with only one corporation.

Before we proceed with the use of sampling distributions in the logic of inferential statistics, we need to summarize more carefully what a sampling distribution is. The concept of the sampling distribution is one of the most important in inferential statistics. We will spend the rest of this chapter examining the concept in general and a number of specific sampling distributions that are commonly used by sociologists.

12.2 THE CONCEPT OF THE SAMPLING DISTRIBUTION

There are three distinct types of distributions of data: (1) one which characterizes the distribution of elements of a population (the **population** distribution), (2) one which characterizes the distribution of elements of a sample drawn from a population (a **sample** distribution), and (3) one which describes the expected behavior of a large number of simple random samples drawn from the same population (a **sampling** distribution). The third type of distribution differs from the other two in a number of ways which will be discussed below, but an outstanding basic difference is that the units which are distributed are summary measures of whole samples of values rather than individual values of characteristics of single cases.

There are many different sampling distributions, depending upon: a) the population being sampled, b) the size of the sample (N), c) the method of sampling (we will consider only simple random samples here), and d) the statistic (function of sample scores) in which we are interested.

A sampling distribution is *a theoretical probability distribution of sample statistics* (*e.g.*, sample means or sample proportions). A sampling distribution may be generated by taking all possible random samples (each with at least one element different) with a fixed N from a population, computing a statistic for each sample, and plotting the distribution of these statistics around the parameter which they estimate. For example, consider a population of 222,000,000 individuals in the United States in 1980. Assume that we draw a simple random sample of 1000 from that population and compute the mean age of the individuals in the sample.* Assume further that we repeat this procedure until we have exhausted all of the unique simple random samples of 1000 in the population. The *distribution of the means of all of these samples* would constitute a **sampling distribution of means.**

Of course the sociologist engaged in research does not actually generate a sampling distribution in this way except in cases of limited distributions of possible samples such as the seven best-buddy choices discussed earlier. Usually he(she) draws one sample, computes a statistic from the sample, and uses knowledge about the nature of its sampling distribution to generalize to the corresponding parameter of the population.

Every statistic, whether it be a frequency, a proportion, a mean, a variance, or whatever, has a sampling distribution. For a statistic to be useful in making inferences about parameters its sampling distribution must be known. One important job for the statistician who develops a new statistical technique is to specify its sampling distribution (or distributions). This has, in fact, been done for the commonly used statistics. These sampling distributions are usually

---

*Of course this is a purely hypothetical example because it would not be feasible to draw a simple random sample of the United States population.

presented in tabular form and are included in statistics books (as they are in this one; see Appendix I, Tables B, C, D, and E). As an example, the normal distribution is the appropriate sampling distribution for several different statistical techniques when large samples are used (e.g., means, proportions, runs, etc.). The chi-square distribution is the appropriate sampling distribution for variances and the chi-square (χ^2) technique. Student's t-distribution is the appropriate sampling distribution for means from small samples when the population variance is unknown. The point is that there are many statistical techniques and many sampling distributions (although there are not as many sampling distributions as there are techniques). It is necessary to match the technique with the appropriate sampling distribution in order to generalize from sample data to population.

A sampling distribution is usually a **univariate distribution** and as such may be described in terms of its central tendency, variability, and form.* We will examine each of these characteristics of a sampling distribution, in turn.

12.2.1 Central Tendency

In referring to the central tendency of a sampling distribution it is customary to speak of the **expected value of a statistic.** The expected value is *the average value a statistic assumes for its sampling distribution* (e.g. the arithmetic mean of the sampling distribution of the statistic). That is, the value that we would obtain if we added all of the statistics in the sampling distribution and divided by their number. An expected value is indicated by the capital letter E. The expected value of the sample mean, therefore, is designated $E\ (\overline{X})$. If the average or expected value of a statistic is, in fact, the parameter which it estimates, then the statistic is said to be an **unbiased estimator** of the parameter. We will establish later in this chapter that the sample mean is an unbiased estimator of the population mean [*i.e.,* $E\ (\overline{X}) = \mu$].

The fact that a statistic is an unbiased estimator does not give us any notion of just how closely a particular statistic from a single sample estimates its parameter. It merely assures us that *on the average* such statistics will equal the parameter. Still, the information that an estimator is unbiased is useful because it tells us that any difference that exists between a particular statistic and its parameter is attributable to random error rather than systematic bias in the statistic itself. *Since scientific generalizations are based upon replications of a study rather than a single, isolated study, when we make repeated estimates of a parameter it is important that our estimation errors not be systematic.*

12.2.2 Variability

A second important characteristic of a sampling distribution is the extent to which the sample statistics vary around their parameter. Variability of scores in population distributions or sample distributions can be measured by techniques

*There are multivariate sampling distributions of interest in more specialized research.

such as ranges, variances, and standard deviations. When we measure the variability of sample statistics around their parameters, we refer to the summarizing measurement as a **standard error.*** There are standard errors for frequencies, proportions, means, medians, variances, correlation coefficients, and so forth. Standard errors, in general, measure the random variation of statistics around the parameter that they estimate. The size of the standard error is dependent, in part, upon the size of the sample from which the statistic is computed. In accordance with *the* **law of large numbers,** *as sample size increases, the standard error decreases*. The larger the *N*, the more closely statistics will cluster around their parameter.†

12.2.3 Form

The form of a sampling distribution refers to the shape of the particular curve that describes the distribution. It is important to determine, for instance, whether the distribution is symmetrical or skewed, leptokurtic, mesokurtic, or platykurtic. Is the distribution of the statistic normal or *J*-shaped? The particular form which a sampling distribution assumes is a significant factor to be considered in generalizing from statistics to parameters. For example, we will see later in this chapter that under certain conditions the sampling distribution of means is platykurtic. Ideally, it is possible to describe a sampling distribution with a mathematical formula (a density function) such as the formula for the normal distribution.

We will not attempt to discuss all of the known sampling distributions in this chapter. Rather, we will examine a few of the more commonly used ones in order to obtain a feeling for what sampling distributions are. Other sampling distributions will be discussed and described as they are needed in later chapters.

In this chapter we will direct our attention to the central tendency, variability, and form of sampling distributions. In the next chapter we will direct our attention to the question of how sampling distributions enter into the estimation of parameters. In the remaining chapters we will see how sampling distributions are used in tests of hypotheses.

12.3 THE BINOMIAL SAMPLING DISTRIBUTION

The distribution of the number of corporations in samples of four retail enterprises that we examined earlier is a particular instance of a sampling distribution known as the **binomial probability distribution.** The binomial probability distribution is *the appropriate sampling distribution for dichotomized, nominal scale*

---

*The standard deviation computed on a sampling distribution of a statistic is called the *standard error* to distinguish it from other standard deviations.
†There are some exceptions which are not of interest in the usual research situation where a population mean could potentially be computed.

data for which the observations are independent (e.g. the distribution of the statistic, the *number* or *percent* of cases in a certain category out of a sample of size N). For example, if the observations are independent, such measurement categories as male—female, success—failure, white—nonwhite, in-group—out-group tend to have binomial sampling distributions. Binomial sampling distributions are expressed in terms of frequencies of occurrence of each of the dichotomous categories. Thus, the distribution of corporations—non-corporations was composed of the possible simple random samples of size 4 for which the number of corporations varied from 4 to 0.

Formula (12.2) allows us to compute directly the probability of occurrence of each of the terms of a **binomial sampling distribution.** Barring measurement error and other non-random errors, the probabilities computed from the formula are exact.

12.3.1 Central Tendency

Because the binomial sampling distribution is *an exact distribution*, the N and P upon which the distribution is based are parameters. Given these parameters it is possible to compute a measure of central tendency that is also a parameter. The mean of a binomial distribution is given by the following formula:

(12.3)
$$\mu_B = NP$$

where N is the size of the sample for each of the samples in the distribution, and P is the proportion of outcomes of interest (*e.g.*, the outcome of interest earlier was the proportion of corporations).

The resulting mean is a frequency of outcomes of interest and may be interpreted as the number of such outcomes to be expected, on the average.

Using Formula (12.3) we may compute the mean of the sampling distribution of percent corporations in samples of 4 retail establishments.

$$\mu_B = NP = (4)(0.3) = 1.2$$

Therefore, the central tendency of this sampling distribution is represented by a mean of 1.2 corporations. Such a low number of expected corporations is, of course, consistent with the results of our earlier examination of this sampling distribution.

12.3.2 Variability

It is also possible to compute a parameter as the measure of sampling error of the binomial sampling distribution. The formula for the *standard error* is as follows:

(12.4)
$$\sigma_B = \sqrt{NPQ}$$

where N and P have the same meanings as above and $Q = 1 - P$.

This *standard error* measures the variability of sample frequencies of outcomes of interest, among themselves and around the mean of the sampling distribution.

Computing the standard error for the sampling distribution of the number of corporations in samples of 4 retail establishments with Formula (12.4) gives the following result:

$$\sigma_B = \sqrt{NPQ} = \sqrt{(4)\,(.3)\,(.7)} = \sqrt{0.8400} = 0.917$$

This, then, is a measure of the variability of frequencies of corporations from simple random samples of size 4 from a population of retail establishments composed of 30 percent corporations.

12.3.3 Form of the Distribution

The form of the binomial sampling distribution is dependent upon N and P. When $P = 0.5$ the distribution will be symmetrical. As N approaches infinity the

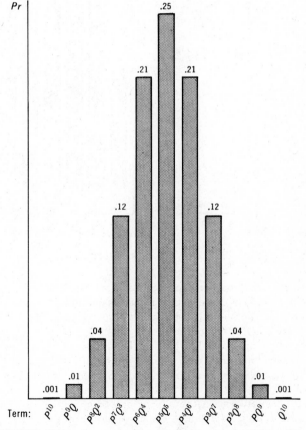

FIGURE 12.1a BINOMIAL DISTRIBUTION WITH $P = 0.5$ AND $N = 10$

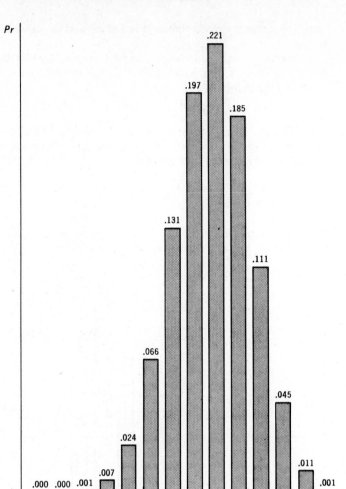

FIGURE 12.1b BINOMIAL DISTRIBUTION FOR $P = 0.4$ AND $N = 13$

binomial distribution approaches the normal distribution. Even when P is not equal to 0.5, as N becomes larger the binomial distribution will begin to look more like the normal distribution. As a matter of fact, as long as both NP and NQ are equal to or greater than 5, the normal curve is a satisfactory approxima- tion to the binomial distribution. Thus, if $P = Q = 0.5$, the normal approximation is satisfactory when N is as small as 10. If $P = 0.4$, N would have to be at least 13 to satisfy the rule of thumb. For a $P = 0.3$, N would have to be at least 17, and so forth. Figures 12.1a, 12.1b, and 12.1c portray the graphs of these three bino- mial distributions. Although the graphs of Figure 12.1b and 12.1c are not sym- metrical, they do resemble the normal distribution.

When the normal distribution closely approximates the binomial distribu- tion, the standard error of the binomial distribution takes on a meaning similar

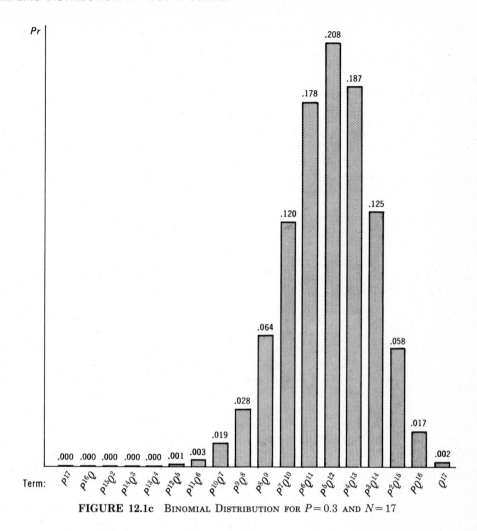

FIGURE 12.1c Binomial Distribution for $P = 0.3$ and $N = 17$

to that of the standard deviation. That is, approximately 68 percent of the sample frequencies can be expected to fall within one standard error of the mean, approximately 95 percent within two standard errors, and approximately all of the sample frequencies within three standard errors. When the binomial distribution departs from normality the standard error does not carry this interpretation.

It is obvious from an examination of Table 12.2 that the binomial sampling distribution in question has a definite skew toward sampling corporations. That is, the bulk of the samples in the distribution are heavily weighted in favor of sampling relatively few corporations. The size of the standard error for this distribution, 0.917, reflects variability of sample frequencies larger than the mean since the mean, itself, is so low. In a skewed distribution such as this one the interpretation of the standard error is obfuscated.

Whenever P deviates from 0.5, if N is not large, the binomial sampling distribution will be skewed. This is not a problem, however, since the binomial distribution yields exact probabilities.

12.3.4   Use of the Normal Approximation to the Binomial Distribution

As long as the sample N is small it is advisable to use the binomial sampling distribution directly because it does give exact probabilities. As N gets larger, however, the binomial distribution becomes more and more cumbersome to work with. The advantage of being able to compute exact probabilities is soon outweighed by the amount of work involved in computing them. If a computer and computer time are available, it is possible to program the computer to do the necessary computations. Another alternative is to make use of published tables of the binomial probability distribution, several of which give the necessary probabilities for all N's up to and including 50 (*e.g.* Nat'l Bur. Standards, 1949).

A third alternative is to use the normal curve as an approximation to the binomial sampling distribution. We will examine the normal approximation to the binomial distribution to determine how probabilities are computed and to compare those probabilities with the exact probabilities given by the binomial sampling distribution itself.

Although the sample size in our earlier example does not meet the rule of thumb for using the normal approximation to the binomial distribution, let us assume we had drawn a larger sample, say, a sample of size 17. First we must find the mean and standard error of the binomial sampling distribution of retail establishments where the sample size is 17 and the probability of drawing a corporate form of social organization is .30, as before. Thus $N = 17, P = .3$, and $Q = (1.0 - .3) = .7$. Given these data we can compute the mean and standard error as follows:

$$\mu_B = NP = (17)(.3) = 5.1$$

$$\sigma_B = \sqrt{NPQ} = \sqrt{(17)(0.3)(0.7)} = \sqrt{3.57} = 1.889$$

To illustrate the similarity of the normal curve approximation to the binomial distribution, let us look at the sample outcome of getting 4 corporations out of the sample of 17 retail establishments.

The 4 corporations is a **frequency** and frequencies are discrete values. The **normal curve**, on the other hand, is a *continuous distribution for which probabilities are represented by areas under the curve*, so it is not possible to use the normal curve to find a probability for a value like 4. What must be done is to **correct for continuity** by treating the integer 4 as though it were a continuous score with a lower boundary of 3.5 and an upper boundary of 4.5. It is possible to find an area under the normal curve between 3.5 and 4.5 to represent the desired probability. Using the following standard score formula, this is just what we shall do.

(12.5)
$$z = \frac{X - \mu_B}{\sigma_B}$$

> **BOX 12.2**   REVIEWING NORMAL CURVE AND STANDARD SCORE
>
> The discussion which follows draws upon your knowledge and understanding of the normal curve and the concept of a standard score. Both the normal curve and the standard score are extremely important for any discussion of inferential statistics. You will find that they will be referred to again and again throughout this volume. If you feel that you need to refresh your memory on these topics at this point, you should review Sections 5.4.7 and 5.5.2.

where X first takes the value of the lower boundary of the frequency and then the upper boundary.

Solving for the two z's we get the following:

$$z_1 = \frac{3.5 - 5.1}{1.889} = -0.847$$

and

$$z_2 = \frac{4.5 - 5.1}{1,889} = -0.318$$

To find the probability associated with the occurrence of a simple random sample with 4 corporations and 13 non-corporations, it is necessary to compute the area under the normal curve lying between these two z scores.

The area between a z score of -0.847 and a z score of -0.318 under the normal curve is approximately $0.3023 - 0.1255 = 0.1768$ (see Appendix Table B); thus this represents our approximation to the desired binomial probability.

Using Formula 12.2 to compute the exact binomial probability for this kind of sample outcome (4 corporations and 13 non-corporations), we get:

$$\frac{N!}{(N-r)!\,r!}\,P^r\,Q^{N-r} \text{ or } \frac{17!}{(17-4)!\,4!}\,(.3)^4\,(.7)^{17-4}$$

$$= 2380\,(.0081)\,(.0096889) = .1867826$$

The normal curve approximation, 0.1768, is slightly lower but approximately the same as the exact binomial probability, 0.1867826 (the calculations for the normal approximation include rounding error). If the sample size had been larger and normal curve calculations carried out to more decimal places, the approximation would have been closer. When N is large, it is much more convenient to use the normal approximation to the binomial distribution than it is to find probabilities from the binomial itself. The approximated probabilities are generally close enough to those computed from the binomial to do the job.*

We could use the normal curve approximation to find the probability of each possible sample outcome. Usually, however, only those outcomes of imme-

---

*In Chapter 15 we will present an alternative approach in which frequencies are converted to proportions.

diate interest are calculated. In this instance, we were interested in the probability of getting a sample with four corporations. Sometimes we may be interested in the probability of getting samples as extreme or more extreme than 4 out of 17 (*e.g.* those with 3, 2, 1, or no corporations).

12.4 SAMPLING DISTRIBUTIONS OF MEANS

Thus far we have used the binomial sampling distribution and have applied the normal approximation to it to evaluate various patterns of dichotomous events. In each case the statistic we examined in the light of its sampling distribution was a frequency (or a *sum* of 0's and 1's).

Frequently, the sociologist focuses his(her) attention on **means.** He(she) might be interested in the mean income of a group of professionals, or the mean number of days of work lost because of strikes, or the mean size of high-socio-economic-status families. *The mean is a much used statistic because it yields the greatest amount of information about the central tendency of a relatively symmetrical distribution of scores.*

Whenever the sociologist draws a simple random sample and computes a mean he(she) does it in order to generalize about the central tendency of the population from which the sample comes. If for example, he(she) drew a simple random sample of workers and computed the mean number of years they had worked, his(her) purpose would be to make some statement about the mean number of years worked by all workers in the population from which the sample was drawn. In order to be able to do this he(she) must know the nature of the sampling distribution of means.

In this section our purpose is to determine the central tendency, variability, and form of the sampling distribution of means so that we will have the information necessary for making generalizations.

To begin with, we will explore further the potential of the sample mean as unbiased estimator of the population mean. We will also examine sample variances and standard deviations as unbiased estimators since they play an integral part in the measurement of sampling error.

12.4.1 Unbiased Estimators: Sample Means

The sample mean is considered a good estimator of the population mean because it is an unbiased estimator. In Section 12.2.1, we have seen that unbiased estimators are those that, on the average, equal the parameter being estimated.

We will now use a simplified example to illustrate the fact that the sample mean is an unbiased estimator of the population mean: Assume that in a community there are ten block clubs that have each existed for the following number of years: 1, 4, 5, 7, 7, 9, 9, 10, 13, 15. If we consider these ten block clubs as a *population*, then the population mean for years of existence is 80/10 = 8.

Column 1 of Table 12.3 lists all of the possible simple random samples of

TABLE 12.3 Sample Means, Variances, and Standard Deviations of Simple Random Samples, $N = 2$, from a Population Consisting of Ten Block Clubs with Longevity Scores of 1, 4, 5, 7, 7, 9, 9, 10, 13, 15

(1) Sample	(2) $\overline{X}$	(3) $\Sigma(X-\overline{X})^2$	(4) $\Sigma(X-\overline{X})^2/N$	(5) $\Sigma(X-\overline{X})^2/N-1$	(6) $\sqrt{\Sigma(X-\overline{X})^2/N-1}$
1,1	1	0	0	0	0
1,4	2.5	4.5	2.25	4.5	2.12
4,1	2.5	4.5	2.25	4.5	2.12
1,5	3	8	4	8	2.83
5,1	3	8	4	8	2.83
1,7	4	18	9	18	4.24
7,1	4	18	9	18	4.24
1,7	4	18	9	18	4.24
7,1	4	18	9	18	4.24
1,9	5	32	16	32	5.66
9,1	5	32	16	32	5.66
1,9	5	32	16	32	5.66
9,1	5	32	16	32	5.66
1,10	5.5	40.5	20.25	40.5	6.36
10,1	5.5	40.5	20.25	40.5	6.36
1,13	7	72	36	72	8.49
13,1	7	72	36	72	8.49
1,15	8	98	49	98	9.90
15,1	8	98	49	98	9.90
4,4	4	0	0	0	0
4,5	4.5	0.5	0.25	0.5	0.71
5,4	4.5	0.5	0.25	0.5	0.71
4,7	5.5	4.5	2.25	4.5	2.12
7,4	5.5	4.5	2.25	4.5	2.12
4,7	5.5	4.5	2.25	4.5	2.12
7,4	5.5	4.5	2.25	4.5	2.12
4,9	6.5	12.5	6.25	12.5	3.54
9,4	6.5	12.5	6.25	12.5	3.54
4,9	6.5	12.5	6.25	12.5	3.54
9,4	6.5	12.5	6.25	12.5	3.54
4,10	7	18	9	18	4.24
10,4	7	18	9	18	4.24
4,13	8.5	40.5	20.25	40.5	6.36
13,4	8.5	40.5	20.25	40.5	6.36
4,15	9.5	60.5	30.25	60.5	7.78
15,4	9.5	60.5	30.25	60.5	7.78
5,5	5	0	0	0	0
5,7	6	2	1	2	1.41
7,5	6	2	1	2	1.41
5,7	6	2	1	2	1.41
7,5	6	2	1	2	1.41
5,9	7	8	4	8	2.83
9,5	7	8	4	8	2.83
5,9	7	8	4	8	2.83
9,5	7	8	4	8	2.83
5,10	7.5	12.5	6.25	12.5	3.54

TABLE 12.3   *(Continued)*

(1) Sample	(2) $\overline{X}$	(3) $\Sigma(X-\overline{X})^2$	(4) $\Sigma(X-\overline{X})^2/N$	(5) $\Sigma(X-\overline{X})^2/N-1$	(6) $\sqrt{\Sigma(X-\overline{X})^2/N-1}$
10,5	7.5	12.5	6.25	12.5	3.54
5,13	9	32	16	32	5.66
13,5	9	32	16	32	5.66
5,15	10	50	25	50	7.07
15,5	10	50	25	50	7.07
7,7	7	0	0	0	0
7,7	7	0	0	0	0
7,7	7	0	0	0	0
7,9	8	2	1	2	1.41
9,7	8	2	1	2	1.41
7,9	8	2	1	2	1.41
9,7	8	2	1	2	1.41
7,10	8.5	4.5	2.25	4.5	2.12
10,7	8.5	4.5	2.25	4.5	2.12
7,13	10	18	9	18	4.24
13,7	10	18	9	18	4.24
7,15	11	32	16	32	5.66
15,7	11	32	16	32	5.66
7,7	7	0	0	0	0
7,9	8	2	1	2	1.41
8,7	8	2	1	2	1.41
7,9	8	2	1	2	1.41
9,7	8	2	1	2	1.41
7,10	8.5	4.5	2.25	4.5	2.12
10,7	8.5	4.5	2.25	4.5	2.12
7,13	10	18	9	18	4.24
13,7	10	18	9	18	4.24
7,15	11	32	16	32	5.66
15,7	11	32	16	32	5.66
9,9	9	0	0	0	0
9,9	9	0	0	0	0
9,9	9	0	0	0	0
9,10	9.5	0.5	0.25	0.5	0.71
10,9	9.5	0.5	0.25	0.5	0.71
9,13	11	8	4	8	2.83
13,9	11	8	4	8	2.83
9,15	12	18	9	18	4.24
15,9	12	18	9	18	4.24
9,9	9	0	0	0	0
9,10	9.5	0.5	0.25	0.5	0.71
10,9	9.5	0.5	0.25	0.5	0.71
9,13	11	8	4	8	2.83
13,9	11	8	4	8	2.83
9,15	12	18	9	18	4.24
15,9	12	18	9	18	4.24
10,10	10	0	0	0	0
10.13	11.5	4.5	2.25	4.5	2.12
13,10	11.5	4.5	2.25	4.5	2.12
10,15	12.5	12.5	6.25	12.5	3.54

TABLE 12.3 *(Continued)*

(1) Sample	(2) $\overline{X}$	(3) $\Sigma(X-\overline{X})^2$	(4) $\Sigma(X-\overline{X})^2/N$	(5) $\Sigma(X-\overline{X})^2/N-1$	(6) $\sqrt{\Sigma(X-\overline{X})^2/N-1}$
15,10	12.5	12.5	6.25	12.5	3.54
13,13	13	0	0	0	0
13,15	14	2	1	2	1.41
15,13	14	2	1	2	1.41
15,15	15	0	0	0	0
Total =	800		780	1560	313.94

size two that may be drawn from the population of ten longevity scores, and in column 2, the mean for each sample. There are 100 such samples. If the sample mean is an unbiased estimator of the population mean, then the mean of the 100-sample means should equal the population mean. As Table 12.3 shows, the sum of the 100-sample means is 800; and the mean of the sample means $[\,E(\overline{X})\,]$ is 800/100 = 8, which is, in fact, equal to the population mean.

Just because an estimator is unbiased, there is no implied assurance that a single estimate (sample mean) will be equivalent to the population mean. As a matter of fact, examination of the sample means in Table 12.3 indicates that only 10 percent of them are equal to the population mean.

In general, unbiased estimators are more desirable even though there are instances where biased estimators have known sampling distributions or can be corrected. Biased sampling procedures will result in biased estimations as well.*

12.4.2 Unbiased Estimators: Sample Variances and Standard Deviations

The formula for the **population variance** is as follows:

(12.6)
$$\sigma^2 = \frac{\Sigma(X-\mu)^2}{M}$$

BOX 12.3 REVIEWING VARIANCE AND STANDARD DEVIATION

The material in this and in the following section requires that you be comfortable with the concepts of *variance* and *standard deviation*. Considerable stress is placed on them and on other measures of variability in the field of descriptive statistics. It is not unusual for students to become confused about the measures of variability, particularly concerning the variance and the standard deviation. If you have not quite mastered their meaning and use, it would be advisable for you to take time now to review some basics of descriptive statistics. (See Chapter 5.)

---

*The impression that unbiased estimators are always "best" is not always true since there are sometimes conflicting criteria for what an investigator may wish to define as a "best" estimator.

where X is a score, μ is the population mean, and M is the size of the population.

The **variance** is *an average squared deviation of the scores of a distribution from their mean.* In Formula (12.6), for the population variance, the population mean is used as the point of origin from which the deviations of the scores are measured. When only the sample data are available, the population mean (μ) is unknown. We must, therefore, substitute our best estimate of the population mean for that parameter. Since the sample mean is an unbiased estimator of the population mean it may be used in lieu of the population mean. If we replace the population M with the sample N in the denominator we obtain the following formula for the **sample variance:**

(12.7)
$$\text{(biased)} \; s^2 = \frac{\Sigma(X-\overline{X})^2}{N}$$

where X is a sample score, $\overline{X}$ is the sample mean, N is the sample size, and s^2 is the sample variance.

The problem that arises with Formula (12.7) is that it yields a *biased* estimator of the population variance [*i.e.,* $E(s^2) \neq \sigma^2$]. Using Formula (12.6) for the population variance (σ^2) on the longevity scores for the ten block clubs we get a variance of 15.6 and a standard deviation (σ) of 3.95.

In column 3 of Table 12.3 we have computed the sum of squared deviations of the sample scores from the sample mean $[\Sigma(X-\overline{X})^2]$ for each of the 100 samples with $N=2$, and in column 4 we have computed the sample variance, $s^2 = \Sigma(X-\overline{X})^2/N$, for each. If these sample variances were unbiased estimators of the population variance, we would be able to add them, divide by 100 and get a variance of $\sigma^2 = 15.6$. The total for column 4 is 780. Note, however, that dividing this sum by 100 does not give 15.6; it gives 7.80. In this example the average sample variance seriously *underestimates* the population variance.

These estimators are biased because we used the sample mean $(\overline{X})$ in place of the population mean (μ) in the formula for the sample variance. According to the second property of the mean, *the sum of squared deviations of a set of scores around their mean is minimal.* That is, the sum of squared deviations around the mean is smaller than the sum of squared deviations of the scores around any other value. The set of sample scores was used to compute the sample mean in each case and in the numerator of Formula (12.7) the sum of squared deviations of the sample scores was taken around the sample mean, so the second property of the mean applies. The expression $\Sigma(X-\overline{X})^2$ is minimal. If the sample mean is exactly equal to the population mean, then no harm is done. Whenever there is any difference between $\overline{X}$ and μ, $\Sigma(X-\overline{X})^2$ will be too small. It is not likely that $\overline{X}$ will be exactly equal to μ; it will usually differ somewhat (in 100 samples only 10 percent of the sample means were equal to the population mean). The sample variance computed from $s^2 = \Sigma(X - \overline{X})^2/N$ will tend, on the average, to *underestimate* the population variance.

Fortunately, it is possible to correct the formula for s^2 and come up with an unbiased estimator. Instead of increasing the size of the numerator (which tends to be too small) the correction is made by decreasing the size of the denominator to $N-1$. It has been found that the expected value of the uncorrected sample variance $[E(s^2)]$ is equal to $[(N-1)/N] \sigma^2$. That is, it is, on the average, too small by $(N-1)/N$. If each sample variance is divided by this factor so that

(12.8)
$$\text{(corrected) } s^2 = \frac{N}{N-1} \frac{\Sigma(X-\overline{X})^2}{N} = \frac{\Sigma(X-\overline{X})^2}{N-1}$$

the estimator will be unbiased.

In column 5 of Table 12.3 we have recalculated the 100 sample variances using the correction factor. The sum of the 100 variances is 1560. Dividing the sum by 100, we obtain an average of 15.6, equivalent to the population variance.

Because it is an unbiased estimator, the corrected sample variance is the preferred measure of variability in inferential statistics. The corrected formula is used in this book.

Even with correction, the sample standard deviation (s) is a biased estimator of the population standard deviation. Remember that the population standard deviation for the ten longevity scores is 3.95. In column 6 of Table 12.3 we have computed the 100 sample standard deviations using the corrected formula, that is:

(12.9)
$$\text{(corrected) } s = \sqrt{\frac{\Sigma(X-\overline{X})^2}{N-1}}$$

The sum of these 100 sample standard deviations is 313.94. When this sum is divided by 100 the average is 3.14, not 3.95. The sample standard deviation is, on the average, still too small.

TABLE 12.4 CORRECTION FACTORS FOR UNBIASED ESTIMATION OF THE STANDARD DEVIATION

N	a_1	a_2	N	a_1	a_2
2	1.77245	1.25331	34	1.02275	1.00760
3	1.38198	1.12838	35	1.02209	1.00738
4	1.25331	1.08540	36	1.02145	1.00717
5	1.18942	1.06385	37	1.02086	1.00697
6	1.15124	1.05094	38	1.02029	1.00678
7	1.12587	1.04235	39	1.01976	1.00660
8	1.10778	1.03624	40	1.01925	1.00643
9	1.09424	1.03166	41	1.01877	1.00627
10	1.08372	1.02811	42	1.01831	1.00612
11	1.07532	1.02527	43	1.01788	1.00597
12	1.06844	1.02296	44	1.01746	1.00583
13	1.06272	1.02103	45	1.01706	1.00570
14	1.05788	1.01940	46	1.01668	1.00557
15	1.05373	1.01800	47	1.01632	1.00545
16	1.05014	1.01679	48	1.01597	1.00533
17	1.04700	1.01574	49	1.01564	1.00522
18	1.04423	1.01481	50	1.01532	1.00511
19	1.04176	1.01398	60	1.01272	1.00425
20	1.03956	1.01324	70	1.01088	1.00363
21	1.03758	1.01257	80	1.00950	1.00317
22	1.03579	1.01197	90	1.00843	1.00281
23	1.03416	1.01142	100	1.00758	1.00253
24	1.03267	1.01093	110	1.00688	1.00230

TABLE 12.4 *(Continued)*

N	a_1	a_2	N	a_1	a_2
25	1.03130	1.01047	120	1.00630	1.00210
26	1.03005	1.01005	130	1.00582	1.00194
27	1.02888	1.00965	140	1.00540	1.00180
28	1.02783	1.00931	150	1.00503	1.00168
29	1.02682	1.00897	160	1.00472	1.00158
30	1.02590	1.00866	170	1.00445	1.00149
31	1.02503	1.00836	180	1.00420	1.00141
32	1.02423	1.00810	190	1.00395	1.00130
33	1.02347	1.00784	200	1.00378	1.00127

Source: (Bolch, 1968: 27).

Because it is a biased estimator, the sample standard deviation has been used less in inductive statistics than the variance. It is used, however, as we shall soon see. A number of statisticians have pointed out that there is a further correction that can be made to s to make it an unbiased estimator (Cureton, 1968; Bolch, 1968; Jarrett, 1968; and Brugger, 1969). The size of the correction factor depends upon the sample size. Table 12.4 gives the correction factors for various N's between 2 and 200. If the standard deviation has been computed using the $N-1$ correction, the table entry under a_2 is multiplied by s to give the unbiased estimate. If the $N-1$ correction factor has not been used, then the correction is $a_1 s$.

To use this correction on our sample standard deviations in column 6 of Table 12.3, we would multiply each s by $a_2 = 1.25331$. Since a_2 is a constant we can merely multiply it by $E(s) = 3.14$ to get the corrected average standard deviation. Thus,

$$(1.25331)(3.14) = 3.94$$

We did not get exactly 3.95 using the correction factor, but that was due to the standard deviations in column 6 of Table 12.3 being decimal fractions rounded to two decimal places. The difference between 3.95 and 3.94 amounts to compounded errors in rounding.

12.4.3 A Note on Computational Formulae for s^2 and s

The corrected formulae for s^2(12.8) and s(12.9) with which we have been working are useful definitional formulae because they express clearly the operations performed. However, they are not very desirable computationally. Altering the form of the numerator of each [usually called the "sum of squares" and symbolized by Σx^2, where $x = (X - \overline{X})$] results in formulae which are much more convenient for computing. It can be shown algebraically that

(12.10)
$$\Sigma(X - \overline{X})^2 = \Sigma X^2 - (\Sigma X)^2/N$$

Substituting into Formulae (12.8) and (12.9), for the sample variance and standard deviation, we arrive at the following computational formulae:

(12.11)
$$s^2 = \frac{\Sigma X^2 - (\Sigma X)^2/N}{N-1}$$

and

(12.12)
$$s = \sqrt{\frac{\Sigma X^2 - (\Sigma X)^2/N}{N-1}}$$

where X is a sample score and N is the sample size.

12.4.4 Standard Error of the Mean

The measure of sampling error that indicates the magnitude of the deviations of sample statistics around their parameter is called the **standard error.** Table 12.3 constitutes a sampling distribution of sample means based on simple random samples of size two from the population of longevity scores of ten block clubs

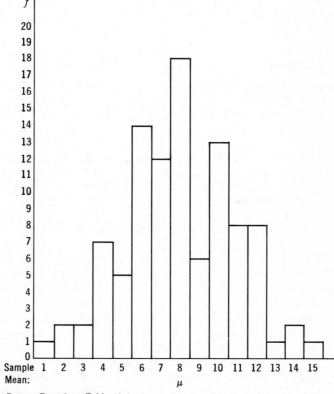

Source: Data from Table 12.4

FIGURE 12.2 HISTOGRAM OF THE SAMPLING DISTRIBUTION OF MEANS OF RANDOM SAMPLES OF SIZE TWO FROM THE POPULATION OF LONGEVITY SCORES OF TEN BLOCK CLUBS (SAMPLE MEANS ROUNDED TO NEAREST EVEN INTEGERS)

(*i.e.,* column 2 does). Figure 12.2 depicts a histogram of this distribution with all of the sample means rounded to the nearest even integer. Note that the sample means cluster around the population mean of 8. *The standard error of the mean is a measure* (the standard deviation) *of the variability of sample means around their population mean.*

If we treated the 100 sample means in this sampling distribution as though they were scores, we could devise a formula very similar to Formula (12.9) for the standard deviation to measure the variability of these 100 "scores" around the population mean. Such a measure of variability is called a "standard error" rather than a "standard deviation." *While the standard deviation measures the variability of scores around their mean, the standard error of the mean measures the variability of sample means around the population mean.*

Taking the formula for the population standard deviation:

$$\sigma = \sqrt{\frac{\Sigma(X-\mu)^2}{M}}$$

we merely substitute $\overline{X}$ for X and n_s for M to convert it into a formula for the standard error of the mean:

(12.13)
$$\sigma_{\overline{X}} = \sqrt{\Sigma(\overline{X}-\mu)^2/n_s}$$

where $\overline{X}$ is a sample mean, μ is the population mean, n_s is the number of samples, and $\sigma_{\overline{X}}$ is the standard error of the mean (a parameter).

We will use this latter formula to compute the standard error of the mean for the sampling distribution of 100 sample means in Table 12.3. Table 12.5 reproduces the 100-sample means from Table 12.3 and shows the deviations and squared deviations of the sample means from the population mean (8) which are necessary to compute the standard error of the mean. Notice that the sum of the squared deviations of the sample means from the population mean, $\Sigma(\overline{X}-\mu)^2$, is 780. We can substitute this sum into the formula along with $n_s = 100$ and solve for the standard error of the mean:

$$\sigma_{\overline{X}} = \sqrt{780/100} = \sqrt{7.80} = 2.79$$

The standard error 2.79 can be interpreted in the same way as the standard deviation is interpreted (since it is a standard deviation), if the sampling distribution is normal or nearly normal. Given a normal sampling distribution, about 68 percent of the sample means in the sampling distribution can be expected to fall within one standard error of the population mean (8 ± 2.79).

We have illustrated for one sample that the sample mean is an unbiased estimator of the population mean and that the variability of sample means around the population mean can be assessed using the standard error of the mean. To generalize from a sample mean to a population mean, however, it is necessary to know more specifically the form of the sampling distribution. In this respect, we are aided by a powerful theorem known as the **central limit theorem** that provides us with information about the form of the sampling distribution of means.

TABLE 12.5 Deviations and Squared Deviations of 100 Sample Means from a Population Mean (8) of a Population Consisting of Ten Block Clubs

$\overline{X}$	$\overline{X}-\mu$	$(\overline{X}-\mu)^2$	$\overline{X}$	$\overline{X}-\mu$	$(\overline{X}-\mu)^2$	$\overline{X}$	$\overline{X}-\mu$	$(\overline{X}-\mu)^2$
1	−7	49	7.5	−0.5	0.25	12	4	16
2.5	−5.5	30.25	7.5	−0.5	0.25	10	2	4
2.5	−5.5	30.25	9	1	1	11.5	3.5	12.25
3	−5	25	9	1	1	11.5	3.5	12.25
3	−5	25	10	2	4	12.5	4.5	20.25
4	−4	16	10	2	4	12.5	4.5	20.25
4	−4	16	7	−1	1	13	5	25
4	−4	16	7	−1	1	14	6	36
4	−4	16	7	−1	1	14	6	36
5	−3	9	8	0	0	15	7	49
5	−3	9	8	0	0		$\Sigma(\overline{X}-\mu)^2=780$	
5	−3	9	8	0	0			
5	−3	9	8	0	0			
5.5	−2.5	6.25	8.5	0.5	0.25			
5.5	−2.5	6.25	8.5	0.5	0.25			
7	−1	1	10	2	4			
7	−1	1	10	2	4			
8	0	0	11	3	9			
8	0	0	11	3	9			
4	−4	16	7	−1	1			
4.5	−3.5	12.25	8	0	0			
4.5	−3.5	12.25	8	0	0			
5.5	−2.5	6.25	8	0	0			
5.5	−2.5	6.25	8	0	0			
5.5	−2.5	6.25	8.5	0.5	0.25			
5.5	−2.5	6.25	8.5	0.5	0.25			
6.5	−1.5	2.25	10	2	4			
6.5	−1.5	2.25	10	2	4			
6.5	−1.5	2.25	11	3	9			
6.5	−1.5	2.25	11	3	9			
7	−1	1	9	1	1			
7	−1	1	9	1	1			
8.5	0.5	0.25	9	1	1			
8.5	0.5	0.25	9.5	1.5	2.25			
9.5	1.5	2.25	9.5	1.5	2.25			
9.5	1.5	2.25	11	3	9			
5	−3	9	11	3	9			
6	−2	4	12	4	16			
6	−2	4	12	4	16			
6	−2	4	9	1	1			
6	−2	4	9.5	1.5	2.25			
7	−1	1	9.5	1.5	2.25			
7	−1	1	11	3	9			
7	−1	1	11	3	9			
7	−1	1	12	4	16			

12.4.5 The Central Limit Theorem

The central limit theorem states that *if repeated simple random samples of size N are drawn from a normally distributed population with mean μ and standard deviation σ, the means of such samples will be normally distributed with mean μ and standard error σ/√N.*

Moreover, *if the N of each sample drawn is large, regardless of the shape of the population distribution, the sample means will tend to distribute themselves normally with mean μ and standard error σ/√N.**

In effect, the central limit theorem tells us that the means of simple random samples drawn from a population that is itself normally distributed will result in a sampling distribution that is normal, even though the sample N is small.

The second part of the theorem adds that, as long as the sample N is large, even though the population is not normally distributed, we will still tend to get a normal sampling distribution of means. What is meant by "large" has not been firmly established. When N is as great as 30, the normal distribution may be a fairly good approximation to the sampling distribution of means. A more conservative estimate of a large N is 100 or more. Naturally, the larger the N, the better is the sampling distribution approximated by the normal curve.

Although the distribution of scores for the ten block clubs is certainly not normally distributed and the sample size is only 2, we will see whether the normal distribution gives us a reasonable approximation to the sampling distribution of 100 sample means. We have already determined that the mean of the sampling distribution is 8 and the standard error is 2.79. To answer a question such as: "How many of the sample means will fall between 5 and 11, inclusive?" we can substitute these values into a standard score formula

(12.14)
$$z = \frac{\overline{X} - \mu}{\sigma_{\overline{X}}}$$

treating 5 and 11 as though they were continuous scores and using 4.5 and 11.5 in Formula (12.14) to obtain the following:

$$z_1 = \frac{4.5 - 8}{2.79} = -1.25$$

$$z_2 = \frac{11.5 - 8}{2.79} = +1.25$$

The area under the normal curve between $z = -1.25$ and $z = +1.25$ is 0.7988 (see Appendix Table B). Since there are 100 sample means, this indicates we would expect, if our sampling distribution was in fact normal, to find 79.88 or 80 means between 5 and 11, inclusive. If we actually count the number of sample means between 5 and 11 in Table 12.5, we find that there are 76. Thus, in this case at least, even with an N as small as 2, the normal approximation to the sampling distribution of means did not err greatly. *Be aware, however, that this illustration does not prove that the normal approximation will work in other situ-*

---

*If the population distribution has a finite mean and variance as it would have in most social science research applications.

ations with such small N's! It is our position that a normal sampling distribution should not be assumed when sampling from a non-normal population unless N is at least 100.

Another important fact brought out by the central limit theorem is that the standard error of the sampling distribution is given by $\sigma/\sqrt{N}$. In our block club example we obtained the standard error of the mean directly from the sampling distribution using Formula (12.13). The resulting standard error was 2.79. Presumably we should get the same standard error by using the following formula as well:

(12.15)
$$\sigma_{\bar{x}} = \sigma/\sqrt{N}$$

To show this, we need to go back to the longevity scores for the ten block clubs and use the population standard deviation 3.95 that we had computed from them earlier. Dividing this standard deviation by the square root of N we get the following:

$$\sigma_{\bar{x}} = \frac{3.95}{\sqrt{2}} = \frac{3.95}{1.414} = 2.79$$

Thus, once we know the population standard deviation, we can find the standard error of the mean with Formula (12.15). *Since the sampling distribution is a theoretical distribution, we would not actually need to generate it as we did from the population of block clubs.* We only reproduced that sampling distribution for purposes of illustration. Ordinarily we would not have the data available (nor is it necessary to go to the trouble of trying to create them) to compute the standard error directly with Formula (12.13). *If the population standard deviation is known, the standard error should be computed using Formula (12.15).*

The problem, however, with using Formula (12.15) to compute the standard error of the mean is that we often do not know σ. In research typically conducted by sociologists, both the population mean and standard deviation are likely to be unknown. These frequently are the parameters that the sociologist is interested in estimating from his(her) sample data. If σ is unknown, the best estimate of it available would be the sample standard deviation (s). If s is corrected using

---

BOX 12.4 STANDARD SCORE FORMULAE AND THE NORMAL DISTRIBUTION

If you examine carefully Formula (12.14), you will notice that it is but another version of the basic standard score formula that is used in descriptive statistics (see Chapter 5). This is the second variation of the basic formula that we have had occasion to use. The first variation appeared as Formula (12.5) in Section 12.3.4 of this chapter. Still other versions will be presented in later sections of this volume. It should be becoming progressively more clear to you that the normal distribution is *the most important* theoretical distribution for the field of statistics.

Table 12.4, it will be an unbiased estimator. Substituting s for σ the following formula is used:

(12.16)
$$s_{\bar{X}} = s/\sqrt{N}$$

where $s_{\bar{X}}$ is a statistic that estimates $\sigma_{\bar{X}}$. This estimate is also known as the standard error of the mean; the difference being that $\sigma_{\bar{X}}$ is the parameter and $s_{\bar{X}}$ is the statistic.

12.4.6   Student's t-Distribution

When $\bar{X}$ is a random variable as it is in a sampling distribution (*i.e.* sample means from random sampling), generalizations may be made by converting $\bar{X}$ to a standard score by means of Formula (12.14) and by evaluating the results on the normal curve. Since $\bar{X}$ is a random variable in Formula (12.14) for z, the numerator is also a random variable. The denominator, however, is a constant. If σ is unknown and we must substitute s for it in Formula (12.14), the formula becomes:

(12.17)
$$t = \frac{\bar{X} - \mu}{s/\sqrt{N}}$$

In this case the denominator as well as the numerator is a random variable because s is a statistic rather than a parameter. If N is large (100 or more), t is approximately equal to z. This is so because the greater the N, the closer s estimates σ. With a large N, then, the normal distribution is a good approximation to the distribution of t.*

When N is small, the t-distribution may diverge greatly from the normal distribution.† Because both the numerator and the denominator are variables in Formula (12.17) for t while only the numerator is a variable for z, the t-distribution tends to have greater dispersion than the normal distribution. Over a number of different samples if $\bar{X}$ is the same, z will be the same because the denominator is a constant; but this is not the case with t. The variability of the t-distribution is related to the size of N. The smaller the N, the more variable the t-distribution. There is, then, a whole series of t-distributions—a different one for each N. The family of t-distributions is presented in Appendix Table C. In Appendix Table C the first column is labelled df, which is the sign for degrees of freedom. When the mean from one sample is evaluated using the t-distribution, $df = N - 1$ (we will have more to say about the concept of degrees of freedom in Section 12.4.7). Each row of the table represents values of a different t-distribution with $N - 1$ degrees of freedom.

---

*Whenever the sample is large we will make a practice of designating Formula (12.17) as z and using the normal curve to evaluate it.
†The t-distribution was studied and introduced as a sampling distribution by W. S. Gossett (1876–1937), a mathematical consultant for the Guiness Brewery in Dublin, Ireland. He found that, with small N's, the use of the sample standard deviation (s) results in a sampling distribution of means that is not normal. He published his findings in 1908 under the pen name "Student" because of a previous agreement with Guiness.

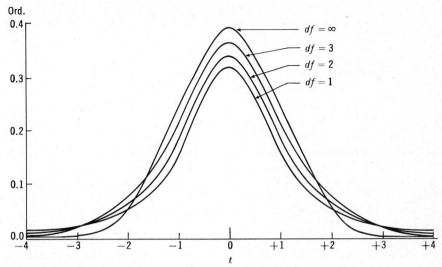

FIGURE 12.3 STUDENT'S *t*-DISTRIBUTIONS WITH 1, 2, 3, AND ∞ *df*.

These *t*-distribution values were derived based upon the assumption that the numerator of the *t* ratio is normally distributed and the denominator is independent of the numerator. Since the sample mean $(\overline{X})$ appears in the numerator and the sample standard deviation (s) appears in the denominator and these two statistics are usually related, it would seem unlikely that the denominator would be independent of the numerator. There is one instance, though, in which the denominator is independent of the numerator. This prevails when simple random samples are drawn from a normally distributed population of data. Such an assumption necessarily limits the instances when the *t*-distribution will be applicable to sociological research.

The *t*-distributions are symmetrical distributions, but they are flatter (more platykurtic) than the normal curve and their tails approach the base line at a slower rate. The smaller the *N*, the flatter the curve and the more slowly the tails approach the base line. If you look at the entries for the *t*-distributions in Appendix Table C, you will notice that when $N=2$ $(df=1)$, the standard scores beyond which a total of 5 per cent of the area under the curve falls are −12.706 and +12.706. In the normal curve 5 per cent of the area lies beyond standard scores of −1.96 and +1.96. When *N* is increased to 3 $(df=2)$, the standard scores cutting off the 5 per cent in the tails drop to −4.303 and +4.303; for an *N* of 4 they become −3.182 and +3.182, etc. Figure 12.3 presents graphs of these *t*-distributions and the *t*-distribution with $df=\infty$. This latter distribution is, in fact, the normal distribution since it is the limiting distribution for the Student's *t*. When *N* reaches 100, the standard scores cutting off 5 percent in the tails of the *t*-distribution become −1.98 and +1.98, which are very close to the standard scores for the normal curve. Thus, when *N* is equal to or greater than 100 the normal curve may be substituted for the *t*-distribution, as a sampling distribution. In the following chapters we will demonstrate how the *t*-distribution can be used for estimating parameters and testing hypotheses.

We will digress for a moment from our discussion of sampling distributions now and examine in greater detail the concept of degrees of freedom.

12.4.7 Degrees of Freedom

The concept of **degrees of freedom** is one that arises frequently in connection with sampling distributions. It is relevant in particular to statistics with which families of sampling distributions are associated rather than a single sampling distribution. The standard normal curve is a single sampling distribution; thus the concept of degrees of freedom does not apply to it. Such sampling distributions as the t-distribution, the chi-square distribution, and the F-distribution are actually families of sampling distributions requiring the concept of degrees of freedom to determine which member of the family of distributions is applicable in a given situation.

Degrees of freedom, in general, relate to the number of restrictions (or parameters estimated) *which are put on data.* Initially, there are as many degrees of freedom as there are cases or categories of data. Each time a restriction is imposed upon the data one degree of freedom is sacrificed.

For example, when we evaluate the ratio,

$$\frac{\overline{X} - \mu}{s/\sqrt{N}}$$

using the t-distribution, $df = N - 1$. For the t-distribution the number of cases (N) determines the number of degrees of freedom which are initially available. The one degree of freedom is sacrificed in the computation of the sample standard deviation (s). This degree is lost because of the restriction that the sum of deviations of scores around their mean must equal 0 [*i.e.,* $\Sigma(X - \overline{X}) = 0$]. Given this limitation $N - 1$ of the scores in the sample may take any value, but once values have been assigned to $N - 1$ of them, the remaining score is determined. For example, if we had five scores whose mean was 10, four of the scores could take any values, but the fifth would have to have a value such that the sum of the deviations of all five from 10 would equal 0. Thus, values such as 1, 3, 14, and 20 might be assigned to the first four scores, but the fifth score would then have to have a value of 12. Or if the first four scores had values of 7, 11, 2, and 16 the fifth score would have to be 14.

When the appropriate number of degrees of freedom is established for a particular application of a statistical technique, then the specific sampling distribution that applies to that situation is also established.

Because the formula for degrees of freedom varies from one sampling distribution to another and from one statistical technique to another, it is difficult to devise definitive statements about finding degrees of freedom. In general, the explanation of the concept given here applies to each use, however. Throughout this book when degrees of freedom are relevant, the appropriate formula will be given.

12.4.8 Recapitulation

We have discussed how the sampling distribution of means is affected by normal and non-normal population distributions, sample size, and knowledge (or lack of it) about the population standard deviation. To bring the discussion into focus, in Diagram 12.1 we have recapitulated the essential points to be considered in choosing the appropriate sampling distribution and standard error.

12.5 SAMPLING DISTRIBUTIONS OF VARIANCES

By the nature of sociology there are sometimes situations in which variability may be as relevant or even more relevant than central tendency to the problem that is being investigated. For example, if a sociologist is engaged in the study of political activities of organizations, the variability of the characteristics of members may be an important factor in determining the amount of political activity in which the organization engages. It might be fruitful, for instance, to test the hypothesis that the greater the homogeneity of members, the more politically active the organization.

Peter Blau has suggested that a number of empirical findings regarding formal organizations may be accounted for by two basic generalizations: "(1) increasing organizational size generates differentiation along various lines at decelerating rates; and (2) differentiation enlarges the administrative component in organizations to effect coordination" (Blau, 1970:201). These generalizations suggest the importance of examining measures of differentiation such as variances.

Blau contends that differentiation of responsibilities characteristic of larger organizations simultaneously leads to intra-unit homogeneity and inter-unit heterogeneity (Blau, 1970:217). Though Blau primarily had job specialization in mind when he mentioned intra-unit homogeneity and inter-unit heterogeneity, there is no reason to believe that a number of variables which lend themselves to interval/ratio level measurement (*e.g.*, age, income, years of education, etc.) might not be significant indices as well.

At any rate, the point being made is that sociologists often need to examine variability of their data and make generalizations that specifically address themselves to the influence of variability on social situations. When focusing his (her) attention on the variability of sample data, the sociologist must have knowledge of the sampling distributions of measures such as variances.

It has been found that sample variances (s^2's) based upon simple random samples from a normally distributed population can be expressed as **chi-square scores** (χ^2's) which have a **chi-square sampling distribution** with $N-1$ degrees of freedom. The formula for converting sample variances to χ^2 scores is as follows:*

(12.18)

$$\chi^2 = \frac{(N-1)s^2}{\sigma^2}$$

where σ^2 is the population variance and s^2 is the sample variance.

*Note that the population variance (σ^2) is frequently unknown. We will show in Chapter 13 that a variation of this formula can be used even though the population variance is unknown.

DIAGRAM 12.1 Determining the Sampling Distribution and Standard Error of the Mean

1. If the population distribution is normal and the population standard deviation is *known*, regardless of whether N is large or small:
    Sampling Distribution: Normal Curve
    Standard Error, use Formula (12.15): $\sigma_{\bar{x}} = \sigma/\sqrt{N}$
2. If the population distribution is normal and the population standard deviation is *unknown*, when N is large (100 or more):
    Sampling Distribution: Normal Curve
    Standard Error, use Formula (12.16): $s_{\bar{x}} = s/\sqrt{N}$
3. If the population distribution is normal and the population standard deviation is *unknown*, when N is small (less than 100):
    Sampling Distribution: Student's t with $df = N - 1$
    Standard Error, use Formula (12.16): $s_{\bar{x}} = s/\sqrt{N}$
4. If the population distribution is not normal and the population standard deviation is *known*, when N is large:
    Sampling Distribution: Normal Curve
    Standard Error, use Formula (12.15): $\sigma_{\bar{x}} = \sigma/\sqrt{n}$
5. If the population distribution is not normal and the population standard deviation is *unknown*, when N is large:
    Sampling Distribution: Normal Curve
    Standard Error, use Formula (12.16): $s_{\bar{x}} = s/\sqrt{N}$

12.5.1 The Chi-Square Distribution

The chi-square sampling distribution is not one distribution but, like the t-distribution, a whole family of continuous distributions. The shape of a particular chi-square distribution depends upon the number of degrees of freedom involved. The chi-square distribution table (Appendix Table D) is set up with degrees of freedom (df) in the first column and selected probabilities (areas in the tails of the distribution) in the other columns. χ^2 scores may vary from 0 to infinity.

Figure 12.4 illustrates the curves of chi-square distributions for 1, 2, 3, 4, 5, 8, and 10 degrees of freedom. Notice that for 1 degree of freedom ($N=2$) the chi-square distribution is essentially a single tailed J-curve with the bulk of the χ^2 scores close to 0. Only 5 percent of the χ^2 scores in this particular distribution are as large or larger than 3.841.

Another interesting fact about the chi-square distribution with 1 degree of freedom is that its distribution is the same as that for z^2. Notice, for instance, that $(1.96)^2 = 3.841$. A z score of 1.96 cuts 2.5 percent of the area off one tail of the normal distribution, while a χ^2 of 3.841 cuts 5 percent off of the right tail of the chi-square distribution.

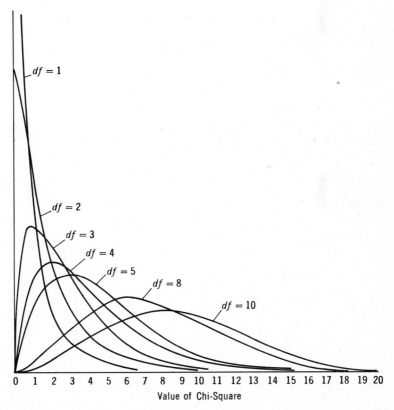

FIGURE 12.4 Chi-Square Distributions for 1, 2, 3, 4, 5, 8, and 10 Degrees of Freedom

Notice in Figure 12.4 that a chi-square distribution with 2 degrees of freedom ($N=3$) also forms a J-curve, but it approaches the baseline more slowly than the curve for 1 degree of freedom. For the curve with 3 degrees of freedom a tail falls on the left as well as on the right. The left tail starts at 0. Notice also that the mode of the curve for 3 degrees of freedom is at a χ^2 score of 1. It can be said, in general, that the mode of a χ^2 curve (with the exception of the curve with 1 degree of freedom) will be at $\chi^2 = df - 2$.

As the number of degrees of freedom (or sample size in the case of variances) increases, the chi-square distributions get less skewed. By the time degrees of freedom reach 10, the distribution is much less skewed, although it is not quite symmetrical. The table of chi-square distributions usually goes only to 30 degrees of freedom. For small samples, therefore, the chi-square distributions can be used directly to evaluate sample variances. When the number of degrees of freedom exceeds 30, the chi-square distribution can be approximated by the standard normal distribution with a standard error of 1 and

(12.19)
$$z = \sqrt{2\chi^2} - \sqrt{2(df) - 1}$$

An alternative approach is possible with a large N because, as N gets large sample *standard deviations* tend to distribute themselves in an approximately normal distribution with a mean of σ and a standard error of $\sigma/\sqrt{2(N-1)}$. Since s tends to come close to σ when N is large, if σ is unknown, the unbiased s may be substituted for it without introducing serious error.

It will be explained in later chapters how these sampling distributions of variances and standard deviations can be used in (1) estimating parameters (Chapter 13, Section 13.4) and (2) testing hypotheses (Chapter 15, Sections 15.7 and 15.8).

All we have said about the sampling distributions of variances and standard deviations is based upon the assumption that the population from which we sample is normally distributed. Unless N is extremely large the error introduced when the population is not normally distributed may be considerable.

12.6 SUMMARY

In this chapter we have examined the concept of the sampling distribution. We found that, in general, statistics distribute themselves around their parameters in predictable patterns with measureable degrees of variability. The predictable patterns are known as sampling distributions and the measures of variability are known as standard errors. We have studied the central tendency, variability, and form of sampling distributions for frequencies, means, variances, and standard deviations.

The sampling distributions of these particular statistics by no means exhaust those of interest and importance to sociologists. The ones discussed in this chapter serve as typical examples of the kinds of sampling distributions with which sociologists work. The sampling distributions of other statistics will be discussed in the chapters that follow.

BOX 12.5 SOME IMPORTANT DISTINCTIONS

It is very important that you be able to distinguish among a population distribution, a sample distribution, and a sampling distribution. If this is not clear to you, go back and review all units under Section 12.2.

The central limit theorem was discussed in connection with the sampling distribution of the mean, but it has a more general applicability than has been indicated thus far. Whenever the normal approximation to the sampling distribution of a statistic can be used with a large sample, it is because the central limit theorem is operative. We have found thus far that the normal approximation can be used with frequencies, variances, standard deviations, and means when N is large. In the following chapters we will find that the normal approximation applies to sampling distributions of other statistics as well.

The concept of the sampling distribution is really the pivotal concept for inferential statistics. It is at the very heart of the logic of statistical induction. In Chapter 13 we will see how sampling distributions are used in estimating parameters and in the remaining chapters we will see how they are used to test statistical hypotheses.

CONCEPTS TO KNOW AND UNDERSTAND

sampling distribution
expected value of a statistic
unbiased estimator
standard error
law of large numbers
binomial sampling distribution
normal sampling distribution
mean of a binomial distribution
standard error of a binomial
    distribution

sampling distribution of means
Student's t-distribution
standard error of the mean
second property of the mean
central limit theorem
large N
degrees of freedom
sampling distribution of variances
chi-square distribution

QUESTIONS AND PROBLEMS

1. Explain the relationship between a *sample* distribution and a *sampling* distribution.

2. What is the relationship between the standard error of the mean and the standard deviation?

3. Explain the meaning of the statement that the normal distribution is the limiting case of the binomial probability distribution.

4. Explain what is meant by the statement that the sample mean is an unbiased estimator of the population mean.

5. Why is the variance preferred to the standard deviation as a measure of variability in inferential statistics?

6. Prove algebraically that: $\Sigma(X-\bar{X})^2 = \Sigma X^2 - (\Sigma X)^2/N$.

7. Why is $N-1$ instead of N used as the denominator of the formula for the sample variance?

8. Of what significance for inferential statistics is the central limit theorem?

9. How does Student's t-distribution differ from the normal distribution? When may the normal distribution be used instead of the t-distribution?

10. Explain what is meant by degrees of freedom as a concept applied to Student's t-distribution.

11. Draw a biased sample from the ten block clubs described in Table 12.3. For example, draw samples with the restriction that the extreme scores (e.g. 1 and 15) can not appear in any sample. Then show that the mean of random samples of size two drawn in this biased way will be a biased estimate of the population mean. Hint: identify all of the possible samples of size two under this biased sampling procedure, compute each sample's mean and describe this distribution of sample means.

12. Create a sampling distribution. Select a dichotomous variable (e.g. sex) and a small population (e.g. members of your immediate family). Using the formula for computing binomial probabilities, compute the sampling distribution. For this problem, use sample size $N = 2$. Portray the sampling distribution in table form and also in the form of a histogram (carefully drawn). Now, create and portray the sampling distribution for the same situation where a larger sample size is used (but one less than the size of the population!). Comment on differences between these distributions.

13. Empirically set up a sampling distribution. You could do this in one of two ways. If you have access to a computer, identify a large data set (e.g. some general social survey which is already stored in computer readable form). Then identify a computer program which permits you to draw simple random samples of cases from this data set. Decide on a sample size (for example, size 30) and have the computer draw repeated random samples of this size (say, about 30 or so samples, if time permits). In each case, the computer should be instructed to summarize statistically the data from the sample on the variable(s) in which you are interested. For example, if you were interested in age of individual, have the computer compute the mean and standard deviation of ages of the, say, 30 individuals included in each random sample. Then, summarize, statistically, the results of your sampling. By making an overall description of all of the cases in the data set you would have, in effect, the population parameter. Compare the parameter with your sampling distribution summary. If you have resources it would be helpful to

repeat the problem using different sample sizes. If you do not have access to a computer, then draw and summarize small samples ($N = 10$, for example) from the data provided below. Complete a sampling distribution and describe that sampling distribution, statistically. Then use the formulas given in this chapter to compute the mean and standard error of a sampling distribution of the kind you have empirically constructed. Comment on similarities of your empirical description and the computed one. Be sure that you use the table of random numbers in the back of the book to draw your simple random samples from the data below.

LISTING OF PERCENT OF POPULATION LIVING IN URBAN PLACES OF 2500 OR LARGER IN 1970 FOR STATES*

Line	State	% Urban	Line	State	% Urban	Line	State	% Urban
01	Maine	50.8	18	N. Dak	44.3	35	Ark	50.0
02	N. H.	56.4	19	S. Dak	44.6	36	La	66.1
03	Vt	32.2	20	Nebr	61.5	37	Okla	68.0
04	Mass	84.6	21	Kans	66.1	38	Tex	79.7
05	R. I.	87.1	22	Del	72.2	39	Mont	53.4
06	Conn	77.4	23	Md	76.6	40	Idaho	54.1
07	N.Y.	85.6	24	D.C.	100.0	41	Wyo	60.5
08	N.J.	88.9	25	Va	63.1	42	Colo	78.5
09	Pa	71.5	26	W. Va	39.0	43	N. Mex	69.8
10	Ohio	75.3	27	N.C.	45.0	44	Ariz	79.6
11	Ind	64.9	28	S.C.	47.6	45	Utah	80.4
12	Ill	83.0	29	Ga	60.3	46	Nev	80.9
13	Mich	73.8	30	Fla	80.5	47	Wash	72.6
14	Wis	65.9	31	Ky	52.3	48	Oreg	67.1
15	Minn	66.4	32	Tenn	58.8	49	Calif	90.9
16	Iowa	57.2	33	Ala	58.4	50	Alaska	48.4
17	Mo	70.1	34	Miss	44.5	51	Hawaii	83.1

*U.S. Bureau of the Census, Statistical Abstract of the United States: 1975 (96th edition). Washington, D.C., 1975, p 20, Table 21.

GENERAL REFERENCES

Games, Paul A., and George R. Klare. *Elementary Statistics: data analysis for the behavioral sciences* (New York, McGraw-Hill Book Company), 1967.

Chapter 8 is an unusually detailed discussion of sampling distributions. It stresses how to distinguish clearly between population distributions, sample distributions, and sampling distributions.

Hays, William L., and Robert L. Winkler. *Statistics: Probability, Inference, and Decision* (New York, Holt, Rinehart and Winston), 1971.

This book presents a more technical, more mathematically oriented discussion of sampling distributions. In Chapter 6 the normal distribution is shown to be the parent distribution of the t-distribution, the chi-square distribution, and the F-distribution.

OTHER REFERENCES

Blau, Peter M. "A Formal Theory of Differentiation in Organizations," *American Sociological Review,* vol. 35, no. 2 (April, 1970), p 201–218.

Bolch, Ben W., "More on Unbiased Estimation of the Standard Deviation," *The American Statistician*, vol. 22, no. 3 (1968), p. 27.

Brugger, R. M., "A Note on Unbiased Estimation of the Standard Deviation," *The American Statistician*, vol. 23, no. 4 (1969), p. 32.

Cureton, E. E., "Unbiased Estimation of the Standard Deviation," *The American Statistician*, vol. 22, no. 1 (1968), p. 22.

Jarrett, R. F., "A Minor Exercise in History," *The American Statistician*, vol. 22, no. 3 (1968), p. 25.

National Bureau of Standards, Applied Mathematics Series 6, *Tables of the Binomial Probability Distribution* (Washington, D.C., United States Government Printing Office), 1949.

13
Point and Interval Estimates of Parameters

When the stock market crashed in 1929 and the depression engulfed the country, the government found itself in the dark about the magnitude of the unemployment problem. At that time the Census Bureau policy was to collect labor statistics using the "gainful worker" concept. Respondents were interviewed to determine the occupations at which they were usually employed. They could report the occupations with which they identified, whether or not they were actually working at them. The use of the gainful worker concept thus made it possible for retired and institutionalized persons to report occupations. Although such information can be very useful to sociologists interested in the self images of workers, potential workers, and ex-workers, it is not an effective way of determining how many people are in the labor force—how many are actually working, and how many are unemployed.

When the federal government took on an active role in regulating the economy, it realized that a more accurate measurement of employment and unemployment, taken at frequent intervals, was necessary. Accordingly, beginning with the 1940 census, the Census Bureau adopted the "labor force" concept. The labor force concept is based principally on each respondent's actual activity during the time surveyed. The respondent is categorized as working, looking for work, or doing something else.

Today, rather than collect labor force statistics only once every ten years as part of the census, the Bureau of Labor Statistics surveys a sample of the population each month. During the week of the 19th day of the month, members of more than 50,000 households in almost 500 sample areas are interviewed to determine their work activities for the previous week. On the basis of these interviews, monthly estimates are made of the size of the labor force and the percentage of those in the labor force who are unemployed. These estimates are now considered to be so vitally indicative of the state of the national economy that they reach front-page coverage in the nation's newspapers and radio and television news bulletins.

Although it is not generally known, these much publicized figures are not parameters. They are estimates of parameters based on sample data. Nonetheless, the federal government takes these estimates seriously enough to use them as a basis for national policy. Certainly, these estimates are efficient enough to warrant the government's concern. The estimated unemployment rate seldom has a standard error larger than four-tenths of one per cent.

Parameter estimation is not merely of interest and importance to the policy makers. It also plays an important role in science. Although the labor force estimates are utilized primarily for policy making, they are also of interest to the sociologist in the course of his(her) study of American society. After all, work is one of the primary sociological variables.

Sociologists not only make use of parameter estimates made by agencies of the federal government; in the course of their research, they also frequently engage in estimation of parameters from sample data they have collected themselves. For example, Nancy Morse and Robert Weiss conducted a study of a national sample of employed men to determine what meanings work had for them (Morse and Weiss, 1955). In response to a hypothetical question, "If by some chance you inherited enough money to live comfortably without working, do you think you would work anyway or not?" 80 per cent of the 393 men who responded said that they would continue to work (Morse and Weiss, 1955:191–192). Because Morse and Weiss were interested in the meanings of work for *all* employed men in the United States, their 80 per cent figure can be looked on as a parameter estimate. Of course, in their study, they went on to ask other questions and to estimate other parameters. The point to consider is that they conducted a sociological study whose import depends primarily upon results of estimation procedures.

Sociologists have sometimes been prone to forget the important role played by estimation procedures in scientific research. Labovitz has argued, for instance, that tests of significance are non-utilitarian for sociology (Labovitz, 1970; see also Morrison and Henkel, 1970). In the process of affirming his case against tests of significance he concludes,

> In a crude sense, it seems that many researchers have already rank ordered the importance of statistical techniques. Usually, a statistic is first developed and reported (*e.g.*, a rank correlation measure like d_{yx}), and then a claim is made that its sampling distribution will be developed later. The arguments in this paper would logically extend this rank order among techniques to the point of forgetting the development of the sampling distribution altogether (Labovitz, 1970:147).

Such an extreme position ignores the fact that sampling distributions are necessary for estimating parameters as well as testing hypotheses. Although Labovitz recalls some important arguments against tests of statistical hypotheses, his arguments do not apply equally as well to estimation procedures.

13.1 POINT VERSUS INTERVAL ESTIMATES

When the Bureau of Labor Statistics' estimate of the percent unemployed is released through the mass media each month, a single value is reported. *When a single value based upon sample data is used as an estimate of a parameter the estimate is known as a* **point estimate.** The daily newspaper may report that the unemployment rate for the month of July was 5 percent. What this actually means is that 5 percent of those surveyed by the Bureau of Labor Statistics interviewers reported themselves unemployed. Assuming that the survey sample is a random sample,* this statistic is used as a point estimate of the parameter — the percentage of the total labor force that is unemployed.

Point estimates are used in the reports to the mass media because they are easiest for the general public to comprehend. Ideally, a point estimate should be based upon an unbiased estimator of the parameter (which this one is), error in estimation should be random; *and* the estimate should be an **efficient** one (*i.e.,* the standard error should be small so that the statistics cluster closely about the parameter they estimate). Unfortunately, a point estimate does not carry with it any indication of how closely it estimates the parameter — there is no probability figure attached to it, as such. Therefore, although it is a popular way of presenting an estimate, both in the sociological literature and in the public media, it leaves something to be desired.

There are real benefits to be derived from making an **interval** rather than a point estimate of a parameter. *An interval estimate is an estimate which consists of a range of values rather than a single value.* One advantage of the interval estimate is that the width of the interval tells something about how efficient the estimate is. Another advantage is that we can attach a probability figure to the estimate. Because of these and other advantages to be pointed out later, interval estimates are generally preferred to point estimates for sociological research.

An interval used to estimate a parameter is known as a **confidence interval** and *the end points of the interval are known as* **confidence limits.** The use of the term *confidence* relates to the fact that probability figures may be associated with interval estimates, and the probabilities give us a notion of how much confidence we can have in our estimation procedure. We may select any probability level that we wish to associate with our estimates. Two probability levels that are commonly used, however, are the 95 and 99 percent levels.

In this chapter we will discuss estimation procedures for several different statistics. Again, our discussion of various statistics will be selective rather than exhaustive. We have chosen to examine statistics that are commonly used and that concisely illustrate estimation procedures. With estimation procedures for various statistics being generally similar, your ability to understand the procedures for a few techniques should enable you to apply them to other techniques as well.

---

*Although the sampling plan is a probability sample, it is more complex than a simple random sample. It is, in fact, a variation of the cluster sample design.

13.2 ESTIMATING PROPORTIONS

Since we began this chapter with examples of the use of sample percentages as point estimates, we will turn first to a consideration of interval estimates for proportions (or for percentages). Actually, frequencies, proportions, and percentages represent three alternative ways of presenting the same data; therefore, they are interchangeable. Computations are usually performed on frequencies or proportions, rather than percentages, because they are more convenient to work with.

In order to make an interval estimate of a proportion we need to know the central tendency, variability, and form of the sampling distribution of proportions. We will turn our attention first to these characteristics of the sampling distribution, then we will demonstrate how interval estimates of proportions are made. We will compute interval estimates of the unemployment rate and of the proportion of men who would continue working even if they did not need the money.

13.2.1 Sampling Distributions of Proportions

If the data with which we are working are dichotomized, if simple random sampling is involved, and if the measurements are independent, then the binomial distribution is the appropriate sampling distribution to use, whether the data are in frequency, proportion, or percentage form.

When the sample size N is small, the easiest way to analyze data of this type is to work with frequencies rather than proportions and use the binomial sampling distribution directly. When N is large and, the proportion, P is not too extreme so that the rule of thumb applies (both NP and NQ are equal to or greater than 5), then the normal curve may be used as an approximation to the bi-

---

BOX 13.1 SYMBOLS FOR STATISTICS AND PARAMETERS

The distinction between samples and populations is central to inferential statistics. The symbols used generally reflect this distinction as well. In the text different symbols are used for sample statistics or estimates and population values themselves. You should be alert to this distinction and extend this illustrative list for yourself.

Sample Statistic		Population Parameter
$\overline{X}$	arithmetic mean	μ
p	proportion p	P
$q = 1 - p$	proportion q	$Q = 1 - P$
s	standard deviation	σ
$s_{\overline{x}}$	standard error of the mean	$\sigma_{\overline{x}}$
s_p	standard error of proportion	σ_p

nomial sampling distribution.* In such cases the frequencies can be used directly with $\mu_B = NP$ and $\sigma_B = \sqrt{NPQ}$ to compute standard scores. Or alternatively, proportions can be computed, converted to standard scores, and evaluated on the normal sampling distribution. If we use the latter alternative, then we need to know how to compute the mean and standard error of proportions.

13.2.2 The Mean Proportion

It can be demonstrated that *a proportion is a mean for its distribution of scores.* Remember that the mean is merely the sum of scores of a distribution divided by the number of scores (*e.g.*, $\overline{X} = \Sigma X/N$). We will use the Morse and Weiss data to demonstrate that a proportion is a mean. Of the 393 men who responded to the question about continuing to work, 314 said they would and 79 said they would not (Morse and Weiss, 1955:192). Suppose that we assign a score of 1 to every man who says he would continue to work and a score of 0 to every man who says he would not. If we add all of the scores, the 1's and the 0's, and divide by the number of respondents (393) we will get a mean. Since the 314 1's equal 314, and the 79 0's equal 0, the sum of the scores is 314 and the mean is $314/393 = 0.80$. Therefore, the proportion 0.80 is the mean.

Earlier we found that the mean of a binomial distribution is simply N multiplied by P and this mean is for a frequency count. This could be changed to a proportion simply by dividing NP (the average frequency of occurrence of events of interest) by N, yielding the mean of the sampling distribution of proportions. Substituting f for NP, we have:

(13.1)
$$\overline{P} = P = f/N$$

In the case of the Morse and Weiss data, the 0.80 is a sample proportion (mean) which is an unbiased estimator of the population proportion. That is, p estimates P.

13.2.3 Standard Error of a Proportion

If the population proportion (P) is known, the **standard error of the proportion** can be computed from the following formula:

(13.2)
$$\sigma_p = \sqrt{PQ/N}$$

where σ_p is a parameter.

Formula (13.2) for the standard error of a proportion is directly related to the formula for the standard error of the binomial distribution (standard error of a frequency). The standard error of the binomial distribution is as follows:

(13.3)
$$\sigma_B = \sqrt{NPQ}$$

---

*If N is large but P is extreme (either very small or very large), the Poisson Distribution may be used as the sampling distribution. Since this situation rarely occurs in sociology, the Poisson Distribution has not received much attention (but see Pierce, 1970).

Since a frequency is converted to a proportion by dividing by N, we can divide the standard error of a frequency by N to get the standard error of a proportion. Thus:

$$\sigma_B/N = \sqrt{NPQ}/N = \sqrt{NPQ/N^2} = \sqrt{PQ/N} = \sigma_p$$

Where it can be assumed that the normal curve is a good approximation to the sampling distribution of P, then σ_p can be interpreted like a standard deviation. That is, the statement can be made that approximately 68 per cent of the sample proportions in the sampling distribution will be within one standard error of the population proportion.

If the population proportion is not known, our best estimator of it is the sample proportion (p). Since the N has to be large in order for us to use the normal approximation to the binomial sampling distribution in the first place, the sample proportion (p) from such a sample should estimate the population proportion (P) fairly closely, and the error introduced by substituting p for P in the standard error formula should not be serious. We shall distinguish the standard error based on the sample proportion from the one based on the population proportion by calling the former s_p and the latter σ_p. The s_p is the statistic that estimates the parameter σ_p.

The formula for the standard error based on the sample proportion is as follows:

(13.4)
$$s_p = \sqrt{pq/N}$$

Given this knowledge of mean proportions and standard errors of proportions, it is now possible to compute confidence intervals.

13.2.4   Interval Estimates

Although the Bureau of Labor Statistics makes point estimates of the unemployment rates for the mass media, in their technical reports they give the information necessary to make interval estimates. As was mentioned earlier, the standard error of the proportion unemployed runs about four-tenths of one per cent (0.004). Assume that in a given month the sample estimate of the proportion unemployed is 0.06. We will use this information to compute a 95 percent confidence interval.

Since the N is large (over 50,000 households in the survey), we can expect the sampling distribution of the proportion unemployed to be approximated by the normal sampling distribution. The sample proportions will distribute themselves normally about their expected value $[E(p) = P]$. Moreover, 95 per cent of the sample p's in the sampling distribution can be expected to fall under the normal curve between standard scores of -1.96 and $+1.96$ since these z scores bound the middle 95 per cent of the normal curve (see Appendix Table B).

Because the parameters P and σ_p are unknown the standard score formula appropriate to this situation is as follows:

(13.5)
$$z = \frac{p - E(p)}{s_p}$$

where $E(p)$ is the expected value of p.

Since the sample proportion (p) is an unbiased estimator of P, $E(p) = P$. Of course, this is the very value that we wish to estimate. Rather than use p as a point estimate of this value, we will make an interval estimate.

Since 95 per cent of the sample proportions in a sampling distribution can be expected to fall within ±1.96 standard errors of P, if we knew P, we could construct a confidence interval around it and see whether our observed sample proportion is in the interval. Since we do not know P what we shall do instead is construct the confidence interval about the observed p (an unbiased estimator of P).

If we take Formula (13.5) for z and manipulate it algebraically to solve for $E(p)$, we can come up with a formula for confidence limits that looks like the following:

$$E(p) = p \pm z(s_p)$$

Let's substitute cl (confidence limits) for the unknown $E(p)$ in the formula because when we use the formula we will actually get lower and upper limits of a confidence interval rather than a single figure representing $E(p)$. The formula for finding the confidence limits of proportions will then be as follows:

(13.6)
$$cl = p \pm z(s_p)$$

To find the confidence limits we substitute the appropriate values of p and s_p into the formula along with the appropriate z scores. For example, to find the 95 percent confidence limits for proportion unemployed we substitute $p = 0.06$, $s_p = 0.004$ and solve the equation with $z = -1.96$ and $z = +1.96$ in order to get the lower and upper limits of the interval:

$$cl_1 = 0.06 - 1.96(0.004) = 0.06 - 0.008 = 0.052$$
$$cl_2 = 0.06 + 1.96(0.004) = 0.06 + 0.008 = 0.068$$

Thus, the 95 percent confidence interval for the proportion unemployed has limits of 0.052 and 0.068. This gives one a somewhat different impression of the unemployment rate than the simple point estimate of 0.06.

13.2.4a   THE MEANING OF CONFIDENCE

But what does the 95 percent probability figure mean? Does it mean that there is a 95 percent chance that the unemployment rate for the total population (the parameter) is between 0.052 and 0.068? This kind of interpretation of the probability associated with a confidence interval is a common error. Actually, the parameter (P) which we are trying to estimate is a fixed value. Its value at a given point in time does not fluctuate. Therefore, the probability that P is between 0.052 and 0.068 is 1 or 0. If it is, in fact, between these two limits, then the probability is 1; if it is not, the probability is 0. What fluctuates from sample to sample is p (the statistic). Since, by necessity, we constructed the interval around p instead of P, the position of the interval with regard to the parameter depends upon the location of the particular p we have used from the sampling distribution. Since 95 percent of the p's in the sampling distribution will be within ± 1.96 standard errors of the population proportion, and we used $\pm 1.96\ s_p$ to construct the interval around p, any p within that range will result in a confidence interval that will include P. Any p not within the range of ± 1.96 standard errors of the parameter will result in a confidence interval that will not include the parameter. Figure 13.1 presents pictorially this situation to help clarify it. Because p_5 in Figure 13.1 was not in the range of ± 1.96 standard errors of P, its interval missed the parameter. Other p's such as p_1, p_2, etc. were within the range, so their intervals do include P.

The 95 percent probability figure thus does not apply to a particular confidence interval (technically speaking, probability never applies to a particular event; it always applies in the long run). What it means is that if we made a very large number of interval estimates (such as those illustrated in Figure 13.1), each based on a sample p, 95 percent of the confidence intervals would include the parameter and only 5 percent of them would miss.

With respect to the confidence interval for proportion unemployed, the 95 percent probability figure means that if a great many independent interval estimates of the proportion unemployed were made instead of just one, 95 percent of those interval estimates would include the parameter and 5 percent would not. The confidence we have in a single interval estimate comes from an intuitive feeling that it is likely to be one of the 95 percent that include the parameter rather than one of the 5 percent that do not.

As we have stated earlier (Section 13.1) the probability figures commonly used for finding confidence intervals are the 95 and 99 percent levels. The 95 percent level leaves 2.5 percent of the area of the sampling distribution in each tail of the curve, and the 99 percent level leaves 0.50 percent in each tail. The assumption is that a sample proportion (p) that deviates from the parameter

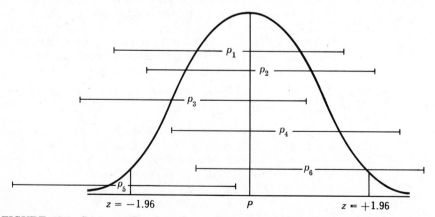

FIGURE 13.1 SAMPLING DISTRIBUTION OF PROPORTIONS ILLUSTRATING DISTRIBUTION OF CONFIDENCE INTERVALS OF SAMPLE p'S AROUND P.

(P) by so much that it falls into one of these tails represents an unusual occurrence. And, as we shall see later (in Chapter 14), a difference great enough to throw the sample proportion into one of the tails is considered a critical difference. Therefore, the tails of the sampling distribution beyond the 95 or 99 percent levels are usually labeled *the critical regions*. The symbol α is sometimes used to refer to this probability – the proportion of the area under the curve that is in the tails of the distribution. We will have occasion to use this notation later in this chapter (Section 13.4.1).

13.2.4b ANOTHER EXAMPLE

Although Morse and Weiss did not report interval estimates of parameters in their article, they did record the necessary data to make such estimates. We will take their finding that 80 percent of the respondents in their sample would continue to work and compute a confidence interval for it. Their $p = 0.80$ and $N = 393$. (Morse and Weiss, 1955:192, Table 1.) Therefore, s_p may be computed as follows:

$$s_p = \sqrt{pq/N} = \sqrt{(0.80)(0.20)/393} = \sqrt{0.00041} = 0.02$$

The 95 percent confidence limits for $p = 0.80$ and $s_p = 0.02$ can then be computed as follows:

$$cl = p \pm 1.96\, s_p$$
$$cl_1 = 0.80 - 1.96\,(0.02) = 0.80 - 0.0392 = 0.76$$
$$cl_2 = 0.80 + 1.96\,(0.02) = 0.80 + 0.0392 = 0.84$$

Therefore, the 95 percent confidence interval has limits of 0.76 and 0.84. Notice that the width of this interval is 0.08 while the width of the interval for the unemployment rate was only 0.016. Although both are 95 percent confidence intervals, the standard error is much smaller for the employment data. Remember that the labor force survey is based on more than 50,000 interviews while the Morse and Weiss data were based on only 393 interviews. The principle involved is that the smaller the standard error, the narrower the confidence interval will be and the more efficient it will be at estimating the parameter. Of course, the obvious way of decreasing the standard error is to increase the sample size. According to **the law of large numbers,** *the larger the sample, the smaller the standard error.*

Both of the examples we have used were 95 percent confidence intervals. Although there is nothing sacred about the 95 percent level, convention says that it is a reassuring level. Under some circumstances, the 95 percent level may not be considered adequate. It may be desirable to use the 99 percent level instead. In such a case it will be necessary to substitute $z = \pm 2.58$ for $z = \pm 1.96$ in the formula for confidence limits (see the normal curve table, Appendix Table B).*

---

*We should note that it is possible to have confidence limits which are not equally distant from the parameter estimate where the sampling distribution of some statistic is not symmetrical (and the excluded proportion under the curve is divided between the two tails of the distribution.) It is also possible to compute only one confidence limit depending upon the shape of the curve and the problem of interest.

We will compute 99 percent confidence limits for the unemployment data for the sake of comparison. Substituting into the formula we get the following:

$$cl_1 = 0.06 - 2.58(0.004) = 0.06 - 0.0103 = 0.050$$

$$cl_2 = 0.06 + 2.58(0.004) = 0.06 + 0.0103 = 0.070$$

Thus, the 99 percent confidence limits are 0.050 and 0.070 as compared to 0.052 and 0.068 for the 95 percent limits. The effect of increasing the probability from 95 to 99 percent was to increase the width of the confidence interval from 0.016 to 0.020. The increase was not drastic because the standard error is very small.

There are no hard and fast rules for determining what is a tolerable width for a confidence interval. It depends upon the research problem and the judgment of the researcher. For example, public opinion pollsters regularly make predictions of election outcomes. They draw samples of registered voters to ask them for whom they will vote. Although most pollsters use quota samples rather than random samples, they make estimates of what their sampling error would be if their samples were random. These rough estimates allow them to compute confidence intervals for their sample estimates, and they do surprisingly well with their predictions.

With two candidates in contention, it is important to predict which candidate will get the majority of the votes. During the 1960 election campaign when John F. Kennedy and Richard M. Nixon were the presidential candidates, the contest was so close that the lower limit of the confidence interval dipped below the 50 per cent mark. Therefore, the pollsters declared the election "too close to call." On the other hand, in the 1964 campaign when Lyndon B. Johnson and Barry M. Goldwater were the candidates, Johnson was so heavily favored the pollsters predicted a Johnson victory without the slightest hesitation. The point estimate for a Johnson victory was about 64 per cent; therefore, it was possible to use 99 per cent confidence limits without fear of dipping below the 50 per cent mark.

In the Nixon–Humphrey campaign of 1968 Humphrey closed the gap in the last days of the campaign, and the result was again in doubt. However, the Nixon–McGovern campaign of 1972 was similar to the 1964 presidential campaign and pollsters predicted a Nixon victory with some degree of confidence (the predictions were generally within 2 or 3 percentage points of the final result, and one poll missed the percentage of the vote that went to Nixon by less than 1 percent).

Ideally, a researcher would routinely use the 99 percent confidence interval and take a sample large enough to minimize sampling error. However, such mundane matters as availability of funds often dictate some compromise in the design of a research project.

13.3 ESTIMATING MEANS

It has become increasingly obvious that illness and health are not solely physical or medical problems. They are only definable and meaningful within a social context. As a matter of fact, from the behavioral perspective, social definitions of illness and health are more predictive than the objective facts. Those who are recognized as being ill by others and see themselves as being ill are likely to play sick roles.* On the other hand, those who are physically ill but are not so defined by others nor by themselves are not as likely to behave as though they are ill.

Sociological variables such as life styles, occupational patterns, and socio-economic statuses are relevant for the study of the incidence of diseases. Also of relevance, sociologically, are studies of attitudes toward medical care and of behavior related to the treatment of illness. The sociologist interested in illness and health has available to him(her) an excellent source of national health data for the United States because of the continuing surveys conducted by the U.S. Public Health Service. The data are from nationwide household interviews conducted by the U.S. National Health Survey. The survey is a continuous one with a random sample of the civilian non-institutionalized population of the United States being interviewed each week.

These data allow for an exploration of sociological and social psychological factors that influence attitudes and behavior towards illness and health (U.S. Dept. of H.E.W., Series B and C). For example, it is known that older people and people from low income families have more health problems than do younger or more affluent people (Lerner and Anderson, 1963; U.S. Senate, 1963). It is reasonable to expect that persons who are ill will seek help from physicians more often than persons who are not ill. Therefore, older persons and those from low income families should have relatively high mean numbers of physician visits per year. We will look at the National Health Survey data to see whether this is the case. In the process of looking at these data, we will also learn the procedures used for estimating population means.

The random sample from which the physician visit data were collected included approximately 115,000 persons from 36,000 households. The population covered by the survey was the civilian population of the United States living at the time of the interviews (U.S. Dept. of H.E.W., 1960).†

The Public Health Service report on physician visits is stated in terms of point estimates, but standard errors are also provided so we will use them to compute interval estimates of population means. A physician visit is defined as "consultation with a physician, in person or by telephone, for examination, diagnosis, treatment, or advice." (U.S. Dept. H.E.W., 1960:49.)

Because of the nature of the data, it is convenient to compare physician visits of persons 65 years of age or older with physician visits of persons of all ages. The mean number of visits per year for persons of all ages was 5.0 while that for persons 65 or older was 6.8 (U.S. Dept. H.E.W., 1960:17, Table 7). We will convert these point estimates to 95 per cent interval estimates of their respective parameters. The appropriate sampling distribution when N is large is the normal distribution and the standard score formula, $z = (\overline{X} - \mu)/s_{\bar{x}}$, can be

---

*For a description of the set of behavior expectations attaching to the "sick role," see Parsons, 1951:436.
†Note that these data do not include persons in hospitals, rest homes, etc. Therefore, there is a bias in favor of the healthier people.

manipulated algebraically to come up with the following formula for confidence limits:

(13.7)

$$cl = \overline{X} \pm z(s_{\bar{x}})$$

where cl is a confidence limit (substituted for μ in the formula), $\overline{X}$ is the sample mean, $s_{\bar{x}}$ is the sample standard error of the mean (our best estimate of the population standard error and close when N is larger), and z is the standard score.

Note the similarity between this formula and the one for finding confidence limits for proportions [Formula (13.6)].

The standard error of the mean number of physician visits for persons of all ages is given as approximately 0.075 and that for persons 65 and over as approximately 0.265 (U.S. Dept. H.E.W., 1960:45–47). We can substitute the means and standard errors into the formula accordingly, and solve for the confidence limits as follows:

For persons of all ages:

$$cl_1 = 5.0 - 1.96(0.075) = 5.0 - 0.147 = 4.85$$
$$cl_2 = 5.0 + 1.96(0.075) = 5.0 + 0.147 = 5.15$$

For persons 65 years of age or older:

$$cl_1 = 6.8 - 1.96(0.265) = 6.8 - 0.519 = 6.28$$
$$cl_2 = 6.8 + 1.96(0.265) = 6.8 + 0.519 = 7.32$$

The 95 percent confidence interval for persons of all ages, then, is 4.85 to 5.15, and that for persons 65 or older is 6.28 to 7.32. Notice that the two intervals do not overlap. This would allow us to say, with some confidence, that the mean number of physician visits per year for persons 65 or older is higher than that for persons of all ages in the United States. These data do tend to support the notion that those with higher rates of illness seek help from physicians more often.

The standard error for persons of all ages is smaller than that for persons 65 or older; consequently, the confidence interval is narrower for the former group. This is reasonable because the sample of persons all ages is larger than the sample of persons 65 or older (the standard deviations may differ as well).

Now compare the mean number of physician visits per person per year of persons from low income families (under $2,000 per year) with that for persons from higher income families ($7,000 or more per year).* The mean for physician visits from low income families was 4.6 and from higher income families, 5.7 (U.S. Dept. H.E.W., 1960:20, Table 11). The standard errors of the means were approximately 0.16 and 0.15, respectively (U.S. Dept. H.E.W., 1960:45–47). The 95 per cent confidence limits may be found as follows:

---

*Note that the comparison of mean physician visits for those 65 and over was with mean physician visits for people of all ages – including people 65 and over. Thus, we were comparing an estimate of a population mean with an estimate of the mean of a subpopulation of that same population. In this case we are working with estimates of means for two independent groups – those from low income families and those from higher income families.

For persons from families with incomes under $2,000:

$$cl_1 = 4.6 - 1.96\,(0.16) = 4.6 - 0.3136 = 4.29$$

$$cl_2 = 4.6 + 1.96\,(0.16) = 4.6 + 0.3136 = 4.91$$

For persons from families with incomes of $7,000 or more:

$$cl_1 = 5.7 - 1.96\,(0.15) = 5.7 - 0.2940 = 5.41$$

$$cl_2 = 5.7 + 1.96\,(0.15) = 5.7 + 0.2940 = 5.99$$

Thus, the 95 percent confidence interval for persons from families with yearly incomes of $2,000 or less is 4.29 to 4.91; and that for persons from families with yearly incomes of $7,000 or more is 5.41 to 5.99. Again, notice that the two intervals do not overlap. This would give us some confidence that persons in the United States from higher income families make more physician visits, on the average, than persons from low income families. This finding is not consistent with our expectation. Although persons from low income families tend to be ill more than persons from high income families, they make fewer physician visits, on the average.

How might we explain the fact that physician visits for older persons occur more frequently than for persons of all ages while physician visits for persons from low income families are fewer than they are for persons from higher income families?

One possible explanation might be that persons from lower income families have less money to spend on medical care than do older persons; however, older persons, too, typically have limited incomes.* Another possible reason might be that older persons have free medical care more readily available to them than do persons from lower income families. But, on the other hand, free medical care is frequently available to persons from lower income families as well. A third possibility is that persons from families with higher incomes make more physician visits than persons from lower income families because they go to physicians for regular medical examinations while lower income persons do not. U.S. National Health Survey data estimate, however, that the mean number of physician visits per year for general checkups is 0.4 for persons from families with less than $2,000 yearly incomes and 0.5 for persons from families with incomes of $7,000 or more (U.S. Dept. H.E.W., 1960:35, Table 26).

Still another possible explanation is that persons from low income differ from persons from higher income families with respect to their attitudes and behavior toward physicians. It may be that persons from higher income families will consult a physician for complaints that are less serious than those necessary to provoke a physician visit by persons from lower income families. If this is the case, the implications for medical care are significant.

These possible explanations deserve further exploration. We will not examine them further here. Our analysis of the data has convinced us that there is a pattern of results that calls for explanation. If we were seriously engaged in

---

*Take note of the fact that the mean physician visits for those 65 and over was the highest of the four means compared, while that for persons from low income families was the lowest of the four means.

research on this subject, we would pursue the answer to our questions by engaging in additional research on the topic. We have satisfied our immediate purpose of demonstrating how parameter estimates may be used to analyze data and of explaining the procedures for making such estimates.

13.3.1   Estimates Based upon Small Samples

Joseph Harry has studied the influence of family life cycle on social participation in a large outdoor leisure organization (Harry, 1970).* He sought to test Wilensky's notion that the high point in social participation is delayed until middle age due to intense involvement in family affairs when children are young (Wilensky, 1966). As one aspect of his study he computed the mean annual attendance at activities of the organization by females whose youngest child was between 0 and 5 years of age and females whose youngest was between 6 and 17. He found that for those 48 females in his sample whose youngest was between 0 and 5 the mean attendance was 3.29 with a standard deviation of 5.06. For those 58 females whose youngest was between 6 and 17 the mean attendance was 6.78 with a standard deviation of 8.99.

Although Harry himself did not make interval estimates of population means, we will use his data to illustrate how such estimates would be made. His data are not ideal for making such estimates because the distributions of attendance are positively skewed. Nevertheless, we will use them for illustrative purposes.

Remember that when sample size is small and we have to use the sample standard deviation to estimate the population standard deviation, the sampling distribution to use is Student's t with $N-1$ degrees of freedom. Therefore, the formula for finding confidence limits is as follows:

(13.8)
$$cl = \overline{X} \pm t(s_{\bar{x}})$$

where t replaces z signifying that the Student's t-distribution will be used as the sampling distribution rather than the normal distribution.

Substituting Harry's data into Formula (13.8) to solve for the 95 percent confidence limits we get the following:

For females with youngest child between 0 and 5:

$$cl_1 = 3.29 - 2.011\,(5.06/\sqrt{48}) = 3.29 = 2.011\,(0.73) = 3.29 - 1.4680 = 1.82$$
$$cl_2 = 3.29 + 2.011\,(0.73) = 3.29 + 1.4680 = 4.76$$

where $t = 2.011$ with 47 df for the 95 percent confidence interval (see Appendix Table C).

For females with youngest child between 6 and 17:

$$cl_1 = 6.78 - 2.003\,(8.99/\sqrt{58}) = 6.78 - 2.003\,(1.18) = 6.78 - 2.3635 = 4.42$$

---

*Harry's data do not represent, strictly speaking, a random sample of the organization he studied; but since he chose to treat them as a sample, we will treat them as such as well and use them to demonstrate how interval estimates can be made from small samples.

BOX 13.2   RECONSIDERING KINDS OF SAMPLES

This is a good time to stop and remind yourself that the sampling distributions we have been using in this chapter to find confidence limits are generated based on the assumption of simple random sampling. The data from the examples we have used were not based upon simple random samples, unfortunately. It is difficult to find examples in sociological literature of confidence intervals computed from simple random samples. Random samples found in this material are more likely to be cluster samples than simple random samples. You should recall from Chapter 11 that cluster samples generally have larger standard errors than simple random samples; therefore, the confidence intervals we have computed in this chapter are probably somewhat narrower than they should be. This fact needs to be considered in interpreting results of the analysis of the data.

$$cl_2 = 6.78 + 2.003\,(1.18) = 6.78 + 2.3635 = 9.14$$

where $t = 2.003$ with 57 df for the 95 percent confidence interval (see Appendix Table C).

The 95 percent confidence limits for mean attendance of females whose youngest child is between 0 and 5, therefore, are 1.82 and 4.76 while those for females whose youngest child is between 6 and 17 are 4.42 and 9.14. Notice that the two confidence intervals overlap, but only slightly.

The fact that the two confidence intervals do overlap slightly complicates the interpretation of the findings. If there was no overlap between the two confidence intervals, we might be willing to conclude that the participation rates for the two categories of women differed. Given that there is overlap, the results should be considered inconclusive. The problem is that with such small samples the sampling error as measured by the standard errors was large. If this aspect of the study were to be repeated with larger samples, the results might be more conclusive.

This example illustrates neatly a possible difficulty of working with relatively small samples. Even though one mean was twice as large as the other, the two confidence intervals overlapped because of the relatively large standard errors. The slight overlap in the two confidence intervals neither gave us confidence that the two sample means estimate two different population means nor that they estimate the same population mean.

13.4   ESTIMATING VARIANCES

Since reading ability correlates with success in and out of school, educators, social scientists, and the public have become concerned about the effectiveness of reading programs in elementary schools. In California reading tests are given periodically to elementary school students, and test results are made public to inform educators and taxpayers of the kind of job their schools are doing. Mean grade placement scores on reading tests are reported for each school for each grade tested. For example, first, second, and third grade students from the 434 elementary schools in the Los Angeles City School District were tested for reading ability in May, 1969.* Mean grade placement scores and percentile scores were computed for each of the three grades for each school in the district. These grade placement scores were then compared with national norms. The national norm for first graders in May of the school year is 1.8, for second graders, 2.8; and for third graders, 3.8.

When brought into comparison with the national norms, grade placement scores varied widely among the 434 schools. In some schools the students in all three grades tested were reading, on the average, well below the national norms. In other schools students in all three grades were reading well above the national norms. In still other schools, students in some grades read above the norms while in other grades they read below.

These reading scores can provide answers to a number of significant questions. Questions such as: How do first graders in Los Angeles City Schools compare with national grade placement norms on reading tests? How do second graders compare? How do third graders compare? Do second graders deviate more from the national norms than do first graders? Do third graders deviate more than second graders? etc.

In addition to these questions about central tendencies, we might wish to raise questions about the variability of grade placement scores from school to school. For instance, we might ask: Do grade placement scores for second graders vary more from school to school than they do for first graders? Do third graders' scores vary more from school to school than second graders' scores?

If these latter questions can be answered in the affirmative, there must be variables operating that account for the fact that the differences between schools (or at least, in their students' reading skills) become more pronounced for students advancing in each grade through the third grade. Of course, at this level of analysis we might not know which variables are responsible for the increasing differences; that would require more intensive study. But, we could establish that such differences exist and that more intensive study of their existence is warranted.

We will use these reading scores as a vehicle for learning how to make interval estimates of variances and of standard deviations. And, in the process, we will see whether there are signs of increasing differences between schools in the reading skills of the students at different grade levels.

13.4.1   Small Sample Estimates

We will start by drawing a simple random sample of 30 schools from the 434, measuring the variability of the grade placement scores for these 30 schools,

---

Los Angeles Times, 16 December, 1969, Part 1, p. 25.

and using these measurements to estimate parameters.* Table 13.1 presents the mean grade placement scores for first, second, and third graders for a random sample of 30 schools.

Using the formula for the sample variance, Formula (12.11), we have computed the variances for the mean grade placement scores for first, second, and third graders. These variances are 0.10, 0.29, and 0.38, respectively. If we wished, we could use these as point estimates of the corresponding parameters. We will, however, make interval instead of point estimates.

TABLE 13.1   READING TEST MEAN GRADE PLACEMENT SCORES FOR FIRST, SECOND, AND THIRD GRADERS FROM A SAMPLE OF 30 LOS ANGELES ELEMENTARY SCHOOLS, MAY, 1969

School Number	First Grade	Second Grade	Third Grade
1	1.4	1.8	2.6
2	1.7	2.7	3.7
3	2.0	3.0	4.0
4	1.5	1.8	2.6
5	1.7	2.7	3.7
6	2.3	2.7	3.7
7	1.6	3.0	3.6
8	1.9	2.6	3.5
9	1.5	1.9	2.7
10	1.4	2.0	2.8
11	1.4	2.1	2.9
12	1.5	2.2	2.8
13	1.6	2.4	3.5
14	1.6	2.8	3.7
15	2.1	3.5	4.0
16	1.5	2.2	3.5
17	1.4	1.8	2.5
18	1.8	2.9	4.1
19	2.2	3.2	3.5
20	2.0	2.8	4.8
21	1.4	2.0	2.1
22	2.2	3.4	3.4
23	1.7	2.2	3.3
24	2.3	3.5	4.0
25	2.4	3.0	3.4
26	1.4	1.8	2.3
27	1.7	2.5	2.9
28	1.6	2.2	2.9
29	1.6	2.1	3.0
30	1.4	1.9	2.7
Total	= 51.8	74.7	98.2

Source: Based upon data from *The Los Angeles Times,* September 30, 1969.

*Although grade placement scores on the reading tests are available for all of the 434 elementary schools in the Los Angeles district, we will work here with data from samples as well as with population data in order to demonstrate how population variances and standard deviations may be estimated.

It was stated in the last chapter (Section 12.5.1) that when the sample size is small—if sample variances are based upon simple random samples from a normally distributed population—they can be expressed as chi-square (χ^2) scores which have a chi-square sampling distribution with $N-1$ degrees of freedom. The formula for converting variances into χ^2 scores was given as follows:

$$\chi^2 = \frac{(N-1)s^2}{\sigma^2}$$

In order to find the lower and upper confidence limits, we will want the inequality:

$$\chi^2_{\alpha/2} < \sigma^2 < \chi^2_{1-\alpha/2}$$

where α (alpha) refers to the area in the tails of the sampling distribution beyond the limits of the confidence interval. Thus, for the 95 percent confidence interval α would equal 0.05. Half of this, or 0.025, would be located in each tail of the sampling distribution as the inequality suggests.

By elementary algebra, the standard error formula for variances, above, can be converted into an expression for finding a confidence interval:

(13.9)
$$\frac{(N-1)s^2}{\chi^2_R} < \sigma^2 < \frac{(N-1)s^2}{\chi^2_L}$$

where the two χ^2 values in the denominator are values read from the table of areas under the chi-square distribution which define the area below which half of α would fall (e.g. χ^2_L) and above which half of α would fall (e.g. χ^2_R).

We will substitute the data for first graders into the inequality and find the 95 percent confidence limits. First, we must find the proper χ^2 values to substitute into Formula (13.9). If we are looking for the 95 percent confidence limits, χ^2_R will be the value that cuts 0.025 off the right tail of the chi-square distribution, with 29 degrees of freedom ($N - 1$). Looking in the chi-square table (Appendix Table D), we find that this value is 45.722. The other value, χ^2_L, will be the one that cuts 0.025 off the left tail of the distribution, with 29 degrees of freedom. This entry will be in the chi-square table under the column headed 0.975 (the area to the right of the given χ^2). The value is 16.047 (see Appendix Table D). We may now substitute these χ^2 values into the inequality along with $N-1=29$ and $s^2=0.10$ as follows:

$$\frac{(29)(0.10)}{45.722} < \sigma^2 < \frac{(29)(0.10)}{16.047}$$

$$\frac{2.92}{45.722} < \sigma^2 < \frac{2.92}{16.047}$$

$$0.06 < \sigma^2 < 0.18$$

Thus, the 95 percent confidence limits for the variance of grade placement scores of first graders are 0.06 and 0.18. Notice that these limits are not symmetrical around the sample variance of 0.10. This is because the chi-square distribution with 29 degrees of freedom is not a symmetrical distribution.

Now we will use the same procedure to find the 95 percent confidence limits for second and third graders for the sample of 30 schools:

For second graders:

$$\frac{(29)(0.29)}{45.722} < \sigma^2 < \frac{(29)(0.29)}{16.047}$$

$$\frac{8.39}{45.722} < \sigma^2 < \frac{8.39}{16.047}$$

$$0.18 < \sigma^2 < 0.52$$

For third graders:

$$\frac{(29)(0.38)}{45.722} < \sigma^2 < \frac{(29)(0.38)}{16.047}$$

$$\frac{11.1}{45.722} < \sigma^2 < \frac{11.1}{16.047}$$

$$0.24 < \sigma^2 < 0.69$$

The χ^2 values in the denominators were the same for these two cases as they were for the first case because in all three cases we were solving for the 95 percent confidence limits with 29 degrees of freedom.

Later we will look at the actual parameters computed from the data for all 434 schools, but, for now, we will look only at these three interval estimates and their possible interpretations. Note that the confidence intervals for first and second graders barely overlap, those for second and third graders have considerable overlap, and those for first and third graders do not overlap at all. We can say with some confidence that there is more school-to-school variability for third graders than there is for first graders in grade placement scores.* Beyond that the results are inconclusive.

13.4.2   Large Sample Estimates

Now we will draw a simple random sample of 100 schools from the population and make interval estimates using large sample procedures. Table 13.2 presents mean grade placement scores for a simple random sample of 100 schools. Using the formula for the sample variance [Formula (12.11)], we find that the variances for first, second, and third graders are 0.09, 0.31, and 0.54, respectively.

In the last chapter (end of Section 12.5.1) we stated that when the size of the sample exceeds 30, the chi-square distribution can be approximated by a normal distribution with $z = \sqrt{2\chi^2} - \sqrt{2(df) - 1}$ and a standard error of 1. Using the facts that $\chi^2 = (N-1)s^2/\sigma^2$ and $df = N-1$, we can substitute into the formula for z above and algebraically set up the following inequality for a confidence interval of the *standard deviation*, given that N is greater than 30:

*One must be careful in making multiple comparisons. When such comparisons are not independent, the usual sampling distributions are not applicable. For example, if we compare first graders with second graders, second graders with third graders, and first graders with third graders, the comparisons are not all independent and the probabilities of occurrence overlap. If variance two (for second graders) is greater than variance one (for first graders), and variance three (for third graders) is greater than variance two, it implies that variance three is greater than variance one. It is proper to compare variance one with variance two and variance one with variance three. These are independent comparisons.

(13.10)
$$\frac{s\sqrt{2N-2}}{\sqrt{2N-3+z}} < \sigma < \frac{s\sqrt{2N-2}}{\sqrt{2N-3-z}}$$

Note that this confidence interval is for the *standard deviation* rather than the variance. However, since the variance is the square of the standard deviation – if we wish to estimate the population variance rather than the standard deviation – once we have established the confidence limits for the standard deviation they may be squared to find the limits for the variance.

To find the 95 percent confidence limits using this inequality, we substitute $z=1.96$ into the formula and the appropriate s. Doing that for the first, second, and third graders' grade placement scores from the sample of 100 schools results in the following:

For first graders:

$$\frac{0.30\sqrt{2(100)-2}}{\sqrt{2(100)-3+1.96}} < \sigma < \frac{0.30\sqrt{2(100)-2}}{\sqrt{2(100)-3-1.96}}$$

$$\frac{0.30(14.07)}{14.04+1.96} < \sigma < \frac{0.30(14.07)}{14.04-1.96}$$

$$\frac{4.22}{16.00} < \sigma < \frac{4.22}{12.08}$$

$$0.26 < \sigma < 0.35$$

or

$$0.07 < \sigma^2 < 0.12$$

For second graders:

$$\frac{0.56(14.07)}{16.00} < \sigma < \frac{0.56(14.07)}{12.08}$$

$$\frac{7.88}{16.00} < \sigma < \frac{7.88}{12.08}$$

$$0.49 < \sigma < 0.65$$

or

$$0.24 < \sigma^2 < 0.42$$

For third graders:

$$\frac{0.73(14.07)}{16.00} < \sigma < \frac{0.73(14.07)}{12.08}$$

$$\frac{10.27}{16.00} < \sigma < \frac{10.27}{12.08}$$

$$0.64 < \sigma < 0.85$$

or

$$0.41 < \sigma^2 < 0.72$$

TABLE 13.2 READING TEST MEAN GRADE PLACEMENT SCORES FOR FIRST, SECOND, AND THIRD GRADERS FROM A SAMPLE OF 100 LOS ANGELES ELEMENTARY SCHOOLS, MAY 1969

School Number	First Grade	Second Grade	Third Grade	School Number	First Grade	Second Grade	Third Grade
1	1.5	2.0	2.6	51	1.6	2.1	3.3
2	1.4	1.9	3.1	52	1.6	1.9	2.9
3	2.6	3.3	4.0	53	1.8	2.7	3.2
4	1.9	2.8	3.7	54	1.4	1.9	2.9
5	1.6	2.3	2.6	55	1.9	2.9	3.8
6	1.7	2.3	2.9	56	1.6	2.0	3.3
7	1.8	2.9	3.8	57	1.4	2.0	2.6
8	1.8	2.6	3.5	58	1.6	1.8	2.3
9	1.6	2.2	2.9	59	2.0	3.4	5.3
10	1.8	3.0	3.7	60	1.7	2.9	3.8
11	1.6	2.6	3.5	61	1.8	2.4	3.4
12	1.4	2.0	2.6	62	1.6	2.5	3.3
13	1.4	1.9	2.9	63	1.6	2.6	3.4
14	1.4	1.8	2.6	64	1.6	2.0	2.3
15	1.5	2.1	2.6	65	1.9	2.6	3.4
16	1.4	1.7	2.1	66	1.5	1.9	2.1
17	2.4	3.0	3.4	67	1.5	1.8	2.6
18	1.5	2.0	2.9	68	2.4	3.8	4.4
19	2.0	2.7	4.5	69	1.5	1.8	2.7
20	2.6	4.0	5.9	70	1.8	3.5	4.5
21	1.7	2.3	3.6	71	1.7	2.4	3.6
22	1.7	2.3	2.9	72	1.5	2.2	3.5
23	1.7	2.3	2.9	73	1.4	1.8	2.5
24	1.4	1.9	2.2	74	1.9	2.6	3.5
25	1.4	1.9	2.1	75	1.6	2.4	3.6
26	1.7	2.4	3.0	76	1.6	2.5	3.0
27	1.7	2.7	3.7	77	1.7	2.6	3.8
28	1.5	2.5	3.5	78	1.7	3.1	4.0
29	1.8	2.9	3.7	79	1.4	2.0	2.1
30	1.8	2.8	3.4	80	1.4	1.7	2.1
31	1.8	2.9	3.7	81	1.7	2.9	4.1
32	2.6	4.0	5.9	82	2.6	3.3	4.0
33	1.6	2.4	3.5	83	1.7	3.6	4.4
34	1.9	3.1	3.7	84	1.7	2.9	2.9
35	1.6	2.7	3.7	85	1.5	1.8	2.6
36	1.6	2.4	3.2	86	1.7	2.3	3.2
37	1.4	1.8	3.0	87	1.5	2.0	3.1
38	2.0	2.8	4.5	88	1.7	3.6	3.3
39	1.5	2.5	3.1	89	1.6	2.7	3.0
40	1.3	2.0	2.4	90	2.3	2.7	3.7
41	1.7	2.5	3.4	91	1.6	2.7	3.6
42	1.6	2.5	3.3	92	1.5	2.1	2.8
43	1.8	2.6	3.4	93	1.8	2.6	3.7
44	1.5	2.0	2.8	94	2.3	4.0	4.7
45	1.6	2.0	3.0	95	1.5	2.0	2.7
46	2.2	3.3	4.2	96	1.4	1.9	2.7
47	2.3	2.7	3.7	97	1.6	2.0	3.0
48	1.8	2.9	3.7	98	2.3	3.0	3.7

(continued)

TABLE 13.2 *(Continued)*

School Number	First Grade	Second Grade	Third Grade	School Number	First Grade	Second Grade	Third Grade
49	1.7	2.6	3.4	99	1.8	2.9	4.0
50	1.7	2.6	3.4	100	2.0	3.0	3.8

Source: Based upon data from *The Los Angeles Times*, September 30, 1969.

Now we will examine these confidence intervals to see whether they allow us to make statements that we were not willing to make on the basis of the confidence intervals that we computed from the sample of size 30. Again, we have evidence that the schools are less variable in grade placement scores for first graders than they are for third graders. The evidence now, however, is somewhat more conclusive than it was for the smaller sample. Furthermore, we are now able to resolve that the scores for second graders are more variable than they are for first graders. The comparison of estimates from a sample of 30 schools with estimates from a sample of 100 demonstrates the advantage that may be gained by taking a larger sample whenever feasible.

In the last chapter (in Section 12.5.1) we presented an alternative way of approximating the sampling distribution of standard deviations when the sample size is large. It was pointed out that if N is very large,* the standard deviation tends to distribute itself normally with a mean of σ and a standard error of $\sigma/\sqrt{2N-1}$. This approximation lends itself to the following inequality for computing a confidence interval for the standard deviation:

(13.11)
$$s\left[1-\frac{z}{\sqrt{2(N-1)}}\right] < \sigma < s\left[1+\frac{z}{\sqrt{2(N-1)}}\right]$$

Although a sample of 100 might better be characterized as moderately large rather than as very large, we will compute 95 percent confidence intervals for the first, second, and third graders using this inequality and compare the results with those previously obtained.

For first graders:

$$0.30\left(1-\frac{1.96}{14.07}\right) < \sigma < 0.30\left(1+\frac{1.96}{14.07}\right)$$

$$0.30(0.8607) < \sigma < 0.30(1.1393)$$

$$0.26 < \sigma < 0.34$$

or

$$0.07 < \sigma^2 < 0.12$$

For second graders:

$$0.56\left(1-\frac{1.96}{14.07}\right) < \sigma < 0.56\left(1+\frac{1.96}{14.07}\right)$$

---

*Large enough so that $(2N-2)/(2N-3)$ is very near 1. Also, z should be small in relation to $\sqrt{2N-3}$.

$$0.56\,(0.8607) < \sigma < 0.56\,(1.1393)$$

$$0.48 < \sigma < 0.64$$

or

$$0.23 < \sigma^2 < 0.41$$

For third graders:

$$0.73\left(1 - \frac{1.96}{14.07}\right) < \sigma < 0.73\left(1 + \frac{1.96}{14.07}\right)$$

$$0.73\,(0.8607) < \sigma_i < 0.73\,(1.1393)$$

$$0.63 < \sigma < 0.83$$

or

$$0.40 < \sigma^2 < 0.69$$

Even though N was not *very* large, the confidence intervals computed by this alternative procedure differ very little from those computed previously. The interpretation of the results would not be changed because of the small differences in confidence limits obtained. The first large sample procedure is generally to be preferred over the alternative just presented because the assumptions behind it are less restrictive.

13.4.3   Comparisons of Sample Estimates and Parameters

Since the data for first, second, and third graders' mean grade placement scores are available for all 434 schools, it is possible to compare the interval estimates with the actual parameters in order to see how good the estimates were. Table 13.3 presents a summary of the parameters and the point and interval estimates of variances from samples of size 30 and 100 using the three different estimation procedures. Also included for the sake of comparison are their population means and their point and interval estimates.

Notice, first, that all of the interval estimates of means actually include the parameters they are intended to estimate; that is, the population means, in every case, are in the intervals. The same cannot be said for the interval estimates of variances. While all of the estimates for second and third graders include their parameters, those for the first graders do not. The interval estimate based on the sample of 30 comes closest to the parameter but does not quite include it.

How might this failure of the intervals for first graders to include the parameter be explained? Could it be a coincidence that both the sample of 30 and the one of 100 produced statistics with unusual values? To test this notion, two other simple random samples of size 30 were drawn and confidence intervals were computed. Both of these likewise failed to include the parameter. The probability of three independently drawn simple random samples producing confidence intervals that miss the parameter is $(0.05)^3 = 0.000125$.

TABLE 13.3 PARAMETERS AND SAMPLE ESTIMATES OF VARIANCES AND MEANS OF GRADE PLACEMENT SCORES

Source of Data	Means	Variances	Interval Estimates of Means	Interval Estimates of Variances
Population	*(Parameters)*			
First Grade	1.68	0.06		
Second Grade	2.46	0.36		
Third Grade	3.22	0.45		
Sample of 30 Schools	*(Point Estimates)*			
First Grade	1.73	0.10	$1.61 < \overline{X} < 1.85$	$0.06 < \sigma^2 < 0.18$
Second Grade	2.49	0.29	$2.29 < \overline{X} < 2.69$	$0.18 < \sigma^2 < 0.52$
Third Grade	3.27	0.38	$2.00 < \overline{X} < 4.54$	$0.24 < \sigma^2 < 0.69$
Sample of 100 Schools				
First Grade	1.72	0.09	$1.66 < \overline{X} < 1.79$	$0.07 < \sigma^2 < 0.12$
Second Grade	2.51	0.31	$2.40 < \overline{X} < 2.62$	$0.24 < \sigma^2 < 0.42$
Third Grade	3.34	0.54	$3.20 < \overline{X} < 3.48$	$0.41 < \sigma^2 < 0.72$
Sample of 100 Schools *Alternative Process*				
First Grade				$0.07 < \sigma^2 < 0.12$
Second Grade				$0.23 < \sigma^2 < 0.41$
Third Grade				$0.40 < \sigma^2 < 0.69$

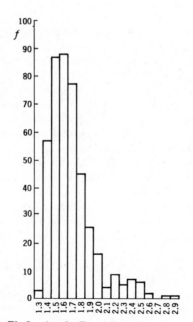

Source: Based upon data from *The Los Angeles Times*, September 30, 1969.

FIGURE 13.2a DISTRIBUTION OF FIRST GRADE MEAN GRADE PLACEMENT SCORES ON READING TESTS FOR A POPULATION OF 434 ELEMENTARY SCHOOLS IN LOS ANGELES (MAY 1969).

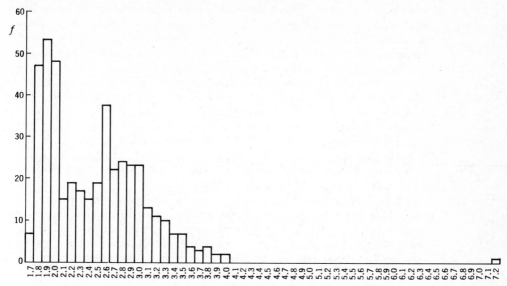

Source: Based upon data from *The Los Angeles Times*, September 30, 1969.

FIGURE 13.2b DISTRIBUTION OF SECOND GRADE MEAN GRADE PLACEMENT SCORES ON READING TESTS FOR A POPULATION OF 434 ELEMENTARY SCHOOLS IN LOS ANGELES (MAY 1969).

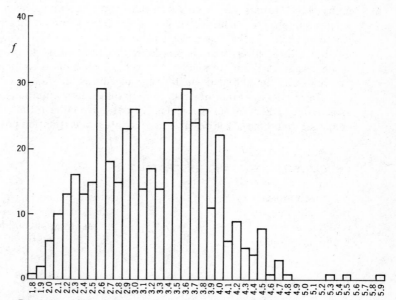

Source: Based upon data from *The Los Angeles Times*, September 30, 1969.

FIGURE 13.2c DISTRIBUTION OF THIRD GRADE MEAN GRADE PLACEMENT SCORES ON READING TESTS FOR A POPULATION OF 434 ELEMENTARY SCHOOLS IN LOS ANGELES (MAY 1969).

There is another explanation that is more feasible. It was stressed in the last chapter (at the end of Section 12.5.1) that sampling distributions of variances are sensitive to the assumption that the population being sampled is normally or nearly normally distributed. To test whether this assumption had been violated by the reading data we computed standard measures of skewness (as in Chapter 5) for the populations of grade placement scores for first, second, and third graders. The skewness for first graders was +1.62, that for second graders, +0.23, and that for third graders, +0.37. A skewness score greater than +1.00 indicates relatively extreme skewness. Extreme skewness was present in the distribution of first grade scores but not for second or third grade scores. These three distributions are graphed in Figures 13.2a, 13.2b, and 13.2c. Apparently, the amount of skewness in the distribution for first graders invalidates the computation of confidence intervals for those data.

This illustrates well the importance of the assumption of a normally distributed population for interval estimates of variances. Violation of the assumption of normality is, by the way, much less serious for the interval estimation of means as evidenced by the fact that the interval estimates of the population mean for first grade data did, in every case, include the parameter being estimated.

13.5 SUMMARY

In this chapter we have studied procedures for making interval estimates of proportions, means and variances. Interval estimates may be computed for many other statistics as well. As a matter of fact, whenever a statistic has a known sampling distribution, it is possible to compute an interval estimate (cf. Noether, 1972). In the chapters that follow we will have occasion to consider a few other interval estimation procedures as they relate to the subject matter being discussed. We will consider, too, the use of interval estimates as an alternative to traditional hypothesis testing procedures.

Interval estimation has received, perhaps, less attention than it deserves as a tool for the analysis of sociological data. We believe that this chapter has demonstrated some of the possible uses of interval estimation procedures.

CONCEPTS TO KNOW AND UNDERSTAND

point estimates
interval estimates
confidence interval
confidence limits

critical regions
standard error of a proportion
the meaning of confidence
efficient estimate

QUESTIONS AND PROBLEMS

1. Explain the difference between a point estimate and an interval estimate of a parameter. Which is preferable? Why?

2. Under what circumstances may the normal distribution be used as an approximation to the binomial probability distribution?

3. What is the relationship between the standard error of the binomial probability distribution and the standard error of a proportion?

4. Explain what the probability figure (*e.g.*, 95 percent) associated with an interval estimate means.

5. What is the relationship between sample size and the magnitude of the standard error?

6. What are the steps that can be taken to decrease the width of a confidence interval?

7. How important is the normal distribution assumption for making interval estimates of variances? What are the consequences of violating the assumption?

8. If you have access to a computer and a data file of sample survey data, select three variables of interest to you. One variable should be dichotomous and another should be interval level. The third variable should be one which would divide the data set into two to four interesting subgroups (e.g. large, medium, small families; sex; age groups; geographic location; etc.). Now, find point and interval estimates of proportions means and variances (or standard deviations) for each of the subgroups defined by the third variable, using the two variables you selected for study. State what each point and interval estimate means in your own words. Then, compare and contrast the various measures across the subgroups. Show your work and indicate the formulae you used in your computations.

9. If you do not have access to a computerized data set, use data from problem 13 in Chapter 12. The U.S. overall percent of persons living in urban places of 2500 or larger in 1970, was 73.5. Using careful random sampling procedures (e.g. using the table of random numbers), draw a sample of size N=10 from the numbered listing of 51 states. Using the percentage urban as an interval level score, compute the sample mean percent urban and set up confidence limits around the mean. Do this for the sample variance in percent urban. Finally, dichotomize the sample data, using the overall U.S. average (73.5) as the cutting point. Any state drawn in the sample with a higher figure call "highly urban" and any state in the sample with a figure at or below the U.S. average call "low urban." Then compute a sample percent urban and set up confidence limits around this percent (or treat it as a proportion). Then, state in your own words what these numbers mean. Compare them to the parameters.

10. Extend problem 9 by computing the same statistics on a larger sized sample as well. Use, say, a sample of size 35. What differences are there in procedure? Justify any similarity or difference in outcome.

11. Again, problem 9 could be extended by drawing random samples from subgroups of states defined by region. Compare and contrast the confidence

intervals and point estimates based upon samples drawn from the separate regions. Census regional classification of states is given below. You may wish to combine these categories into larger categories for purposes of drawing a sample. Numbers refer to line numbers in problem 13, Chapter 12.

North East   (01 through 06)
Middle Atlantic   (07 through 09)
East North Central   (10 through 14)
West North Central   (15 through 21)
South Atlantic   (22 through 30)
East South Central   (31 through 34)
West South Central   (35 through 38)
Mountain   (39 through 46)
Pacific   (47 through 51)

GENERAL REFERENCES

Blalock, Hubert M., Jr. *Social Statistics*. Revised ed. (New York, McGraw-Hill Book Company), 1979.

   Chapter 12 of this book treats clearly the subject of point and interval estimates.

   Blalock argues convincingly that confidence intervals are implicit tests of a whole range of hypotheses.

Walker, Helen M., and Joseph Lev. *Statistical Inference* (New York, Holt, Rinehart and Winston), 1953.

   Chapter 8 deals with inferences concerning variances and standard deviations. This is a topic that is ignored in many other statistics books.

OTHER REFERENCES

Harry, Joseph, "Family Localism and Social Participation," *American Journal of Sociology*, vol. 75, no. 5 (March, 1970), pp. 821–827.)

Labovitz, Sanford, "The Nonutility of Significance Tests: The Significance of Tests of Significance Reconsidered," *Pacific Sociological Review*, vol. 13, no. 3 (Summer, 1970), pp. 141–147.

Lerner, Monroe, and Odin W. Anderson, *Health Progress in the United States, 1900–1960* (Chicago, University of Chicago Press), 1963.

Morrison, Denton E., and Ramon E. Henkel, *The Significance Test Controversy* (Chicago, Aldine), 1970.

Morse, Nancy C., and Robert S. Weiss, "The Function and Meaning of Work and the Job," *American Sociological Review*, vol. 20, no. 2 (April, 1955), pp. 191–198.

Noether, Gottfried E., "Distribution-Free Confidence Intervals," *The American Statistician*, vol. 26, no. 1 (February, 1972), pp. 39–41.

Parsons, Talcott, *The Social System* (Glencoe, Ill, The Free Press), 1951.

Pierce, Albert, *Fundamentals of Nonparametric Statistics* (Belmont, Calif., Dickenson Publishing Co., Inc.), 1970.

U.S. Department of Health, Education, and Welfare, Public Health Service, *Health Statistics from the U.S. National Health Survey*, Series B and C (Washington, D.C., U.S. Government Printing Office).

U.S. Department of Health, Education, and Welfare, Public Health Service, *Health Statistics from the U.S. National Health Survey: Volume of Physician Visits, United States, July 1957–June 1959*, Series B-No. 19 (Public Health Service Publication No. 584-B19, Washington, D.C.), August, 1960.

[U.S. Senate], *Developments in Aging 1959 to 1963: A Report of the Special Committee on Aging, United States Senate* (Washington, D.C., U.S. Government Printing Office), 1963.

Wilensky, Harold L., "Work, Careers, and Social Integration," in *Comparative Social Problems*, S. N. Eisenstadt (ed.) (New York, The Free Press), 1966.

VI Inferential Statistics: Hypothesis Testing

14 The Logic of Hypothesis Testing

Suicide is obviously an individual act with psychological overtones. Emile Durkheim, the famous French sociologist, maintained, however, that the regularity and predictability of suicide rates over time could not be explained by psychological variables. He was convinced that suicide rates were explainable, rather, in terms of "social facts." He said,

> If, instead of seeing in them only separate occurrences, unrelated and to be separately studied, the suicides committed in a given society during a given period of time are taken as a whole, it appears that this total is not simply a sum of independent units, a collective total, but is itself a new fact *sui generis*, with its own unity, individuality, and consequently its own nature—a nature, furthermore, dominantly social (Durkheim, tr. 1951:46.).

With the collective nature of the suicidal act in mind, Durkheim attributed the regularity of its rates for various populations to the social fact of group solidarity, or the lack thereof. Thus he theorized that people lacking support from group solidarity are most vulnerable to suicide. His theory did not seek to explain the dynamics of individual suicides; it sought, rather, to explain suicide *rates* in terms of the differential vulnerability of various cohorts of people.

Although Durkheim's work was completed more than 75 years ago, it is considered a sociological classic because it illustrates the proper relationship between theory and data. In spite of soft spots in his methods, his general theoretical notions about suicide as a sociological phenomenon are still relatively sound.*

From his theory Durkheim derived some specific hypotheses about the relationship between social solidarity and suicide rates. Even though he never

*Douglas, 1967, critiques Durkheim's work and reviews further studies of suicide.

defined social solidarity in a rigorous manner, he did indicate various indices of it that could be observed and measured. For instance, he felt that social solidarity varied from one religious persuasion to another. He reasoned that social solidarity was highest among Jews, next among Catholics, and lowest among Protestants. He attributed these differences to differences in the degree to which the lives of individual members were dominated by their religions. Because Protestantism allowed for more free inquiry and placed more responsibility on the shoulders of the individual, it fostered less social solidarity than the other two. Accordingly, Durkheim hypothesized that suicide rates would be highest for Protestants and lower for Catholics and Jews. In order to test his hypothesis, he examined suicide statistics for various European countries. In general, he found that his hypothesis was supported by the data. For example, in the states of Germany suicide rates varied in direct proportion to the number of Protestants and in inverse proportion to the number of Catholics (Durkheim, tr. 1951:153). Moreover, those European countries that were predominantly Catholic (*e.g.*, Portugal, Spain, and Italy) had low suicide rates as compared with high rates for predominantly Protestant countries, while the rates for mixed Catholic–Protestant countries were intermediate (Durkheim, tr. 1951:152). When the rates of Protestant, Catholic, and Jewish groups were compared, Protestant rates were consistently higher than those of the other two, and Jewish rates were generally lower than those for Catholics (Durkheim, tr. 1951:155).*

Another index of social solidarity examined was marital status. Durkheim reasoned that the married enjoyed more social solidarity than the unmarried or widowed; thus, suicide rates would be lower for them. Again, the data tended to support his hypothesis. Rather consistently, for each age category, married persons had lower rates than unmarried or widowed persons (Durkheim, tr. 1951: 176–177). Furthermore, suicide rates for married persons with children were consistently lower than those for married persons without children (Durkheim, tr. 1951:186 ff.).

Durkheim went on to examine several other variables which he took to be indices of social solidarity, in each case comparing suicide rates from several different sources. The data generally supported his theory.

The sketchy, incomplete picture of Durkheim's study of suicide presented here obviously does not do justice to the man and his work. It is sufficient, however, to show how he related his theory to data.†

Durkheim's basic approach was to develop theory to account for the regularity and predictability of suicide rates in Europe in the latter part of the 19th century. His major explanatory concept was social solidarity. He *selected* a number of measurable variables to be indices of social solidarity, *deduced* a number of specific hypotheses relating these indices to suicide rates, *gathered* all available data bearing on his hypotheses, and *examined* them to see whether

---

*It should be noted that Durkheim's statistics relating suicide rates and religion were for countries or areas of countries and not for individuals. If predominantly Protestant countries have high suicide rates, it does not necessarily follow that those who are committing suicide are Protestants. (*See also* Robinson, 1950.)

†Although the statistical techniques known to us were not available to Durkheim, on a raw level he duplicated the reasoning underlying modern statistics. Sir Francis Galton invented correlation earlier (Galton, 1886), but the technique was not generally known nor understood at the time that Durkheim conducted his study.

they lent support to his predictions. He *concluded* that the data, on the whole, supported his hypotheses and theory. Therefore, he offered his theory as a fruitful explanation of suicide rates.

The point of this exercise in social science was to develop a theoretical explanation for a social phenomenon useful for predicting the same class of phenomena in the future. Durkheim did not *prove* his theory any more than any scientific theory is ever proven. He did demonstrate, however, that his findings were useful for predicting suicide rates. This is the way science uses theory. A theory is never proven; it is demonstrated to be useful or not useful, and is used or revised. If a competing theory is developed which predicts better, it is substituted for the previous theory and is used until it is replaced, in turn, by another theory that is an even better predictor.

The process of developing theoretical explanations of social behavior is what sociology is all about. Science is, after all, a continual interplay between theory and data. The examination of data in a systematic manner gives rise to theory; the theory gives direction to the collection of new data; and new data either give additional support to the theory or contribute to its refutation. The game of science is concerned with the selection of the most useful theory from a number of competing ones. The *application* of scientific knowledge by practitioners and policy makers involves the practical use of the most fruitful theory currently in vogue.

When Durkheim developed his theory he had in mind an explanation that would transcend particular populations. He did not, however, use sampling techniques nor did he worry whether his data were representative of either general or special populations. The data he used were descriptive of specific geographic areas at specific points in time. He did examine sets of data descriptive of a number of *different* special populations. In effect, he replicated his hypothesis tests. A very important principle of science is that hypothesis tests should be repeated independently a number of times to gauge their soundness. The impressive aspect of Durkheim's work was the consistency with which these separate sets of data upheld his hypotheses.

14.1   STATISTICS AND HYPOTHESIS TESTING

The concepts that appear in sociological theories are not usually defined in measurable form. They are generally defined in terms of other concepts equally abstract. When such is the case, the hypotheses linking the concepts are not directly testable. The usual procedure is to specify measurable indices of the concepts, frame hypotheses relating the indices, and test these **"working hypotheses"** against empirical data.

For example, Robert E. Clark conducted a study designed to test the **general hypothesis** that incidence of mental disorders varies with occupational status (Clark, 1949). As indices of mental disorders Clark used diagnostic categories assigned to patients in mental hospitals in the Chicago area. He looked separately at rates for alcoholic psychoses, senile psychoses, paresis, manic-depressive psychoses, and schizophrenia among patients whose occupations were known. He used two indices of occupational status: a measure of the prestige of the occupation on the North–Hatt scale,* and the median income for the occupation in the Chicago area at the time of the study.

From the one general hypothesis Clark framed several working hypotheses, relating his indices of mental disorders to his indices of occupational status. When he tested his working hypotheses against the data, he found that his general hypothesis had to be qualified. While alcoholic psychoses, senile psychoses, paresis, and schizophrenia were inversely related to occupational status (the higher the occupational status, the lower the incidence of the disorder), manic-depressive psychoses were unrelated (Clark, 1955:440). Although Clark's substantive findings are interesting in themselves, we are concerned primarily with the procedures used to test the general hypothesis. Clark selected indices of his theoretical concepts and recast his general hypothesis in terms of indices, thus deriving working hypotheses; then he subjected his working hypotheses to empirical test and, on the basis of these tests, drew conclusions about the general hypothesis.

When population data are available, the indices serve as parameters; and decisions about hypotheses can be made by examining the parameters. When Durkheim compared suicide rates in Catholic Bavaria with those in Protestant Prussia he merely had to take note of the fact that the Prussian rates were higher. Since these rates were parameters for his populations, the differences between them were actual differences (assuming that the data were error free). Therefore, descriptive statistics allow for direct tests of hypotheses without further complications.

Unfortunately, population data are not usually available. We are faced with the necessity of examining sample data and making generalizations about the population. The procedure for testing hypotheses with sample data is as follows: (1) *specify indices of the concepts included in the general hypothesis*, (2) *derive working hypotheses which link the indices*, (3) *draw a sample of data from the population to which the hypotheses apply*, (4) *use statistical techniques to analyze the sample data*, (5) on the basis of that analysis, *decide whether the data support the working hypotheses*, and finally, (6) *decide whether the general hypothesis fruitfully describes the population*.

When hypotheses are tested using sample data, the complication introduced is that parameters must be estimated, since they cannot be examined directly. For example, if Durkheim's comparisons of suicide rates in Catholic Bavaria and Protestant Prussia had been based on sample data rather than population data, he would have been faced with the necessity of estimating suicide rates from his sample data, and then deciding whether the estimated parameters actually differed.

We found in Chapter 13 that estimating parameters involves the use of probability theory applied to sampling distributions. As you will see later, there is a close relationship between interval estimates of parameters and statistical tests of hypotheses. The difference between them is a difference in orientation rather than kind.

*For a description of the North-Hatt Scale see Reiss, 1961.

> **BOX 14.1** SAMPLING DISTRIBUTION
>
> If the concept of the sampling distribution is not yet quite clear to you, perhaps you should go back and review Chapter 12, particularly the units under Section 12.2. An understanding of the concept of sampling distribution is essential to the discussion that follows.

14.1a The Statistical Hypothesis. When we use sample data to test hypotheses, it is necessary to introduce a third type of hypothesis—the **statistical hypothesis.** *We start with a general hypothesis, which we translate into working hypotheses, and from our working hypotheses we derive statistical hypotheses.* Statistical hypotheses make statements about population parameters, but are tested by examination of statistics computed from sample data. As a result of the outcomes of tests of statistical hypotheses, we decide what conclusions about the working hypotheses are warranted and, in turn, these decisions help us to make decisions about the general hypothesis.

Again, we will rely heavily on sampling distributions to help us make decisions. The questions we will ask, however, will be somewhat different than those raised in estimating parameters. We will examine the following questions: Is it reasonable to conclude that the statistic we have computed from the sample is an estimate of a specific, given parameter? Are the statistics we have computed separate sample estimates of a common parameter or do they estimate different, distinct parameters? In each case primary concern will be with making decisions about hypotheses that refer to parameters or to relationships between parameters. In classical hypothesis testing we must choose between two competing hypotheses.

14.2 TESTING STATISTICAL HYPOTHESES

Thus far our discussion of hypothesis testing has been fairly abstract. It might be helpful at this point to take a concrete example, run through the process involved in testing an hypothesis, then analyze the procedure involved. In the process of presenting the example we will introduce the concepts that play an integral part in the testing of statistical hypotheses.

According to folk lore, "Birds of a feather flock together." However, there is as well a contradictory assertion that "opposites attract." Sociologists have conducted numerous studies on the factors involved in the selection of mates and friends. They have generally found more evidence to support the first contention that individuals select others with similar backgrounds, interests, attitudes, and values as their mates or friends (Newcomb, 1960). Although the proof for the opposite reasoning is less convincing, there are some sociologists who support the notion that opposites do attract (Winch, 1955; Ktsanes, 1955; but also see Bowerman and Day, 1956).

Several years ago Herman Loether made a study of the effects of propinquity (nearness) and homogeneity (similarity of personal characteristics and roles) on the choosing of "best buddies" in the military (Loether, 1960).* Questionnaires were administered to 182 men stationed at a radar site. Among other things, the respondents were asked to name their "best buddy" on the site. Best buddies were listed by 126 of the respondents. The assumption was made that the other 56 did not write down their best buddies' names because they did not have any. The best-buddy choices of the 126 were then analyzed for any evidence of propinquity and homogeneity (rather than a random choice process). It was theorized that Air Force men would select as their best buddies men with whom they had come into close contact and with whom they had shared similar personal characteristics and roles.†

One aspect of the study was to find out whether the subjects tended to choose their best buddies from within their own work sections. Sufficient data were available to analyze 15 of the 20 work sections at the base. Table 14.1 summarizes these data.

The general hypothesis was recast into a number of working hypotheses relating the indices of propinquity and homogeneity to best-buddy choices. These hypotheses, in turn, were reformulated as statistical hypotheses and subjected to test. We will limit ourselves to considering only one of the working

TABLE 14.1   BEST-BUDDY CHOICES BY WORK SECTION

Work Section	Percent of Total Population	Number of Choices	In-Group Choices	Out-Group Choices
1	5%	7	3	4
2	4	3	1	2
3	9	7	5	2
4	2	3	1	2
5	7	10	9	1
6	6	6	4	2
7	4	6	5	1
8	9	13	13	0
9	24	35	25	10
10	6	7	4	3
11	7	9	3	6
12	2	4	0	4
13	1.7	2	0	2
14	1.7	2	0	2
15	4	5	2	3

Source: Loether, 1960:20, Table 1.

*Propinquity has been studied numerous times as a factor in friendship formation (*cf.* Byrne, 1961). These revised excerpts from Loether (1960) are reprinted by permission of Sage Publications, Inc.
†We will assume that the data from the study constitute simple random samples in order to use them for this example.

hypotheses, namely: there is a positive relationship between common work group membership and best-buddy choice. Since data were available for fifteen different work sections, it was possible to derive fifteen statistical hypotheses from the one working hypothesis. For example, one hypothesis dealt with predicting the outcome of best-buddy choices in work section 3. In a sense, the tests of these fifteen statistical hypotheses represented fifteen independent tests of the working hypothesis: that there is a positive relationship between common work group membership and best-buddy choice. The procedure used is not quite the same as that of drawing separate and independent samples and replicating the study; rather, it is a process of dividing the sample into subsamples and running separate tests of the working hypothesis on the subsamples. This procedure is sometimes called pseudoreplication. Ideally, pseudoreplication involves random selection of the subsamples from the overall sample in order to make multiple estimates of relevant parameters. In this case the subsamples were selected on the basis of work group affiliations of respondents.*

We will focus our attention for the time being on work section 3. The workers in section 3 constituted 9 percent of all the men at the base. Seven members of work section 3 made best-buddy choices—5 of fellow workers and 2 of men from other work sections. However, this particular outcome is only one of several possible outcomes for a set of seven best-buddy choices. For example, there might have been 4 in-group and 3 out-group choices, or 3 in-group and 4 out-group choices. As a matter of fact, the following range of choices is possible:

> 7 in-group and 0 out-group
> 6 in-group and 1 out-group
> 5 in-group and 2 out-group
> 4 in-group and 3 out-group
> 3 in-group and 4 out-group
> 2 in-group and 5 out-group
> 1 in-group and 6 out-group
> 0 in-group and 7 out-group

In a random sample of seven choices it would be possible to get any one of these eight distinct in-group—out-group choice patterns. Since in Loether's survey only 9 percent of the men at the installation belonged to work section 3 and 91 percent didn't, each of the eight possibilities would not be expected to occur with equal frequency. If the men choosing best buddies were asked to disregard work-section membership, it would be more likely that a larger number of them would choose men outside their own work section, since then there would be so many more men from which to choose best buddies. The question we need to consider is, how probable is the particular combination of in-group—out-group choices observed for section 3? If we can establish that a random choice process (i.e., disregarding work-section membership) did not likely account for the 5 in-group and 2 out-group best-buddy choices, then we will have some evidence for the contention that these men had selected their fellow workers on the basis of propinquity.

*For an extended discussion of pseudoreplication see Finifter, 1972.

14.2.1   The Null Hypothesis

Framing a statistical hypothesis in a positive manner, we would come up with some such hypothesis as the following: There is a significant tendency for members of work section 3 to choose their best buddies from among their fellow workers. The kind of sample data we would consider as evidence of a significant tendency would have to be stated explicitly so that we could test the hypothesis.

Since statistical inference is based upon probability theory, tests of statistical hypotheses are probabilistic rather than absolute. It is not possible to prove or disprove statistical hypotheses in an absolute sense. The best that can be achieved is an estimate of their truth or falsity.

It so happens that the rejection of a statistical hypothesis is much more clear-cut than is its acceptance. Therefore, the usual procedure is to frame a statistical hypothesis contrary to that which we are hoping to prove. Such an hypothesis is known as a **null hypothesis.** If sample data warrant rejection of the null hypothesis, that is regarded as evidence for its alternatives—those our theory predicted and those we proposed as explanations in the first place. The null hypothesis gets its name from the fact that it is the hypothesis to be nullified by statistical test.

The advantage of using the null hypothesis is that it serves as a basis for selecting a specific sampling distribution, that is, the sampling distribution that would be found if the null hypothesis were, in fact, true. This sampling distribution is then used to determine whether sample data warrant rejection of the null hypothesis in favor of some set of alternatives to it.

As concerns work section 3 we wish to demonstrate that there is a tendency for members to choose best buddies from within the group. Since work section 3 constitutes 9 percent of the total population of the air site, if names of best buddies were selected at random, we would expect about 9 percent of them to be those of members from work section 3. The null hypothesis could be formulated as follows: The proportion of in-group best-buddy choices will not differ significantly from 9 percent of the total number of choices made by members of work section 3. If this null hypothesis were true, we would have a sampling distribution with an expected value of 9 percent in-group choices. What we do is assume that the null hypothesis is true, generate a sampling distribution from the null hypothesis, draw a random sample, collect the relevant sample data, compute the relevant statistic, and decide whether it is reasonable to assume that the statistic came from the given sampling distribution. If the probability that the statistic came from the given sampling distribution is as small as or smaller than some predetermined level, we reject the null hypothesis in favor of its alternatives. If the probability is not as small as or smaller than the predetermined level, we fail to reject the null hypothesis.

Notice that we *fail to reject* the null hypothesis rather than accept it. If we accepted the null hypothesis, we would be saying, in effect, that it is true. However, if we fail to find reason to reject the null hypothesis, it does not necessarily follow that it is true. We are saying, rather, that the data we collected did not provide us with sufficient basis for concluding that the null hypothesis is false. Perhaps, our data collection was merely inadequate.

By the same reasoning, if we do reject the null hypothesis, it does not follow that we accept its alternatives. All we imply by rejecting the null hypothesis is that some set of its alternatives is more probable than the null hypothesis itself.

Note, also, that both the null hypothesis and its alternatives apply not to sample data but to population data — not to statistics but to parameters. We test the null hypothesis with sample data and generalize from statistics to parameters.

14.2.2   One-Tailed and Two-Tailed Tests of Hypotheses

If the null hypothesis claims that the proportion of in-group choices will not differ significantly from 9 percent, then the alternative hypothesis is that it will differ. Symbolically, the null hypothesis and the set of its alternatives can be stated as follows:

$$H_0 : P = 0.09$$

$$H_a : P \neq 0.09$$

where H_0 is the null hypothesis; H_a, the set of alternatives; and P the proportion of in-group choices in the work section 3 population.

Actually, the alternatives to the null hypothesis consist of a whole range of possible hypotheses that can be divided meaningfully into two distinct classes. First, there is the set for which the proportion of in-group choices is greater than 9 percent (this is, in fact, the one for which we are interested in finding evidence). We might call this set Alternative 1 and represent it symbolically as follows:

$$H_1 : P > 0.09$$

Secondly, there is a set for which the proportion of in-group choices is less than 9 percent. We might call this Alternative 2 and represent it symbolically as follows:

$$H_2 : P < 0.09$$

When a null hypothesis has two distinct sets of alternatives as this one does, *the test is called a two-sided or* **two-tailed test** of the hypothesis. This terminology is used because each set of alternatives to the null hypothesis is represented by a separate side or tail of the sampling distribution.

We are not interested in both sets of alternatives to the null hypothesis for the test of work section 3 data. We are interested in demonstrating that members of work section 3 have a tendency to make *more* in-group choices than chance would lead one to expect. We have labelled this particular set of outcomes Alternative 1 (H_1). Since we defined the null hypothesis earlier as that hypothesis that includes all except that which we are interested in demonstrating, the appropriate null hypothesis would not be $H_0 : P = 0.09$; rather it would be $H_0 : P \leq 0.09$. The null hypothesis and its set of alternatives appear as follows:

$$H_0 : P \leq 0.09$$

$$H_1 : P > 0.09$$

When there is only one set of alternatives to the null hypothesis as there is in this case, *the test of the null hypothesis is said to be a one-sided or* **one-tailed test.** Here we anticipate a significant deviation from chance in only one tail of the sampling distribution. The null hypothesis can only be rejected if the sample data produce a sample proportion of in-group best-buddy choices enough larger than the hypothesized 9 percent to meet some predetermined criterion for rejection. If it turned out that the sample proportion of in-group choices were considerably smaller than 9 percent, it would not be possible to reject the null hypothesis because that possibility is included as part of the null hypothesis.

Whenever research is done to test theory, the theory will generally suggest which set of alternatives to the null hypothesis the sample data are expected to support. The appropriate null hypothesis will be for a one-tailed test. Whenever exploratory research is undertaken, and there is no theory to suggest direction of relationship, the two-tailed test of the null hypothesis is ordinarily employed. As we shall see later, the one-tailed test enjoys certain advantages that make it the preferred test. The two-tailed test should be reserved for those cases where the one-tailed test is not feasible.

14.2.3   Significance Levels and Critical Regions

Once we have decided upon a one-tailed or a two-tailed test and have framed the null hypothesis and its alternatives, we must decide on criteria for rejecting the null hypothesis.

In order to establish criteria for rejecting the null hypothesis, we must specify the applicable sampling distribution. In order to compute a sampling distribution, we need a specific value of P. The null hypothesis for the one-tailed test, however, specifies a whole range of possible values (e.g. those equal to or smaller than .09). However, if we would use the limiting value (in this case the highest value, .09, included in the null hypothesis) and find that we could reject the null hypothesis for this limiting value, then we would automatically know that we could have rejected any other hypothesis included in the one-sided null hypothesis because the sample outcome would be shown to be even less likely for these more extreme hypotheses. On the other hand, if the sample outcome is not deemed unusually rare when testing the limiting value of the null hypothesis, then, at least for one value included under the one-sided null hypothesis, the hypothesis could not be rejected. The procedure, then, is to use the specific limiting value under the one-sided null hypothesis to construct the appropriate sampling distribution.

For work section 3 data, the appropriate sampling distribution is the binomial probability distribution with an expected value of 9 percent in-group choices. This expected value is determined by the null hypothesis. The value is set at 9 percent because that figure represents the percentage that work section 3 members constitute of the total population (see Section 14.2.1). The specific shape of the sampling distribution will depend upon the size of the sample from which the statistic is computed. If we can specify the sample size before data are collected, we can determine exactly which outcomes will justify rejecting the null hypothesis. If sample size depends upon the outcome of data collection, then we must set up the criteria for rejecting the null hypothesis in a more general way. The latter took place when the criteria were established for rejection of the null hypotheses of the best-buddy study. We had decided, therefore, that any sample *outcome* so infrequent as to occur by chance in the specified sampling distribution 5 percent of the time, or less, would warrant rejecting the null hypothesis. This 5 percent figure is known as a level of significance – in this case, *the 5 percent level of significance.*

When we choose to use the 5 percent level of significance as a criterion for rejecting the null hypothesis, we are saying, in effect, if the outcome we observe is so unusual that it would be likely to occur in the sampling distribution less than 5 times out of 100 by chance, then we will consider the rejection of the null hypothesis warranted.

There is nothing sacred about the 5 percent level of significance (Skipper, *et al.*, 1967; Labovitz, 1968). It has merely been defined by convention to be a reasonably rare, chance, occurrence. The more stringent 1 percent level of significance is also used in sociology although it is less frequently used than the 5 percent level. It is reasonable to assume that the popularity of the 5 percent level relates to the relatively unsophisticated measurement levels of sociological data. We will have more to say later about the choice of an appropriate level of significance.

Once we have decided upon the level of significance necessary to reject the null hypothesis we are ready to sample and collect data.

We will now turn to the data from work section 3 (Table 14.2). The sampling distribution has been worked out using procedures for computing a binomial sampling distribution discussed in Chapter 12. Here $P = .09$, $Q = .91$, and the sample size is $N = 7$, since that is the number of best-buddy choices made in work section 3. Figure 14.1 is a graphic representation of the same distribution.

The set of alternatives to the null hypothesis is in one tail. Our only interest is whether there will be more in-group choices than chance would lead one to expect. Since we have decided to use the 5 percent level of significance, we must

TABLE 14.2   SAMPLING DISTRIBUTION OF BEST-BUDDY CHOICES IN WORK SECTION 3 ($N = 7$, $P = 0.09$)

In-Group – Out-Group Choices	Probability*
	In Critical Region Pr = 0.019+
7 in-group, 0 out-group	0.00000005
6 in-group, 1 out-group	0.000003
5 in-group, 2 out-group	0.0001
4 in-group, 3 out-group	0.00173
3 in-group, 4 out-group	0.0175
	Boundary of Critical Region for Pr $\leq$ 0.05
2 in-group, 5 out-group	0.1061
1 in-group, 6 out-group	0.3578
0 in-group, 7 out-group	0.5168

Source: Based on data from Loether, 1960.
*These probabilities are approximate due to errors in rounding.

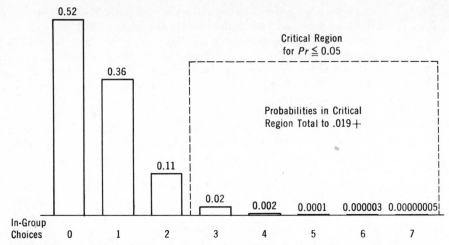

FIGURE 14.1 Histogram of Probability Distribution of In-Group and Out-Group Best-Buddy Choices with $N = 7$ and $P = 0.09$.

examine the sampling distribution and determine how many extreme outcomes are likely to occur by chance 5 percent or less of the time. It is obvious from Table 14.1 that the probability of 3 or more in-group choices out of 7 meets this criterion $(0.0175 + 0.00173 + 0.0001 + 0.000003 + 0.00000005 = 0.019+)$. If we included the probability of 2 in-group choices, we would exceed 5 percent. Therefore, if we find 3 or more in-group choices in the sample of 7 work section 3 members, we will reject the null hypothesis and fail to reject its alternatives.

All outcomes that fall in the area of the tail of the distribution where rejection of the null hypothesis is warranted are said to be in the **critical region.** Thus, *the critical region is that area in the tail (or tails) of a sampling distribution that meets the preestablished criteria for rejection of the null hypothesis.**

BOX 14.2 CHECKING BACK

Let us pause for a moment and take stock. Some rather abstract ideas have been presented to you thus far in Chapter 14. We have discussed the null hypothesis and its alternatives, one- and two-tailed tests of hypotheses, critical regions, and levels of statistical significance. If all of these ideas are still not clear to you, before you proceed further you had better go back and reread the chapter up to this point. In the next few pages we will be discussing some more very important abstruse concepts. It is advisable to read this entire chapter slowly and thoughtfully. If you can master the basic logic involved in testing statistical hypotheses, the rest of the volume will come easily to you.

---

*Actually, the binomial distribution is a discrete rather than a continuous distribution; therefore, the critical region is made up of a finite set of points (outcomes) rather than an area.

When a one-sided test is called for, the critical region is located in one tail of the sampling distribution. When a two-sided test is in order, the critical region is divided in half and located in both tails of the sampling distribution. In effect we are dividing up the area under the sampling distribution into an area where observed sample outcomes favor the null hypothesis and another area (or areas) in which sample outcomes favor the alternative hypothesis.

If we wished to test the null hypothesis regarding work section 3 at the 1 percent level rather than the 5 percent level, we would require 4 or more in-group choices out of 7, since that critical region would include 0.18 percent or 0.0018 of the area in the tail (this in addition to the term for 3 in-group and 4 out-group choices would lead to a critical region that would exceed 1 percent).

14.2.4 Decision Errors

The *decision making process in scientific research is based upon the* **principle of uncertainty.** Whenever we make a decision about a null hypothesis, whether the decision is to reject or fail to reject, we make our decision in the face of uncertainty. We set up criteria for the rejection of the null hypothesis and reject if the criteria are met, or fail to reject if the criteria are not met. If we reject the null hypothesis, we are saying, in effect, that we believe it to be false. It is possible, however, to observe an outcome that warrants rejection of the null hypothesis according to our criteria even though the null hypothesis is true. Any outcome included in the sampling distribution can occur as a result of chance (that is what the sampling distribution is all about). Some outcomes are, however, much less probable than others. It is these less probable outcomes that we include in our critical regions. When we get an outcome in the critical region and reject the null hypothesis, we are saying that we choose to believe that we got that outcome, not as a result of chance, but because the null hypothesis was false. This could, perhaps, be due to the systematic operation of some variable that interests us. If we consistently reject the null hypothesis whenever we reach the critical region, over the long haul, we will be wrong 5 percent of the time in rejecting it at the 0.05 level of significance. We never know for sure whether a particular outcome was due to chance. That is where the uncertainty principle enters the picture. If the observed outcome falls in the critical region, and the critical region only includes 5 percent of the total possible occurrences in the sampling distribution, it is reasonable to suppose that we did not get such a result by chance. For example, of the 7 best-buddy choices made by members of work section 3, 5 were in-group choices. Since this outcome is well within the critical region ($Pr < 0.0001$), we reject the null hypothesis and conclude that there is a tendency for members of that work section to choose each other as best buddies.

14.2.4a Type-I Error. Our conclusion might be *wrong;* the null hypothesis might be *true.* If this is the case and we have rejected the null hypothesis because our criteria for rejection were satisfied by the data, then we made what is known as a **type-I error.** *A type-I error is committed if the null hypothesis is rejected when it is, in fact, true.* Type-I errors can be committed because it is possible to observe sample outcomes that diverge widely from the hypothesized parameter because of the vicissitudes of random sampling variation.

The *potential probability* of committing a type-I error is no greater than the

level of significance adopted. If the test is made at the 5 percent level of significance, then the potential probability of a type-I error is 5 percent. This is so because we reject the null hypothesis whenever we observe a sample outcome that
falls in the critical region; and indeed, 5 percent of the time the sample outcomes
can be expected to fall in the critical region even when the null hypothesis is true.

Five percent is the maximum possible type-I error but is not the error
that will necessarily be risked each time the null hypothesis is tested. In the
test regarding work section 3, 5 out of 7 in-group choices had a probability of
occurring by chance of less than 0.0001; thus, for this particular test, that figure
represents the probability of a type-I error. The risk of a 5 percent type-I error
would result from a finding that actually was right on the boundary of the critical region and therefore had a probability of 5 percent.

14.2.4b   Type-II Error.   Whenever the sample outcome does not fall into
the critical region, in accordance with our predetermined rules for decision making we will fail to reject the null hypothesis. If the sample outcome falls short of
the critical region because the null hypothesis is true, then we have been correct
in failing to reject it. *If, on the other hand, the null hypothesis is false and we fail
to reject it, then we are guilty of a type-II error.*

Notice that the **type-II error** arises whenever the sample outcome fails to
reach the critical region, while the type-I error arises whenever the sample outcome falls in the critical region. It should be obvious that our decision rules do not
subject us to risks of type-I and type-II errors simultaneously. Thus, when we
reject the null hypothesis, we are subject to type-I error; when we fail to reject
the null hypothesis, we are subject to type-II error. (Figure 14.2 diagrams the
relationship between decisions, type-I, and type-II errors.)

We found that the maximum possible type-I error was equal to the area
(or proportion of outcomes) in the critical region—5 percent in the case of the 5
percent level of significance. Apparently the maximum possible type-II error is
equal to the area in the sampling distribution that is not in the critical region—
95 percent in the case of the 5 percent level of significance. However, it is not
quite as simple as all that. The maximum possible type-II error depends upon the
actual value of the parameter of interest and the sample size. This information,
unfortunately, is not generally available to us.

We will turn once again to the data from work section 3 and examine the
type-II error. Since the members of work section 3 constituted 9 percent of the
personnel of the base, we set up the sampling distribution with an expected value
of 9 percent in-group choices. The critical region consisted of 3 or more in-group
choices out of a total of 7. Because of the nature of the distribution, 1.9 percent
rather than 5 percent of the possible outcomes fell in the critical region. Therefore, 98.1 percent of the possible outcomes fell short of the critical region. Since
the sample outcome was 5 in-group and 2 out-group choices, we might *assume*
that the actual value of the parameter (the population proportion) is 71 percent
(5/7ths) instead of 9 percent. Table 14.3 gives the probabilities for the various outcomes of 7 best-buddy choices, first assuming the parameter to be 9 percent and
then assuming it to be 71 percent. Figure 14.3 is a graphic presentation of these
two probability distributions. The sampling distribution used to test the null
hypothesis is based upon an expected value of 9 percent. Thus, we include in the
critical region 3 or more in-group choices. Notice in Table 14.3 that the proba-

Null Hypothesis	Decision Made	
	Reject	*Fail to Reject*
True	Type-I Error	Correct Decision
False	Correct Decision	Type-II Error

FIGURE 14.2 POSSIBLE OUTCOMES OF STATISTICAL DECISIONS.

bility of 3 or more in-group choices is 0.019, and the probability of 2 or fewer such choices is 0.981 when 9 percent is used as the expected value. If the parameter is really 71 percent in-group choices, then the probability of not reaching the critical region is approximately 0.024 (2.4 percent) rather than 0.981 (98.1 percent). The probability of a type-II error, then, would only be about 2.4 percent.

We were only able to come up with a figure of 2.4 percent for type-II error by assuming that we knew the parameter for in-group choices to be 71 percent. Of course, we do not know at all that this is the case—that is the crucial problem involved in attempting to compute the probability of a type-II error. If we actually knew the parameter we would not need to use inductive statistics to make statements about it. Since we really do not know the parameter, we are unable to get any clear idea of how large the type-II error is for a particular test of the null hypothesis.

14.2.4c Power of the Test. The complement of the type-II error is known as the **power of the test.** *Power of the test is defined, therefore, as $1 - \Pr$ (type-II error).* Since type-II error stems from *failing to reject* the null hypothesis

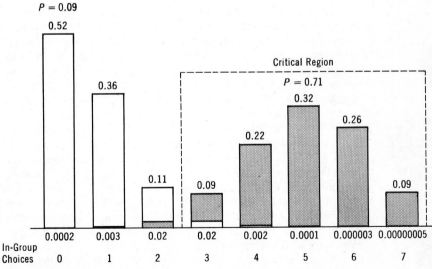

Source: Based on data from Loether, 1960.

FIGURE 14.3 HISTOGRAM OF PROBABILITY DISTRIBUTIONS OF IN-GROUP AND OUT-GROUP BEST-BUDDY CHOICES WITH EXPECTED VALUES OF 9 PERCENT AND 71 PERCENT ($N = 7$).

TABLE 14.3 Probability Distributions of In-Group and Out-Group Best-Buddy Choices with Expected Values of 9 Percent and 71 Percent ($N = 7$)

In-Group—Out-Group Choices	Probabilities when $P = 0.09$		Probabilities when $P = 0.71$	
7 in-group, 0 out-group	0.00000005		0.09	
6 in-group, 1 out-group	0.000003		0.2598	
5 in-group, 2 out-group	0.0001	} 0.019+	0.3179	} 0.972+
4 in-group, 3 out-group	0.00173		0.2169	
3 in-group, 4 out-group	0.0175		0.0877	
	Boundary of Critical Region			
2 in-group, 5 out-group	0.1061		0.0212	
1 in-group, 6 out-group	0.3578	} 0.981	0.0029	} 0.024+
0 in-group, 7 out-group	0.5168		0.00017	

Source: Based on data from Loether, 1960.

when it is *false,* the *power* of the test refers to the probability of actually *rejecting* the null hypothesis when it is *false.* (Power of the test is represented in the lower left cell of Figure 14.2.) In the example above, the probability of a type II error was 2.4 percent, so the power of the test would be 97.6 percent. In that particular situation the test of the null hypothesis would be very efficient.

The dilemma facing us in trying to make use of the concepts of type-II error and the power of the test is that in any one test of a null hypothesis we do not know the value of the parameter and, thus, are unable to compute a type-II error or the power of the test.

However, even though we cannot compute the probability of a type-II error or the power of the test for a particular test of the null hypothesis, we can assume various values for the parameter and compute probabilities for a type-II error and the power that would exist for each. Table 14.4 gives the distributions of type-II errors and power generated by assuming various actual values of the parameter. Figure 14.4 represents graphically the same distributions. The critical region was based on the null hypothesis—that the proportion of in-group choices would be 0.09 and the condition that type-I error should not exceed 0.05 (as it turned out the potential type-I error used was 0.019). The test is a one-tailed test of significance.

Note that in Table 14.4 when the actual value of P is close to the hypothesized value, 0.09, the probability of type-II error is large and the power of the test is minimal. This happens because the sampling distribution around the actual parameter is very nearly superimposed upon the sampling distribution generated from the null hypothesis (*i.e.,* the one having an expected value of 0.09). When the actual value of the parameter is 0.09, there is only one sampling distribution, and the null hypothesis is true with the probability of a type-II error being 0. The power of the test has no meaning at that particular point because the null hypothesis is true.

As the actual value of P varies more from the hypothesized value of 0.09, the probability of type-II error decreases and the power of the test increases. The greater the difference, the greater the power and the smaller the probability of type-II error. This occurs because the sampling distribution around the actual

TABLE 14.4 Distributions of Type-II Errors and Power for Various Values of P when the Critical Region Contains 3 or More In-Group Choices out of 7 (for Sampling Distribution with Expected Value of $P = 0.09$)

Actual Value of P	Pr (Type-II Error)	Power of Test
0.1	0.97	0.03
0.2	0.86	0.14
0.3	0.65	0.35
0.4	0.42	0.58
0.5	0.23	0.77
0.6	0.10	0.90
0.7	0.03	0.97
0.8	0.00+	0.99+
0.9	0.00+	0.99+
1.0	0.00	1.00

Source: Based on data from Loether, 1960.

value of P overlaps increasingly less with the sampling distribution posited by the null hypothesis. If you refer back to Figure 14.3 you will see that the overlap of the hypothesized sampling distribution with $P = 0.09$ and the sampling distribution around the assumed parameter of 0.71 is very small. It is understandable that there would be little confusion between the two sampling distributions, and, thus, there would be little tendency to err in making a decision about the null hypothesis when the hypothesized parameter and the actual parameter are substantially different in value.

What is the point of generating theoretical distributions for type-II error and power of the test such as those in Table 14.4 and Figure 14.4?* These theoretical distributions are actually very useful to the sociologist in planning his statistical analysis. For example, if there is a choice of alternative statistical techniques that could be used to analyze a body of data, and information about their respective power distributions is available, that information may be used as a basis for deciding which technique to use. Other things being equal, it would be reasonable to select uniformly the most powerful test (if one exists). Furthermore, it can generally be demonstrated that the power distribution for a one-tailed test will be more efficient than that for a two-tailed test. This useful bit of information comes from a comparison of theoretical power distributions such as the one we have been examining.

14.2.5   Minimizing Decision Errors

In testing a null hypothesis it is desirable, if possible, to minimize the probabilities of committing both type-I and type-II errors. It is easy enough to control the maximum probability of a type-I error by selecting the size of the critical region. If we decide to test the null hypothesis at the 5 percent level of significance and

---

*The distribution of type-II errors is sometimes referred to as the Operating Characteristics *(OC)* Curve, and the distribution of power of the test is referred to as the Power Curve or Power Function.

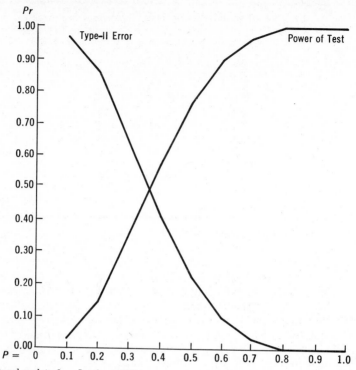

Source: Based on data from Loether, 1960.

FIGURE 14.4 DISTRIBUTIONS OF TYPE-II ERROR AND POWER FOR VARIOUS VALUES OF P WHEN THE CRITICAL REGION CONSISTS OF 3 OR MORE IN-GROUP CHOICES OUT OF 7.

set the critical region accordingly, we are saying, in effect, that we are not willing to risk a type-I error larger than 5 percent.

The size of the type-I error we are willing to risk depends upon the seriousness of committing such an error as compared to a type-II error. For example, if the null hypothesis states that couples who have had premarital counseling do not have lower divorce rates than the average for all couples, and we reject that hypothesis in favor of the alternative that their divorce rates are lower,

---

BOX 14.3 DECISION ERRORS AGAIN

This Section 14.2.4 on decision errors is usually not an easy one to understand the first time around. If you are confused at this point, don't give up. Read the section over again carefully. Many students tend to glance at tables and figures without really reading them. The tables and graphs included in this section are an integral part of the explanation, so don't take them lightly. Study them carefully and they will help you to comprehend what is being said.

the consequences of making a type-I error will probably not be considered serious. If engaged couples were subjected to premarital counseling because of our findings and our findings were spurious, it would probably do them no harm. On the other hand, if we failed to reject the null hypothesis, and it was false (a type-II error), failure to institute a premarital counseling program might lead to the dissolution of marriages that might have been "saved." In this case, a type-I error might be considered less serious than a type-II error.

On the other hand, if our null hypothesis states that capital punishment does not deter people from committing crimes of violence, and rejection of the hypothesis will lead to capital punishment, a type-I error would have serious consequences while a type-II error might not be considered as serious because of the alternative of life imprisonment. In a case such as this we would probably want to minimize the probability of type-I error by reducing the critical region to 1 percent or less.

Unfortunately, reducing the critical region to 1 percent increases the proportion of the distribution outside the critical region to 99 percent. This has the potential of increasing type-II error. What we wish to accomplish is the reduction of type-I and type-II error simultaneously.

There are two ways of reducing the probability of type-I error without increasing the probability of type-II error. The first way is to increase the sample size. As sample size increases, sampling error decreases. The smaller the sampling error, the more sensitive a statistical test is to differences in parameters and the less likelihood of confusing actual differences with differences attributable to chance.

The second way to reduce the probability of a type-I error is to replicate the study. If a study is conducted and the statistic in question just reaches the 5 percent level of significance, the null hypothesis can be rejected in favor of its alternative. The probability of a type-I error is 0.05. There are 5 chances out of 100 that the statistic reached the critical region as a result of sampling error. If the same study is conducted a second time with an independent sample and with the same result as the first time, the probability of a type-I error would again be 0.05. However, the probability that two type-I errors were committed in succession in two independent tests of the same hypothesis would be $(0.05)^2 = 0.0025$. If the study were conducted a third time with the same result, the probability of a type-I error would be $(0.05)^3 = 0.000125$. As you can see, the probability of type-I error gets increasingly smaller with replication.*

There are, then, three ways of reducing the probability of a type-I error: (1) reduce the size of the critical region, (2) increase the sample size, and (3) replicate the study. The first method is not very satisfactory since it has the effect of increasing the probability of a type-II error. The second and third are preferable because they do not increase the probability of a type-II error — as a matter of fact, they decrease the probability of a type-II error. Increasing sample size reduces type-II error because the reduced sampling error makes the statistical technique more sensitive to actual differences between parameters.

Moreover, replication reduces type-II error as well as type-I error. If the

---

*For a discussion of the dangers of confusing type-I errors with theoretically significant findings see Sterling, 1959.

probability of making a type-II error was 0.42, then the probability of making two type-II errors in a sequence of two independent studies would be $(0.42)^2 = 0.18$.

Of course, type-II error can be reduced by increasing the area in the critical region. This, however, has the effect of increasing the probability of a type-I error.

There is one other way of reducing the probability of a type-II error without affecting the probability of a type-I error. As was mentioned at the beginning of this unit (Section 14.2.5) a one-tailed test of significance has more power than a two-tailed test. This implies that the type-II error is smaller for a one-tailed than for a two-tailed test. Certainly, one must have sound reasons for using a one-tailed rather than a two-tailed test. Otherwise, it is foolhardy to place the whole critical region in one tail of the sampling distribution.

It is possible, then, to reduce the probability of type-II error four different ways: (1) increase the size of the critical region, (2) increase the sample size, (3) replicate the study, and (4) use a one-tailed test of significance. The last three steps can be taken without increasing the probability of a type-I error and, incidentally, methods 2 and 3 also reduce the probability of a type-I error. More generally one could look for more powerful statistical tests as well.

14.2.6   Steps in Testing Statistical Hypotheses

To recapitulate the process of testing statistical hypotheses here is a list of the steps involved:

1. Select an hypothesis from a theory to be tested. If possible, select a crucial hypothesis; that is, one that is a keystone of the theory. This makes for an efficient and economical test of the theory.

2. Derive a working hypothesis or hypotheses from the general hypothesis. This is done by specifying measurable indices of the concepts of the general hypothesis and of the relationships between the indices.

3. State the null hypothesis or hypotheses. The null hypothesis is derived from the working hypothesis and is contradictory to the working hypothesis.

4. Decide which statistical technique will be used to test the null hypothesis.

5. Choose the appropriate sampling distribution. The sampling distribution that is appropriate is suggested by the null hypothesis and is determined by the nature of the data to be collected and by the statistical technique to be used.

6. Define the critical region. This includes deciding on the level of significance to be used and on whether a one-tailed or two-tailed test should be conducted.

The two-tailed test of significance should be used primarily for exploratory research. The major motivation for doing research should be to test theory, and theory should lead to directional hypotheses. Therefore, one-tailed tests should be used more frequently by sociologists than two-tailed tests.

7. Draw a random sample of observations. This includes the process of designing the sample and deciding on the size of the sample.

8. Compute a statistic from the data, and locate it on the sampling distribution.

9. If the statistic is in the critical region, reject the null hypothesis; if not, fail to reject.

10. Decide what bearing the results have on the working hypothesis and the general hypothesis.

11. Replicate the study. It is only through the repetition of a study that we develop a real feeling for how fruitful or unfruitful a theory is.

The fact that these steps are listed from 1 to 11 should not be construed to mean that the steps are carried out in that order. They are listed in this way to suggest a flow of logic rather than a step-by-step guide for accomplishing the work. There is no simple, orderly sequence of steps involved in testing a statistical hypothesis. The order in which the steps are carried out may vary, and, frequently, several steps are carried out simultaneously.

It should also be noted that a rather comprehensive understanding of theory and research methods is a necessary prerequisite to carrying out these steps successfully. Statistical analysis, theory, and research methods are so integrally interrelated that they cannot really be separated.

14.2.7   The Importance of Replication

Scientists have long recognized the value of replicating studies. Some experiments in the physical and life sciences are repeated again and again until the researchers are convinced of the validity of their findings.

Unfortunately, replication receives all too little attention from sociologists. Sociologists seem to be reluctant to repeat studies that have already appeared in their professional literature. Most researchers seem to be determined to do something unique. The effect of not replicating studies is, of course, to perpetuate type-I errors in the guise of scientific generalizations. For example, suppose that a sociologist does a study, tests a null hypothesis at a given significance level, rejects the null hypothesis, concludes that he has a valid finding, and publishes his results. Suppose further that his rejection of the null hypothesis is based upon a type-I decision error (he rejects the null hypothesis when it is in fact true). Once his study appears in print, other sociologists will be less likely to conduct similar studies because "the research has already been done." The spurious findings of the original study may be taken at face value and may be given the status of a scientific generalization (Sterling, 1959:30–34). Perhaps this explains why many so-called principles of sociology have little predictive value and little basis in fact.

Much more attention needs to be given to replication if sociologists are to develop theoretical models which are useful predictive and explanatory instruments. It is only when we do replicate studies that we begin to realize how tenuous many of our *research findings* are.

14.2.8   An Additional Example

To illustrate the steps in the test of a statistical hypothesis we will use data from another work section of the air force study, section 9, and follow the process from beginning to end. There are enough observations involved in work section 9 to justify the use of the normal approximation to the binomial distribution. This will make possible the use of a different statistical technique and a different sampling distribution than that used in analyzing the data from work section 3. This should supply a more general perspective of the process involved in testing statistical hypotheses.

   1. *Select an hypothesis from a theory to be tested.* The hypothesis of interest is that homogeneity and propinquity are factors in the choice of a best buddy.
   2. *Derive a working hypothesis from the general hypothesis.* A number of working hypotheses may be derived from the general hypothesis. We will concentrate on one. We will use as an index of propinquity common membership in a work section on the air base studied. Best-buddy choice will be determined by a subject's response to a request to name his best buddy. The working hypothesis is that men will tend to name as their best buddies members of their own work section.
   3. *State the null hypothesis.* The null hypothesis is that subjects who belong to work section 9 will not select as their best buddies members of work section 9 any more than chance would lead one to expect (again, this is only one of several possible null hypotheses that can be derived from the working hypothesis). More specifically,

$$H_0: P[\text{best buddy in same work group}] \leq 0.24$$
$$H_1: P[\text{best buddy in same work group}] > 0.24$$

   4. *Decide which statistical technique will be used to test the null hypothesis.* The statistical technique we will use will be the test of significance of a sample frequency against the mean frequency for the population.
   5. *Choose the appropriate sampling distribution.* Since the size of the sample will be large enough to use the normal approximation of the binomial sampling distribution (see Section 12.3.4), and the proportion of the population in work section 9 is 0.24, we will use the normal sampling distribution with an expected value (mean) equal to $0.24N$.
   6. *Define the critical region.* Since we wish to show that there are significantly more in-group best-buddy choices in work section 9 than chance would lead one to expect, we will use a one-tailed test of significance with the critical region on the side of the high number of in-group choices. In light of our confidence in our theory, we will use a 1 percent level of significance, putting the critical region in the right tail of the distribution. A standard score of $+2.33$ will bound the critical region.

7. *Draw a random sample of observations.* The sample of observations for members of work section 9 consists of 35 best-buddy choices — 25 in-group choices and 10 out-group choices.

8. *Compute a statistic from the data and locate it on the sampling distribution.* First we must find the mean and standard error for frequencies using the following formulae:

$$\mu_B = NP$$

where P is the proportion that the work section is of the population, N is the number of best-buddy choices made by the members of the group, and μ_B is the mean frequency for a binomial distribution.

$$\sigma_B = \sqrt{NPQ}$$

where $Q = 1 - P$ and σ_B is the standard error of a frequency in a binomial distribution.

For our data these parameters are:

$$\mu_B = (35)(0.24)$$

$$= 8.4$$

$$\sigma_B = \sqrt{(35)(0.24)(0.76)}$$

$$= 2.53$$

Now we substitute these parameters into the z score formula along with the observed frequency of in-group choices (25), corrected for continuity ($25 - 0.5 = 24.5$):

$$z = \frac{X - \mu_B}{\sigma_B}$$

$$= \frac{24.5 - 8.4}{2.53}$$

$$= +6.36$$

9. *If the statistic is in the critical region, reject the null hypothesis.* The z score on the boundary of the critical region is $+2.33$, therefore, any computed z score as large or larger will be in the critical region. The computed z score of $+6.36$ is well within the critical region; so we will reject the null hypothesis in favor of the set of alternatives.* The set of alternatives is that members of work section 9 do tend to select their fellow workers as best buddies more often than chance would lead one to expect.

10. *Decide what bearing the result has on the working hypothesis and the general hypothesis.* Rejection of the null hypothesis lends support to the working hypothesis and, thus, to the general hypothesis. The evidence contributes to the contention that propinquity is a factor in the selection of a best buddy.

---

*If you look up the probability attached to a z-score of $+6.36$ in the normal curve table, you will find that it is less than 0.001. It is entirely proper to report the actual probability associated with a finding if you so desire.

> **BOX 14.4**   CORRECTION FOR CONTINUITY
>
> If you do not remember the *correction for continuity*, don't let it throw you. Go back and read the explanation of the correction in Section 12.3.4. We will have occasion to refer to the correction for continuity again in this volume. It is used generally in evaluating a discrete empirical distribution with a continuous sampling distribution.

11. *Replicate the study.* Since data are available for fifteen work sections, analysis of each work section constitutes a pseudoreplication of the study. In the original study the number of in-group choices was statistically significant for ten of the work sections. These results were considered supportive of the working hypothesis and the general hypothesis. We would want, however, to do additional studies of the same kind with different populations to demonstrate the general applicability of the theory.

14.2.9   Confidence Intervals and Tests of Hypotheses

An alternative to the traditional method of testing null hypotheses described in this chapter is to use confidence intervals. Some authors have pointed out that the use of confidence intervals to test hypotheses enjoys some advantages over the traditional approach (Blalock, 1972; Natrella, 1960). First we will consider how confidence intervals may be used to test hypotheses; then we will examine the advantages of using this approach.

We have just examined the test of the null hypothesis — that members of work section 9 do not tend to select members of their own work section as best buddies any more than chance would lead one to expect. We rejected this null hypothesis because we got a z score of +6.36 that was well into the critical region. Instead of computing the z score we could have computed 99 percent confidence limits around the population mean of 8.4. Using the population mean and the standard error (2.53) we can substitute into the following formula with $z = \pm 2.58$ (which bounds the middle 99 percent of the normal sampling distribution; see Appendix Table B) and get the required limits:

$$cl = \mu_B \pm z(\sigma_B) = 8.4 \pm 2.58 \ (2.53)$$

(14.1)
$$cl_1 = 8.4 - 6.53 = 1.87$$

$$cl_2 = 8.4 + 6.53 = 14.93$$

Therefore, the 99 percent confidence limits are 1.87 and 14.93, inclusive. Any number of in-group best-buddy choices that falls between these limits is a reasonable expectation for a population with a mean of 8.4. Any number of choices falling outside these limits may be considered unusual. The observed number of in-group choices was 25. This result is outside of the confidence interval and quite far above the upper limit, so we would feel justified in rejecting the null hypothesis.

The following advantages are offered for this approach to testing hypotheses over the traditional approach: (1) The approach divides all of the possible hypotheses into two sets, those that are tenable because they are in the confidence interval and those that are not because they are outside the interval. It is possible, therefore, to test simultaneously a whole set of hypotheses. (2) In rejecting the null hypothesis we not only discover what outcome is unlikely, but by placing confidence limits around the observed sample outcome (25) we have available to us, at the same time, a whole set of likely outcomes to offer as a parameter estimate. (3) The width of the confidence interval gives us some notion of how confident we can be about rejecting the null hypothesis. The narrower the confidence interval, the more assurance we can have in rejecting. Also important, of course, is the distance of the observed outcome from the confidence limits.

In the example above, the interval was not as narrow as we might have liked; but the observed outcome did fall quite a distance from the upper limit of the interval.

The significance of the width of the confidence interval is that it gives one a notion of how powerful the test of the null hypothesis is. Narrow confidence intervals tend to be found in those cases where the power curve is efficient (Natrella, 1960:21).

14.3   SUMMARY

Chapter 14 has introduced you to the general process by which statistical hypotheses are tested. We have explained the use of the null hypothesis and pointed out how it can be used to specify the sampling distribution. A distinction was made between one-tailed and two-tailed tests, and the applications of each were discussed. Significance level was defined, and its relationship to critical regions explained. Type-I errors, type-II errors, and the power of the test were discussed; and techniques for reducing the probabilities of making decision errors were suggested. We listed the eleven steps in the testing of a statistical hypothesis and gave an example of how these steps are actually carried out. Finally, an alternative approach to hypothesis testing was discussed.

It is important to note that the inductive process in statistics is merely an aid to the researcher in making substantive decisions. Ultimately, the researcher must decide whether his(her) data support or fail to support the theoretical notions he(she) is investigating. Statistics is a crutch for the researcher to lean on in the process of making decisions. However, statistics alone are not capable of making substantive decisions.

Furthermore, tests of statistical significance are a crude way of determining whether there is anything that can be generalized from samples to populations. Tests of null hypotheses constitute a primitive but necessary step in the process of scientific investigation. The significant analysis, from the theoretical standpoint, is the descriptive analysis that is used. Such an analysis addresses itself to the meaning of data. Tests of null hypotheses help us determine whether such meanings can be generalized to populations.

Since tests of null hypotheses are an integral part of the generalization process, in the chapters that follow we will consider a number of statistical techniques that are used to test null hypotheses. In each case, the general hypothesis-testing process will be the one described here; but the detailed procedures will vary from one application to another.

CONCEPTS TO KNOW AND UNDERSTAND

general hypothesis
working hypothesis
statistical hypothesis
null hypothesis
one-tailed and two-tailed tests
level of statistical significance
critical region
type-I error
type-II error
power of the test
replication

QUESTIONS AND PROBLEMS

1. Why is Durkheim's study of suicide considered a classic example of sociological research? Is the generalization of Durkheim's findings justified? If yes, why? If not, why not?

2. What are the steps involved in using sample data to test statistical hypotheses?

3. Explain the logic involved in using the null hypothesis.

4. Explain the difference between a one-tailed and a two-tailed test of significance. Under which circumstances would you use a two-tailed test? Under which circumstances would you use a one-tailed test?

5. Explain the meaning of the concept *critical region*.

6. Essentially, what does *statistical significance* mean?

7. Distinguish between a type-I and a type-II error. How would you go about minimizing the probabilities of committing both types of errors?

8. What is the relationship between the type-II error and the power of the test? How is the concept *the power of the test* used in statistical analysis?

9. Why is it important to replicate a study? See if you can find examples of studies in sociology that have been replicated. Ask your physics or biology professor how often replication is used in his discipline.

10. Explain the relationship between finding a confidence interval and testing a statistical hypothesis. What are the advantages of using a confidence interval to test a statistical hypothesis?

11. Using information from Chapter 12, verify the probabilities given in Table 14.2. Show how you arrived at your figures.

12. Suppose we had decided to run a two-tailed test with the null hypothesis that $P = 0.24$ and a level of significance of .05 for work section 9 in Table 14.1. The sampling distribution is shown below. Indicate where the critical region would be. Laying out your work according to the steps discussed in this chapter, indicate what your conclusion would be. Would you reach the same conclusion using the normal approximation to the exact binomial sampling distribution?

BINOMIAL SAMPLING DISTRIBUTION WITH $N = 35$, $P = 0.24$

Binomial Term	Probability*	Binomial Term	Probability*
P^{35}	0	$4537567650P^{17}Q^{18}$	0.0009442
$35P^{34}Q$	0	$4059928950P^{16}Q^{19}$	0.0026752
$595P^{33}Q^2$	0	$3247943160P^{15}Q^{20}$	0.0067771
$6545P^{32}Q^3$	0	$2319959400P^{14}Q^{21}$	0.0153294
$52360P^{31}Q^4$	0	$1476337800P^{13}Q^{22}$	0.0308909
$324632P^{30}Q^5$	0	$834451800P^{12}Q^{23}$	0.0552903
$1623160P^{29}Q^6$	0	$417225900P^{11}Q^{24}$	0.0875429
$6724520P^{28}Q^7$	0	$183579396P^{10}Q^{25}$	0.1219765
$23535820P^{27}Q^8$	0	$70607460P^9Q^{26}$	0.1485611
$70607460P^{26}Q^9$	0	$23535820P^8Q^{27}$	0.1568145
$183579396P^{25}Q^{10}$	0	$6724520P^7Q^{28}$	0.1418798
$417225900P^{24}Q^{11}$	0	$1623160P^6Q^{29}$	0.1084484
$834451800P^{23}Q^{12}$	0.0000002	$324632P^5Q^{30}$	0.0686839
$1476337800P^{22}Q^{13}$	0.0000010	$52360P^4Q^{31}$	0.0350806
$2319959400P^{21}Q^{14}$	0.0000048	$6545P^3Q^{32}$	0.0138860
$3247943160P^{20}Q^{15}$	0.0000213	$595P^2Q^{33}$	0.0039975
$4059928950P^{19}Q^{16}$	0.0000842	$35PQ^{34}$	0.0007446
$4537567650P^{18}Q^{17}$	0.0002982	Q^{35}	0.0000674

*Probabilities are approximate due to errors of rounding.
Source: Based upon data from Loether, 1960.

13. Select another work group from Table 14.1 (not group 3 or 9) and run a one-tailed test as was illustrated in this chapter. Show your work and lay it out according to the steps discussed in this chapter.

GENERAL REFERENCES

Blalock, Hubert M. *Social Statistics,* Revised Edition. (New York, McGraw-Hill Book Company), 1979.
 In this excellent intermediate-level book, Blalock devotes Chapter 10 to a discussion of the steps involved in the test of a statistical hypothesis.
Edwards, Allen. *Statistical Methods for the Behavioral Sciences.* (New York, Holt, Rinehart and Winston), 1954.
 This is another clearly written intermediate statistics book. The discussion of the concept of the power of the test is one of the best to be found anywhere.

Siegel, Sidney. *Nonparametric Statistics for the Behavioral Sciences.* (New York, McGraw-Hill Book Company), 1956.
   The first three chapters of this book are especially recommended as an introduction to the subject of hypothesis testing.

OTHER REFERENCES

Blau, Peter M. "A Formal Theory of Differentiation in Organizations," *American Sociological Review,* vol. 35, no. 2 (April, 1970).

Bowerman, Charles E., and Barbara R. Day, "A Test of the Theory of Complementary Needs as Applied to Couples During Courtship," *American Sociological Review,* vol. 21, no. 5 (October, 1956), pp. 602–605.

Byrne, D., "The Influence of Propinquity and Opportunities for Interaction on Classroom Relationships," *Human Relations,* vol. 14 (1961), pp. 63–69.

Clark, Robert E., "Psychoses, Income, and Occupational Prestige," *American Journal of Sociology,* vol. 54 (1949), pp. 433–440.

Douglas, Jack D., *The Social Meaning of Suicide* (Princeton, N.J., Princeton University Press), 1967.

Durkheim, Emile, *Suicide: A Study in Sociology,* 1897. Translated by John A. Spaulding and George Simpson (Glencoe, Ill., The Free Press), 1951.

Finifter, Bernard M., "The Generation of Confidence: Evaluating Research Findings by Random Subsample Replication," in Herbert L. Costner, editor, *Sociological Methodology 1972* (San Francisco, Jossey-Bass), 1972, pp. 112–175.

Galton, Francis, *Hereditary Stature,* Royal Society Proceedings, XL, 1886.

Ktsanes, Thomas, "Mate Selection on the Basis of Personality Type: A Study Utilizing an Empirical Typology of Personality," *American Sociological Review,* vol. 20, no. 5 (October, 1955), pp. 547–551.

Labovitz, Sanford, "Criteria for Selecting A Significance Level: A Note on the Sacredness of .05," *American Sociologist,* vol. 3 (1968), pp. 220–222.

Loether, Herman J., "Propinquity and Homogeneity as Factors in the Choice of Best Buddies in the Air Force," *Pacific Sociological Review,* vol 3, no. 1 (Spring, 1960), pp. 18–22.

Natrella, Mary G., "The Relation Between Confidence Intervals and Tests of Significance," *The American Statistician,* vol. 14, no. 1 (February, 1960), pp. 20–22.

Newcomb, Theodore M., "Varieties of Interpersonal Attraction," in D. Cartwright and A. Zander (eds.), *Group Dynamics: Research and Theory* (Evanston, Ill., Row), 1960, pp. 104–119.

Reiss, Albert J., Jr., *Occupations and Social Status* (New York, The Free Press of Glencoe), 1961.

Robinson, William S. "Ecological Correlations and Behavior of Individuals," *American Sociological Review,* vol. 15, no. 3 (June, 1950), pp. 351–357.

Skipper, J.K., Jr., *et al,* "The Sacredness of .05; A Note Concerning the Uses of Significance in Social Sciences," *American Sociologist,* vol. 2 (February, 1967), pp. 16–19.

Sterling, T.D., "Publication Decisions and Their Possible Effects on Inferences Drawn from Tests of Significance – or Vice Versa," *Journal of The American Statistical Association,* vol. 54 (1959), pp. 30–34.

Winch, Robert F., "The Theory of Complementary Needs in Mate-Selection: Final Results on the Test of the General Hypothesis," *American Sociological Review,* vol. 20, no. 5 (October, 1955), pp. 552–554.

15 *Testing Hypotheses about Central Tendency and Dispersion*

Sociology is concerned with the study of social systems such as groups and organizations. People are of interest not as individuals, but as members of social systems.* Lazarsfeld and Menzel point out that there are three types of properties of social systems (or, as they call them, collectivities) which are of interest sociologically (Lazarsfeld and Menzel, 1969).

15.1 PROPERTIES OF SOCIAL SYSTEMS

First, there are **analytical properties,** which are characteristics of members that are relevant to the social system. These include any characteristics that influence the operation of the system. Such properties might include the average age of the members of an organization, the percentage of members who are married, etc. Second, there are **structural properties,** which are based upon the ways in which members are related to each other in the system. The emphasis here is on the network of roles that constitutes the structure of the system. Included might be information about who interacts with whom, who initiates action for whom, etc. Finally, there are **global properties,** which are not based upon information about the properties of members and are not analyzable in terms of relations between members. Global properties are products or outcomes of the operations of social systems that cannot be further reduced to constituent components. These might include the number of bachelor's degrees produced by

---

*It should be noted that the members of a social system need not be individuals. They may be other social systems (subsystems) or they may be combinations of social systems and individuals.

a university in a year, the number of automobiles produced by an auto manufacturer, the number of games won by a baseball team, etc.

Studies of the structural properties of social systems often do not generally lend themselves to analysis by the more commonly used statistical techniques unless they can be reduced to numerical indices, *e.g.* centrality, connectivity, etc. As a matter of fact, researchers often have to develop new analytical strategies to handle their data (Procter and Loomis, 1951). Studies of global and analytical properties, however, provide data that can be and are readily analyzed in terms of central tendency, variability, and form.

It is, perhaps, a commentary on the state of current sociology that more attention is given to analysis of ancillary analytical properties than to the more central global and structural properties of social systems. It appears that a majority of the studies in the sociological literature focus upon analytical properties. A sizeable percentage of these involve tests of hypotheses about central tendencies. For example, sociologists seek answers to questions such as the following: Do graduates of private colleges have annual incomes that are higher on the average than those of graduates of American colleges in general? Are more draftees from southern states rejected because of nutritionally related diseases than draftees from other parts of the country? Is the average age of members of the Democratic party lower than that for members of the Republican party?, etc.

Studies dealing with global properties of social systems often are focused on questions of central tendencies too. For example, such questions as the following may be asked: Do companies with diversified product lines tend to have higher average volumes of sales than companies with less diversified product lines? Do lower-income families spend larger proportions of their incomes for food than higher-income families? Do industrialized countries have higher population density than non-industrialized countries?, etc.

Popular as tests about central tendency are in sociology, the sociologist can hardly afford to ignore questions about variability. The relative homogeneity or heterogeneity of social systems may be crucial to certain theoretical questions. Although sociologists have given much less attention to questions of variability than they warrant, there are many significant questions begging answers. For example, with respect to analytical properties, sociologists might investigate such questions as the following: Are income differentials greater in large cities than they are in small cities? Is the age structure of the population of central cities more or less variable than that of the suburbs?, etc. And with respect to global properties, such questions as the following might be investigated: Does per capita expenditure on education vary more among states with growing populations than among states with stable or declining populations? Does size of membership vary more among the locals of craft unions than the locals of industrial unions? etc.

These questions suggest just a few of the kinds of studies that sociologists undertake in which they may be interested in testing hypotheses about central tendency or variability. This chapter will be devoted to a consideration of the techniques available for testing hypotheses of these kinds. We will specifically discuss the one-sample case for means, proportions, and variances; the case of two independent samples for means, proportions, and variances; and the k-sample case for means.

15.2   THE ONE-SAMPLE CASE FOR MEANS

The Bureau of Labor Statistics of the United States Department of Labor carries out area wage surveys annually in selected metropolitan areas to provide information on earnings for standard work weeks in occupations common to a variety of industries. The data are used to generalize to all metropolitan areas in the United States. A standard of pay is computed on the basis of the average weekly salary (or hourly rate) for the occupations in all standard metropolitan areas combined; then the rates for particular metropolitan areas are expressed as percentages of the standard rate. For example, the standard rate of pay for clerical office workers in all industries is set at 100. The rate for the Chicago metropolitan area in the year from March, 1968 to February, 1969 was 104 while for Des Moines it was 88 (U.S. Dept. Labor, 1970).

These data can be used to seek answers to questions relevant to an understanding of various labor markets. For purposes of illustration, attention will be directed to clerical office workers, although the data are also available for skilled maintenance and unskilled plant workers.

One question we might explore is this: What are the consequences of size of metropolitan area for the pay scale of clerical office workers? That is, how do the pay scales compare to the standard in large metropolitan areas (areas with populations of 1,000,000 or more), moderate-sized metropolitan areas (250,000 to 1,000,000 population), and small metropolitan areas (less than 250,000 population)?

We could argue that the results should come out in either of two ways. First, we might suspect that the pay scales will be most like the standard in larger metropolitan areas because of the concentration of industries that employ many clerical office workers and the resulting competition for labor. Consequently, the pay scales might be below standard for smaller metropolitan areas because of scarcity of jobs and an abundant labor supply. Then too, the higher costs of living found in larger metropolitan areas might be associated with higher pay scales.

On the other hand, it may be that the abundant labor supply in the large metropolitan areas leads to low pay scales because of the many potential workers available, while in smaller metropolitan areas a shortage of labor may lead to higher pay scales.

To see which of these alternatives is supported by the data we might compare each of the average pay scales, for large, moderate, and small metropolitan areas, with the standard of 100 for all metropolitan areas.*

In this case we do not have a specific theory from which to work. We will conduct an exploratory analysis of the data to see which alternative is supported. The following three null hypotheses will be tested:

$$H_0 (1): \mu_S = 100$$

where μ_S is the mean pay rate for clerical office workers in *small* metropolitan areas.

$$H_0(2): \mu_M = 100$$

*For purposes of illustration we will assume that the survey data are based upon a simple random sample of metropolitan areas.

where μ_M is the mean pay rate for clerical office workers in *moderate* sized metropolitan areas.

$$H_0(3): \mu_L = 100$$

where μ_L is the mean pay rate for clerical office workers in large metropolitan areas.

Each of these hypotheses will be tested by the single sample mean test. Each sample mean will be compared with the expected value of 100 to see whether it is significantly different. Since the study is exploratory, the null hypotheses were set up for a two-tailed test.

It should be noted that, although the data used to test these three null hypotheses are independent for each test, the hypotheses themselves are interrelated in that we are comparing pay scales for metropolitan areas of three different sizes. That is, once we have tested the three null hypotheses, we wish to put the results of those three tests together and make some general statements about the relationship between size of metropolitan area and level of pay scale for clerical office workers.

Also note that in order to justify the use of the one-sample test for means to test these hypotheses we must have reason to believe that we can satisfy, sufficiently, those assumptions of the technique that are discussed below in Section 15.2.2.

Table 15.1 presents data from the *Handbook of Labor Statistics* that may be used to test the null hypotheses. Because the three null hypotheses will be tested with samples of 24, 49, and 16 metropolitan areas, respectively, the appropriate sampling distributions will be Student's t-distributions. The appropriate formula for degrees of freedom for the one-sample case is $df = N - 1$.* If the null hypotheses are tested at the 5 percent level of significance using two-tailed tests, the t's that bound the critical regions are $t = \pm 2.069$ for large metropolitan areas, $t = \pm 2.010$ for moderate-sized metropolitan areas, and $t = \pm 2.131$ for small metropolitan areas. These t scores were found by entering Table C (in the Appendix) with 23, 48, and 15 degrees of freedom, and finding the entries for the 5 percent level of significance – two-tailed test.

Using the survey data in Table 15.1 we compute sample means for large, moderate, and small metropolitan areas, find standard errors for the means, and solve for t's. Then we compare the computed t's with those that bound the critical regions of the sampling distributions and decide whether rejection of the null hypotheses is warranted.

In order to compute the t scores for tests of the hypotheses we use the following variation of the standard score formula:

(15.1)
$$t = \frac{\bar{X} - \mu}{s_{\bar{X}}}$$

where $\bar{X}$ is the sample mean, μ is the population mean, and $s_{\bar{X}}$ is the estimated standard error of the mean based on sample data (see Formula (12.16), Chapter 12).

Tables 15.2, 15.3, and 15.4 give step-by-step accounts of testing the three

---

*See Section 12.4.7 for a general explanation of the concept of degrees of freedom.

TABLE 15.1 PAY RATES (MARCH 1968-FEBRUARY 1969) FOR CLERICAL OFFICE WORKERS IN ALL INDUSTRIES FOR SAMPLES OF METROPOLITAN AREAS*

Large Metropolitan Areas (1 million or more population)	Moderate Metropolitan Areas (250,000 to 1 million)		Small Metropolitan Areas (less than 250,000)
96	99	99	94
100	102	104	83
100	97	91	86
104	98	97	87
99	88	95	97
96	108	99	86
101	94	93	85
100	96	107	100
98	92	112	86
94	94	97	94
99	93	91	93
102	112	91	99
104	93	86	94
92	101	89	91
102	91	101	106
115	88	93	90
97	97	95	
99	85	106	
93	90	104	
98	85	88	
111	95	99	
105	90	93	
109	91	93	
107	95	105	
	92		

Source: U.S. Dept. Labor, 1970.
*These rates are in terms of percentages of the standard rate of 100.

null hypotheses and present the results. The data in Table 15.1 were used to get the sums necessary for computing the statistics in Tables 15.2, 15.3, and 15.4.

We will focus our attention on the test of the first null hypothesis which is summarized in Table 15.2. This null hypothesis states that the pay scales of clerical office workers in small metropolitan areas will not differ, on the average, from the standard scale of 100 (the expected mean).

There are two sets of alternatives to the null hypothesis. The first set of alternatives is that μ_S is actually greater than 100 ($\mu_S > 100$) and the second is that μ_S is actually less than 100 ($\mu_S < 100$). According to the previously established criteria for testing the null hypothesis, we will reject the null hypothesis in favor of the first set of alternatives if $t \geq +2.131$ or, we will reject the null hypothesis in favor of the second set of alternatives if $t \leq -2.131$. If t does not reach either of these two critical regions, we will fail to reject the null hypothesis.

Using the data in Table 15.1 for small metropolitan areas we compute the sum of scores (ΣX) and the sum of squared scores (ΣX^2). These sums are used to

TABLE 15.2   One-sample Test for Means: Pay Scales for Small Metropolitan Areas

$H_0(1): \mu_S = 100$

Alternatives to H_0:

   $H_1: \mu_S > 100$ designate this set of alternatives if
   $t \geq +2.131$

   $H_2: \mu_S < 100$ designate this set of alternatives if
   $t \leq -2.131$

Computations:

0.025   0.025

-2.131 μ $+2.131$

$$\Sigma X = 1471 \qquad \Sigma X^2 = 135{,}855 \qquad \overline{X} = \frac{\Sigma X}{N} = \frac{1471}{16} = 91.938$$

$$s = \sqrt{\frac{\Sigma X^2 - (\Sigma X)^2/N}{N-1}} = \sqrt{\frac{135{,}855 - (1471)^2/16}{15}} = \sqrt{40.996} = 6.4$$

[Correcting the standard deviation using Table 12.4 of Chapter 12 we get:

   $(s)(a_2) = (6.4)(1.01679) = 6.51$]

$s_{\overline{x}} = s/\sqrt{N} = 6.51/\sqrt{16} = 1.63$

$$t = \frac{\overline{X} - \mu}{s_{\overline{x}}}$$

$$= \frac{91.938 - 100}{1.63}$$

$$= \frac{-8.062}{1.63} = -4.95$$

Conclusion:

   Therefore, rejection of the null hypothesis in favor of the set of alternatives $H_2: \mu_S < 100$ is warranted.

---

compute the sample mean $(\overline{X})$ and the sample standard deviation (s). Because the sample standard deviation is a biased estimator, we use Table 12.6 of Chapter 12 to correct it. Then, we compute the sample estimate of the standard error of the mean $(s_{\overline{x}})$.

Substituting into Formula (15.1), we compute t. As we can see from Table 15.2, $t = -4.95$, which is well into the critical region in the left tail of the sampling distribution. Therefore, we reject the null hypothesis in favor of the set of alternatives that the sample mean is significantly lower than the population mean of 100. We would conclude from this result that pay scales for clerical office workers in small metropolitan areas are lower than the standard rate.

Before we consider the tests of hypotheses two and three, let us analyze the process used to test hypothesis one. We determined that the proper sampling distribution to use is Student's t-distribution, with 15 degrees of freedom. This is so because when N is as small as 16, sample means tend to distribute themselves in a Student's t-distribution with $N-1$ degrees of freedom. We decided to do a two-tailed test of the null hypothesis at the 5 per cent level of significance, so we split the critical region evenly between the two tails of the sampling dis-

TABLE 15.3 One-sample Test for Means: Pay Scales for Moderate-sized Metropolitan Areas

$H_0(2): \mu_M = 100$

Alternatives to H_0:

$H_1: \mu_M > 100$ designate this set of alternatives if $t \geq +2.010*$

$H_2: \mu_M < 100$ designate this set of alternatives if $t \leq -2.010*$

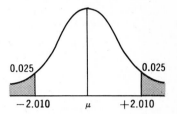

0.025   0.025

−2.010 μ +2.010

Computations:

$\Sigma X = 4694 \qquad \Sigma X^2 = 451{,}758 \qquad \overline{X} = \dfrac{\Sigma X}{N} = \dfrac{4694}{49} = 95.796$

$s = \sqrt{\dfrac{\Sigma X^2 - (\Sigma X)^2/N}{N-1}} = \sqrt{\dfrac{451{,}758 - (4694)^2/49}{48}} = \sqrt{43.58} = 6.60$

[Correcting the standard deviation using Table 12.4 of Chapter 12 we get:

$(s)(a_2) = (6.6)(1.00522) = 6.63]$

$s_{\overline{X}} = s/\sqrt{N} = 6.63/\sqrt{49} = .95$

$t = \dfrac{\overline{X} - \mu}{s_{\overline{X}}}$

$= \dfrac{95.796 - 100}{.95}$

$= \dfrac{-4.204}{.95}$

$= -4.43$

Conclusion:

Therefore, rejection of the null hypothesis in favor of the set of alternatives $H_2: \mu_M < 100$ is warranted.

*These values were arrived at through interpolation from the values in Appendix Table C.

tribution, putting 2½ percent in each tail. We found that the critical regions in the tails were bounded by t scores of +2.131 on the right and −2.131 on the left.

Given that the area under the curve in the sampling distribution consists of sample means drawn from simple random samples of the population for which $\mu = 100$ is the parameter, we then elected to use the 5 percent level of significance and agreed to call any sample mean that fell into one of the critical regions a "rare" or unusual sample mean.

The standard error of the mean measures the variability of sample means around the population mean in a sampling distribution. The formula used to compute t is a ratio between sampling error (as measured by the standard error) and the observed difference between the sample mean and the expected population mean. If the observed difference is at least 2.131 times as large as the standard error, then the resulting t score will be in one or the other of the critical regions, and the null hypothesis will be rejected. The probability that the t score will fall into one of the critical regions by chance is 5 percent. We are thus risking a maximum probability of type-I error of 5 percent by rejecting the null

TABLE 15.4 ONE-SAMPLE TEST FOR MEANS: PAY SCALES FOR LARGE METROPOLITAN AREAS

$H_0(3): \mu_L = 100$

Alternatives to H_0:

$H_1: \mu_L > 100$ designate this set of alternatives if
$t \geq +2.069$

$H_2: \mu_L < 100$ designate this set of alternatives if
$t \leq -2.069$

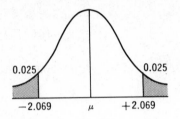

0.025   0.025

-2.069 μ $+2.069$

Computations:

$$\Sigma X = 2421 \qquad \Sigma X^2 = 244{,}943 \qquad \overline{X} = \frac{\Sigma X}{N} = \frac{2421}{24} = 100.875$$

$$s = \sqrt{\frac{\Sigma X^2 - (\Sigma X)^1/N}{N-1}} = \sqrt{\frac{244{,}943 - (2421)^2/24}{23}} = \sqrt{31.51} = 5.61$$

[Correcting the standard deviation using Table 12.4 of Chapter 12 we get:

$(s)(a_2) = (5.61)(1.01093) = 5.67]$

$s_{\overline{x}} = s/\sqrt{N} = 5.67/\sqrt{24} = 1.157$

$$t = \frac{\overline{X} - \mu}{s_{\overline{x}}}$$

$$= \frac{100.875 - 100}{1.157}$$

$$= \frac{.875}{1.157} = +.76$$

Conclusion:
     Therefore, fail to reject the null hypothesis.

---

hypothesis whenever the computed t is in one of the critical regions. Since this is an exploratory analysis, a type-I error of 5 percent is considered an acceptable risk. Of course, it is only through replication of the study that the sociologist can develop confidence that the finding is not due to chance.

     This same process was used in the tests of hypotheses two and three, relating to moderate-sized and large metropolitan areas. The steps in the test of these hypotheses are summarized in Tables 15.3 and 15.4. In the case of moderate-sized metropolitan areas, the null hypothesis was rejected in favor of the set of alternatives that the pay scales are lower than the standard. This is essentially the same finding as that for small metropolitan areas. The null hypothesis was not rejected, however, for large metropolitan areas. The mean pay scale in those areas was very near the standard rate.

     Comparing the results of the tests of the three hypotheses, an apparent pattern emerges. The means were significantly lower than the standard rate for small and moderate-sized metropolitan areas. Meanwhile, for large metropolitan areas, the mean rate was very much like the standard rate. The three means were 91.9 for small metropolitan areas, 95.8 for moderate-sized areas, and 100.9 for large areas. From this we might observe that the smaller the metropolitan areas, the lower the mean pay scales for clerical office workers. The findings of this exploratory analysis could form the basis for the design of further studies to determine the factors that explain why rates are lower in smaller areas.

15.2.1   Use of the Normal Approximation to the *t*-distribution

In the example just given, the Student's *t*-distribution was used as the sampling distribution to evaluate the null hypotheses. This was necessary because the N's were small and s was used as an estimate of the unknown σ in the formula for the standard error of the mean.* Note that the use of Student's *t*-distribution is based upon the assumption that the population being sampled is normally distributed.

Given that the normal population assumption is a reasonable one, and the additional condition that the sample N is large, the normal sampling distribution may be used to approximate the *t*-distribution.† In this case *large* will be defined as 100 or more cases. If the normal approximation is used, the process of analyzing the data is the same, but the standard scores bounding the critical regions are taken from the normal curve instead of one of the *t*-distributions. Formula (15.1) is still used to compute the standard score, but the standard score is designated z instead of t to indicate that the normal curve is the sampling distribution being used.

15.2.2   Assumptions of the One-Sample Test for Means

In order to justify using the one-sample test for means, it is necessary to make the following assumptions: (1) there is interval-level measurement of the variable; (2) there are independent observations; (3) the population being sampled is normally distributed; and (4) the data being tested are from a simple random sample of the population. If these assumptions are met, the test is a powerful one. If the assumptions are violated, one must be very careful about the kinds of interpretations of data that are made. The test, however, is fairly robust. That is, small violations of assumptions have little impact on conclusions.

15.3   THE TWO-SAMPLE CASE FOR MEANS

Although the one-sample case for means is used occasionally by sociologists, (*e.g.*, Verwaller, 1970) the two-sample case is found much more frequently in the literature.

Basically, the two-sample case involves first drawing two random samples that differ on some dichotomous variable and then comparing their central ten-

---

*See Section 12.4.6 for an explanation of the effects of substituting s for σ in computing the standard error.

†An examination of the last entry in the table of the *t*-distribution under the .05 column indicates that the normal curve is the limiting case of Student's *t*-distribution. However, when degrees of freedom reach 60, the t score, which cuts $2\frac{1}{2}$ percent of the distribution out of each tail of the curve, is already down to 2.00; therefore, it is fast approaching the limit of 1.96.

dencies on a second variable to see whether any difference they show is large enough to be called unusual or rare. For example, the sociologist might wish to compare the mean incomes of a sample of college graduates *vs.* the mean incomes of a sample of high school graduates, hoping to see whether there is a systematic difference, and whether such a difference can be generalized to the populations from which the samples were drawn. The solution to the problem is again a comparison of observed differences with expected differences and the key to the comparison is a sampling distribution and a standard score with the appropriate standard error term in its denominator.

15.3.1   The Sampling Distribution of Differences

In order to arrive at a standard error for comparing two sample means, we must consider the sampling distribution that is generated by repeatedly drawing a pair of simple random samples from a common population and by finding the difference between their means. Just for the sake of discussion, assume that we draw pairs of simple random samples of size N from a normally distributed population, continuing until we exhaust all of the unique samples of size N in the population. Furthermore, assume that we compute the differences between all of those pairs of means, add up all of the difference scores, and divide by the number of difference scores to find the mean difference. Because all of the simple random samples are drawn from the same population, and all of the sample means are estimates of the same population mean, the differences found between pairs of sample means are merely random differences; the sum of the differences will be zero, and the mean difference will be zero. Furthermore, the differences will distribute themselves normally around the mean difference of zero. Thus, we will have a normal sampling distribution of differences between pairs of sample means. *Of course, we do not actually go through the work of generating the sampling distribution of differences.* Knowing what would happen if we did generate such a distribution gives us the necessary information to use it without having to generate it ourselves.

15.3.1a **Standard Error of the Difference.** The measure of variability of the differences between pairs of sample means around their mean difference of zero* is known as the **standard error of the difference.** Its interpretation

> **BOX 15.1** SOME BASIC CONCEPTS
>
> It will help you considerably to understand this section if you feel comfortable with the following concepts: standard deviation, variance, and standard error of the mean. You should know how they differ and how they are related. The material covered in Sections 12.4.2; 12.4.3; 12.4.4; 12.4.5 in Chapter 12 discusses these three concepts in some detail. If you feel the need to refresh your memory, go back and review those sections and then proceed with this section.

*The expected value of the mean difference is zero only if $\mu_1 = \mu_2$. It is possible to test an hypothesis where the mean difference is specified at some other value than zero.

is the same as that of the standard deviation or the standard error of the mean. That is, approximately 68 percent of the differences between pairs of sample means will fall within one standard error of the expected mean difference, assuming a normal distribution. Let the expression, $\sigma_{\bar{x}_1-\bar{x}_2}$, be the symbolic name for the standard error of the difference between pairs of sample means; then, the formula for the standard error of the difference (for independent samples) is as follows:*

(15.2)

$$\sigma_{\bar{x}_1-\bar{x}_2} = \sqrt{\frac{\sigma^2}{N_1}+\frac{\sigma^2}{N_2}}$$

or,

$$= \sqrt{\sigma^2\left(\frac{1}{N_1}+\frac{1}{N_2}\right)}$$

where σ^2 is the population variance; N_1 is the number of cases in the first sample; and N_2 is the number of cases in the second sample.

The problem with this formula is that it requires that the population variance be known. It is very seldom true that the sociologist working with sample data will know the population variance. Given that two samples have been drawn, there will be two sample variances, s_1^2 and s_2^2, available, however. Presumably the two sample variances are independent, unbiased estimates of the unknown population variance.

An assumption of the test of significance of difference between means is that the populations from which the two samples are drawn have the same standard deviation. This assumption is especially pertinent given our hypothesis that the two samples were drawn from the same underlying population. Yet, typically, we have two sample estimates s_1 and s_2 of that common population standard deviation σ and the sample estimates are not identical to each other, presumably differing by chance.

Instead of having two sample variances as separate estimates of the same parameter, it is efficient to combine them into a more reliable single estimate of the parameter. The single estimate will be more reliable because it will be based on N_1+N_2 cases, whereas the individual estimates are based on N_1 and N_2 cases, respectively. To combine the two separate sample estimates, s_1^2 and s_2^2, into a single pooled estimate which we will call simply s^2, we may use the following formula:

(15.3)

$$s^2 = \frac{\Sigma x_1^2 + \Sigma x_2^2}{N_1 + N_2 - 2}$$

where

$$\Sigma x_1^2 = (N_1-1)s_1^2 = \Sigma(X-\overline{X}_1)^2$$

and

$$\Sigma x_2^2 = (N_2-1)s_2^2 = \Sigma(X-\overline{X}_2)^2$$

Once s^2 is found, it can be substituted for the population variance in the formula for the standard error of the difference. (Formula 15.2), giving us the following sample estimate of the standard error:

---

*The general formula for the standard error of the difference includes a correlation term, but in the case of independent samples the correlation is zero and the term disappears.

(15.4)
$$s_{\bar{x}_1 - \bar{x}_2} = \sqrt{s^2\left(\frac{1}{N_1} + \frac{1}{N_2}\right)}$$

where $s_{\bar{x}_1 - \bar{x}_2}$ is the sample estimate of the standard error of the difference. It is also known as the *standard error of the difference*.

In order to pool the sample variances using Formula (15.3) we must have some indication that the variances are homogeneous. As we shall see later in this chapter (Section 15.8), it is possible to test for the homogeneity of sample variances. If the test for homogeneity of variances leads us to conclude that they are heterogeneous, they cannot be pooled, and we must use the following formula for the standard error of the difference instead:

(15.5)
$$s_{\bar{x}_1 - \bar{x}_2} = \sqrt{\frac{s_1^2}{N_1} + \frac{s_2^2}{N_2}}$$

Since we substituted s^2 for σ^2 in Formula (15.4) for the standard error of the difference, we are faced with the same situation we encountered with the sampling distribution for the one-sample case. That is, instead of having a constant, σ^2, in the numerator of each fraction in the formula and a variable (N_1 or N_2) in the denominator, we have variables in both the numerators and the denominators.* Thus, we find that the sampling distribution is the symmetrical but platykurtic Student's t-distribution rather than the normal distribution. The appropriate formula for degrees of freedom in the two-sample case is $df = N_1 + N_2 - 2$ if the population standard deviations are equal.† However, when $N_1 + N_2 \geq 100$, the normal sampling distribution is a good approximation to the Student's t-distribution.

With Formula (15.4), given the data from two random samples, we can get an estimate of the sampling error for the sampling distribution of differences between pairs of sample means. Armed with this estimate, we can test hypotheses about empirically determined differences between sample means using the following test statistic:

$$t = \frac{(\overline{X}_1 - \overline{X}_2) - (\mu_1 - \mu_2)}{S_{\bar{x}_1 - \bar{x}_2}}$$

where $\overline{X}_1$ and $\overline{X}_2$ are the two sample means, μ_1 and μ_2 reflect any expected difference between population means (in this case the expected difference is zero so that term can be ignored), $S_{\bar{x}_1 - \bar{x}_2}$ is the standard error of the difference between means.

15.3.2   An Example from the Literature

Eldon L. Wegner and William H. Sewell conducted a study of the relationship between the type of college attended and the probability of graduation from that college (Wegner and Sewell, 1970). They explored four specific questions:

*Note that $s^2(1/N_1 + 1/N_2) = s^2/N_1 + s^2/N_2$
†If $\sigma_1 \neq \sigma_2$ (tested by a test for homogeneity) the test is only approximately a t-distribution with df a weighted average of $N_1 - 1$ and $N_2 - 1$.

BOX 15.2 DEGREES OF FREEDOM AGAIN

Notice that the appropriate degrees of freedom formula for this two-sample case is $df = N_1 + N_2 - 2$. The formula for the standard error of the difference essentially combines the standard error of the mean for the first sample with the standard error for the second sample. This can be seen most clearly in Formula (15.5). Actually, the standard errors are squared (thus, $[s_1/\sqrt{N_1}]^2 = s_1^2/N_1$), summed, and the square root of their sum is taken. The combination is handled this way because standard errors are not additive, but their squares are.

Since the standard errors of the two samples are combined in the formula for the standard error of the difference, the degrees of freedom associated with each sample standard error are also combined. That is, the degrees of freedom for the first sample, $N_1 - 1$, are added to the degrees of freedom for the second sample, $N_2 - 1$, to arrive at the proper degrees of freedom for the standard error of the difference; i.e., $(N_1 - 1) + (N_2 - 1) = N_1 + N_2 - 2$.

Perhaps this would be a good point to go back to Section 12.4.7 and review the general discussion of the concept of degrees of freedom.

(1) Do different types of colleges recruit students with different characteristics, and are those characteristics related to the probability of graduating from college? (2) Are there institutional differences in graduation rates which cannot be accounted for by the background characteristics of students or by differential rates of students transferring or students with extended undergraduate careers? (3) From what types of schools do students of different intelligence and socioeconomic levels experience the greatest probability of graduating? (4) Does the process of selection into institutions influence the probability of completing college beyond the effects of different college contexts? Are students of different intelligence and socioeconomic status distributed among institutions in accordance with their chances of graduating from them? (Wegner and Sewell, 1970:667.)*

To illustrate how to do a test of differences between means from two independent samples we will focus on the first part of Wegner and Sewell's first question: Do different types of colleges recruit students with different characteristics?

The data for the study came from a one-third probability sample of high school seniors in the state of Wisconsin in 1957. In June 1964 a follow-up mail questionnaire identified those men who had attended college between 1957 and 1964. Data used by Wegner and Sewell came from these cases. Those who went to college attended 126 different colleges or universities. The authors divided these colleges and universities into different types and compared characteristics of those attending the various types.* They looked at rank in high school class, rank in intelligence, family socioeconomic status, and level of occupational aspiration.

We will concentrate on the family socioeconomic status of those men who went either to state colleges or "good" liberal arts colleges. Family socioeconomic status was measured on a scale which combined items involving the father's occupation, parents' educational level, and family economic resources. The potential range of scores on the scale was from 1 to 99.

A question for which we might seek an answer is whether these state colleges and "good" liberal arts colleges have selected different kinds of students as far as family socioeconomic status is concerned. Since tuition charges and other expenses involved in going to college are generally lower for state college students than for students of liberal arts colleges, it is reasonable to hypothesize that those men who attended the "good" liberal arts colleges would, on the average, come from families with higher socioeconomic status scores than those who attended the state colleges. This hypothesis can be stated symbolically as follows:

$$H_1 : \mu_1 > \mu_2$$

where μ_1 is the mean family socioeconomic status score of those men who attended the "good" liberal arts colleges and μ_2 is the mean family socioeconomic status score of those who attended state colleges. The null hypothesis covering the other alternatives to this hypothesis would be as follows:

$$H_0 : \mu_1 \leqq \mu_2$$

although the null hypothesis we actually test is the limiting case of $\mu_1 = \mu_2$.

The test of the null hypothesis will be a one-tailed test with H_1 as the only set of alternatives. Assume that we wish to test the null hypothesis at the 1 percent level of significance. If the sample sizes (designated N_1 and N_2) are large enough, a normal approximation to the sampling distribution can be used, and the standard score that bounds the critical region in the right tail of the curve will be +2.33. If the samples are small, Student's t-distribution will be used with $N_1 + N_2 - 2$ degrees of freedom.

It so happens that the authors had data on 394 men who attended state colleges and on 93 who attended the "good" liberal arts colleges, so it is possible to use the normal approximation.

In Table 15.5 we have presented the essential steps in the test of the null hypothesis using the Wegner and Sewell data. The standard score for evaluating the data against the normal curve is as follows (assuming no expected difference between μ_1 and μ_2):

(15.6)
$$z = \frac{\overline{X}_1 - \overline{X}_2}{s_{\bar{x}_1 - \bar{x}_2}}$$

*The types of colleges used by Wegner and Sewell were (1) high-prestige state university, (2) urban state university, (3) state colleges, (4) Catholic urban university, (5) out-of-state universities (that is, out of the state of Wisconsin), (6) "good" liberal arts colleges, (7) other liberal arts colleges, and (8) other four-year colleges. The distinction between "good" liberal arts colleges and other liberal arts colleges was based upon a quality-of-colleges scale developed by Wegner and Sewell (see Wegner and Sewell, 1970:669). Data from this study used by permission of the University of Chicago. (Copyright 1970 by the University of Chicago: all rights reserved.)

TABLE 15.5 Two-sample Test for Means: Type of College and Socioeconomic Status of Students

$H_0 : \mu_1 \leqq \mu_2$

Alternative to H_0:

$H_1 : \mu_1 > \mu_2$ designate this set of alternatives if
$z \geqq +2.33$

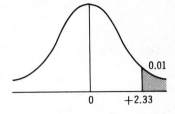

0.01

0   +2.33

Sample 1 (Good liberal arts colleges	Sample 2 (State colleges)
$\overline{X}_1 = 50$	$\overline{X}_2 = 38$
$N_1 = 93$	$N_2 = 394$
$s_1^2 = 169$	$s_2^2 = 121$
$\Sigma x_1^2 = 15{,}548$	$\Sigma x_2^2 = 47{,}553$

Pooled variance equals:

$$s^2 = \frac{\Sigma x_1^2 + \Sigma x_2^2}{N_1 + N_2 - 2} = \frac{15{,}548 + 47{,}553}{93 + 394 - 2} = 130.11$$

Standard error of difference equals:

$$s_{\bar{x}_1 - \bar{x}_2} = \sqrt{s^2\left(\frac{1}{N_1} + \frac{1}{N_2}\right)}$$
$$= \sqrt{130.11(.010753 + .00253807)}$$
$$= \sqrt{1.73} = 1.32$$

Standard score equals:

$$z = \frac{\overline{X}_1 - \overline{X}_2}{s_{\bar{x}_1 - \bar{x}_2}} = \frac{50 - 38}{1.32} = +9.09$$

Conclusion:
Therefore, reject the null hypothesis in favor of the set of alternatives $H_1 : \mu_1 > \mu_2$

---

where $\overline{X}_1$ is the sample mean for men who attended "good" liberal arts colleges and $\overline{X}_2$ is the sample mean for men who attended state colleges.

This is another variation of the familiar standard score formula as we have seen. If $\overline{X}_1$ is larger than $\overline{X}_2$ and the difference between them is at least 2.33 times the expected difference, we will reject the null hypothesis in favor of its set of alternatives.*

In using this technique to test the null hypothesis we must assume that the conditions are such that the statistical model underlying the technique (this model is discussed in Section 15.3.3) is reasonably satisfied.

As you can see from Table 15.5, the computed z score is +9.09, which is well into the critical region in the right tail of the sampling distribution. Therefore, the null hypothesis is rejected in favor of the set of alternatives to it. The conclusion is that the "good" liberal arts colleges selected students with higher socioeconomic status, on the average, than did the state colleges.

Note that the probability of the type-I error being risked is not 1 percent when the computed z score is +9.09; rather it is equivalent to that area beyond

*Note that, in general, hypotheses are stated in terms of parameters but tested in terms of statistics.

+9.09 in the tail of the sampling distribution (for a one-tailed test). The area beyond a z score of +9.09 is very small, less than 0.00001; therefore, the probability of committing a type-I error is negligible.

15.3.3   Assumptions of the Test for Differences between Means

The test for differences between means is a powerful statistical tool for testing null hypotheses. Its use is limited, however, by the following set of assumptions: (1) the level of measurement of the variable being studied is at least interval scale, (2) the observations are independent, (3) the samples are independent, (4) the samples are simple random samples from a normally distributed population, and (5) the sample variances are homogeneous. This last assumption is particularly important because the technique is sensitive to differences in variability as well as differences in central tendency. If variances are not equal or nearly so, it is difficult to establish that statistically significant differences are due to differences in central tendency alone.

15.4   THE K-SAMPLE CASE FOR MEANS OF INDEPENDENT SAMPLES

Thus far we have examined techniques for testing statistically significant differences between a sample mean and a population mean, and between means from two independent samples. It is conceivable that a researcher may be interested in testing for differences between means from more than two independent samples (that is, k samples where k is any number greater than two). For instance, it may be desirable to compare the means from three samples to see if they differ.

It might at first seem that the way to handle such a situation would be to take the means in pairs and then test the various combinations of pairs to see if they differ significantly. Thus mean one would be tested against mean two, mean two against mean three, and mean one against mean three, using the test for differences between means described in the previous section. The problem with this approach is that the various tests of pairs of sample means would not all be independent tests, and the resulting probabilities would be overlapping. For example, if mean one is tested against mean two and mean two is tested against mean three, the test of mean one against mean three is not independent of the other two tests. In such a case the usual sampling distributions are not applicable.

15.4a   Analysis of Variance.   What must be done instead is a test that will make possible a simultaneous comparison of more than two sample means, and will have a known sampling distribution. The **analysis of variance** is such a test. Analysis of variance (sometimes called ANOVA) is, in effect, an extension of the test for differences between means to cover the case where more than two samples (k samples) are involved. It has a known sampling distribution — the **F-distribution** (see Section 15.4.2). Although the analysis of variance is a test of differences in central tendencies, it is based upon a comparison of variances — thus, the name.

The basis for the analysis is a comparison of the amount of heterogeneity within samples with the amount between samples. It stands to reason that if subjects are exposed to the same conditions as others in their own group, but to conditions that differ from those to which subjects in other groups are exposed (and those conditions make a difference), then subjects within groups will be more alike than subjects between groups.

This is the kind of reasoning that is behind the use of the concept of social class by many sociologists. Those who defend the concept and use it generally assume that there is homogeneity within classes and heterogeneity between classes. For example, they claim that "upper-class" persons think and act similarly, but differently from persons who belong to other classes (Warner *et al.*, 1949). On the other hand, critics of the class concept argue that the concept is unfruitful because there is as much heterogeneity within classes as there is between classes. (See Faris, 1954, and Svalastoga, 1964.)

Perhaps the best way to make the analysis of variance understandable and to explain the procedures used in doing such analysis is to work through an example from the literature.

15.4.1   An Example from the Literature

Mark and Schwirian have explored "the general relationship between urban central-place function and community-population growth in a region rapidly improving in transportation." (Mark and Schwirian, 1967:31.)

They assert that during a "given time period, the general relationship between urban central-place function and community-population growth is a function of *both* the level of regional industrialization, and of the particular ecological niche or position of the community in the total pattern of intercommunity relations." (Mark and Schwirian, 1967: 31.)

The central-place model assumes (1) that cities service their surrounding agricultural populations and are centrally located within their trade areas, (2) that within regions a hierarchy of central places emerges because of cities' differing ecological positions and functions, and (3) that distance between cities is a function of size and mode of transportation in vogue at the time of location (Mark and Schwirian, 1967:30). The point is made that central-place function is a community building activity under certain conditions, but that as conditions change, central-place function ceases to be a community building activity. When industrialization takes place, according to the theory, central cities usurp the trade of the hinterland centers as transportation improves, and it is easier to get to the larger population centers. The authors hypothesize "that during regional industrialization, while there is a rapid population growth and expansion of central-place function for certain ecological classes of cities, central-place function is no longer a significant community-building activity."(Mark and Schwirian, 1967:33.)

In order to test their hypothesis they collected data on urban incorporated places in Iowa for 1950 and 1960. They classified the Iowa urban places into five ecological types: central cities (of metropolitan areas), suburbs, metropolitan neighbors (independent communities at the periphery of metropolitan areas), competitive trade centers (hinterland towns in competition with other hinterland towns for retail trade), and non-competitive trade centers (hinterland towns without competition from other centers). Among these five types of communities the authors studied differences in population growth and central-place function as measured by per capita retail trade, while holding constant the stage of industrial development.

We will focus on that part of their analysis in which they examined the relationship between type of community and central-place function. For reasons to be explained a little later, in our examination of the relationship between community type and central-place function, we will look only at central cities, metropolitan neighbors, competitive trade centers, and non-competitive trade centers. We will not include suburbs. Using per-capita retail sales as an index of central-place function, and comparing data for 1950 with 1960, we will seek to answer the question: Do the four community types differ in mean changes in central-place function between 1950 and 1960? The authors' theory would lead us to expect the mean increase in per-capita retail sales to be greater for central cities than for either competitive or non-competitive trade centers, with metropolitan neighbors in an intermediate position. This is so because it is at the industrialization stage that the central cities are supposed to profit at the expense of the trade centers. The null hypothesis will be that the mean increases will be the same for all four types of community. That is:

$$H_0: \mu_1 = \mu_2 = \mu_3 = \mu_4 = \mu$$

where μ_1 is the mean change for central cities, μ_2 is the mean change for metropolitan neighbors, μ_3 is the mean change for competitive trade centers, μ_4 is the mean change for non-competitive trade centers, and μ is the mean change for all urban areas. In using the analysis of variance to test this hypothesis, we assume that the conditions for its use are reasonably satisfied (see Section 15.4.4).

If the analysis of variance leads us to believe that there is a difference between any, some, or all of the means for the four types of community, we will reject the null hypothesis in favor of the set of alternatives that two or more of the means are significantly different.

Table 15.6 is adapted from Table 2 of the Mark and Schwirian article. We have deleted the data for suburbs, recomputed the mean for all urban places, and

TABLE 15.6 COMMUNITY ECOLOGICAL POSITION AND MEAN CHANGE IN URBAN CENTRAL-PLACE FUNCTION INDEX, 1950–1960

Community Ecological Position	Change in Index, 1950–1960			
	$\overline{X}$	s^2	n	Σx^2
Central Cities	41.2	81	6	405
Metropolitan Neighbors	34.4	75.69	19	1362.42
Competitive Trade Centers	32.7	179.56	48	8439.32
Non-Competitive Trade Centers	32.7	176.89	14	2299.57
Total Urban Places	33.7	150.18	87	12915.12

computed the variances and sums of squared deviations for all four community ecological types and total urban places. The sums of squared deviations were computed with the following formula:

(15.7)
$$\Sigma x^2 = (n-1)s^2$$

The first four variances in the table $(81, 75.69, 179.56, \text{and } 176.89)$ measure the variability in changes in per capita retail sales of the four types of communities. For example, $s_1^2 = 81$ reflects the variability of changes in per capita retail sales among the 6 urban places classified as central cities. The variance for total urban places (150.18) measures the variability of changes, in per capita retail sales of each of the 87 urban places, around the mean change for all urban places (33.7).

15.4.1a   Total Sum of Squares.   The **sum of squares** (short for the sum of the squared deviations of scores around their mean) for total urban places $(14{,}277.52)$, interestingly enough can be partitioned into two separate sources of variation — the variation among urban places within each of the four ecological types (called the sum of squares within groups), and the variation of the mean changes of each of the four ecological types around the mean change for all urban places (called the sum of squares between groups). Let Σx_t^2 be the total sum of squares; Σx_w^2, the within-groups sum of squares; and Σx_b^2, the between-group sum of squares. We can say then that:

(15.8)
$$\Sigma x_t^2 = \Sigma x_w^2 + \Sigma x_b^2$$

The **total sum of squares** may be computed by taking the deviation of each change score for each of the 87 urban places from the mean change of all 87 (mean change for the total), squaring these deviations and summing them. That is:

(15.9)
$$\Sigma x_t^2 = \sum_{i=1}^{87} (X_i - \overline{X}_t)^2$$

where X_i is the change score of the ith urban place, $\overline{X}_t$ is the mean change for all 87 urban places, and where $\Sigma_{i=1}^{87}$ means the squared deviations are summed for all 87 places.

15.4.1b   Within-group Sum of Squares.   The **sum of squares within groups** may be computed by finding the sum of squared deviations of the change scores within each sample (community type) from their sample mean, and then summing these sums of squares for all k of the samples. That is:

(15.10)
$$\Sigma x_w^2 = \Sigma x_1^2 + \Sigma x_2^2 + \Sigma x_3^2 + \cdots + \Sigma x_k^2$$

where Σx_1^2 is the sum of squares for the first sample, Σx_2^2 is the sum of squares for the second sample, etc., and k is the number of samples.

For the data in Table 15.6 we would find the within-group sum of squares by substituting into the formula and summing as follows:

$$\Sigma x_w^2 = 405 + 1362.42 + 8439.32 + 2299.57$$
$$= 12{,}506.31$$

Since the variances were already computed for these data we computed the sum of squares by using Formula (15.7). However, if we were working with raw data we would compute the sum of squares with the following computational formula:

(15.11)
$$\Sigma x^2 = \Sigma X^2 - \frac{(\Sigma X)^2}{n}$$

15.4.1c Between-group Sum of Squares. The sum of squares between groups is computed by taking the deviation of each sample mean from the total mean, squaring the deviation, multiplying the squared deviation by the sample n, and summing these products for all k of the samples. That is

(15.12)
$$\Sigma x_b^2 = \sum_{i=1}^{k} n_i (\overline{X}_i - \overline{X}_t)^2$$

where n_i is the number of cases in the ith sample, $\overline{X}_i$ is the mean for the ith sample, $\overline{X}_t$ is the total mean, and k is the number of samples.

For the data at hand, the sum of squares between groups may be computed as follows:

$$\Sigma x_b^2 = 6(41.2 - 33.7)^2 + 19(34.4 - 33.7)^2 + 48(32.7 - 33.7)^2$$
$$+ 14(32.7 - 33.7)^2$$
$$= 6(7.5)^2 + 19(.7)^2 + 48(-1)^2 + 14(-1)^2$$
$$= 337.5 + 9.31 + 48 + 14$$
$$= 408.81$$

Since it is true that $\Sigma x_t^2 = \Sigma x_w^2 + \Sigma x_b^2$, we can check the results by adding the sum of squares within groups to the sum of squares between groups to see whether their sum is equal to the total sum of squares.

$$12,915.12 = 12,506.31 + 408.81$$

15.4.1d Within-groups and Between-groups Variances. In order to test the null hypothesis it is necessary to compute the variance between groups and the variance within groups. The between-groups variance is then divided by the within-groups variance to get the computed F. The between-groups vari-

---

BOX 15.3 WATCH THE SUMS OF SQUARES

The reasoning involved here is not at all complicated. However, if you do not follow the explanation carefully, you may not understand the total sum of squares, the between-groups sum of squares, and the within-groups sum of squares. As you read this material pay close attention to Table 15.6 and relate the numbers there to the formulae for sums of squares. It may be very helpful to you to take paper and pencil and actually substitute the numbers from Table 15.6 for the symbols in the formulae for the within-groups and the between-groups sums of squares.

ance is computed by dividing the between-groups sum of squares by $k-1$ degrees of freedom where k is the number of samples:

(15.13)
$$s_b^2 = \Sigma x_b^2 / (k-1)$$

The within-groups variance is computed by dividing the sum of squares within groups by $N-k$ degrees of freedom where N is the number of scores in the total sample (*i.e.*, $N = n_1 + n_2 + n_3 + \cdots + n_k$):

(15.14)
$$s_w^2 = \Sigma x_w^2 / (N-k)$$

15.4.2   The F-Distribution

If the null hypothesis that two sample variances are independent estimates of a common population variance ($s_1^2 = s_2^2 = \sigma^2$) is true, then the ratio of the two sample variances has an F sampling distribution. The F-distribution, like the chi-square (χ^2) distribution and the t-distribution, is actually a family of related sampling distributions rather than a single distribution. Consequently, the F table (see Appendix Table E) is set up with degrees of freedom to represent individual distributions from the family. In the table of F, however, there are entries for degrees of freedom across the tops of the columns and down the edge of the rows. The df_1 across the top of the table refers to the number of degrees of freedom for the variance in the *numerator* of the ratio, and the df_2 down the first column of the table refers to the number of degrees of freedom for the variance in the *denominator* of the ratio.

The first section of the F table gives values of F that cut 5 percent of the area out of the right tail of the distribution, and the second section gives values of F that cut 1 percent of the area out of the right tail. The F-distributions are such that the critical region usually appears only in the right tail. If the two variances that form the ratio are equal, the F will equal 1. The F values in the critical region will always be greater than 1. The larger the F values, the farther they are into the right tail of the distribution. In general the distributions in the family of F-distributions are skewed to the right. The shape of a particular F-distribution varies with different values of df_1 and df_2.

15.4.3   Test of the Null Hypothesis

In order to test the null hypothesis for the Mark and Schwirian data we must compute F from the following formula:

(15.15)
$$F = s_b^2 / s_w^2$$

where s_b^2 is the between-groups variance and s_w^2 is the within-groups variance.

The between-groups variance always appears in the numerator of the F ratio, and the within-groups variance in the denominator. The F ratio essentially compares the observed differences between the sample means of the groups being studied (as measured by the between-groups variance) with an error term

TABLE 15.7 SUMMARY TABLE OF ANALYSIS OF VARIANCE FOR COMMUNITY TYPE AND CENTRAL-PLACE FUNCTION

Source of Variation	Σx^2	df	s^2	F
Between-Groups	408.81	3	136.27	0.90
Within-Groups	12506.31	83	150.68	
Total	12915.12			

consisting of measures of variability of scores within the groups being studied (the within-groups variance). Since we are interested in rejecting the null hypothesis, and the null hypothesis can only be rejected when differences appear between the sample means for the various groups studied, we always divide the error term into the between-groups variance. Only when the between-groups variance is *larger* than the within-groups variance will there be a possibility of rejecting the null hypothesis.

The computed F ratio is evaluated against the tabled F value that bounds the critical region of the sampling distribution. To find that particular F value for the 5 percent level of significance, we enter the first section of the F table with $df_1 = k - 1$ or 3 and $df_2 = N - k$ or 83. The $k - 1$ degrees of freedom are those for the between-groups variance, and the $N - k$ degrees of freedom are those for the within-groups variance. Since the F table does not have an entry for $df_2 = 83$, we will use the entry for 80 degrees of freedom, thus erring on the side of conservatism, because the tabled F will be slightly larger than the one for 83 degrees of freedom would have been. Looking in the table with 3 and 80 degrees of freedom, we find that the F value that bounds the 5 per cent critical region is 2.72. Thus, if the computed F value is as large or larger than 2.72, we will reject the null hypothesis that all of the sample means are estimates of a common population mean, and we can conclude that at least two of them estimate different parameters. Table 15.7 is a summary table of the analysis of variance for the community type and central-place function data. Figure 15.1 is a graph of the F-distribution of interest. Note that the computed F ratio is 0.90, which is not in the critical region. Therefore, rejection of the null hypothesis is not warranted. An examination of the four sample means in Table 15.6 would seem to indicate that the mean change for central cities was higher than for the other three community types, although the F-test does not seem to indicate significant differences in any of the means.

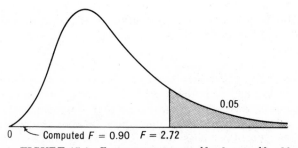

FIGURE 15.1 F-DISTRIBUTION FOR $df_1 = 3$ AND $df_2 = 80$.

It should be reiterated that analysis of variance does not zero in on specific differences between means. It tests a whole set of sample means simultaneously, and rejection of the null hypothesis merely indicates that there is a statistically significant difference between at least two means in the set. However, there could be differences between more than two means, or even between all of the means in the set. Edwards discusses a number of techniques that allow for tests of all possible contrasts between two or more sample means, without having to worry about any confusion that might result from overlapping probabilities arising from interdependent comparisons.* These techniques are based upon joint sampling distributions that are appropriate for such multiple comparisons.

15.4.4   Assumptions of One-Way Analysis of Variance

The analysis of variance discussed here is the simplest case, called the **one-way analysis of variance.** The technique may be elaborated in a number of ways that we will not discuss. (See Edwards, 1960.) These variations on the simple case naturally make more restrictive assumptions. For example, it is generally necessary to be able to assign subjects randomly to different experimental conditions. This random assignment is most easily handled in a laboratory situation. Because sociologists usually engage in survey research and field studies in which it is not possible to assign subjects randomly to conditions, the one-way analysis of variance is most applicable.

The one-way analysis of variance makes basically the same assumptions as the test for differences between means. Interval measurements are assumed. It is assumed that the observations are independent. The samples should be independent simple random samples from a normally distributed population. And, the population variances are assumed to be equal.

If the sample N's are large, deviation from the assumption of a normally distributed population will not result in serious errors of interpretation. The assumption of homogeneous variances is crucial, however, if we are interested in establishing that differences between samples are due to differences in central tendency. The analysis of variance is sensitive to differences in variability as well as differences in central tendency. Therefore, if variances are not homogeneous, it will be difficult to sort out differences in variability from differences in central tendency.

The variance for suburbs in the Mark and Schwirian study was, by statistical test (see Section 15.8), significantly larger than the variances of the other four types of communities. Since this presents problems in the interpretation of the results, we excluded the data on suburbs from our analysis of their data.

Diagram 15.1 (on the adjacent pages) summarizes the essential features of tests of hypotheses about means for the one-sample, two-sample, and k-sample cases. Study the Diagram and refer back to our discussions of these techniques for more detailed information.

---

*These techniques will not be discussed here, but they can be found in Edwards (1960).

DIAGRAM 15.1 Tests of Hypotheses about *Means*

> THE ONE-SAMPLE CASE

Problem: To determine whether a given sample mean is an estimate of a known or theoretical population mean.

Assumptions:
1. Interval level of measurement
2. Independent observations
3. Normally distributed population
4. A simple random sample of that population

Null and Alternate Hypotheses:

H_0:μ = (some specified value)
H_1:$\mu \neq$ (specified value)

or,

$\mu <$ (this specified value) or $\mu >$ (this specified value)

Standard Error of the Mean:

$\sigma_{\bar{X}} = \sigma/\sqrt{N}$ if σ is known.

$s_{\bar{X}} = s/\sqrt{N}$ if σ is unknown.

Standard Score Formulae:

$z = \dfrac{\overline{X} - \mu}{\sigma_{\bar{X}}}$ if $\sigma_{\bar{X}}$ is known.

$z = \dfrac{\overline{X} - \mu}{s_{\bar{X}}}$ if $\sigma_{\bar{X}}$ is unknown but N is large.

$t = \dfrac{\overline{X} - \mu}{s_{\bar{X}}}$ if $\sigma_{\bar{X}}$ is unknown and N is small.

Sampling Distribution:
1. The normal curve if σ is known or N is large (100 or more).
2. Student's t-distribution with $N-1$ degrees of freedom if σ is unknown and N is small.

Step 1: List assumptions.

Step 2: Set up the null hypothesis. Specify whether the test is one-tailed or two-tailed. Specify the level of significance. Determine critical region.

Step 3: Draw a simple random sample from the population and collect data.

Step 4: From the sample data, compute the mean, standard deviation, and standard error of the mean.

Step 5: Substitute into the appropriate standard score formula and compute the standard score. If the standard score is in the critical region, reject the null hypothesis; if not, fail to reject it.

DIAGRAM 15.1 *(Continued)*

THE TWO-SAMPLE CASE

Problem: To determine whether the means of two independent samples estimate a common parameter.

Assumptions:
1. Interval level of measurement
2. Independent observations
3. Normally distributed population
4. Two independent simple random samples
5. Homogeneity of sample variances

Null and Alternate Hypotheses:

$$H_0 : \mu_1 = \mu_2$$
$$H_1 : \mu_1 \neq \mu_2$$

or,

$$\mu_1 < \mu_2 \text{ or } \mu_1 > \mu_2$$

Standard Error of the Difference:

$$\sigma_{\bar{X}-\bar{x}} = \sqrt{\sigma^2 \left(\frac{1}{N_1} + \frac{1}{N_2} \right)} \quad \text{if } \sigma \text{ is known.}$$

$$s_{\bar{X}-\bar{x}} = \sqrt{s^2 \left(\frac{1}{N_1} + \frac{1}{N_2} \right)} \quad \text{if } \sigma \text{ is unknown and the sample variances are homogeneous.}$$

Formula for finding pooled variance:

$$s^2 = \frac{\Sigma x_1^2 + \Sigma x_2^2}{N_1 + N_2 - 2}$$

Standard Score Formulae:

$$z = \frac{\overline{X}_1 - \overline{X}_2}{\sigma_{\bar{x}_1 - \bar{x}_2}} \quad \text{if } \sigma \text{ is known.}$$

$$z = \frac{\overline{X}_1 - \overline{X}_2}{s_{\bar{x}_1 - \bar{x}_2}} \quad \text{if } \sigma \text{ is unknown but } N_1 + N_2 \geqq 100.$$

$$t = \frac{\overline{X}_1 - \overline{X}_2}{s_{\bar{x}_1 - \bar{x}_2}} \quad \text{if } \sigma \text{ is unknown and } N_1 + N_2 < 100.$$

Sampling Distribution:
1. The normal curve if σ^2 is known or if $N_1 + N_2 \geqq 100$.
2. Student's *t*-distribution with $N_1 + N_2 - 2$ degrees of freedom if σ is unknown and $N_1 + N_2 < 100$.

Step 1: List assumptions.

Step 2: Set up the null hypothesis. Specify whether the test is one-tailed or two-tailed. Specify the level of significance. Determine critical region.

DIAGRAM 15.1 *(Continued)*

Step 3: Draw two simple random samples and collect data.

Step 4: From the sample data compute means, variances, the pooled variance, and the standard error of the difference.

Step 5: Substitute into the appropriate standard score formula and solve. If the standard score is in the critical region, rejection of the null hypothesis is warranted. If not, rejection is not warranted.

THE K-SAMPLE CASE

Problem: To determine whether the means of k independent samples (where k is any number greater than two) estimate a common parameter.

Assumptions:

1. Interval level of measurement
2. Independent observations
3. Normally distributed population
4. Independent simple random samples
5. Homogeneous sample variances

Null and Alternate Hypotheses:

$$H_0 : \mu_1 = \mu_2 = \mu_3 \ldots = \mu_k$$

$$H_1 : \mu_1 \neq \mu_2 \neq \mu_3 \ldots = \mu_k$$

The Sums of Squares:

Total sum of squares:

$$\Sigma x_t^2 = \sum_{i=1}^{N} (X_i - \overline{X_t})^2$$

Between-groups sum of squares:

$$\Sigma x_b^2 = \sum_{i=1}^{k} n_i (\overline{X_i} - \overline{X_t})^2$$

Within-groups sum of squares:

$$\Sigma x_w^2 = \Sigma x_1^2 + \Sigma x_2^2 + \cdots \Sigma x_k^2$$

where $\quad \Sigma x_1^2 = \sum\limits_{i=1}^{n_i} (X_i - \overline{X_1})^2, \quad$ etc.

As an alternative: $\quad \Sigma x_w^2 = \Sigma x_t^2 - \Sigma x_b^2$

Variances:

Between groups:

$$s_b^2 = \Sigma x_b^2 / (k-1)$$

Within groups:

$$s_w^2 = \Sigma x_w^2 / (N-k)$$

553

DIAGRAM 15.1 *(Continued)*

F Ratio:

$$F = s_b^2/s_w^2$$

Sampling Distribution:
 The F-distribution with $df_1 = k-1$ and $df_2 = N-k$.

Step 1: List assumptions.

Step 2: Set up the null hypothesis. Specify the level of significance. Determine critical region.

Step 3: Draw simple random samples and collect data.

Step 4: From the sample data compute total sum of squares, sum of squares between groups, and sum of squares within groups. Compute the within-groups variance and the between-groups variance.

Step 5: Substitute into the F ratio and solve. Enter the F table with $k-1$ and $N-k$ degrees of freedom. If the computed F is as large or larger than the tabled F, reject the null hypothesis. If not, fail to reject.

15.5   THE ONE-SAMPLE CASE FOR PROPORTIONS

Much of the data with which sociologists work lend themselves most readily to presentation as proportions or percentages.* The occasion arises, therefore, to test hypotheses about proportions. In this section we will examine procedures used to test a sample proportion against a known or theoretical population proportion to determine whether the sample proportion is a likely estimate of the population proportion. In the following section we will examine procedures used to test for differences between proportions from two independent samples.

Coombs *et al.,* conducted a study of the disadvantaged status of premaritally pregnant couples (Coombs *et al.,* 1970). They sought answers to three principal questions:

1. Are those couples whose first child was premaritally conceived (hereafter referred to as PMP couples) drawn distinctively from lower-status groups, or are they distinctive in other ways in the social strata to which they belonged before marriage?
2. Does the social background of the family of origin of the PMP couples account for their disadvantaged position after marriage?
3. Is being premaritally pregnant similar in its effect to having a child born a short interval after marriage, although not premaritally conceived? (Coombs *et al.,* 1970:801.)†

The data were from a longitudinal study of the family growth of white, married couples in the Detroit area who were interviewed after the birth of a first, second, or fourth child. A random sample was drawn of all white, married women in the Detroit area who had a first, second, or fourth live birth during the month of July, 1961. The data for this study were from the 1053 respondents who were married only once.

Among other variables investigated was the educational level of the wife's father. The proportion of premaritally pregnant couples among all of the respondents in the sample was .198. Coombs *et al.* present data on the proportion of premaritally pregnant couples in which the wife's father had attended college (.123). We will explore the possibility that this proportion differs significantly from the overall proportion of .198. The authors suggest it is commonly hypothesized that there is an inverse relationship between incidence of premarital conceptions and status of the couple's parents (Coombs *et al.,* 1970:804). Taking our cue from this suggestion we will hypothesize that the sample proportion (p) for wives whose fathers have attended college estimates a parameter significantly lower than the proportion of all premaritally pregnant couples (P_u). That is:

$$H_1 : P_c < P_u$$

Therefore, the null hypothesis will be as follows:

$$H_0 : P_c \geq P_u$$

where P_c is the population proportion (parameter) for premaritally pregnant

---

*A percentage is easily converted to a proportion by dividing it by 100. Since proportions are more convenient to use for computations, it is common to do tests of hypotheses for proportions rather than percentages.

†Reprinted by permission of the publisher; copyright 1970 by The University of Chicago; all rights reserved.

wives whose fathers attended college, and P_u is the population proportion of all premaritally pregnant couples.

Since there were 122 women in the study whose fathers had attended college and P_u was .198, the rule of thumb applies,* and we can use the normal distribution as an approximation of the sampling distribution of the sample proportion p. If we do a one-tailed test of the null hypothesis at the .05 level of significance, the z score that bounds the critical region will be -1.64. Consequently, if the standard score computed from the sample data is equal to or less than -1.64, we will reject the null hypothesis in favor of its set of alternatives.

As was pointed out in Chapter 13 (Section 13.2.3), the standard error of a proportion when P_u is known (as it is in this case) is given by the following:

(15.16)
$$\sigma_p = \sqrt{\frac{P_u Q_u}{N}}$$

Given the standard error, a standard score can be computed from the following:

(15.17)
$$z = \frac{p - P_u}{\sigma_p}$$

Assuming that all the necessary conditions of the statistical model can be met (see next section), we can use Formula (15.17) to test the null hypothesis. All we need do is substitute $P_u = .198$, $Q_u = 1 - P_u = .802$, and $N = 122$ into Formula (15.16) to solve for σ_p. Then, p, P_u and σ_p can be substituted into Formula (15.17) to compute z. Table 15.8 traces this process of testing the null hypothesis step by step.

As can be seen from the table, σ_p turns out to be .036. The computed stan-

---

TABLE 15.8 One-sample Test for Proportions: Premaritally Pregnant Couples and Educational Level of Wife's Father

$H_0 : P_c \geq P_u$

Set of alternatives to H_0:

$H_1 : P_c < P_u$ Accept this set of alternatives if
    $z \leq -1.64$.

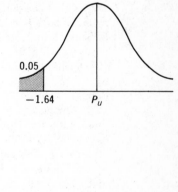

$P_u = .198$
$Q_u = .802$
$p = .123$
$N = 122$

$$\sigma_p = \sqrt{\frac{P_u Q_u}{N}} = \sqrt{\frac{(.198)(.802)}{122}} = \sqrt{.0013} = .036$$

$$z = \frac{p - P_u}{\sigma_p} = \frac{.123 - .198}{.036} = \frac{-.075}{.036} = -2.08$$

Therefore, reject the null hypothesis and accept the set of alternatives $H_1 : P_c < P_u$.

---

*Both NP and NQ are equal to or greater than 5. See Chapter 12.3.3

dard score comes to −2.08, which is well into the critical region beyond −1.64, so rejection of the null hypothesis is warranted. As a matter of fact, the probability of getting z as small as −2.08 by chance is less than .02 (see the normal curve table, Appendix Table B). It is reasonable to conclude from these results that wives whose fathers attended college were less likely than all wives represented by the sample to be premaritally pregnant. This finding offers some support to the original hypothesis that premarital pregnancy is inversely related to social status.

15.5.1   Assumptions of the One-sample Test for Proportions

The one-sample test for proportions is really just another version of the test for frequencies which makes use of the binomial probability distribution. Therefore, the assumptions are basically the same. It is assumed that the observations are independent, and that the data constitute a simple random sample from the population. If it is possible to meet the rule of thumb that NP and NQ are both equal to or greater than 5, then the binomial sampling distribution can be approximated by the normal distribution.

It is not necessary to assume that the population is normally distributed. As a matter of fact, the shape of the population distribution need not be specified. Though most appropriate for nominal-level data, the test can also be applied to ordinal or interval-level data. At the ordinal and interval levels the test is wasteful of information, however.

15.6   THE TWO-SAMPLE CASE FOR PROPORTIONS

Sociologists have been concerned for some time with assessing the consequences of matriarchal family structures in black families for children. It is sometimes argued that the dominance of black mothers leads to poor academic achievement by black boys and increased emphasis upon the education of black girls. Denise B. Kandel conducted a study of a sample of adolescents (black and white) from a large working-class urban high school to test some of these notions (Kandel, 1971). She examined interaction within comparable black and white families, both intact and mother-headed, and examined the consequences for the educational aspirations of adolescents and parents.

Among other things, she compared mothers' educational aspirations for their boys and girls by race and type of family. She reports proportions (percentages) of black mothers and white mothers from intact and mother-headed families who aspire to college educations for their boys and girls. In order to determine whether the differences between levels of aspirations of black and white mothers are significant, we will test her data for significance of the difference between proportions from independent samples. To justify the use of this test, the assumptions related to it must be met reasonably well (see Section 15.6.3). In making such a test we must first determine the appropriate sampling distribution and the measure of sampling error.

15.6.1   The Sampling Distribution of Differences between Proportions

If we were actually to generate a sampling distribution of differences between proportions, we would do it by drawing pairs of simple random samples from a population, computing a proportion for each sample, and taking the difference between each pair of proportions. If we were to do so until all pairs of unique samples of size N were exhausted, and then sum all of the differences, the differences would sum to zero and the mean difference would be zero. This is so because any differences between pairs of proportions, if from simple random samples of the same population, are nothing more than random differences. If we were to plot all of the differences on a curve, they would distribute themselves approximately normally about the mean difference of zero, with a standard error equal to the following:

(15.18)
$$\sigma_{p_1 - p_2} = \sqrt{PQ\left(\frac{1}{N_1} + \frac{1}{N_2}\right)}$$

where P is the population proportion, $Q = 1 - P$, N_1 is the number of cases in the first sample and N_2 is the number of cases in the second sample.

This standard error is known as the **standard error of the difference between proportions.**

Unfortunately, when we have data from only two samples, P and Q are unknown parameters. The null hypothesis states that $P_1 = P_2 = P$. If the sample proportions, p_1 and p_2, are estimates of P_1 and P_2, then by the null hypothesis they are estimates of the common parameter, P. Thus, we have two independent estimates of P, one based on N_1 cases and the other on N_2 cases. We can get a more reliable common estimate of P by pooling p_1 and p_2 using the following formula:

(15.19)
$$p = \frac{N_1 p_1 + N_2 p_2}{N_1 + N_2}$$

We justify this pooling procedure by assuming, for the time being, that the null hypothesis is true.

The pooled estimate of Q can be arrived at by subtraction, since $q = 1 - p$. These pooled estimates can be substituted into the formula for the standard error of the difference in place of the unknown parameters, P and Q. The resulting standard error will be an estimate of $\sigma_{p_1 - p_2}$. Its formula is as follows:

(15.20)
$$s_{p_1 - p_2} = \sqrt{pq\left(\frac{1}{N_1} + \frac{1}{N_2}\right)}$$

This standard error has the same interpretation as the other standard errors we have talked about so far. That is, in a *normal* sampling distribution, approximately 68 per cent of the differences between pairs of sample proportions will be within one standard error of the mean difference of zero.

With the use of Formula (15.20), and the knowledge that the sampling distribution of differences between proportions will be approximately normal, we are ready to test the null hypothesis.

15.6.2 Testing the Null Hypothesis

We will return to the Kandel study now and use her data to illustrate how to test for significance of difference between proportions. We will compare the educational aspirations that black and white mothers of intact families had for their boys and girls.

According to Kandel, empirical evidence on this question is sparse, and past definitions of relevant concepts have been ambiguous. In a sense, then, her study was an exploration of this and related questions. Since Kandel does not explicitly set up one-tailed tests of the null hypotheses, we have set up two-tailed tests at the .05 level of significance. Table 15.9 presents the essential steps in the test of the null hypothesis for boys, and Table 15.10 does the same for girls.

To compute the standard score from the sample data we use the following formula:

(15.21)
$$z = \frac{p_1 - p_2}{s_{p_1 - p_2}}$$

As can be seen from an examination of Tables 15.9 and 15.10, the null hypotheses with respect to educational aspirations for both boys and girls were rejected in favor of their sets of alternatives.* Black mothers had higher aspirations for their children than white mothers. Furthermore, the difference between white and black mothers was much more pronounced for girls than for boys. While 78 per cent of white mothers and 90 per cent of black mothers aspired to college educations for their sons, only 50 per cent of white mothers as compared to 86 per cent of black mothers aspired to college educations for their daughters.

15.6.3 Assumptions of the Test for Differences between Proportions

In order to justify use of the test for differences between proportions, we must assume that the observations are independent and that the samples are independent. It is also assumed that the samples are simple random samples of the population. The assumption that the null hypothesis is true serves as a rationale for getting a pooled estimate of the population proportion and using it to compute the standard error.

It is not necessary that the population being sampled be normally distributed; in fact, it is not necessary to specify its form at all. The technique is most appropriately used for nominal-scale data, but it can also be used for ordinal or interval data, although it is wasteful of information for such data.

---

*When the computed standard score falls near the boundary of the critical region, but within the region, it may be advisable to correct for continuity. This is suggested because the normal distribution is a continuous distribution, whereas proportions are discrete. The correction is accomplished by recomputing the proportions used in the numerator of the z-score formula. Since the proportions are arrived at by f/N, the proportions are converted back to these fractions. Then the f for the smaller of the two proportions is increased by 0.5 and the f for the larger of the two is decreased by 0.5. The proportions are recomputed with these corrections. The effect will be to reduce the numerator of the z-score formula slightly. The corrections treat the frequencies as though they were continuous rather than discrete scores.

TABLE 15.9 TWO-SAMPLE TEST FOR PROPORTIONS: EDUCATIONAL ASPIRATIONS WHICH MOTHERS OF INTACT FAMILIES HELD FOR THEIR SONS

$H_0 : P_1 = P_2 = P$

Sets of Alternatives to H_0:

$H_1 : P_1 > P_2$ designate this set of alternatives if $z \geq +1.96$.

$H_2 : P_1 < P_2$ designate this set of alternatives if $z \leq -1.96$.

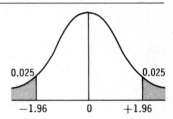

	Sample 1 White Mothers	Sample 2 Black Mothers
	$p_1 = .78$	$p_2 = .90$
	$q_1 = .22$	$q_2 = .10$
	$N_1 = 276$	$N_2 = 49$

Pooled estimates of P and Q:

$$p = \frac{N_1 p_1 + N_2 p_2}{N_1 + N_2} = \frac{(276)(.78) + (49)(.90)}{276 + 49} = \frac{259}{325} = .80$$

$$q = 1 - p = 1 - .80 = .20$$

Standard error of the difference:

$$s_{p_1 - p_2} = \sqrt{pq \left(\frac{1}{N_1} + \frac{1}{N_2} \right)} = \sqrt{(.80)(.20) \left(\frac{1}{276} + \frac{1}{49} \right)} = \sqrt{(.16)(.024)} = \sqrt{.0038} = .06$$

Standard score:

$$z = \frac{p_1 - p_2}{s_{p_1 - p_2}} = \frac{.78 - .90}{.06} = \frac{-.12}{.06} = -2.00$$

Therefore, reject the null hypothesis in favor of the set of alternatives $H_2 : P_1 < P_2$.

Source: Data from Kandel, 1971.

Diagram 15.2 summarizes the essential features of the one-sample test for proportions and the two-sample test for proportions.

15.7 ONE-SAMPLE CASE FOR VARIANCES

It is sometimes of interest to test a sample variance against a known or hypothesized population variance to see whether it is an estimate of the given population variance. To illustrate how this kind of problem is handled, we will use data on accidental deaths during 1967 from a random sample of 25 of the United States.*

*The sample data were drawn from U.S. Bureau of the Census (1971:59, Table 74; and 56, Table 70). We could have used the data for all of the states, but we drew a random sample of 25 states in order to illustrate the process of testing a sample variance against a known or theoretical population variance. Note that in this case we drew a random sample without replacement, so we need to recognize the fact that we do not have a simple random sample when we interpret the results. The decision to sample without replacement in this instance was based upon substantive rather than statistical grounds. It was felt desirable for the problem being investigated to get the widest possible representation of states without resorting to a total enumeration of the population.

TABLE 15.10   Two-sample Test for Proportions: Educational Aspirations which Mothers of Intact Families Held for Their Daughters

$H_0 : P_1 = P_2 = P$

Sets of Alternatives to H_0:

$H_1 : P_1 > P_2$ designate this set of alternatives
if $z \geq +1.96$.

$H_2 : P_1 < P_2$ designate this set of alternatives
if $z \leq -1.96$.

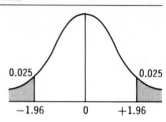

0.025   0.025

−1.96   0   +1.96

Sample 1	Sample 2
White Mothers	*Black Mothers*
$p_1 = .50$	$p_2 = .86$
$q_1 = .50$	$q_2 = .14$
$N_1 = 290$	$N_2 = 52$

Pooled estimates of P and Q:

$$p = \frac{N_1 p_1 + N_2 p_2}{N_1 + N_2} = \frac{(290)(.50) + (52)(.86)}{290 + 52} = \frac{190}{342} = .56$$

$$q = 1 - p = 1 - .56 = .44$$

Standard error of the difference:

$$s_{p_1 - p_2} = \sqrt{pq\left(\frac{1}{N_1} + \frac{1}{N_2}\right)} = \sqrt{(.56)(.44)\left(\frac{1}{290} + \frac{1}{52}\right)} = \sqrt{(.246)(.023)} = \sqrt{.0057} = .075$$

Standard score:

$$z = \frac{p_1 - p_2}{s_{p_1 - p_2}} = \frac{.50 - .86}{.075} = \frac{-.36}{.075} = -4.80$$

Therefore, reject the null hypothesis in favor of the set of alternatives $H_2 : P_1 < P_2$.

Source: Data from Kandel, 1971.

We will compare the variability of these sample data with the variability of deaths from all causes to see whether deaths from accidents are more or less variable from state to state. Table 15.11 lists the 25 sample states and their accidental death rates for 1967. The variance for these is 180.15. The question is whether this sample variance can be seen as an estimate of population variance for the overall death rates. To answer this question we will set up a test of the following null hypothesis:

$$H_0 : \sigma_A^2 = \sigma_D^2 = 91.20$$

where σ_A^2 is the population variance for accidental death rates and σ_D^2 is the population variance for overall death rates.* There are two sets of alternatives to this null hypothesis:

*Since the overall death rates were not comparable to the rates for accidental deaths in units of measurement, it was necessary to transform the overall rates into comparable units and compute a transformed σ^2. The conversion was accomplished by using the ratio of the sample mean for accidental deaths to the population mean for all causes of deaths ($\overline{X}_A / \mu_D = 0.069$) as a conversion factor. Therefore, the σ^2 for the null hypothesis represents what the population variance for overall death rates would be, if those rates were in units comparable to those for accidental deaths.

561

DIAGRAM 15.2   Tests of Hypotheses about Proportions

THE ONE-SAMPLE CASE

Problem: To determine whether a given sample proportion is an estimate of a known or theoretical population proportion.

Assumptions:
1. Independent observations
2. Simple random sample of the population
3. NP and NQ are both equal to or greater than 5

Null and Alternative Hypotheses:

$H_0:P=$ (some specified value)
$H_1:P \neq$ (that specified value), or
$P >$ (that value) or $P <$ (that value)

Standard Error:

$$\sigma_p = \sqrt{\frac{PQ}{N}}$$

Standard Score Formula:

$$z = \frac{p-P}{\sigma_p}$$

Sampling Distribution:
  Normal curve

Step 1: List assumptions.
Step 2: Set up the null hypothesis. Specify whether the test is one-tailed or two-tailed. Specify the level of significance.
Step 3: Draw a simple random sample from the population and collect data.
Step 4: From the sample data compute the sample proportion. Compute the standard error of the proportion.
Step 5: Substitute into the standard score formula and solve for z. If the computed z is in the critical region, reject the null hypothesis. If not, fail to reject.

DIAGRAM 15.2 *(Continued)*

THE TWO-SAMPLE CASE

Problem: To determine whether two sample proportions from two independent random samples estimate a common population proportion.

Assumptions:

    1. Independent observations

    2. Independent simple random samples

Null and Alternative Hypotheses:

$H_0 : P_1 = P_2$
$H_1 : P_1 \neq P_2$, or
    $P_1 > P_2$ or $P_1 < P_2$

Standard Error of the Difference:

    Pool sample estimates p_1 and p_2 get p; find q by subtraction $(q = 1 - p)$, and substitute into following standard error formula:

$$s_{p_1 - p_2} = \sqrt{pq\left(\frac{1}{N_1} + \frac{1}{N_2}\right)}$$

Standard Score Formula:

$$z = \frac{p_1 - p_2}{s_{p_1 - p_2}}$$

Sampling Distribution:

    The normal curve with a mean difference between proportions of zero.

Step 1: List assumptions.

Step 2: Set up the null hypothesis. Specify whether the test is one-tailed or two-tailed. Specify the level of significance.

Step 3: Draw two simple random samples and collect data.

Step 4: From the samples compute sample proportions. Compute the standard error of the difference between proportions.

Step 5: Substitute into the standard score formula and compute z. If the computed z is in the critical region, reject the null hypothesis. If not, fail to reject.

TABLE 15.11   ACCIDENTAL DEATH RATES FOR RANDOM SAMPLE OF 25 STATES—1967*

State	Death Rate
Maine	62.9
Massachusetts	49.1
New York	44.6
New Jersey	41.5
Pennsylvania	48.6
Ohio	52.4
Illinois	51.9
Michigan	52.3
North Dakota	62.9
South Dakota	71.8
Kansas	59.7
North Carolina	67.6
South Carolina	68.4
Florida	65.2
Alabama	70.7
Arkansas	66.5
Oklahoma	69.0
Texas	60.7
Montana	88.3
Idaho	74.2
Wyoming	79.0
Washington	65.2
Oregon	72.3
Alaska	93.4
Hawaii	41.1

Source: Based on data from U.S. Bur. Census, 1970:56, Table 70; 59, Table 74.
*Rates per 100,000 estimated mid-year population in each area.

$$H_1 : \sigma_A^2 > \sigma_D^2$$

and

$$H_2 : \sigma_A^2 < \sigma_D^2$$

or more simply:

$$H_1 : \sigma_A^2 \neq \sigma_B^2$$

Assuming the necessary conditions have been met (see Section 15.7.1), the null hypothesis may be tested using the following formula:

(15.22)
$$\chi^2 = \frac{(N-1)s^2}{\sigma^2}$$

The value in Formula (15.22) has a chi square sampling distribution with $df = N - 1$ for random samples of size N drawn from a normal population with variance σ^2. If we use the .01 level of significance for the test, the χ^2 values that bound the critical regions (with $df = 24$) will be 45.558 for the set of alternatives H_1, and will be 9.886 for the set of alternatives H_2 (see Appendix Table D).

If accidental death rates were comparable in variability to the overall

---

TABLE 15.12 ONE-SAMPLE TEST FOR VARIANCES: ACCIDENTAL DEATH RATES FROM A
SAMPLE OF TWENTY-FIVE STATES*

$H_0 : \sigma_A^2 = \sigma_D^2$

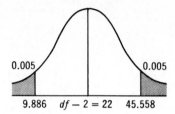

Sets of Alternatives to H_0:

$H_1 : \sigma_A^2 > \sigma_D^2$ designate this set of alternatives
       if $\chi^2 \geq 45.558$

$H_2 : \sigma_A^2 < \sigma_D^2$ designate this set of alternatives
       if $\chi^2 \leq 9.886$

0.005   0.005

9.886 $df - 2 = 22$ 45.558

$N = 25$

$df = N - 1 = 24$

$\sigma_D^2 = 91.20$, the hypothesized population variance based on a transformation of the
population variance for overall death rates.
$s^2 = 180.15$ the sample variance of accidental death rates for the 25 randomly drawn
states.

$$\chi^2 = \frac{(N-1)s^2}{\sigma_D^2} = \frac{(24)(180.15)}{91.20} = 47.41$$

Therefore, reject the null hypothesis at the .01 level of significance and in favor
of the set of alternatives $H_1 : \sigma_A^2 > \sigma_D^2$.

---

*Note that in the pictured curve of the chi-square distribution, the point at the mode of the distribution
is marked $df - 2 = 22$. As was mentioned in Chapter 12 (Section 12.5.1), with the exception of the curve
for $df = 1$, the mode of a chi-square distribution will be at $df - 2$.

death rates, the population variance for accidental death rates would be 91.20
(the transformed rate for overall deaths). We will use this figure for the hypothesized population variance and see whether the observed sample variance is significantly different. Table 15.12 summarizes the test of the null hypothesis.

As was mentioned previously, the sample variance for the 25 states is
180.15. Substituting this and the population variance (91.20) into the χ^2 formula
we get the following:

$$\chi^2 = \frac{(N-1)s^2}{\sigma^2} = \frac{(24)(180.15)}{91.20} = 47.41$$

Since this χ^2 is into the critical region in the right tail of the distribution,
we reject the null hypothesis in favor of the set of alternatives H_1, which implies
that the sample variance estimates a population variance larger than that for
the overall death rates. This result suggests the need for further study to determine why accidental death rates are more variable.

15.7.1 Assumptions of the One-sample Test for Variances

The one-sample test for variances is based on the following set of assumptions:
(1) that the level of measurement is at least interval scale, (2) that the population is normally distributed, (3) that the data are from a simple random sample of
that population, and, thus, (4) that the observations are independent.

Especially important is the assumption that the population is normally distributed. Variances are sensitive to deviations from normality and serious errors of interpretation can result. The assumption is less crucial, however, as N gets larger. Therefore, if there is some doubt about the normality of a population distribution, it is wise to take as large a sample as is feasible.

15.8   THE TWO-SAMPLE CASE FOR VARIANCES

With growing concern over the population explosion, it is of special interest to attempt to isolate those factors which influence trends in birth rates. Some of these factors have already been identified, though they are not necessarily fully understood. For example, it is usually the case that birth rates tend to decline as countries become industrialized and urbanized. This finding may be attributed, in part, to such factors as changing attitudes about the desirability of having a large family as opposed to being able to afford a higher standard of living. Chil-

TABLE 15.13   LIVE BIRTHS (per 1000 population) FOR RANDOM SAMPLES OF HIGH-LITERACY AND LOW-LITERACY COUNTRIES

High-Literacy Countries		Low-Literacy Countries	
Country	Birth Rate	Country	Birth Rate
Sweden	13.7	Cambodia	41.4
New Zealand	26.3	Brazil	45.0
Singapore	42.8	Madagascar	45.0
Puerto Rico	33.7	Ghana	54.0
Ceylon	44.0	Congo	44.0
Ireland	21.1	Guatemala	50.0
Greece	16.3	Nicaragua	48.5
Bulgaria	18.7	Haiti	46.5
Hong Kong	38.3	Iran	45.0
Albania	35.3	Ruanda-Urundi	42.0
Hungary	17.8	Peru	45.0
W. Germany	16.8	India	43.2
Finland	19.9	Jordan	47.4
U.S.S.R.	25.3	Egypt	50.8
E. Germany	13.9	Angola	45.0
Portugal	23.6	Malaya	44.4
Austria	16.8	Tunisia	43.2
Netherlands	21.2	Sarawak	33.0
Panama	39.9	Algeria	45.0
Israel	27.9	Sudan	51.0
$\overline{X}_1 = 25.66$		$\overline{X}_2 = 45.47$	
$N_1 = 20$		$N_2 = 20$	

Sources: Based on data from United Nations, 1962, and United Nations, 1963.

dren, who were once assets in the rural setting as cheap sources of labor, have become financial liabilities in the urban setting.

Also related to level of the birth rate is the level of education of a country's population. Among other things, it appears that more education leads to more knowledge about and more use of birth control techniques. If literacy rate is taken as an index of level of education, then it stands to reason that countries whose populations are more literate should have lower birth rates, on the average, than countries with less literate populations.

Table 15.13 presents data on the birth rates (live births per 1,000 population) of a random sample of 20 countries in which a majority of the population is literate and 20 countries in which a majority of the population is illiterate.* For convenience we will refer to these two sets of countries as "high-literacy" countries and "low-literacy" countries. As you can see from the means of the two samples, the birth rates are considerably higher in the low-literacy countries than in the high-literacy countries. If put to a formal test (two-sample case for means from independent samples, Section 15.3), it will be found that the difference between the sample means is statistically significant.

Another significant question that might be asked about high-literacy versus low-literacy countries is whether there are differences in the between-country birth-rate variability. If we were to find that the variability in birth rates was either larger or smaller for high-literacy countries than for low-literacy countries, that would tell us that there is something about the social dynamics of changes in birth rates that requires further explanation. For example, if it so happens that the variance for birth rates is larger for low-literacy countries than it is for high-literacy countries, that would seem to indicate that there are several diverse combinations of variables that influence birth rates in such countries. On the other hand, if the variance turns out to be larger for high-literacy countries, that may indicate that the diverse combination of influences is present in high-literacy countries, while conditions influencing birth rates are more uniform in low-literacy countries.

Either finding would lead to fruitful channels of further research in an effort to understand the social processes involved in the birth rate trends. In order to determine whether the two groups of countries do differ in variability of birth rates, and also to demonstrate the test of significance of difference between the two variances, we will test the following null hypothesis:

$$H_0: \sigma_1^2 = \sigma_2^2$$

where σ_1^2 is the variance for birth rates of countries in which a majority of the population is literate and σ_2^2 is the variance for birth rates of countries in which a majority of the population is illiterate. The test will be made by comparing the two sample variances, s_1^2 and s_2^2, to see whether they are likely to be

---

*These samples were drawn from all of those countries for which data were available; therefore, the population of countries is limited to those which keep records. Sources for these rates were United Nations, *Demographic Yearbook, 1961* (New York, 1962) and United Nations, *Compendium of Social Statistics, 1963* (New York, 1963). Here again the random samples were drawn without replacement. The reason for using this procedure is basically the same as it was for the previous example. Because of the nature of the problem and the availability of data, it was considered desirable to get as broad a representation of countries as possible in the samples.

estimates of a common population variance σ^2 (the variance for birth rates of all countries) or estimates of two separate parameters σ_1^2 and σ_2^2. The two sets of alternatives to the null hypothesis which are consistent with the notion that there are two separate population variances are as follows:

$$H_1: \sigma_1^2 > \sigma_2^2$$

and

$$H_2: \sigma_1^2 < \sigma_2^2$$

or,

$$H_1: \sigma_1^2 \neq \sigma_2^2$$

We have chosen to use a two-tailed test of the null hypothesis because our investigation of the variability of birth rates is an exploratory one, and we do not have a theory which suggests which variance should be larger. We also choose to test the null hypothesis at the .05 level of significance for similar reasons.

As we found earlier in this chapter (Section 15.4.2) the F-distribution is the appropriate sampling distribution to test a null hypothesis about two variances. To find the F that bounds the critical region, we enter the F table with $df_1 = N_1 - 1$ and $df_2 = N_2 - 1$, where N_1 is the sample size of the sample with the larger variance, and N_2 is the sample size of the sample with the smaller variance. This is so because in a two-tailed test of this kind, the F ratio is always computed by dividing the smaller variance into the larger one.* For a one-tailed test of significance, the variance which the theory predicts will be larger is put in the numerator of the F ratio. In either case, the critical region for the F-distribution is put in the right tail of the distribution, and the computed F approaches the critical region as it gets larger than 1. Therefore, the computed F will only be significant if the variance in the numerator of the F ratio is enough larger than the variance in the denominator so that the ratio will reach the critical region.

Using the data from Table 15.13 to compute sample variances for the birth rates of high-literacy and low-literacy countries, we get the following:

(15.23)
$$s^2 = \frac{\Sigma X^2 - (\Sigma X)^2 / N}{N - 1}$$

For high-literacy countries:

$$s_1^2 = \frac{15{,}043.93 - (513.3)^2/20}{19} = 98.43$$

For low-literacy countries:

$$s_2^2 = \frac{41{,}715.70 - (909.4)^2/20}{19} = 19.23$$

Assuming that we can meet the conditions necessary for performing an F test for two-sample variances (see Section 15.8.1), we may now substitute the two-sample variances into the F ratio and solve for F as follows:

$$F = s_1^2 / s_2^2 = 98.43 / 19.23 = 5.12$$

---

*You should recall that in the analysis of variance, the between-groups variance was always put in the numerator and the within-groups or error variance in the denominator. This procedure is not necessary. Both tails of the F distribution could be used if tables for both critical values were available.

TABLE 15.14   Two-sample Test for Variances: Data on Birth Rates of High-Literacy and Low-Literacy Countries

$H_0: \sigma_1^2 = \sigma_2^2$

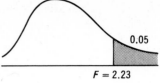

Sets of alternatives to H_0:

$H_1: \sigma_1^2 > \sigma_2^2$ designate this set of alternatives if
$F \geq 2.23$ when $F = s_1^2/s_2^2$

$H_2: \sigma_1^2 < \sigma_2^2$ designate this set of alternatives if
$F \geq 2.23$ when $F = s_2^2/s_1^2$

Sample Group	Variance	N	df
High-literacy countries	98.43	20	19
Low-literacy countries	19.23	20	19

$$F = s_1^2/s_2^2 = 98.43/19.23 = 5.12$$

Therefore, reject the null hypothesis in favor of the set of alternatives $H_1\ \sigma_1^2 > \sigma_2^2$.

To find the F that bounds the 5 percent critical region, we enter Table E (in the Appendix) with $(N_1 - 1) = (20 - 1) = 19\ df_1$ and $(N_2 - 1) = (20 - 1) = 19\ df_2$. However, since the table does not have an entry for $df = 19$, we will use the entry for $df_1 = 15$, thereby erring on the side of conservatism. We find from the table that the F bounding the critical region is 2.23; therefore, our computed F of 5.12 is well into the critical region and rejection of the null hypothesis is warranted in favor of the set of alternatives $H_1:\sigma_1^2 \neq \sigma_2^2$, in the direction of the set of alternatives $H_1:\sigma_1^2 > \sigma_2^2$. The steps in the test of this null hypothesis are summarized in Table 15.14.

We may conclude from these results that the birth rates of the high-literacy countries are more variable than those of the low-literacy countries. This difference in variances deserves more study. Preliminary indications are that those factors related to high birth rates operate more uniformly than the factors related to lower birth rates. Since it was our intention only to demonstrate the use of a statistical technique, we will not pursue this investigation further. It appears, however, to be a subject worth pursuing.

15.8.1   Assumptions of the Two-Sample Test for Variances

The assumptions underlying the two-sample test for variances are very much like those for the one-sample case. An interval level of measurement is assumed. It is assumed that the observations are independent. The samples are assumed to be simple random samples from normally distributed populations.

Again, the assumption of a normal population distribution is a crucial one. Violation of this assumption can lead to serious problems of interpreting results. Large samples will help somewhat in reducing errors of interpretation.

Diagram 15.3 summarizes the step-by-step procedures for testing hypotheses about variances.

DIAGRAM 15.3   Tests of Hypotheses about Variances   569

THE ONE-SAMPLE CASE

Problem: To determine whether a sample variance is an estimate of a known or theoretical population variance.

Assumptions:

1. Interval level of measurement
2. Independent observations
3. Normally distributed population
4. Simple random sample of that population

Test Statistic:

$$\chi^2 = \frac{(N-1)s^2}{\sigma^2}$$

Sampling Distribution:

   χ^2-distribution with $N-1$ degrees of freedom.

Step 1: List assumptions.

Step 2: Set up the null hypothesis. Specify whether the test is one-tailed or two-tailed. Specify the level of significance.

Step 3: Draw a simple random sample from the population and collect data.

Step 4: From the sample data, compute the sample variance.

Step 5: Substitute sample variance and population variance into χ^2 formula and solve. If χ^2 is in the critical region, reject the null hypothesis. If not, fail to reject.

THE TWO-SAMPLE CASE

Problem: To determine whether two sample variances estimate a common population variance.

Assumptions:

1. Interval level of measurement
2. Independent observations
3. Normally distributed populations
4. Simple random samples from those populations

Test Statistic:

$$F = s_1^2 / s_2^2$$

If the null hypothesis is for a one-tailed test, the sample variance, which the alternative hypothesis predicts will be larger, is put in the numerator of the F ratio.

If the null hypothesis is for a two-tailed test, the larger sample variance will always be the numerator.

Sampling Distribution:

   F-distribution with $N_1 - 1$ and $N_2 - 1$ degrees of freedom. Degrees of freedom for variance in numerator of F ratio will be df_1 and in denominator df_2.

Step 1: List assumptions.

Step 2: Set up the null hypothesis. Specify whether the test is one-tailed or two-tailed. Specify the level of significance.

Step 3: Draw two simple random samples and collect data.

Step 4: From the sample data compute the two-sample variances.

Step 5: Substitute the sample variances into the F formula. If the computed F is in the critical region, reject the null hypothesis. If not, fail to reject.

15.9   SUMMARY

In this chapter we have examined techniques for testing hypotheses about central tendency and variability. We discussed procedures for testing hypotheses about means in the one-sample case, the two-sample case, and the k-sample case. For the k-sample case, we discussed the simple one-way analysis of variance. We pointed out that there are many types of analysis of variance procedure, but that most of those are limited in their use to laboratory experimentation. Certain fields of sociology, such as small group research, are amenable to laboratory experimentation, but others are not. Therefore, space was not devoted to a consideration of the more elaborate analysis of variance designs.

Tests of hypotheses of proportions for the one-sample and the two-sample cases were described and discussed. It was explained that these techniques have more general applicability because the assumptions underlying their use are much less stringent than the assumptions underlying tests about means.

We finally discussed tests of hypotheses about variances. We explained procedures for making tests for the one-sample case and the two-sample case. It is important to note that the assumption of a normal population distribution is a crucial one, because variances are sensitive to departures from normality. As a matter of fact, variances are much more sensitive to such departures than are means.*

We began the chapter by discussing global, structural, and analytical-level variables. It was pointed out that structural properties do not lend themselves readily to analysis by the common statistical techniques. They generally require special treatment. We remarked that analytical properties lend themselves most readily to tests about central tendency and variability. Less often are tests such as these used to analyze global properties, but such tests are to be found in the literature.

Much attention was devoted earlier (Parts III and IV) to measures of association. In the next chapter the discussion will focus on techniques for generalizing statements about association from sample data to population data.

CONCEPTS TO KNOW AND UNDERSTAND

analytical properties

structural properties

global properties

sampling distribution of differences
   between means

standard error of the difference
   between means

analysis of variance (ANOVA)

total sum of squares

within-groups sum of squares

between-groups sum of squares

within-groups variance

between-groups variance

F distribution

F ratio

one-way analysis of variance

sampling distribution of differences
   between proportions

standard error of the difference
   between proportions

*There is a test for this assumption of equal variances, called Bartlett's Test. See Wilfrid J. Dixon and Frank J. Massey, Jr., *Introduction to Statistical Analysis*, (New York: McGraw-Hill Book Company, 1957) p 179–180.

QUESTIONS AND PROBLEMS

1. Read several of the abstracts that appear in the workbook or in the *American Sociological Review* and classify the variables studied as analytical, structural, or global.

2. If the formula $t = \dfrac{\overline{X} - u}{s_{\overline{x}}}$ is a ratio, explain the nature of the ratio. That is, what does the numerator represent, what does the denominator represent, and what is the significance of the ratio between the numerator and denominator?

3. How would you interpret the finding that the mean pay scale for clerical office workers was 91.9 for small metropolitan areas, 95.8 for moderate-sized metropolitan areas, and 100.9 for large metropolitan areas? How might these differences be explained?

4. Explain the part played by the null hypothesis in generating the sampling distribution of differences between pairs of sample means.

5. Explain what is measured by the standard error of the difference between means.

6. In a test for significance of difference between sample means, why is the assumption that the variances are homogeneous an important one?

7. What is the rationale behind the analysis of variance? That is, what kind of comparison does the test imply?

8. If an analysis of variance test is conducted to test the null hypothesis: $\mu_1 = \mu_2 = \mu_3 = \mu_4 = \mu$, and the null hypothesis is rejected, what can we say about differences between the means?

9. What justifies the pooling of two sample proportions (p_1 and p_2) to get a common estimate of the population proportion (P)?

10. Which assumption is crucial for testing hypotheses about variances? Why?

11. At this point, one of the most effective ways to firm up your knowledge of the various hypothesis testing techniques is to try them out on real problems of interest to you. It is important to understand thoroughly not just the calculations but the logic and usefulness of the procedures in drawing conclusions about worthwhile ideas or problems of relevance to your subject-matter interest. This can be achieved in any number of ways such as:
   a) Design and conduct (perhaps with others in the class) a manageable survey or observational study of some phenomenon of interest to you. It would be helpful to include in your study, some reading or review of at least the main recent work in your topic area. Consult your course instructors for ideas or use work started in a past term paper. You will want to pay particular attention to the interesting variable/s and how they are measured plus subgroupings of subjects which you will use in making contrasts and comparisons. What do you expect the data to show? How would you explain that? How would you explain results that were

exactly opposite? Then, carefully state your problem, define the relevant population and sampling frame, specify how it is you will draw the sample, set up measures and conduct the data gathering. Try to draw a sample sufficiently large (say 30 or even 100) so that large sample statistical assumptions can be met. You can treat a random small sample from these data as a data set for small sample procedures to contrast with large sample procedures. Be sure that some measures suit themselves to percentage or proportion use and some suit themselves to interval level treatment. Your instructor can show you how to use computer facilities to process your data. Then, test a variety of hypotheses you have using the range of techniques discussed in this chapter. You may be able to use the same data to test hypotheses using techniques in the next chapter. Be sure to clearly state your procedures and conclusions and the reasons behind these.

b) If you do not have resources or time to conduct a study, you could use available data from published articles on topics of interest to you, governmental data (a wealth exists in the government documents section of your library), or data sets already stored on the computer (such as data from an instructor's research or data from national studies). In some instances where, for example, information is given for each of the states in the U.S., you can draw a random sample from these data for your use. In that case, compare your results with the population parameters if they are available to you. Cities, counties, states, organizations, or countries are interesting units of analysis for sociological studies.

12. Some statistical reports, such as the one selectively reproduced below, include information on sampling error. The following table provides number and frequency of deaths for the U.S. by sex, color and age for two years plus the information about sampling error.
   a) Using this information, state confidence intervals for the rate and for the frequency of deaths for the U.S. and for males and females. Pick a confidence interval of your choice for which information is given.
   b) Test the hypothesis that male death rates are not higher than the overall rate.
   c) Test the hypothesis that White and "Other" background are equal in terms of frequency of deaths; in terms of death rate.
   d) Test the hypothesis that, overall, the death rate has not declined between the two years.
   In each instance, organize your work in terms of the steps of the procedure you use, show your work and state conclusions in your own words. There are many other hypotheses you may want to test on these data in terms of change from year to year, comparisons of sex, color and age groups with each other or over time. In general, what conclusions can one draw from the frequencies and rates shown in the table? (Data are from table 5 and page 11 of "Births, Marriages, Divorces, and Deaths for February, 1978", Monthly Vital Statistics Report, volume 27, No. 2, May 18, 1978, Washington, D.C., National Center for Health Statistics. DHEW Publication No. (PHS) 78-1120.)

DEATHS IN THE SAMPLE AND ESTIMATED DEATH RATES, BY AGE, COLOR, AND SEX: UNITED STATES, 12 MONTHS ENDING WITH JANUARY 1977 AND 1978

[Rates on an annual basis per 1,000 estimated population residing in area for specific group]

| Age, color, and sex | 12 months ending with January | | | |
| | 1978 | | 1977 | |
	Number	Rate	Number	Rate
Total	190,541	8.8	191,102	8.9
Male	104,976	10.0	105,366	10.1
Female	85,565	7.7	85,735	7.8
White	167,513	8.9	167,812	9.0
Male	91,867	10.0	92,009	10.1
Female	75,646	7.9	75,803	8.0
All other	23,028	8.0	23,290	8.2
Male	13,109	9.5	13,357	9.8
Female	9,919	6.5	9,933	6.7
Under 1 year	4,589	14.5	4,744	15.7
1–14 years	2,153	0.4	2,136	0.4
15–24 years	4,801	1.2	4,544	1.1
25–34 years	4,496	1.4	4,193	1.3
35–44 years	5,776	2.5	5,811	2.5
45–54 years	14,328	6.1	15,051	6.4
55–64 years	29,313	14.4	29,538	14.7
65–74 years	44,659	30.6	44,578	31.4
75–84 years	49,373	72.4	49,548	73.2
85 years and over	30,995	149.0	30,923	156.3
Not stated	58	. . .	36	. . .

Since the estimates of deaths and death rates presented in this report are based on a sample of the death certificates, they are subject to sampling variability. The percent errors shown in the following table are measures of sampling variability. They show the percent difference between figures based on a sample and those based on a complete count that may result from sampling variation. Values are shown for two different sized samples. The first column refers to the monthly sample size, the second to the annual; cumulative monthly totals shown during the year will be between the two. The percent differences apply to both frequencies and rates. Chances are about 2 out of 3 that the percent difference between a sample estimate and a complete count is less than the percent shown. Chances are about 19 out of 20 that the percent difference due to sampling variation is less than twice the percent shown. In this report a change in an estimated monthly rate from one year to the next is regarded as significant only when the percent change is at least 3 times the percent error of the most recent rate.

| Number of deaths in sample for a group | Percent error of estimate | |
	15,000 total deaths in sample each month	180,000 total deaths in sample each year
5	42.4	42.4
10	30.0	30.0
20	21.2	21.2

(Continued)

(Continued)

Number of deaths in sample for a group	Percent error of estimate	
	15,000 total deaths in sample each month	180,000 total deaths in sample each year
50	13.4	13.4
100	9.5	9.5
200	6.7	6.7
500	4.2	4.2
1,000	2.9	3.0
2,000	2.0	2.1
5,000	1.1	1.3
10,000	0.5	0.9
20,000	...	0.6
50,000	...	0.4
100,000	...	0.2

13. As part of a study of "The Causes and Cost of Racial Exclusion from Job Authority," James R. Kluegel (*American Sociological Review*, 1978, 43, June, p 285–301) developed a measure of job authority from answers by individuals in his sample to a series of questions about kinds of supervisory behavior they exercise. The higher the index the higher the amount of job authority. For our purposes, we will consider the data he gathered from two samples of employed men in urban Wisconsin as a simple random sample. Job authority (and other statistics) are presented below for five occupational categories in his sample. You may want to consult his article to see how he tested his main conclusions.

From these data, test the following hypotheses:
a) That the mean job authority for self-employed managers (or another category you select) is equal to the overall mean job authority of these groups, 1.591.
b) That salaried managers are not higher than clerical workers.
c) That clerical and sales workers are equal.
d) That there is no difference in mean job authority for these five groups.
In each case, show the steps in your work, interpret the results in your own words and discuss the conclusions that can be drawn.

Occupational Category	$\overline{X}$ Job Authority	s	n	Σx^2
Self-employed managers	4.254	1.293	64	105.326
Salaried managers	2.459	1.164	111	149.039
Sales	1.433	1.249	67	102.960
Clerical	.834	1.124	103	128.864
Laborer	.462	.778	158	95.030
Overall (for these groups)	1.591	1.657	503	1380.799

GENERAL REFERENCES

Blalock, Hubert M., Jr., *Social Statistics,* Revised Edition (New York, McGraw-Hill Book Company), 1979.

If you read the sections in Blalock dealing with tests of hypotheses about central tendencies, it will reinforce what you have learned in this chapter.

Edwards, Allen, *Experimental Design in Psychological Research,* Second Edition (New York, Holt, Rinehart and Winston), 1960.

This is an excellent source to read on the analysis of variance. It discusses those more complex analysis of variance applications for laboratory experiments that were not discussed in this chapter.

Freund, John E., *Modern Elementary Statistics,* Third Edition (Englewood Cliffs, N.J., Prentice-Hall), 1967.

In chapter 10 of this book there is a discussion of tests of hypotheses about variability. Reading the material in Freund may help to reinforce what you have learned in this chapter.

OTHER REFERENCES

Coombs, Lolagene C., Ronald Freedman, Judith Friedman, and William F. Pratt, "Premarital Pregnancy and Status Before and After Marriage," *American Journal of Sociology,* vol. 75, no. 5 (March, 1970), pp 800–820.

Faris, Robert E.L., "The Alleged Class System in the United States," *Research Studies, State College of Washington,* vol. 22 (1954), pp. 77–83.

Kandel, Denise B., "Race, Maternal Authority, and Adolescent Aspiration," *American Journal of Sociology,* vol. 76, no. 6 (May, 1971), pp. 999–1018.

Lazarsfeld, Paul F., and Herbert Menzel, "On the Relation Between Individual and Collective Properties," in Amatai Etzioni, editor, *A Sociological Reader on Complex Organizations,* Second Edition (New York, Holt, Rinehart, and Winston), 1969, pp 499–516.

Mark, Harold, and Kent P. Schwirian, "Ecological Position, Urban Central Place Function, and Community Population Growth," *American Journal of Sociology,* vol. 73, no. 1 (July, 1967), pp 30–41.

Proctor, C.H., and C.P. Loomis, "Analysis of Sociometric Data," in M. Jahoda, M. Deutsch, and S.W. Cook, *Research Methods in Social Relations,* vol. II, First Edition (New York, Dryden Press), 1951.

Svalastoga, Kaare, "Social Differentiation," in Robert E.L. Faris, editor, *Handbook of Modern Sociology* (Chicago, Rand McNally), 1964, pp 547–550.

United Nations, *Demographic Yearbook, 1961* (New York, 1962).

———, *Compendium of Social Statistics,* 1963 (New York, 1963).

U.S. Bureau of the Census, "Table 70: Deaths–Total and Rate, 1960 to 1968, and by Race, 1967: States," and "Table 74: Death Rates for the 10 Leading Causes of Death–States: 1967," *Statistical Abstract of the United States: 1970* (Washington, D.C., 1971), p 56 and p 59.

U.S. Department of Labor, Bureau of Labor Statistics, "Table 94. Interarea Pay Comparisons–Relative Pay Levels by Industry Division, 1960–69," *Handbook of Labor Statistics,* Bulletin 1666 (Washington, D.C.) 1970. p. 179.

Verwaller, Darrel J., "Social Mobility and Membership in Voluntary Associations," *American Journal of Sociology,* vol. 75, no. 4 (January, 1970), pp. 481–495.

Warner, W. Lloyd, Marchia Meecker, and Kenneth Eells, *Social Class in America* (Chicago, Science Research Associates), 1949.

Wegner, Eldon L., and William H. Sewell, "Selection and Context as Factors Affecting the Probability of Graduation from College," *American Journal of Sociology,* vol. 75, no. 4, part 2 (January, 1970), pp. 665–679.

16 *Hypothesis Tests of Association*

Lethal aggressive behavior, resulting in suicides and homicides, has long been of interest to sociologists. Durkheim's classical work on suicide was an early study in this tradition. Recently Whitt, Gordon, and Hofley conducted a cross-national study of the relationship of suicide and homicide rates to level of industrialization and religious tradition (Whitt *et al.*, 1972). They postulated that suicide and homicide are alternative aggressive responses to frustration. Frustration is thought to arise in a population as an effect of incongruencies between religious and economic values. The researchers formed a "lethal aggression rate" by adding up suicide and homicide rates for countries, and they hypothesized that the lethal aggression rate would be positively correlated with industrialization in those countries where religious traditions are congruent with industrialization and negatively correlated in those countries where religious traditions are incongruent with industrialization. The proportion of economically active males not engaged in agriculture was used as an index of industrialization.

The researchers argued that Protestantism is congruent with industrialization, and, therefore, in a sample of Protestant countries the correlation between lethal aggression rate and industrialization should be negative. When they analyzed their data for some Protestant countries they found the coefficient of correlation between lethal aggression rate and level of industrialization to be −0.53, a moderate negative correlation. The authors concluded that this result supported their theoretical expectations.

If Whitt, Gordon, and Hofley wished only to reach conclusions about the countries for which they had data, their merely computing the correlation coefficient would have been sufficient. If they desired, on the other hand, *to generalize their finding* beyond the countries for which they had data, they would be obligated to turn to **inferential hypothesis testing procedures.** They were, in fact, interested in generalizing and they did proceed to test inferential statistical hypotheses about the particular coefficient of −0.53 and others which they computed in the course of their study.

Since statistics based on sample data are subject to sampling error, it is necessary to determine whether the association found for the sample can be generalized to the population or whether it may be attributed to sampling error. The usual procedure used is one in which we test the null hypothesis of zero association, using our knowledge about the sampling distribution of the statistic in question. For example, we might wish to test the null hypothesis that the Whitt, Gordon, and Hofley coefficient of −0.53 is a sample estimate of a population coefficient of correlation which is actually 0.

$$H_0 : \rho = 0$$

where rho (ρ) is the population correlation coefficient.

If the data were such that rejection of this null hypothesis was warranted, we might choose to conclude that the correlation found for the sample data would be generalizable to the population from which the sample was derived. If the null hypothesis is not rejected, we are saying, in effect, that the correlation found for the sample data may be attributable to sampling error and is, therefore, not generalizable. The correlation of −0.53 when used to test the null hypothesis does, by the way, warrant rejection of the null hypothesis. A technique to test the null hypothesis regarding a correlation coefficient will be described in Section 16.4.

In this chapter we will examine procedures for testing null hypotheses for various measures of association. First we will look at a technique appropriate to test the null hypothesis about various measures of nominal association. Then we will discuss tests of the null hypothesis for measures of ordinal association. Finally, we will explain how to test the null hypothesis for interval level correlation and regression.

16.1   THE CHI-SQUARE (χ^2) TEST

Sociologists are frequently interested in measuring the association between two nominal scale variables. Many of the variables with which they work are measured by crude instruments, and these measurements lend themselves best to nominal scale interpretation. For example, in studying sex as a variable the sociologist commonly makes a gross distinction between male and female. He(she) seldom attempts to measure degree of maleness or femaleness.

A number of measures of nominal-level association, which are based upon the χ^2 statistic (delta based), are important subjects for study in the field of descriptive statistics (*cf.* Chapter 6). Since the χ^2 statistic can be used to test the null hypothesis of no association, when one of these measures of association is used, a frequent procedure is to test the null hypothesis first, and if it is rejected, then compute a measure of association. The χ^2 technique that is appropriate for this

> **BOX 16.1** INDEPENDENCE
>
> The discussion that follows assumes that you understand the concept of independence and the special multiplication rule of probability related to it. If this material seems hazy to you, it might be wise to go back and review Section 10.2.2f.

kind of application is known as the **chi-square (χ^2) test for independence**. The null hypothesis that is tested is, in fact, that the two variables in question are independent rather than interrelated. If this hypothesis is rejected, then a measure of the extent of association (interrelatedness) is computed.

It should be noted at the outset that chi-square (χ^2) is both the term for a statistical technique used to test hypotheses and the term for a family of sampling distributions. Chi-square sampling distributions were discussed at some length in Chapter 12 (Section 12.5.1), and used in Chapter 13 to compute a confidence interval for the variance (Section 13.4) and in Chapter 15 in the one-sample test for the variance (Section 15.7).

The χ^2 test for independence employs a formula that differs from any seen before in this volume, and it is based upon a different concept of analysis. However, the resulting statistic is evaluated against the same sampling distributions of chi-square.

To illustrate the χ^2 test for independence we will turn to a study of leisure or "not-work" conducted by Neil H. Cheek, Jr. (Cheek, 1971). Cheek has been engaged in a series of studies of "the sociological aspects of the sociocultural behavior of going to parks and zoos" (Cheek, 1971:253). One such study dealt with the relationship between park and zoo going and other aspects of the social institutions of society. His data came from a probability sample of adults in the continental United States. For the article with which we are concerned Cheek concentrated on the 444 males in his sample. Among other things he was interested in learning whether a person's occupation relates to his inclination to visit a local park. He categorized occupations as white-collar or blue-collar and crosstabulated these categories against whether respondents had or had not gone to a park in the preceding two years. Table 16.1 presents Cheek's data for the 444 males.

The χ^2 was used to test the null hypothesis that occupational status is independent of tendency to go or not go to the park. We will follow through the process by which such a test is made, using Cheek's data.

TABLE 16.1 MALE PARK-GOING BEHAVIOR, BY OCCUPATION

Occupation	In Local Park	Not in Local Park	Totals
Blue-Collar	164	61	225
White-Collar	156	63	219
Totals =	320	124	444

Source: Cheek, 1971: Table 1. Reprinted by permission of Sage Publications, Inc.

16.1.1 Computation of Expected Frequencies

If the 444 males for which data are available represent a simple random sample of all males in the population to which we wish to generalize, then we can use the information in Table 16.1 to estimate certain parameters. The four marginal totals of Table 16.1 (320, 124, 225, and 219) give the distributions of the 444 men in the sample on occupational status and on park-going behavior. Thus, 225 of the men had blue-collar occupations; 219 had white-collar occupations; 320 went to the local park, and 124 did not. These marginal frequencies can be converted to proportions by dividing each by $N = 444$. Then, if the sample is representative of the population, these proportions will be estimates of the corresponding parameters. That is, the proportion $320/444 = 0.72$ will be an estimate of the proportion of all men in the population who went to a local park during the two years; and the proportion $124/444 = 0.28$ will be an estimate of all men in the population who did not. Likewise, the proportion $225/444 = 0.51$ will estimate the proportion of all men in the population who are blue-collar workers and $219/444 = 0.49$, the proportion of all men in the population who are white-collar workers.

For the random sample of 444 men the four entries in Table 16.1 represent the observed number of blue-collar workers who had gone to local parks (164), of white-collar workers who had gone to local parks (156), of blue-collar workers who had *not* gone to local parks (61), and of white-collar workers who had *not* gone to local parks (63). The question to which the χ^2 test of independence addresses itself is whether these **observed frequencies** resemble those that would be obtained if occupational status and tendency to go to the park were actually independent of each other.

Earlier, in Chapter 10 (Section 10.2.2f), we found that *the probability of the joint occurrence of independent events is equal to the product of their separate probabilities.* Assuming the null hypothesis to be true, we can use this rule of probability to determine what the entries in Table 16.1 would look like if the two variables in question were actually independent. This second set of entries we will call **expected frequencies.** The four proportions that were computed from the **marginal totals** of Table 16.1 (0.72, 0.28, 0.51, and 0.49) can be treated as probabilities of occurrence of events in the population. Thus, 0.72 represents the probability of men in the population going to the park; 0.28 represents the probability of these men not going; 0.51 represents the probability of these men being blue-collar workers; and 0.49 represents the probability of these men being white-collar workers.

If the null hypothesis were true and occupational status were, in fact, independent of the tendency to go to the park, then by the probability rule for the joint occurrence of independent events—the probability of blue-collar workers going to the park would be $(0.51)(0.72) = 0.367$; the probability of white-collar workers going to the park would be $(0.49)(0.72) = 0.353$; the probability of blue-collar workers *not* going to the park would be $(0.51)(0.28) = 0.143$; and finally, the probability of white-collar workers *not* going to the park would be $(0.49)(0.28) = 0.137$. These latter probabilities (0.367, 0.353, 0.143, and 0.137) can now be used to compute the expected frequencies for the four entries in Table 16.1. For convenience in discussing the entries and marginal totals for a two-by-two table, let them be labelled as they are below:

a	b	$a+b$
c	d	$c+d$
$a+c$	$b+d$	N

Thus 0.367 is the expected proportion for cell a, 0.353 for cell c, 0.143 for cell b, and 0.137 for cell d. These expected proportions may be converted into expected frequencies by multiplying each by $N=444$. The resulting expected frequencies are those that would occur if the two variables in question were independent.

Now, we will recapitulate how we computed the expected frequency for the blue-collar workers who went to the park. First, we assumed that the 444 cases represented a random sample of the population and that the proportions computed from the marginal totals could be used to estimate the corresponding population proportions. Next, we assumed that the null hypothesis of independence was true, allowing us to apply the special multiplication rule of probability for independent events. This justified multiplying 320/444 by 225/444 to get the expected proportion of cases, then we multiplied this expected proportion by N to get the expected frequency. That is:

$$(320/444)(225/444)(444)$$

Note that one of the N's (444) in the denominator of the fractions can be canceled with the final N to give the following result:

$$\frac{(320)(225)}{444}$$

In general, the expected frequencies for each cell can be computed by multiplying the marginal totals for the cell and dividing by N.

To find the expected frequency for cell a the following formula is used:

$$f_e(a) = (a+c)(a+b)/N$$

For cell b the formula is as follows:

$$f_e(b) = (b+d)(a+b)/N$$

For cell c the formula is as follows:

$$f_e(c) = (a+c)(c+d)/N$$

And finally, for cell d the formula is

$$f_e(d) = (b+d)(c+d)/N$$

Using the formula for cell a with the data from Table 16.1 gives the following result:

$$(320)(225)/444 = 162.16$$

In a two-by-two table such as Table 16.1 it is not necessary to compute all four expected frequencies in this way. Because the sums of the various pairs of

TABLE 16.2 Expected Frequencies for Data of Table 16.1

Occupation	In Local Park	Not in Local Park	Totals
Blue-Collar	162.16	62.84	225.00
White-Collar	157.84	61.16	219.00
Totals =	320.00	124.00	444.00

Source: Based on data from Cheek, 1971.

expected frequencies must equal the marginal totals, once we find the expected frequency for one cell we can find the other three by subtracting the known number from the marginal totals. For example, we can get the expected frequency for cell c of Table 16.1 by subtracting the expected frequency for cell a (162.16) from the marginal total $a + c$ (320). Then the expected frequencies for cells b and d can be found, in turn, by subtracting from marginal totals $a + b$ and $c + d$. Table 16.2 presents the expected frequencies (f_e's) we calculated from Table 16.1.

16.1.2   Test of the Null Hypothesis

Once all of the expected frequencies have been computed, they can be compared with the observed frequencies to see whether they are similar. Essentially, χ^2 measures the extent to which the observed frequencies in a contingency table deviate from those frequencies that would be expected if the null hypothesis were true. Symbolically the null hypothesis could be stated as follows: H_0: no association between variables in the population, or, in this case, the population probability of park-going given one is a blue-collar worker, is simply equal to the probability of being a blue-collar worker (i.e. they are independent). The alternative to the null hypothesis would be simply: H_a: there is some association between variables in the population.

To test the null hypothesis, the following formula is used to compute χ^2:

(16.1)
$$\chi^2 = \sum \frac{(f_o - f_e)^2}{f_e}$$

where f_o is the frequency observed for a particular cell and f_e is the frequency expected for the same cell. The large summation sign, Σ (capital sigma), tells us to compute the fractions for each cell and then sum over all cells to get χ^2. In Table 16.3 we have set up the computations necessary to get χ^2 for the Cheek data.

A careful examination of Table 16.3 will reveal that what was done to compute χ^2 was (1) to find the difference between each observed frequency and the corresponding expected frequency for each cell in the table, (2) to square each difference, (3) to divide each squared difference by its respective expected frequency, and (4) to add the resulting quotients. The sum of these quotients is the computed χ^2.

Once χ^2 has been obtained, it must be compared to the χ^2 value in the appropriate chi-square sampling distribution which bounds the desired critical region. If the computed χ^2 is in the critical region, it is appropriate to reject the

TABLE 16.3   CHI-SQUARE (χ^2) COMPUTATIONS FOR DATA FROM TABLE 16.1

Cell	f_o	f_e	$f_o - f_e$	$(f_o - f_e)^2$	$(f_o - f_e)^2/f_e$
a	164	162.16	+1.84	3.3856	0.021
b	61	62.84	−1.84	3.3856	0.054
c	156	157.84	−1.84	3.3856	0.021
d	63	61.16	+1.84	3.3856	0.055
					$\chi^2 = 0.151$

Source: Based on data from Cheek, 1971.

null hypothesis. The computed χ^2 for Table 16.3 was 0.151. To determine whether that value is in the critical region of the sampling distribution we must turn to the table of χ^2 (Appendix Table D).

As we have pointed out in Chapter 12 (Section 12.5.1), the chi-square sampling distribution is actually a family of related distributions; and the table must be entered after the degrees of freedom and level of significance have been determined. The first column of the table lists degrees of freedom and the other column headings list various areas in the right tail of the distribution. The values in the body of the table represent the χ^2 scores that bound the critical regions with the areas noted in the column headings.

Since we have already decided to test the null hypothesis at the 5 percent level of significance, the column headed 0.05 is the one we must look at. We also, however, must determine the number of degrees of freedom for the data we have analyzed so that we will know which of the many chi-square distributions in the table to use.

In the case of the χ^2 test for independence the number of categories in which the data have been classified (on both variables) is the determinant of the number of degrees of freedom. The applicable formula for degrees of freedom is as follows:

(16.2)
$$df = (r-1)(c-1)$$

where r equals the number of categories into which the horizontally listed (row) variable is classified, and c equals the number of categories into which the vertically listed (column) variable is classified. In the example, occupational status is the row variable; and it is divided into two categories—blue-collar and white-collar, thus $r = 2$. The column variable represents visits to the parks, and it is divided into two categories as well—in local park and not in local park, thus $c = 2$. Therefore, the number of degrees of freedom for this case is:

$$df = (2-1)(2-1) = 1$$

A restriction that was placed on the computation of the expected frequencies for the test for independence was that the sums of the expected frequencies had to be equal to the marginal totals. Therefore, when the first expected frequency was found for the two-by-two table, the other three expected frequencies were also determined. The number of degrees of freedom for a two-by-two table is 1. This corresponds to the number of entries in the table that are not determined by the need for the sums of the expected frequencies to agree with the marginal totals. It can be said, in general, that the number of degrees of freedom in a χ^2 test of independence will be equal to the number of cells in the contingency table that are not determined by the marginal totals of the table.

We will return now to the χ^2 table with a 0.05 critical region and 1 degree of freedom to see what the tabular χ^2 value is that bounds the critical region. The tabular value is 3.841, in fact. Therefore, any computed χ^2 as large or larger than 3.841 will be in the critical region and warrant rejection of the null hypothesis. The χ^2 computed from Cheek's data, however, was only 0.151; thus rejection of the null hypothesis is not warranted for these data.

Although failure to reject the null hypothesis is not evidence that the null hypothesis is true, Cheek can say that *his data* do not give him any basis for rejecting the null hypothesis in favor of its set of alternatives that occupational status does make a difference.

16.1.3 χ^2 for Larger Tables

The χ^2 test for independence is, by no means, limited to two-by-two contingency tables. The χ^2 can be computed for tables of any size as long as the assumptions we will discuss below are met. The amount of work involved in computing χ^2 increases as the size of the table increases. But, in any case, the computations should be done by calculator or computer.

To illustrate the computation of χ^2 from a larger table we will look at a study conducted by Diana Crane (Crane, 1965). She studied a sample of scientists (biologists, political scientists, and psychologists) at major and minor universities to explore factors related to productivity and recognition. Note that Crane's data do not constitute a simple random sample of a specified population. She apparently assumed that they were a random sample from some population, and thereby justified her use of a test of statistical significance. For purposes of illustrating the technique in question, we will also make this assumption.

Among other things Crane was interested in finding whether there was an association between scholarly productivity and the current academic affiliation of the scientists in the population. Current academic affiliation was assigned to one of three categories: major university, high minor university, and low minor university, based on the work of Berelson (Berelson, 1960). These categories were cross-classified against high and low productivity and the null hypothesis of independence was tested at the 0.01 level of significance. Crane's data and our test of her null hypothesis are presented in Table 16.4.

In Table 16.4 as in the previous two-by-two Table 16.1 the expected frequencies were computed by multiplying the appropriate marginal totals and dividing by N. Thus, the expected frequency for the high-productivity – major-university entry was found by multiplying the appropriate marginal totals and dividing by N as follows:

$$(72)(54)/150 = 25.92$$

Likewise, the high-productivity – high-minor university entry expected frequency was found by:

$$(36)(54)/150 = 12.96$$

TABLE 16.4 CHI-SQUARE (χ^2) TEST OF PRODUCTIVITY BY CURRENT ACADEMIC AFFILIATION USING CRANE DATA

| | Current Academic Affiliation | | | |
Productivity	Major University	High Minor University	Low Minor University	Totals
High	33 (25.92)	14 (12.96)	7 (15.12)	54
Low	39 (46.08)	22 (23.04)	35 (26.88)	96
Totals	72	36	42	150

Computations for χ^2:

Cell	f_o	f_e	$f_o - f_e$	$(f_o - f_e)^2$	$(f_o - f_e)^2 / f_e$
a	33	25.92	7.08	50.1264	1.93
b	14	12.96	1.04	1.0816	0.08
c	7	15.12	−8.12	65.9344	4.36
d	39	46.08	−7.08	50.1264	1.09
e	22	23.04	−1.04	1.0816	0.05
f	35	26.88	8.12	65.9344	2.45
					$\chi^2 = 9.96$

Source: Adapted from Crane, 1965:704, Table 2.

Since Table 16.4 is a two-by-three contingency table the number of degrees of freedom is $df = (r-1)(c-1) = (2-1)(3-1) = 2$. With 2 degrees of freedom two of the expected frequencies are free to vary and must be computed by multiplying marginal totals and dividing by N. The other expected frequencies, then, may be found by subtracting from the marginal totals. Therefore, once the expected frequencies 25.92 and 12.96 are found as above, all of the others can be found by subtraction.

Once all of the expected frequencies were computed for Table 16.4, the differences between observed and expected frequencies were taken and squared. Each squared difference was divided by its expected frequency, and the quotients were summed to obtain χ^2.

You may have noticed that all of the differences in the two-by-two Table 16.1 of the previous example were the same (although two were negative and two were positive). In the two-by-three Table 16.4 the differences were not all the same. The differences for high and low productivity were the same for pairs in the same column because the productivity variable was dichotomized; but the differences across rows for current academic affiliation were not the same. Only when variables are dichotomized will the differences between observed and expected frequencies be the same. In a two-by-two table, because both variables are dichotomized, the differences for all four entries will be the same.

The computed χ^2 in Table 16.4 was 9.96. Entering the χ^2 table (Appendix

Table D) with 2 degrees of freedom and a 0.01 critical region we find that the χ^2 value bounding the critical region is 9.210. The computed χ^2 is in the critical region so rejection of the null hypothesis was warranted. Crane reached the conclusion that current academic affiliation was associated with scholarly productivity.

In order to determine the nature of the relationship, it is necessary to examine the differences between observed and expected frequencies for the individual cells of the contingency table. The differences reveal that scientists affiliated with major universities produced more scholarly work than chance would lead one to expect and those affiliated with low minor universities produced less. The expected and observed frequencies for high and low productivity were very much alike for those scientists affiliated with high minor universities so those cells contributed practically nothing to the χ^2. The statistically significant result is attributable to the differences between scientists in major universities and those in low minor universities.

Crane also collected data on the universities at which her subjects did their graduate work and found that variable was related to productivity as well. In order to explore further the relationship between productivity and current academic affiliation, it would be possible to use scientist's graduate school as a control variable. Thus, a set of conditional tables could be examined in which prestige level of the scientist's graduate school was controlled at the three levels — major university, high minor university and low minor university — and current academic affiliation was compared with level of productivity. Separate χ^2's could be computed for the three conditional tables to test them for independence. Furthermore, the three resulting χ^2's could be summed to give an overall test of the independence of current academic affiliation and the level of productivity holding graduate school constant. That is, adding the three χ^2's for the conditional tables tests the independence of the two main variables while sorting out the effects of the control variable. To evaluate this pooled χ^2, the degrees of freedom for each of the conditional tables are also summed. Thus, if we were to pool the χ^2's from three two-by-three conditional tables, the resulting χ^2 would be evaluated against a chi-square sampling distribution with 6 degrees of freedom, again, assuming independent random samples.

Note that the value of the pooled χ^2 would not necessarily be the same as the overall χ^2 of 9.96 which was computed from the data of Table 16.4. That χ^2 ignored data on the scientist's graduate school; the pooled χ^2 would reflect the influence of that variable.

BOX 16.2 CONDITIONAL TABLES

If you don't quite understand this discussion about the additive nature of χ^2, perhaps it is because you are unclear about the conditional table concept. If this is the problem, you must review material on conditional tables. (Cf. early portions of Chapter 8.)

16.1.4   Interpretation of a Significant χ^2

The χ^2 computed from Diana Crane's data fell in the critical region but fairly close to the boundary of that region. It is possible for a computed χ^2 to extend quite far into the right tail of the sampling distribution. The larger the χ^2 (depending upon the number of degrees of freedom involved), the farther it will push into the right tail of the sampling distribution.

What does it mean when the computed χ^2 is large? It does *not* mean that the relationship between the variables being investigated is strong. The χ^2 *does not measure strength of relationship*. It merely measures whether there is a relationship, which is not likely to be due to chance. When the value of χ^2 is large, it means that we can be more confident about rejecting the null hypothesis and concluding that the variables are related.

The level of the statistical significance of the χ^2 test is, in fact, independent of the strength of the relationship which exists between the variables studied. It is possible to get a large and very significant χ^2 when the association between the variables of interest is very weak. It is also possible to get a nonsignificant χ^2 when the association between the variables of interest is very high. This latter situation would mean, in effect, that although there is a strong association between the variables for the data at hand, generalization beyond the data would not be warranted.

If Cramer's V* is computed for Crane's data in Table 16.4 the resulting coefficient would be 0.26, which represents roughly a 7 percent association between the two variables. This is a modest association, to say the least; and yet χ^2 was found to be statistically significant at the 0.01 level.

If the sizes of the deviations between the observed and expected frequencies are held constant and the size of the sample is increased, the size of the resulting χ^2 will be decreased. For example, if the differences between observed and expected frequencies in a two-by-two table were 10 and $N = 100$, holding the differences constant at 10 and increasing N to 200 would have the effect of decreasing χ^2. The principle involved is that *deviations of 10 are less likely in smaller samples (e.g., 100 cases) than in larger ones*.

On the other hand, if both the deviations between observed and expected frequencies and the sample size are doubled, then the resulting χ^2 will also be doubled, thereby moving it farther into the critical region of the sampling distribution. The principle to consider here is that *large deviations found in a large sample provide more evidence for rejecting the null hypothesis than proportionately the same deviations from a smaller sample*.

16.1.5   Assumptions of the χ^2 Test for Independence

There are a number of assumptions that are implied by the χ^2 test for independence. *First*, it is assumed that the data being analyzed are a simple random sample of the population. This assumption makes it possible to use the marginal totals to compute expected frequencies and test the null hypothesis of no association.

---

*Cramer's V is a measure of association for nominal scale data discussed in Chapter 6.

Second, it is assumed that the observations are independent, a property that simple random sampling assures. This rules out the use of χ^2 for comparing observations of the same subjects at two different points in time or for comparing observations of matched subjects.

Third, it is assumed that no expected frequency in the contingency table being analyzed will be less than 5. When expected frequencies are too small the chi-square sampling distribution does not represent adequately the distribution of the test statistic:

$$\sum \frac{(f_o - f_e)^2}{f_e}$$

It is assumed that the underlying distribution of the computed chi-square statistic is continuous because the chi-square sampling distribution is a continuous distribution. In fact, the distribution of the chi-square statistic is discrete; however, when degrees of freedom are more than 1, the assumption of continuity is a reasonable one.

There is some disagreement about how reasonable the assumption is for the two-by-two table with 1 degree of freedom (Grizzle, 1967; Mantel and Greenhouse, 1968). It is suggested that when the computed χ^2 is in the critical region but near the boundary of that critical region a correction for continuity should be made. The correction is effected by reducing the absolute differences between observed and expected frequencies by 0.5. The formula for the corrected χ^2 is as follows:

(16.3)
$$\chi_c^2 = \sum \frac{(\,|f_o - f_e|\,-0.5\,)^2}{f_e}$$

The correction has the effect of reducing the magnitude of the computed χ^2. Therefore, a χ^2 that is very near the boundary of the critical region may be made nonsignificant by the correction. If the χ^2 is well within the critical region, the correction will make no difference in the results of the analysis.

In a table that is larger than a two-by-two table it is often possible to combine categories, reduce the size of the table, and increase the size of the expected frequencies. Any such collapsing of categories will reduce the amount of information to be obtained from the analysis and should not be attempted unless the combination of categories has some logical justification. For example, if the expected frequencies had not been large enough in the two-by-three Table 16.4 we just completed analyzing, it would be logically defensible to combine the high minor universities and the low minor universities into a single minor university category if that would serve the purpose of increasing the sizes of the expected frequencies sufficiently. It would not be logically defensible, however, to combine the major universities and the low minor universities and compare them to the high minor universities.*

The sum of all of the expected frequencies must be equal to the sum of all of the observed frequencies. If this is not so, then it would be impossible to get a χ^2 of zero. In testing for independence a χ^2 of zero (representing no difference be-

---

*When the expected frequencies are too small in a two-by-two table, Fisher's Exact Test can be used instead of χ^2. (See, for example, Blalock, 1972.)

tween observed and expected frequencies) must be a distinct possibility. This is assured by the way expected cell frequencies are computed under H_0.

Two conditions that can be relaxed in using the χ^2 test are the measurement requirement and the requirement that the shape of the population distribution must be specified. The χ^2 lends itself to the analysis of nominal-level data. As a matter of fact, it is insensitive to order; and when it is used to analyze ordinal- or interval-level data, information on ordering of categories is discarded.

Neither is χ^2 limited to use with a normally distributed population variable. It can be used no matter what the shape of the population distribution. It is a distribution-free or nonparametric technique.

These two latter points make the χ^2 technique an extremely popular one among sociologists. It should be pointed out, however, that χ^2 should not be used in lieu of another more powerful technique just because it is rather widely applicable. The sociologist should always use the most appropriate and powerful technique called for by the research problem and warranted by the data. Furthermore, χ^2 should be used in conjunction with one of the measures of association appropriate to nominal-level data.

If a measure of association is computed for a body of sample data and that measure does not reveal a level of association strong enough to be theoretically or practically worthwhile, there seems to be little point in computing χ^2. This is because the only function that χ^2 serves is to help the researcher decide whether sample findings can be generalized to the population from which the sample was drawn.

Unfortunately, there seems to be widespread misunderstanding of what a statistically significant χ^2 means. That fact, plus the fact that χ^2 is easy to compute and has a set of assumptions that are not too demanding, has led to a greater use of the technique in the literature than is warranted.

Diagram 16.1 summarizes step-by-step the procedures used in the χ^2 test of independence.

16.2 TEST FOR GAMMA (γ)

A frequently used measure of association for ordinal data is gamma (γ), commonly considered in the field of descriptive statistics.* The *parameter* γ (symbolized by the Greek letter) may be estimated for the total population by using the *statistic G* (expressed as a capitalized Latin letter), as computed from sample data. When G is used as an estimate of the corresponding γ parameter, it is appropriate to test the null hypothesis, $\gamma = 0$, to evaluate the possibility that the computed G is merely due to sampling error.

Armer and Youtz investigated the extent to which a western style education has modernizing effects on African youths (Armer and Youtz, 1971). Interviews were conducted with a random sample of 591 young men in Kano City, Nigeria, to determine the extent of their education and their commitment to "modern value orientations." The six value orientations taken as symptomatic

*Gamma is introduced in Chapter 7.

DIAGRAM 16.1   Chi-Square (χ^2) Test for Independence

Problem: To determine whether two nominal-level variables are related.

Assumptions:
1. Independent observations
2. Nominal-level measurements
3. Simple random sample
4. Underlying continuous distribution of the χ^2 statistic, no expected frequency is less than five.

χ^2 Formula:

$$\chi^2 = \Sigma \ (f_o - f_e)^2 / f_e$$

If the significance of the computed χ^2 is marginal, and $df = 1$, you may use the following corrected formula:

$$\chi_c^2 = \Sigma \ (\ |f_o - f_e| - 0.5)^2 / f_e$$

Sampling Distribution:
  The χ^2-distribution with degrees of freedom equal to $(r-1)(c-1)$.
Step 1: Set up the null hypothesis. Specify the level of significance.
Step 2: Draw a simple random sample from the population and collect the data.
Step 3: Put the sample data in a contingency table and compute the expected frequencies from the marginal totals and N.
Step 4: Compute χ^2. If the computed χ^2 is in the critical region, reject the null hypothesis. If not, fail to reject it.
Step 5: If the null hypothesis is rejected, examine the differences between observed and expected frequencies to determine the nature of the relationship.

TABLE 16.5 LEVEL OF WESTERN EDUCATION AND INDIVIDUAL MODERNITY

Western Education Level	Individual Modernity		Totals
	Low	High	
No Education	194	118	312
Some Primary Education	94	117	211
Some Secondary Education	11	57	68
			N = 591

Computation of G:

$$G = (N_s - N_d)/(N_s + N_d)$$

$$N_s = 194(117 + 57) + 94(57) = 39,114$$

$$N_d = 118(94 + 11) + 117(11) = 13,677$$

$$G = \frac{39,114 - 13,677}{39,114 + 13,677}$$

$$= +0.48$$

of modernity were: independence from family, ethnic equality, empiricism, mastery over nature, futurism, and receptivity to change. A composite measure of modernity based on these value orientations was dichotomized into high and low modernity categories and these were cross-tabulated against three educational levels: no education, some primary education, and some secondary education. Table 16.5 summarizes the data for these cross-tabulations and the computation of G. The sample G formula is as follows:

(16.4)
$$G = (N_s - N_d) / (N_s + N_d)$$

where N_s = the number of pairs of cases ranked in the same way on both variables, and N_d = the number of pairs ranked in the opposite way on the two variables.

As you can see from the computations carried out in Table 16.5, the G for these data was +0.48. Since the motive for sampling the young men of Kano City was to generalize about all of the young men of Kano City (and perhaps beyond), it is appropriate to determine whether the correlation of 0.48 is likely to have come from a population in which the correlation between education and modernity is 0. Therefore, it is appropriate to test the null hypothesis:

$$H_0 : \gamma \leq 0$$

If this null hypothesis can be rejected, there will be reason to believe that the computed sample G reflects an actual population correlation which differs from zero. Notice that the null hypothesis, as stated, is for a one-tailed test. This is because we are interested in the thesis that γ is greater than zero. Generally, tests of the hypothesis of zero correlation or association will be one-tailed tests because the theoretically expected direction of association is specified before the tests of significance are conducted.

Goodman and Kruskal have worked out a normal approximation of the sampling distribution of G which makes test of the null hypothesis possible (Goodman and Kruskal, 1963). They give the following formula for converting G to a standard score:

(16.5)
$$z = (G - \gamma) \sqrt{\frac{N_s + N_d}{N(1 - G^2)}}$$

Assuming that the null hypothesis is true, the γ in the formula will be 0. We can then substitute the necessary sample data into the formula and solve for z. Making these substitutions, we get the following results:

$$z = (0.48 - 0) \sqrt{\frac{39,114 + 13,677}{591(1 - 0.2304)}}$$

$$= (0.48) \sqrt{\frac{52,791}{454.83}}$$

$$= (0.48)(10.77)$$

$$= +5.17$$

If we chose to test the null hypothesis at the 0.01 level, the z score that bounds the critical region of the sampling distribution would be +2.33. Since the computed z score is +5.17, it extends well into the right tail of the sampling distribution, so the null hypothesis can be rejected in favor of the set of alternatives that γ is greater than 0. The interpretation of the results would be that there is a positive relationship between educational level and individual modernity.

Formula (16.5) gives a conservative estimate of the z score. Goodman and Kruskal present another formula that gives a more accurate estimate, but it is very cumbersome to work with (Goodman and Kruskal, 1963:325). It involves the use of computing matrices to find the necessary values for the formula. Formula (16.5) is so much more convenient to handle that it may be used for all but those cases in which rejection of the null hypothesis is in doubt. Formula (16.5) generally underestimates z; thus a test of the null hypothesis which uses that formula and which falls short of reaching the critical region could presumably reach the critical region if the more accurate formula was used. To compare results from Formula (16.5) with results from the more involved but more accurate formula, we computed z for the Armer and Youtz data both ways. The more accurate formula yielded a z of +6.24 as compared to the z of +5.17 computed with Formula (16.5). In this case both formulae resulted in z scores that led to the same conclusion.

The procedure just described for testing the significance of γ [with Formula (16.5)] applies as well to conditional tables for which the G's have been computed. Each conditional table is merely treated as a separate case and each G is used to test the null hypothesis: $H_0: \gamma = 0$. It is entirely possible that an overall γ which is statistically significant may yield a set of conditional γ's, some of which are statistically significant and others of which are not.

16.2.1 Assumptions of the Test for Significance of γ

In using the sample G to test the null hypothesis that $\gamma=0$, it is necessary to make the following assumptions: (1) the measures upon which the sample G is based are independent, (2) the measures of both variables are ordinal in nature, (3) the sample is a simple random sample from the population, and (4) the sample is large enough to justify using a normal approximation to the sampling distribution. It is not necessary to assume that the population is normally distributed, however. The γ is another distribution-free or nonparametric technique.

Diagram 16.2 lists the essential steps in the test of significance for γ.

16.3 TEST FOR TAU (τ)

An alternative to γ for measuring association of ordinal-level data is Kendall's Tau (τ). We shall consider tests of significance for two versions of the τ coefficient, tau-a (τ_a), which is appropriate when there are no tied ranks, and tau-b (τ_b), which takes tied ranks into consideration. These coefficients were introduced in Chapter 7.

The formula for estimating τ_a from sample data is as follows:

(16.6)
$$t_a=\frac{N_s-N_d}{\frac{1}{2}N(N-1)}$$

where the numerator is the same as the numerator for G and N is the sample size.

The formula for estimating τ_b from sample data is as follows:

(16.7)
$$t_b=\frac{N_s-N_d}{\sqrt{(N_s+N_d+T_y)(N_s+N_d+T_x)}}$$

where the numerator is the same as the numerator for G, N_s represents the number of pairs ranked in the same way; N_d, the number of pairs ranked in opposite ways; T_y, the number of pairs with tied ranks on the dependent variable only; and T_x, the number of pairs with tied ranks on the independent variable only.

16.3.1 The Test of Significance for τ_a

Since τ_a is really only appropriate for the measurement of association between two ordinal variables when there are no ties, it is likely to be of limited use to the sociologist. When it is applicable for the data at hand, however, it is relatively easy to test for significance. The null hypothesis to be tested is that $\tau_a=0$.

When $N \geq 10$ the sampling distribution of t_a is approximately normal with a mean of zero and a standard error equal to the following:

(16.8)
$$s_{t_a}=\sqrt{\frac{2(2N+5)}{9N(N-1)}}$$

DIAGRAM 16.2   Test of Significance for Gamma (γ)

Problem: To determine whether the population correlation of two ordinal level variables is different from zero.

Assumptions:
1. Independent observations
2. Ordinal-level of measurement on both variables
3. Simple random sample
4. N is large enough to justify using a normal approximation to the sampling distribution of G.

G Formula:

$$G = N_s - N_d / N_s + N_d$$

Standard Score Formula:

$$z = (G - \gamma) \sqrt{N_s + N_d / N(1 - G^2)}$$

Sampling Distribution:
　　The normal curve as an approximation of the sampling distribution of G.

Step 1: Set up the null hypothesis. Specify the level of significance.

Step 2: Draw a random sample from the population and collect data.

Step 3: Compute G from the sample data.

Step 4: Compute the z score using the sample data. If computed z score is in the critical region, reject the null hypothesis. If not, fail to reject it.

Step 5: If the computed z score does not reach the critical region but is close, use Goodman and Kruskal's alternate formula. (See Goodman and Kruskal, 1963.)

A standard score may then be computed as follows:

(16.9)

$$z = \frac{t_a}{s_{t_a}}$$

Once z has been computed, it can be evaluated against the normal distribution in the usual way.* If the computed z is in the critical region, the null hypothesis can be rejected. If not, fail to reject it.

16.3.2 Assumptions of the Test of Significance for Tau-a (τ_a)

The following assumptions underly the test of significance for τ_a : (1) the observations are independent, (2) both variables are measured on at least the ordinal level, (3) the data constitute a simple random sample from the population, (4) $N \geq 10$, and (5) there are no tied ranks on either variable. It is most appropriate, therefore, for continuous variables. If there are ties, then the limits of t_a will not be −1.0 and +1.0 because the denominator $[\frac{1}{2}N(N-1)]$ will be greater than $N_s + N_d$.

It is not necessary to make any assumptions about the shape of the population distribution. The test is, therefore, distribution-free or nonparametric.

16.3.3 Test of Significance for τ_b

It is much more likely that τ_b will be used by the sociologist rather than τ_a because the data with which he(she) works are likely to include ties. When the situation calls for the statistic t_b, testing the null hypothesis $(\tau_b = 0)$ is possible, but it is much more cumbersome than the test for τ_a. We will compute t_b for the data from the Armer and Youtz study presented in Table 16.5 and test the null hypothesis that the population correlation is zero. This will not only demonstrate how to test for significance, but will also allow us to compare the results of computing G and t_b for the same data. Table 16.6 consists of the same data as were presented in Table 16.5, but in this case t_b has been computed. Notice that t_b only yields a measure of association of 0.27 as compared to a G of 0.48. This difference is due to the more conservative denominator of t_b which takes ties into consideration while G ignores the ties. Also, when the contingency table is not square the upper limit of t_b is less than 1.0.

According to Kendall, the sampling distribution of $S = N_s - N_d$, which is the numerator of the formula for t_b, can be approximated by the normal distribution if N is large (although the approximation is good with N as small as 30) (Kendall, 1955:chapter 4). The sampling distribution has a mean of 0 and an error variance equal to the following:

---

*If the computed z-score falls in the critical region but near the boundary of the region, it should be corrected for continuity. (See Kendall, 1955: Chapter 4, for a discussion of the correction.)

TABLE 16.6 Level of Western Education and Individual Modernity: Test for Tau-b (τ_b)

	Individual Modernity		
Western Education Level	Low	High	Row Totals
No Education	194	118	312
Some Primary Education	94	117	211
Some Secondary Education	11	57	68
Totals = 299		292	N = 591

Computation of t_b:

$$t_b = \frac{N_s - N_d}{\sqrt{(N_s + N_d + T_y)(N_s + N_d + T_x)}}$$

$$N_s = 194(117 + 57) + 94(57) = 39{,}114$$
$$N_d = 118(94 + 11) + 117(11) = 13{,}677$$
$$N_s + N_d = 52{,}791$$
$$T_y = 194(94 + 11) + 94(11) + 118(117 + 57) + 117(57) = 48{,}605$$
$$T_x = 194(118) + 94(117) + 11(57) = 34{,}517$$

$$t_b = \frac{39{,}114 - 13{,}677}{\sqrt{(52{,}791 + 48{,}605)(52{,}791 + 34{,}517)}}$$

$$= \frac{25{,}437}{\sqrt{(101{,}396)(87{,}308)}}$$

$$= \frac{25{,}437}{94{,}089}$$

$$= +0.27$$

Source: Adaptation from Armer and Youtz, 1971:611, Table 2.

(16.10)

$$\sigma_S^2 = 1/18 \left[N(N-1)(2N+5) - \sum_{i=1}^{r} n_i(n_i-1)(2n_i+5) \right.$$
$$\left. - \sum_{j=1}^{c} n_j(n_j-1)(2n_j+5) \right] + \frac{1}{9N(N-1)(N-2)} \left[\sum_{i=1}^{r} n_i(n_i-1)(n_i-2) \right]$$
$$\left[\sum_{j=1}^{c} n_j(n_j-1)(n_j-2) \right] + \frac{1}{2N(N-1)} \left[\sum_{i=1}^{r} n_i(n_i-1) \right] \left[\sum_{j=1}^{c} n_j(n_j-1) \right]$$

where N is the size of the sample, n_i is the ith row total, and n_j is the jth column total. $\Sigma_{i=1}^{r}$ instructs you to sum all rows from the first to the rth where r is the number of rows; and $\Sigma_{j=1}^{c}$ instructs you to sum all columns from the first to the cth where c is the number of columns.

This rather awesome and cumbersome formula is based on nothing but the sample size and marginal totals for rows and columns. In Table 16.7 we have taken the data from Table 16.6, broken down Formula (16.10) into its constituent parts, completed the necessary computations, and used the results of these computations to find σ_S^2.

The normal approximation of the sampling distribution of the S statistic

TABLE 16.7 Computations for σ_S^2 Using Data from Table 16.6

$1/18 = 0.05556$

$N(N-1)(2N+5) = 591(590)(1182+5) = 413{,}895{,}030$

$$\sum_{i=1}^{r} n_i(n_i-1)(2n_i+5) = 312(311)(624+5) + 211(210)(422+5)$$

$\qquad + 68(67)(136+5) = 80{,}595{,}894$

$$\sum_{j=1}^{c} n_j(n_j-1)(2n_j+5) = 299(298)(598+5) + 292(291)(584+5) = 103{,}777{,}014$$

$$\frac{1}{9N(N-1)(N-2)} = \frac{1}{9(591)(590)(589)} = 0.0000000005$$

$$\sum_{i=1}^{r} n_i(n_i-1)(n_i-2) = 312(311)(310) + 211(210)(209) + 68(67)(66)$$

$\qquad = 39{,}641{,}406$

$$\sum_{j=1}^{c} n_j(n_j-1)(n_j-2) = 299(298)(297) + 292(291)(290) = 51{,}105{,}174$$

$$\frac{1}{2N(N-1)} = \frac{1}{2(591)(590)} = 0.0000014$$

$$\sum_{i=1}^{r} n_i(n_i-1) = 312(311) + 211(210) + 68(67) = 145{,}898$$

$$\sum_{j=1}^{c} n_j(n_j-1) = 299(298) + 292(291) = 174{,}074$$

Computations of the error variance:

$\sigma_S^2 = 0.05556(413{,}895{,}030 - 80{,}595{,}894 - 103{,}777{,}014) + 0.0000000005$

$(39{,}641{,}406)(51{,}105{,}174) + 0.0000014(145{,}898)(174{,}074) = 13{,}800{,}745.44$

---

can be used to test the null hypothesis that $\tau_b \leq 0$ or $\tau_b = 0$ by computing the standard score with the following formula:

(16.11)
$$z = \frac{S}{\sigma_S}$$

Notice that the denominator of Formula (16.11) is the standard error of S (σ_S) rather than the error variance (σ_S^2); therefore, we must take into account the square root of the error variance just computed in Table 16.7:

$$\sigma_S = \sqrt{13{,}800{,}745.44}$$

$$= \quad 3714.94$$

Then, substituting into Formula (16.11) we get the following:

$$z = \frac{25{,}437}{3714.94}$$

$$= +6.85$$

If we were testing the null hypothesis with a one-tailed test at the 0.01 level of significance, the z score bounding the critical region would be +2.33. The standard score of +6.85 computed from the data obviously extends well into the right tail of the sampling distribution.* Therefore, rejection of the null hypothesis in favor of the set of alternatives that τ_b is greater than zero would be warranted. This is the same conclusion that was reached when γ was tested for significance. The conclusion would be that those young men who were "better" educated exhibited more signs of individual modernity.

16.3.4 Assumptions of the Test of Significance for τ_b

It is important to note that in using this test of significance for τ_b we are assuming the following: (1) the observations are independent which is assumed when, (2) the data constitute a simple random sample from the population, and (3) both variables are measured on at least the ordinal level. It is not necessary to specify the shape of the population sampled.

The normal approximation to the sampling distribution of S applies best when the N is large and the number of ties is small. It should also be pointed out that the standard error computed is based upon the assumption that all of the samples that constitute the sampling distribution of S have exactly the same number of ties on both variables as were found in the sample actually drawn from the population.

Since the test of significance was made of S instead of τ_b the technique just described could also be used as a test of significance for any of the measures of association for which the numerator is $S = N_s - N_d$.

Diagram 16.3 summarizes the procedures for testing hypotheses about τ.

16.4 TEST FOR PRODUCT MOMENT COEFFICIENT OF CORRELATION AND THE REGRESSION COEFFICIENT

David L. Featherman studied the socioeconomic achievement of a sample of the white male metropolitan population (Featherman, 1971). His sample included 715 men from the "Princeton Fertility Study" for whom longitudinal data were available over a ten-year period. He sought to develop a model to represent the process of socioeconomic achievement through the middle years of the work career. Among the variables he considered were education and income during the 35 to 44 age period. He found that the **product moment coefficient of correlation** for these two variables was 0.478 (Featherman, 1971:297, Table 1).

Since his data represented a random sample of a larger population, we may legitimately ask whether a correlation of 0.478 is likely to indicate the presence of a relationship between education and income for that age period in the population as a whole. The accepted procedure is to test the null hypothesis that the population correlation is zero or less ($H_0: \rho \leq 0$ or $H_0: \rho = 0$). If we can reject the null hypothesis on the basis of the sample data and accept the set of al-

---

*For this test, too, a correction for continuity should be made if the significant z-score is borderline. (See Kendall, 1955: Chapter 4.)

DIAGRAM 16.3   Testing Hypotheses about Tau (τ)

TEST OF SIGNIFICANCE OF τ_a

Problem: To determine whether the association between two ordinal-level variables as measured by t_a is significantly different from zero.

Assumptions:
1. Independent observations
2. Both variables measured on ordinal level
3. Simple random sample of the population
4. $N \geq 10$
5. No tied ranks on either variable

Formula for t_a:

$$t_a = \frac{N_s - N_d}{\frac{1}{2}N(N-1)}$$

Formula for the Standard Error of t_a:

$$s_{t_a} = \sqrt{\frac{2(2N+5)}{9N(N-1)}}$$

Standard Score Formula:

$$z = \frac{t_a}{s_{t_a}}$$

Sampling Distribution:

When N is 10 or larger, the normal approximation to the sampling distribution of t_a may be used.

Step 1: Set up the null hypothesis. Specify the level of significance.

Step 2: Draw a simple random sample from the population and collect data.

Step 3: Compute τ_a from the data, and compute the standard error.

Step 4: Compute the standard score. If the computed standard score is in the critical region, reject the null hypothesis. If not, fail to reject it.

DIAGRAM 16.3   *(Continued)*

TEST OF SIGNIFICANCE OF τ_b

Problem: To determine whether the association between two ordinal-level variables as measured by t_b is significantly different from zero.

Assumptions:

1. Independent observations
2. Both variables measured on ordinal level
3. Simple random sample of population
4. N is large and number of ties is small
5. All samples in the sampling distribution have exactly the same distribution of ties

Sampling Distribution:

The normal approximation to the sampling distribution of S.

Formula for t_b :

$$t_b = \frac{N_s - N_d}{\sqrt{(N_s + N_d + T_y)(N_s + N_d + T_x)}}$$

Error Variance of S:

$$\sigma_S^2 = 1/18 \left[N(N-1)(2N+5) - \sum_{i=1}^{r} n_i(n_i-1)(2n_i+5) - \sum_{j=1}^{c} n_j(n_j-1)(2n_j+5) \right]$$

$$+ \frac{1}{9N(N-1)(N-2)} \left[\sum_{i=1}^{r} n_i(n_i-1)(n_i-2) \right] \left[\sum_{j=1}^{c} n_j(n_j-1)(n_j-2) \right]$$

$$+ \frac{1}{2N(N-1)} \left[\sum_{i=1}^{r} n_i(n_i-1) \right] \left[\sum_{j=1}^{c} n_j(n_j-1) \right]$$

Standard Score Formula:

$$z = S/\sigma_S, \qquad S = N_s - N_d$$

Step 1: Set up the null hypothesis. Specify the level of significance.
Step 2: Draw a simple random sample from the population and collect data.
Step 3: Compute S, t_b , and the error variance from the data.
Step 4: Compute the standard score. If the computed standard score is in the critical region, reject the null hypothesis. If not, fail to reject it.
NOTE: This same test applies to any of the measures of association for which the numerator is S.

ternatives that $\rho > 0$, we will feel justified in concluding that the computed value of 0.478 is not just a quirk of sampling error. A two-tailed test may be appropriate in certain situations. ($H_0: P = 0$)

The sample coefficient of correlation (r) will only have a symmetrical sampling distribution when ρ is actually zero. This is so because r has absolute limits of -1.00 and $+1.00$. When ρ is zero the sample r's can deviate from the parameter an equal distance above and below. However, as ρ deviates from zero the sampling distribution becomes progressively skewed. For example, if ρ were actually $+0.5$, the sample r's could go as low as -1.00 (a distance of 1.5) on the left side of the sampling distribution, but they could go as high as $+1.00$ (a distance of 0.5) on the right side. The sampling distribution, therefore, would be negatively skewed.*

An approach which avoids the problem of the skewed sampling distribution is one which involves transforming r into a statistic that has a normal sampling distribution. The formula for making this transformation from r to Z' is as follows:†

$$Z' = 1.151 \log (1+r)/(1-r) \tag{16.12}$$

The sampling distribution of Z' will be approximately normal even though the sampling distribution of r is not. Furthermore, the standard error of Z' depends only upon N and can be found with the following simple formula:

$$\sigma_{z'} = 1/\sqrt{N-3} \tag{16.13}$$

To test the null hypothesis that $\rho \leq 0$ we will convert r to Z' and find the 95 percent confidence limits around Z'; then we can convert those limits to r's. If the confidence interval does not include 0, the null hypothesis can be rejected at the 0.05 level of significance. If the interval does include 0, we will not reject the null hypothesis. It is not necessary to use Formula (16.12) to convert r to Z' since the Z's and their corresponding r's are tabulated in Appendix Table F at the end of this volume.

Since Featherman's sample consisted of an N of 715, that figure can be substituted into Formula (16.13) to solve for the standard error of Z' as follows:

$$\sigma_{z'} = \frac{1}{\sqrt{N-3}} = \frac{1}{\sqrt{715-3}} = \frac{1}{27.4226} = 0.0365$$

According to Appendix Table F, the Z' that corresponds to an r of 0.478 is 0.5204.

Since we are interested in a one-tailed test (of the hypothesis that the population correlation coefficient is not greater—in the positive direction—than zero), we would reject this hypothesis when the *lower* confidence limit around our estimate of $+r$ does not fall below an r of zero. To find the one-sided, 95% confidence limit we would look for the standard score that cut off 5% of the area under a normal curve in one tail of that distribution (or, in effect, use the standard score for 90%, 2-sided confidence intervals). That value is -1.645 (see the normal

*There is a test of the null hypothesis available which makes use of Student's t as the sampling distribution. That test is most applicable, however, when $H_0: \rho = 0$ because Student's t is a symmetrical sampling distribution (see Hayes, 1963:520–521).
†Note that this Z' is not the standard z score that we have had occasion to use frequently throughout this volume.

BOX 16.3

Notice that Formula (16.14) for finding confidence limits is very similar to the formulae we used for making interval estimates in Chapter 13 of this volume [*e.g.*, Formula (13.6) and Formula (13.7)]. In Chapter 14 we pointed out that confidence intervals may be used to test hypotheses (Section 14.2.9). Formula (16.14) is an example of such an application.

curve table, Appendix Table B). Equipped with this information we can substitute into the following formula for confidence limits and compute the one-sided, 95% confidence limit around Z':

(16.14)
$$cl = Z' - 1.645(\sigma_z)$$

Thus

$$cl_1 = 0.5204 - 1.645(0.0365) = 0.5204 - 0.0600 = 0.4604$$

Now that the lower confidence limit has been computed for Z', it can be converted back to r by returning to Appendix Table F and reading the r that correspond to Z of 0.4604. This r is 0.430. Since this r is above 0 the null hypothesis

SAMPLING DISTRIBUTION AND CONFIDENCE INTERVAL FOR Z'

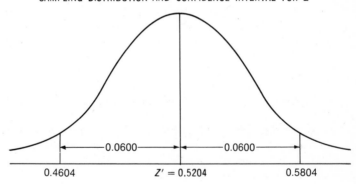

SAMPLING DISTRIBUTION AND CONFIDENCE INTERVAL FOR r

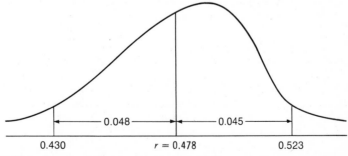

FIGURE 16.1 Sampling Distributions with 90 Percent Confidence Intervals for Z' and r when $r = 0.478$, $N = 715$.

can be rejected in favor of the set of alternatives that $\rho > 0$. If the lower limit had been below 0, rejection of the null hypothesis would not have been warranted.

If we had computed the upper confidence limit $(cl = 0.5204 + 1.645(0.0365) = 0.5804)$ you would note that the confidence limits in terms of Z' (0.4604 and 0.5804) are symmetrical about the Z' of 0.5204. When these limits were reconverted to r's, the limits became 0.430 and 0.523. These limits are not symmetrical about the original r of 0.478. The sampling distribution of r's is slightly negatively skewed. Figure 16.1 represents graphically the sampling distributions of Z' and r for this particular case.

Formula (16.14) includes a standard (z) score of $+1.645$ for the one-sided 95 percent confidence limit. Other z scores may be used in the formula to find other confidence intervals. For example, if a one-sided 99 percent confidence limit was wanted, the z score used would be $+2.33$.* Two-sided standard scores would be ±1.96 and ±2.58, respectively.

16.4.1   Test of Significance for the Regression Coefficient

Since the regression coefficient $\beta = 0$, when the correlation coefficient $\rho = 0$, the null hypothesis can be expressed as follows:

$$H_0 : \beta = \rho = 0$$

Therefore, the procedure of converting r to Z' and finding confidence limits is also a test of significance for the regression coefficient. We need only to test $\rho = 0$, and, if that null hypothesis can be rejected, we know that the regression coefficient also differs from 0.

16.4.2   Assumptions of the Test of Significance Using Z'

In order to test the null hypothesis about ρ and the regression coefficient it is necessary to make the following assumptions: (1) the observations are independent, (2) both variables are measured at the interval level, (3) both variables are normally distributed (bivariate normal distribution), (4) the variables are linearly related, (5) N is large, and (6) the data constitute a simple random sample from some population. If we are not able to satisfy these assumptions, we must proceed with caution and consider carefully whether we are still willing to make generalizations beyond the sample data.

16.4.3   Tests of Significance for Partial and Multiple Correlation Coefficients

The Z'-transformation can also be used to compute confidence intervals for the **partial** and the **multiple correlation coefficients.** The only difference in procedure is that the standard error of Z' has a slightly different formula. The

---

*Whether the sign should be $+$ or $-$ depends on the one-tailed hypothesis being considered.

603

DIAGRAM 16.4a Tests for Product Moment Coefficient of Correlation and the Regression Coefficient

Problem: (1) To determine whether the association between two interval-level variables as measured by r is significantly different from zero. (2) To determine whether the corresponding regression coefficient is significantly different from zero.

Assumptions:

1. Independent observations
2. Both variables measured at interval level
3. Both variables normally distributed (bivariate normal distribution)
4. Variables are linearly related
5. Simple random sample of the population

Z'-Transformation:

$$Z' = 1.151 \log (1+r)/(1-r)$$

Standard Error of Z':

$$\sigma_{z'} = \frac{1}{\sqrt{N-3}}$$

Formula for Confidence Limits of Z':

$$cl = Z' \pm z(\sigma_{z'})$$

Sampling Distribution:

Normal sampling distribution when r is transformed to Z'.

Step 1: Set up the null hypothesis. Specify the level of significance. If 0.05, use the 95 percent confidence interval. If 0.01, use the 99 percent confidence interval, etc.

Step 2: Draw a simple random sample from the population and collect data.

Step 3: Compute r, b (the sample estimate of the regression coefficient), etc.

Step 4: Convert r to Z' using Appendix Table F. Find the standard error of Z'. Find confidence limits for Z'.

Step 5: Convert the confidence limits for Z' to r's. If 0 is included within the limits, fail to reject the null hypothesis. If not, reject the null hypothesis.

DIAGRAM 16.4b   Tests for Partial and Multiple Coefficients of Correlation

Problem: (1) To determine whether the association between two interval-level variables as measured by partial r is significantly different from zero after additional variables have been partialed out. (2) To determine whether the association between one dependent interval-level variable and two or more independent interval-level variables as measured by multiple R is significantly different from zero.

Assumptions:

1. Independent observations
2. All variables measured at interval level
3. All variables normally distributed
4. All variables linearly related
5. A large sample N
6. A simple random sample of the population

Z'-Transformation:

$$Z' = 1.151 \log \frac{1 + \text{partial } r}{1 - \text{partial } r}$$

or

$$Z' = 1.151 \log \frac{1 + \text{multiple } R}{1 - \text{multiple } R}$$

Standard Error of Z':

$$\sigma_{z'} = \frac{1}{\sqrt{N - m - 1}}$$

Formula for Confidence Limits of Z':

$$cl = Z' \pm z(\sigma_{z'})$$

Sampling Distribution:

Normal sampling distribution when partial r or multiple R is transformed to Z'

Step 1: Set up the null hypothesis. Specify the level of significance. If 0.05, use the 95 percent confidence interval. If 0.01, use the 99 percent confidence interval, etc.

Step 2: Draw a simple random sample from the population and collect data.

Step 3: Compute partial r or multiple R.

Step 4: Convert partial r or multiple R to Z' using Appendix Table F. Find the standard error of Z'. Find the confidence limits for Z'.

Step 5: Convert the confidence limits for Z' to partial r's or multiple R's. If 0 is included within the limits, fail to reject the null hypothesis. If not, reject it.

appropriate standard error formula for both partial r and multiple R is as follows:

(16.15)
$$\sigma_{z'} = \frac{1}{\sqrt{N-m-1}}$$

where N is the number of cases in the sample, and m is the total number of variables involved. Thus, for a second-order partial r (e.g., $r_{12\cdot3}$), m would equal 3; and for a multiple R with three independent variables (e.g., $R_{1\cdot234}$), m would equal 4.

Once the standard error of Z' is found, the partial or multiple correlation coefficient can be transformed to Z' using Appendix Table F, and Formula (16.14) can be applied to compute confidence limits.

The assumptions behind the test of significance are the same as for the product moment coefficient of correlation, but they are extended to as many variables as are used in the analysis.

Diagrams 16.4a and 16.4b summarize the essential steps in the tests of null hypotheses concerning the product moment coefficients of correlation, regression coefficients, partial coefficients of correlation, and multiple coefficients of correlation.

16.5   SUMMARY

A study of descriptive statistics must devote considerable time to a discussion of measures of association and their use for finding the meanings of data (as in Chapters 6 and 7). The present chapter was devoted to a discussion of procedures for generalizing these measures beyond the sample data from which they are computed.

First, we considered the *chi-square test for independence*, which serves as a test of significance for a number of the delta-based measures of association. We found that if the null hypothesis of independence could be rejected through chi-square (χ^2) analysis, a measure of association computed could be generalized to the population sampled.

Next, we considered procedures for testing the significance of *gamma* (G), *tau-a* (t_a), and *tau-b* (t_b), measures of association of ordinal variables. The test of significance for t_b was actually accomplished by testing the significance of its numerator $S=N_s-N_d$. It was pointed out that this same test, cumbersome though it was, could also be used for the other measures of association whose numerators are S. These include G, the *tau* coefficients, and Somers' d_{yx}.

Finally, we considered procedures for testing the significance of the *product moment* coefficient of correlation and its associated regression coefficient. We examined the skewed sampling distribution usually associated with r and the reasons for its skewness. We circumvented the problem of skewness by converting r to Z', which has a normal sampling distribution, and by finding a confidence interval. It was pointed out that this test is a simultaneous test for the coefficient of correlation and the regression coefficient. Furthermore, the same test with a slightly altered standard error formula can be used to test the significance of a partial or a multiple coefficient of correlation.

In the next and final chapter of this volume we will attempt to put statistics into context as an integral part of the sociological endeavor.

CONCEPTS TO KNOW AND UNDERSTAND:

chi-square (χ_2) test for independence
observed frequencies
expected frequencies
marginal totals
G
product moment r
partial r
multiple R

t_a
t_b
S statistic
standard error of t_a
error variance of S
Z'-transformation
standard error of Z'

QUESTIONS AND PROBLEMS

1. Explain when and why you would test the null hypothesis that a measure of association or correlation is zero.

2. Why is it important to have a simple random sample of data when you are doing a χ^2 test for independence?

3. If you make a χ^2 test for independence at the 5 percent level of significance with 1 degree of freedom, and the computed χ^2 is 48.6, what does the magnitude of the χ^2 mean?

4. Why are sociologists likely to use τ_b in preference to τ_a? Can you think of a case in which a sociologist might be justified in using τ_a rather than τ_b?

5. Why does the test of significance for the product moment coefficient of correlation simultaneously test the significance of the regression coefficient?

6. When and why is the sampling distribution of product moment r likely to be skewed rather than symmetrical?

7. The sociological journals contain a wealth of studies in which various measures of association are used and their significance tested. Your task is to examine some of these studies in your library's journal collection and, showing your steps and reasoning, compute the following tests of significance.
   a) a chi square test on a table cross classifying nominal variables. You may have to multiply column n's by percentages in order to get cell frequencies.
   b) a test of an ordinal measure of association.
   c) a test of a Pearsonian correlation coefficient.
   d) a test of a multiple correlation coefficient.
   e) a test of a beta coefficient.
   Alternatively, you could compute tests of significance on tables selected from earlier examples in this text (see chapters on descriptive statistics).

GENERAL REFERENCES

Goodman, Leo A., and William H. Kruskal. "Measures of Association for Cross Classifications III: Approximate Sampling Theory." *Journal of the American Statistical Association, 58*: 302 (June 1963), 310–364.

> This technical article includes a discussion of the sampling distribution of G. The relevant part of the article (pp. 322–330) is readable even for the statistically unsophisticated.

Kendall, Maurice G. *Rank Correlation Methods.* (New York, Hafner), 1955.

> This is the authoritative source for the τ measures of association. Kendall discusses their sampling distributions and measuring their sampling error.

Lieberman, Bernhardt, ed. *Contemporary Problems in Statistics.* (New York, Oxford University Press), 1971.

> This book contains many articles of interest for the student of statistics. Section 5 on "The Use and Misuse of Chi-Square" is particularly relevant to this chapter.

OTHER REFERENCES

Armer, Michael, and Robert Youtz, "Formal Education and Individual Modernity in an African Society," *American Journal of Sociology*, vol. 76, no. 4 (January, 1971), pp. 604–626.

Berelson, Bernard, *Graduate Education in the United States* (New York, McGraw-Hill), 1960.

Blalock, Jr., Hubert M., *Social Statistics,* Revised Edition (New York, McGraw-Hill Book Company), 1979.

Cheek, Neil H., Jr., "Toward a Sociology of Not-Work," *Pacific Sociological Review*, vol. 14, no. 3 (July, 1971), pp. 245–258.

Crane, Diana, "Scientists at Major and Minor Universities: A Study of Productivity and Recognition," *American Sociological Review*, vol. 30, no. 5 (October, 1965), pp. 699–714.

Featherman, David L., "A Research Note: A Social Structural Model for the Socioeconomic Career," *American Journal of Sociology*, vol. 77, no. 2 (September, 1971), pp. 293–304.

Grizzle, James E., "Continuity Correction in the χ^2 Test for 2x2 Tables," *American Statistician*, Vol. 21, no. 4 (October, 1967), pp. 28–32.

Hays, William L., *Statistics for Psychologists* (New York, Holt, Rinehart and Winston), 1963.

Mantel, Nathan and Samuel W. Greenhouse, "What Is the Continuity Correction?" *American Statistician,* vol. 22, no. 5 (December, 1968), pp. 27–30.

Whitt, Hugh P., Charles C. Gordon, and John R. Hofley, "Religion, Economic Development, and Lethal Aggression," *The American Sociological Review*, vol. 37, no. 2 (April, 1972), pp. 193–201.

VII Summary and Conclusions

17 *Statistical Analysis in Sociology*

The field of descriptive statistics is concerned with ways of making sense out of the data at hand. Tables, graphs, and statistics (or parameters) are used analytically to investigate the distributions of single variables and the relationships between two or more variables. In Parts II, III, and IV we considered various ways in which data are analyzed and described. Part II focused on techniques for analyzing one variable, Part III considered techniques for analyzing two variables, and Part IV dealt with techniques for handling more than two variables.

In Parts V and VI we attempted to show how inferential leaps are made from descriptive analysis of the data at hand to larger bodies of data. We proceeded to explain the basic logic of statistical inference, building upon the three fundamental topics of (1) probability, (2) sampling, and (3) the sampling distribution. Also, we focused attention upon the two major inferential strategies of estimating parameters and testing statistical hypotheses and considered a number of techniques useful for following these strategies.

It is our intent in this chapter to stand back and look at all that has gone before in terms of how it relates to sociology and to the work of sociologists. It is our contention that, when put into proper perspective, statistics is and will continue to be a valuable tool for the sociologist in his(her) quest for understanding and predicting social behavior.

17.1 THEMES IN DESCRIPTIVE STATISTICS

We have covered most of the primary techniques sociologists use to describe their data statistically. Although sociologists are most often interested in multi-variate description simply because their ideas usually involve more than two variables, we have seen that the differences among techniques unfold from a few basic themes. Starting with simple differentiation of a common collection of cases, we moved into techniques for analyzing two or more groups on the basis of some characteristic that interests us.

17.1.1 Quality of Data

The first basic theme is that *the results of analysis depend heavily on the quality of data.* "Garbage-in-garbage-out" is a snide remark sometimes made about those who unthinkingly "computerize" data. As with any analytic process, no matter how impressive or compact the result, if the original data are biased or measurements invalid or unreliable, no amount of statistical summary will remove these defects. Investigators should force themselves to confront their original data and ask: Are these data worth *any* analysis? The answer should be a firm yes or no, and investigators should firmly accept the result of that binary (yes-no) decision.

17.1.2 The Defined Meaning of Scores

A related theme here has been that the *scores* an investigator examines have a *specific meaning* which has been defined from the start. Scores are always mea-surements from certain kinds of cases collected at specific points in time and place. Scores thus measure variables whose particular meanings have already been defined and limited in advance. Certain information about the state of a case, on the scale of a particular variable, is the meaning a score conveys. In the measurement process some essential elements may not be gathered, and thus some information may be lost. Exactly which components of the meaning of a score are most important to preserve depends upon the research project. Often it seems that measurement problems loom so large that other elements of the meaning of a score shrink beside them. We have emphasized level of measure-ment (nominal, ordinal, interval and ratio), scale continuity (continuous, dis-crete), and role in research (independent, control, intervening, dependent) as useful features of the meaning of scores. We have also devoted considerable space to discussing the different levels of units which may be examined—indi-vidual and group.

17.1.3 Identifying Relevant Data

The third basic theme is that it is positively useful to ignore or drop some data. In fact, we tended to divide data into two parts: the information the investigator wants and needs for a certain problem, and the other detail—perhaps equally sound in every other sense—which he(she) does not want. The general objective of descriptive statistics is to provide tools which permit the investigator to distinguish basic research information from annoying or irrelevant detail. What is basic information, of course, stems directly from the research questions a sociologist is asking, and what is good information to one investigator may be irrelevant or annoying detail to another. Worse, the detail, if it is not carefully controlled and separated from interesting information, may often turn out to be a basic contamination which makes valid comparisons impossible. Many of the statistical techniques are of the sort which permit unwanted sources of variation to be controlled. It is probably painfully clear by now that no complete description of a set of scores is either possible or wise. There are just too many different descriptions that could be made for this to be a useful research strategy.

17.1.4   The Importance of Comparison

A fourth theme here is the central importance of comparison and contrast to research in general, and to the statistical techniques we have covered. Contrasts are made between an individual score and the group from which the score comes. The z-score or standard score was a statistical description which permitted us to make some of the valid comparisons here. Frequently contrasts are made between groups — either groups selected separately for comparison, or subgroups within one general sample. Means and standard deviations, percentages and rates, histograms and scattergrams, correlation coefficients, odds ratios, and path coefficients — all were statistical means for making these contrasts. Finally and importantly, contrasts were made between the actual data on hand and models of the way the data might be or should be. Measures of association, to take one example, were often designed to contrast real data with some abstract model, such as a model of no association or a model of perfect association. In multivariate analysis, an investigator often graphically lays out relationships between variables which he(she) expects on the basis of his(her) theory. Elaboration, standardization, partial correlation and path analysis are techniques which help describe the "fit" between actual data and model. This theme is so essential that one could summarize it by saying that *where there is no contrast to be made, there is no study to be done.* Carried further, where there is no variation between scores, no statistical analysis is meaningful.

17.1.5   The Usefulness of Ratios

A fifth theme has been the importance of the simple ratio as a means for handling and describing contrasts of interest. Virtually all of the statistical descriptions presented in this book are, at root, only ratios. The trick, of course, is to pick the numerator and denominator so that the measure is useful. Often the numerator is a measured sub-part of a whole represented in the denominator. *PRE* measures of association, for example, were ratios of predictive error. How much error might be made in predicting the dependent variable with only its distribution as information, and how much of this potential error can be reduced or eliminated if certain added information on an independent variable is available as a basis of prediction? Most of the association measures we discussed were simply contrasts between different methods for making predictions.

17.1.6   Working with Variation

A sixth theme, mentioned earlier, is that statistics is designed to handle variation in scores — to describe in the presence of variation. Variation, however, comes from many sources, not all of which may be interesting or helpful to consider. Clearly, if there is no variation in the dependent variable, there is nothing to explain. Thus there are relatively few studies of the number of heads humans are born with. Given some variation, we can ask why, and introduce independent variables to help explain. There is also no progress to be made in explanation unless independent variables have some variation. If there were no variation in independent variables, then only one prediction would always be made for the dependent variable, which would hardly help account for differences in scores on the dependent variable. Investigators jealously guard and try to increase variation in the scores of dependent and explanatory variables. Other sources of variation are not desired and a segment of multivariate description dealt with ways to remove statistically unwanted sources of variation in data. For example, if an investigator wants to examine the relationship between ethnicity and moving, then variation in home ownership and rentership has to be removed by standardization or partial correlation coefficients or, perhaps, by elaboration techniques.

17.1.7   The Active Role of the Investigator

Finally, the most important theme is that the investigator's ideas take command of the statistical description from start to finish. Without a research question there is no study, without a knowledge of what is information and what is contaminating detail the proper measures can not be selected, without knowing what is to be measured appropriate descriptions can not be made, and without some sound question or theory the contrasts most useful to answering the question can not be determined. The role of an investigator's ideas and expectations is very clearly seen at key points in his(her) research: (a) in his(her) choice of what is "relevant" as a control variable or independent variable; (b) in the decision to use standardization or partial correlation techniques rather than various forms of elaboration; (c) in setting up a path diagram and the resulting computations to evaluate it; (d) in the choice of measures of association which contrast observed data with specific models of "independence" or "perfect association." Thus, for example, the concern with multivariate descriptive statistics in sociology is a direct result of the way sociologists develop theory.

From start to finish, the organizing core of statistical analysis in sociology is the sociologist's ideas. This, in fact, is the primary reason for a statistics course in sociology or any discipline. The tools of inquiry must be closely woven into the fabric of the inquiry itself. Of course this is not to deny the value of the expert in tools or the tool designer as a separate specialist, any more than it is to disparage the role of computer operator or clerical mathematician or computer designer or professional interviewer. Within a field like sociology, however, the problem is organized around substantive inquiry and not around the formal completeness of systems of logical tautology. Naturally, the more refinements we develop in appropriate kinds of computers and math and formal statistics and data-gathering mechanics, the more efficiently sociologists can get on with the task of using these analytic tools to perform substantive analysis.

17.2   THE PLACE OF STATISTICS IN THE RESEARCH PROCESS

It is frustrating to try to discuss statistics in isolation from theory and research methods. The three are really inseparable. Statistics is only meaningful when it relates to theory and research. Interaction among these three subjects is so important that even leaving one out for the sake of pedagogical simplification creates a hiatus in the flow of logic that is the very essence of science.

For a sociologist to identify himself or herself as a theorist or a researcher and deny the need for a grasp of theory, statistics, and research methods is both unrealistic and unproductive. The sociologist must be informed about and, indeed, must use the full range of conceptual and analytical tools available to him(her) if he(she) wishes to advance his(her) discipline. Durkheim is commonly pointed to as a prototype for sociologists because, in spite of the undeveloped nature of sociology in the nineteenth century, he was able to illustrate the necessary interaction between conceptualization and analysis in his work. Following the manner of the scientific method, he generated theory, collected data bearing on his theory, analyzed his data, and reached conclusions about the implications of his data for his theory.

In broad outline, sociological investigations should progress through these stages:

1. Generate a theory purported to explain some sociological phenomenon.

2. Derive from the theory some hypotheses to be tested against empirical observations.

3. Define relevant concepts to make them observable and measurable.

4. Specify the population to which the theory and hypotheses apply. The population may be either finite or infinite.

5. Design the sample from which the observations will be taken.

6. Specify strategies to be employed in collecting empirical observations that are relevant to testing the hypotheses.

7. Specify or construct instruments to be used for collecting the data.

8. Collect the data.

9. Describe analytically the sample data.

10. Interpret the findings and evaluate their implications for the hypotheses and theory. This is where the decision to make the inferential leap from sample to population is made.

11. Reconsider the hypotheses and theory in light of the findings. This is where any necessary alterations in the theory are made.

12. Replicate the process starting with step 2.

It should be understood that listing these steps is an oversimplification of the process as it exists in reality. The process really has no beginning nor an end. Rather, it is a cyclical process that keeps returning to steps already completed, but hopefully, each time at a higher level representing a progression

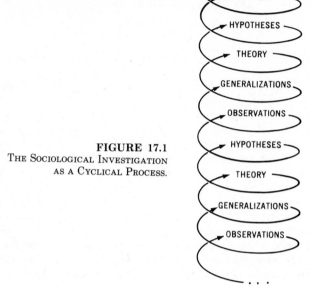

FIGURE 17.1
THE SOCIOLOGICAL INVESTIGATION
AS A CYCLICAL PROCESS.

toward a more adequate explanatory system. Figure 17.1 attempts to portray this cyclical process graphically.

Moreover, it should not be assumed that the steps listed above follow in sequence as neatly as they have been presented here. Sometimes the steps interchange, but, usually, they overlap, or even occur simultaneously. And, not every investigation necessarily includes all of the steps enumerated. This presentation is in fact an idealized version of what can actually take place when the sociologist is at work.

When the researcher gets to the stage of data analysis, statistical tools begin to play a major part. But statistical coɪ.siderations also enter in at various other points in the process. As a matter of fact, if the researcher does not concern himself or herself with statistics until he(she) reaches the data analysis stage, he(she) is likely to have difficulty finding statistical techniques appropriate for his(her) data. The use of statistics must be planned early in the research process if data analysis is to proceed smoothly.

When hypotheses are derived and concepts are defined, it is necessary to begin thinking ahead about the kind of statistical analysis that will come later so that the hypotheses are stated appropriately and the concepts defined in measurable ways. At this point it is appropriate to set up the statistical (null) hypotheses to be tested, to decide whether the tests will be one-tailed or two-tailed, and to select the level of significance (see Chapter 14). Furthermore, it is useful to set up dummy tables (tables without data) and dummy analyses to help anticipate what the data might look like once they are collected. This constitutes a check system to determine whether the kinds of data necessary for the statistical analysis will be obtained.

> **BOX 17.1** Look Back
>
> This discussion assumes that you have a good grasp of such concepts as sampling distribution and sampling error. If you feel the need to refresh your memory turn back and review Sections 12.2; 12.2.1; 12.2.2; 12.2.3; 12.4; 12.4.2; 12.4.4; and 12.4.5.

When the population is specified and the sample is designed it is necessary to decide how large the sample should be. Again statistics is relevant because a determinant of sample size is the amount of sampling error that is tolerable. For example, assume that we wish to test a sample mean against a population mean. The formula for estimating the standard error of the mean from sample data is as follows:*

(17.1)
$$s_{\bar{X}} = \frac{s}{\sqrt{N}}$$

If it is possible to estimate the standard deviation (s), and if we know how large a standard error we are willing to tolerate, then we can turn the above formula around to solve for the sample size.† Thus,

(17.2)
$$N = \frac{s^2}{s_{\bar{X}}^2}$$

As an example of how Formula (17.2) might be used, if we estimated the standard deviation to be 10 and we wished to keep the standard error of the mean down to a maximum of 1, we could substitute into Formula (17.2) and solve for N as follows:

$$N = \frac{100}{1} = 100$$

Therefore, we would draw a random sample with a minimum of 100 cases.

Of course, solving for the appropriate sample size assumes that we have also considered steps 6 and 7 in the research process because we would also have to settle on our strategies for collecting data and have in mind the kinds of data collection instruments we will use. The nature of the data collection instruments is relevant to the amount of sampling error tolerable. For example, if a measurement scale with a ten-point range were to be used, we would need to

---

*You became familiar with this formula earlier when it was identified as Formula (12.16).

†Another useful procedure in practice is to determine how small a difference (for example, in means) an investigator wishes to be able to detect as statistically significant. In the case of single sample means, the difference $\bar{X} - \mu$ which is the smallest difference we wish to treat as meaningful, would be determined. Then, substituting in the formula for the t-test of means, the sample size could be determined:

$$N = \left(\frac{ts}{\bar{X} - \mu} \right)^2$$

where t is the t-value for the desired level of significance. In addition to this estimated size, of course, an investigator would have to take into account non-response rates and missing data which are likely to be a part of his(her) research.

Steps in the Research Process	Statistical Considerations
1. Generation of theory	
2. Derivation of hypotheses	Construct statistical hypotheses
3. Definitions of relevant concepts	
4. Specification of population	Select appropriate sample size
5. Design of sample	
6. Specification of data collection strategies	Determine maximum sampling error tolerable; determine measurement levels; select appropriate statistical tests
7. Construction of data collection instruments	
8. Collection of data	
9. Analysis of data	Perform necessary computations
10. Interpretation of findings	Evaluate statistical results in terms
11. Reconsideration of hypotheses and theory	of amount of information generated
12. Replication of the whole process	

FIGURE 17.2   STEPS IN THE RESEARCH PROCESS AND PARALLEL STATISTICAL CONSIDERATIONS.

keep the sampling error small enough so that an observed difference of ten points or less would result in a computed statistic with a possibility of reaching the critical region. If the sampling error is too large, the statistical test will be insensitive to even the largest difference that can be expected from the data.

When decisions are made about data collection strategies and data collection instruments, the level of measurement of the variables is crucial. The data collection instruments must produce scores compatible with the measurement requirements of the statistical techniques employed to analyze the data.*

Even after the statistical analysis of the data has been completed, a knowledge of statistics plays a direct part in interpreting the findings. We must be thoroughly informed about the statistical tools used to analyze data and the kinds of information that they make available to us. Otherwise, we will not be in a position to express what it all means.

Figure 17.2 summarizes the points we have tried to make here to illustrate the interaction between theory, research methods, and statistics. In the left-hand column of the figure are listed the 12 steps enumerated earlier and in the right-hand column the statistical considerations that most nearly parallel the steps. The attempt to portray these "parallel" statistical considerations in Figure 17.2 demonstrates again the point stressed earlier in this chapter that the steps in research are not mutually exclusive and sequential; rather they overlap and sometimes occur simultaneously.

In a carefully planned program of research most of the important decisions about statistical analysis are made before the data are collected. Decisions made after the data have been collected are usually stopgap measures that may lead to less than desirable solutions to problems arising from poor planning, or no planning at all. Of course, one of the most universally applicable generalizations of science, Murphy's Law, can be expected to hold true even for the most carefully designed research. Murphy's Law states, "If anything can go wrong, it

---

*Throughout this volume the measurement requirements of each technique discussed have been made explicit and have been included among the lists of assumptions.

will." Nevertheless, a well planned project including a carefully thought out statistical analysis is likely to encounter fewer problems than would otherwise be the case.

17.3   PLACING DESCRIPTIVE AND INFERENTIAL STATISTICS IN PERSPECTIVE

Recently sociologists and other behavioral scientists have questioned the necessity and desirability of inferential statistical techniques in sociological and behavioral science research. They have claimed that sociological data are not appropriate for inferential analysis or that inferential analysis gives no more than trivial results[*]. If these criticisms are correct, then it is proper to ask what the justification is for discussing statistical inference.

Our position on this matter is as follows. As sociologists, we are interested in developing generalizations about social behavior. We commonly work with sample data but are not inherently concerned with samples. Rather, we are interested in the populations which they represent, whether those populations are finite or infinite. The logic of inferential statistics gives us a basis for making inferential leaps from samples to populations. When we analyze sample data and perform a test of statistical significance, or estimate a parameter, we are seeking to apply the results of the sample analysis to the population which the sample represents. A test of statistical significance merely supplies us with a probability figure that we can weigh in making the decision whether or how to generalize beyond the sample. The test does not make the decision for us. It is merely a crutch that we may use to help us make the decision. As researchers we are responsible for our decisions, and we cannot blame wrong decisions on statistics. Likewise, we cannot legitimately pass on to others the responsibility of making the decisions for us. It is not proper, in other words, to conduct a test of statistical significance, publish the results of the test, and expect the reader to decide whether the inferential leap is justified. We should report the results of statistical tests to inform the reader of the information used to make decisions, but as researchers we must take responsibility for the decisions reached.

Some critics of tests of significance have complained that statistical significance will almost always be found if the sample size is large enough. It is true that larger samples have smaller sampling errors and are, thus, sensitive to smaller observed differences. It is not necessary, however, to decide in favor of the inferential leap merely because a statistically significant result is obtained. The decision to make an inferential leap should depend heavily upon *the substantive nature of the finding*. Statistical significance is only one criterion that should be considered in the decision process and "significant" does not mean "important."

In the past there have been a number of studies concerned with the subject of selective migration.[†] These studies investigated the question whether migration from farm to city is selective with respect to intelligence — whether there is

---

[*]See Labovitz, 1970 and 1971; Selvin, 1957; Morrison and Henkel, 1970; Rozeboom, 1960.
[†]See Gist and Clark, 1938; Gist *et al.*, n.d.; Pihlblad and Gregory, 1954.

a tendency for the more able persons to migrate to cities and the less able to remain in rural areas. One of the best known of these studies was conducted by Noel P. Gist and Carroll D. Clark (Gist and Clark, 1938). They secured the I.Q. scores of a large number of students who had been enrolled in rural Kansas high schools in 1922–23. In 1935, the year in which they actually conducted the study, they were able to locate 2,544 of these students by place of residence. They found 964 of them had moved to urban areas and 1,580 were still living in rural areas. The researchers computed the mean I.Q. for those who had moved to urban areas and compared it to that for those who remained in rural areas. They found the mean I.Q. of urban migrants to be 98.26 with a standard deviation of 11.52 and of rural nonmigrants to be 94.78 with a standard deviation of 11.36.

Because of their large samples the standard errors of the means of the two groups were small (0.385 for the urban migrants and 0.286 for the rural nonmigrants) and the standard error of the difference was 0.48.* When the standard score is computed using the mean difference of 3.48 in the numerator and the standard error of the difference in the denominator, we get the following results:

$$z = \frac{\overline{X}_u - \overline{X}_r}{s_{\bar{x}_1 - \bar{x}_2}} = \frac{98.26 - 94.78}{0.48} = \frac{3.48}{0.48} = +7.25$$

The probability of occurrence of a z score as large as $+7.25$ by chance is very small (see Appendix Table B). Therefore, Gist and Clark conclude, "The chances are overwhelming that the observed difference in the means of these samples is a true difference and not to be explained as arising from the accidental conditions of sampling."[†]

Gist and Clark should not be criticized for rejecting the null hypothesis of no difference in the means because their data warranted its rejection. On the other hand, they might well be reprehended for conducting a test of significance at all, in the light of the small difference of 3.48 I.Q. points they observed. Taking into consideration the modest reliability of I.Q. scores, a difference of only 3.48 in mean I.Q.'s is substantively unimportant and theoretically uninteresting. Such a difference could certainly not justify the conclusion that the rural nonmigrants are inferior to the urban migrants. If, on the other hand, the researchers had found that the rural nonmigrants had a mean I.Q. of 90 and the urban migrants a mean I.Q. of 120, then their study might be of substantive importance and theoretical interest.

Although the researchers may be correct in concluding that, "the observed difference in the means of these samples is . . . not to be explained as arising from the accidental conditions of sampling," the small difference observed may well have been attributed to the vicissitudes of individual performance on the I.Q. test itself. Furthermore, since it is reasonable to suppose that individuals' scores on an I.Q. test will vary a few points from one administration of the test to another, a mean difference between groups of 3.48 points is an unimportant one. Even if the observed difference is a real difference, in this case it is so small

*When Gist and Clark conducted their study they used "probable errors"—a measure of sampling error that is no longer used. We computed standard errors instead, and computed the standard score with the standard error of the difference between means as the measure of sampling error (See section 15.3.1a).
†Gist and Clark, 1938:45–46.

that it really doesn't make any difference as far as potential achievement of the two groups involved is concerned.

We believe that the critics of inferential statistics should be leveling their criticisms at the users of statistics and at the way in which they use statistics rather than at the techniques themselves. Statistical techniques are tools that help the researcher make decisions about his(her) data. Statistical tools like any other tools can be used or misused. If used properly, they can be a valuable aid to the researcher. If misused, they can lead to serious errors of interpretation.

The critics of statistical inference are correct when they claim that tests of the null hypothesis represent a very primitive level of analysis. Rejection of the null hypothesis merely allows one to assume that whatever is happening at the sample level is probably paralleled by what is happening at the population level (Section 14.2.1). The crucial analysis of data is that which takes place at the sample level to explain the nature of observed relationships. This type of analysis is the subject of descriptive statistics. The techniques of descriptive statistics are those that are designed to make sense of what is happening in the data. It is upon these techniques that the emphasis must be placed if we are to advance sociology as a science. What inferential statistics adds to the process is a logic and a set of techniques for determining the probability that what we have learned about a sample also applies to the population which the sample represents.

17.4   USES AND ABUSES OF STATISTICS

Wisely handled, statistics can be a powerful tool of analysis for the sociologist. He(she) has a whole storehouse of techniques available for him(her) to attack a huge variety of research problems. In this book we have done no more than scratch the surface. The serious student of sociology will go on to take more advanced methods and statistics courses and will consult the literature suggested in the bibliography in order to acquaint herself or himself with the many analytical tools at his(her) disposal.

It should be kept in mind that most of the statistical techniques in the storehouse were not developed by sociologists with sociological applications in mind. For example, the analysis of variance was applied first to agricultural research. It was used to analyze the relative effectiveness of different types of fertilizers. The sociologist-user of statistics often must adapt the techniques to his(her) needs. If the data do not quite fit the assumptions of the technique he(she) elects to use, then he(she) should use the technique with caution, keeping in mind those assumptions he(she) has failed to satisfy, the importance of satisfying those assumptions (*i.e.* robustness), and interpreting his(her) results accordingly.

If you examine the sociological literature for use of the techniques we have discussed, you will usually find that the data analyzed do not quite fit the assumptions of the techniques used. For example, you will frequently find tests of significance run on data from samples that are not quite random samples or means and standard deviations computed on data that are not clearly interval scale data. When statistics are used in such circumstances and the researcher who has done the analysis really understands the techniques he(she) has used and the implications of their use for the analysis of the data, we must assume that he(she) has taken these factors into consideration in interpreting his(her) findings. The general principle is to be conservative in drawing conclusions.

Unfortunately, however, statistical techniques are frequently used by those who are ignorant of the function of the techniques and of the information that can be gleaned from their use. A little knowledge, particularly in statistics, can be a dangerous thing. The person who knows a little about χ^2, for example, may go around blithely computing χ^2 on whatever data he(she) can lay his(her) hands on, completely oblivious to whatever abuses of the technique he(she) may be perpetrating (Section 16.1.4) and, more "sophisticated" researchers may apply the latest statistical fad to any data they can find! Beware of the researcher who thinks that a test of significance is a magic technique for grinding out findings and that the 0.05 level of significance is sacred (Section 14.2.3). He(she) is no less a hindrance to the advancement of a sound, scientific sociology than the person who, through complete ignorance of statistics, says that sociological data are inherently unquantifiable and statistical techniques are completely useless in sociology.

17.5   IS STATISTICS NECESSARY FOR SOCIOLOGY?

When we put the matter into its proper perspective, it is rather meaningless to debate whether statistics is necessary for sociology. Statistics is basically a body of logic and techniques useful for organizing, analyzing, and making sense of information. It is an information processing tool. Sociology, like all scientific disciplines, is an information processing endeavor. We, as sociologists, are interested in making sense out of human social behavior, in understanding that behavior, and in predicting it. The basic postulate of sociology is that human social behavior is understandable and predictable. If we accept the postulate, then we should consider using *any* method available to us which will help us to understand and predict human social behavior.

The use of statistics in sociology does not imply that all data have to be reduced to numbers. Those aspects of data that are quantifiable should be quantified, but so-called "qualitative" data need not be discarded. Quantitative and qualitative data can supplement and complement each other. It is often useful to analyze quantitative data, come up with findings, and return to qualitative data – such as observation narratives – to try to make further sense of the quantitative findings.

The sociologist should focus on his(her) theory and attack significant sociological problems in the best way he(she) knows, using any tools at his(her) disposal to bring light to those problems. He(She) should be problem oriented, not technique oriented. He(She) should choose to investigate not only problems he(she) can handle with the techniques he(she) knows and favors but, rather, significant problems to which he(she) should apply any techniques available that will be helpful to him(her) in his(her) investigation.

17.6 FUTURE DEVELOPMENTS

The increasing availability of computers has shifted the emphasis in sociology from the study of univariate and bivariate problems to the study of multivariate problems. To be efficient predictors, sociological theories generally need to be stated in multivariate terms. Before computers, multivariate statistical techniques were so tedious that they were not commonly used. The computer has now made these techniques accessible and practical. In response to this breakthrough, sociological theories are increasingly becoming multivariate in form. It is becoming increasingly important for the sociologist to be a knowledgeable computer user. Computer technology is racing forward at a breathtaking pace, and the potential uses of computers for sociological analysis stagger the imagination.

Another very promising development in sociology is the gradual but dramatic disappearance of the chasm separating theory and research. Sociology appears to be moving forward by returning to the model which Durkheim set for us in the nineteenth century. The effect of this long overdue marriage of theory and research is the development of theory that is researchable and the appearance of more theory-oriented research. The advent of the computer in sociology and the increasing emphasis upon multivariate analysis have done much to facilitate this development.

There is a third important development that promises to have a significant impact on sociology and on the academic preparation of future sociologists. Traditionally, sociologists have used a structure rather than a process approach to theorizing and researching. Social behavior has been viewed in static terms, and much research has focused on single points in time, much like stopping a movie and studying a single frame. Sociologists are now beginning to realize that what is orderly about social behavior may be the way in which it changes rather than the way in which it resists change. This perspective focuses attention on time series and longitudinal analysis. The shift in statistics is toward the increasing use of stochastic processes and techniques of time series analysis. This emphasis will make the understanding of calculus an important requirement in the academic training of future sociologists. It seems inevitable that process models involving the use of calculus will appear with increasing frequency in the sociological literature.*

Obviously sociology is coming of age. The public and our public leaders are beginning to realize that the pressing problems of today and the forseeable future are those for which solutions are encouched in a knowledge of social behavior. Sociology is in a position to contribute that knowledge. This is an exciting time in which to be a part of it. We believe that those students of sociology who will make important contributions to that knowledge will be those who are well versed in theory, research methods, and statistics.

We hope that this volume has given you some appreciation of the place of statistics in sociology and that it will only mark the beginning of statistical knowledge you will wish to pursue in your preparation for sociology.

*For examples of the use of process models in sociology see Coleman, 1964.

CONCEPTS TO KNOW AND UNDERSTAND

major themes in descriptive statistics

the interaction of theory, research methods and statistics

the stages of sociological investigation

dummy tables and dummy analyses

estimation of sample size

Murphy's Law

responsibility for research decisions

the substantive nature of a finding

abuses of statistics

impact of computers on statistical analysis

integration of theory and research process analysis

QUESTIONS AND PROBLEMS

1. Describe the process through which a sociological investigation should progress. Read a sociological article or monograph and determine to what extent this process was followed.

2. Why is it important to consider statistical analysis at an early stage in planning a study?

3. What is meant by the statement that statistical analysis is merely a crutch upon which the sociologist may lean?

4. Make a clear-cut distinction between statistical significance and substantive importance. Which is more important to the sociologist?

5. What is meant by the statement that the sociologist should be problem oriented rather than technique oriented?

6. Describe the interaction that takes place between the researcher and the research.

GENERAL REFERENCES

Blalock, Hubert M., Jr. *Theory Construction: From Verbal to Mathematical Formulations.* (Englewood Cliffs, N.J., Prentice-Hall), 1969.

   The marriage between theory and research which is currently taking place in sociology has led to the kinds of theory construction activities described in this book.

Kerlinger, Fred N. *Foundations of Behavioral Research.* (New York, Holt, Rinehart and Winston), 1964.

   This excellent text attempts to integrate theory, research methods, and data analysis techniques. It addresses itself to many of the methodological issues which we have raised in this book.

Land, Kenneth C. "Formal Theory." *Sociological Methodology 1971,* edited by Herbert L. Costner, pp. 175–220. (San Francisco, Jossey-Bass), 1971.

   This is an outstanding paper on the formalization of theory in sociology. It shows the relationship between theory, mathematical models, and data. Process models of social behavior are discussed.

OTHER REFERENCES

Coleman, James S., *Introduction to Mathematical Sociology* (New York, The Free Press of Glencoe), 1964.

Gist, Noel P., and Carroll D. Clark, "Intelligence as a Selective Factor in Rural-Urban Migrations," *American Journal of Sociology*, vol. 44 (July, 1938), pp. 36–58.

Gist, Noel P., C.T. Pihlblad, and C.L. Gregory, *Selective Factors in Migration and Occupation*, University of Missouri Studies, XVIII, no. 2, n.d.

Labovitz, Sanford, "The Nonutility of Significance Tests: The Significance of Tests of Significance Reconsidered," *Pacific Sociological Review*, vol. 13 (Summer, 1970), pp. 141–148.

————, "The Zone of Rejection: Negative Thoughts on Statistical Inference," *Pacific Sociological Review*, vol. 14, no. 4 (October, 1971), pp. 373–381.

Lieberman, Bernhardt, editor, *Contemporary Problems in Statistics* (New York, Oxford University Press), 1971.

Morrison, Denton E., and Ramon E. Henkel, *The Significance Test Controversy* (Chicago, Aldine), 1970.

Pihlblad, C.T., and C.L. Gregory, "Selective Aspects of Migration among Missouri High School Graduates," *American Sociological Review*, vol. 19, no. 3 (June, 1954), pp. 314–324.

Rozeboom, William W., "The Fallacy of the Null-Hypothesis Significance Test," *Psychological Bulletin*, vol. 57, no. 5 (1960), pp. 416–428.

Selvin, Hanan, "A Critique of Tests of Significance in Survey Research," *American Sociological Review*, vol. 22, no. 5 (October, 1957), pp. 519–527.

Appendix:
Statistical Tables

TABLE A TABLE OF RANDOM NUMBERS

I.	1-4	5-8	9-12	13-16	17-20	21-24	25-28	29-32
1	74 44	80 91	21 41	40 25	98 81	57 12	30 13	24 93
2	72 59	62 28	26 74	90 62	91 20	70 31	19 10	23 06
3	33 76	63 60	48 79	23 76	28 61	87 65	79 30	38 27
4	78 02	65 54	17 61	60 15	00 81	18 07	66 38	88 33
5	32 46	64 91	77 63	26 35	94 81	54 90	10 70	10 66
6	19 68	47 39	30 75	39 71	13 14	55 59	25 38	79 00
7	06 34	06 19	72 70	53 47	57 70	04 07	54 81	04 98
8	62 23	31 49	29 34	39 38	99 54	66 13	94 08	17 03
9	97 69	04 44	89 07	31 84	66 59	86 21	61 78	60 26
10	71 00	49 40	32 08	38 57	58 59	69 72	31 52	94 75
11	32 40	60 49	51 52	47 01	29 40	59 31	14 60	26 91
12	63 42	42 91	61 34	02 22	44 76	88 32	06 36	71 39
13	09 20	12 10	05 86	08 66	45 84	84 80	69 45	65 08
14	62 50	89 39	48 02	24 64	45 60	20 27	83 65	33 82
15	73 95	05 00	52 98	08 57	74 19	05 58	27 46	23 26
16	62 43	84 11	42 38	74 13	04 57	80 26	28 04	20 52
17	07 00	14 93	55 80	14 27	69 56	24 43	30 82	55 08
18	83 26	52 43	70 61	67 23	07 36	49 78	40 25	09 66
19	14 67	02 60	21 02	64 47	54 86	62 88	81 12	28 29
20	33 60	94 02	89 25	44 73	03 85	01 95	17 85	70 51
21	02 73	10 59	69 74	76 11	57 27	04 63	94 15	74 96
22	75 46	82 03	13 43	64 34	96 68	23 86	38 49	83 98
23	79 06	02 26	58 82	81 17	91 27	44 27	71 89	17 90
24	85 16	88 13	08 97	13 08	73 28	25 34	79 27	16 34
25	37 30	30 55	08 17	55 15	93 34	10 60	00 95	60 26

II.								
1	24 25	56 63	72 28	28 39	91 61	34 21	52 63	73 21
2	50 63	56 77	89 77	80 28	03 14	79 27	86 26	35 33
3	03 65	03 42	90 08	30 90	14 39	85 94	74 39	97 71
4	63 60	97 32	86 61	10 39	07 30	89 99	09 11	21 55
5	54 69	75 00	63 95	78 98	01 93	93 77	57 30	50 82
6	88 05	23 40	46 42	52 55	27 85	28 12	00 51	23 76
7	97 38	25 33	16 78	32 87	47 58	19 34	31 76	44 97
8	03 12	82 62	94 17	66 36	56 10	23 73	50 93	15 55
9	19 66	44 73	79 92	31 66	72 70	27 72	90 95	46 77
10	46 32	77 62	36 51	73 93	57 25	09 15	50 30	50 95
11	40 78	23 70	51 26	87 86	69 22	21 85	62 27	39 01
12	20 66	67 60	26 95	80 64	02 42	97 88	48 34	27 23
13	97 77	58 66	34 10	91 71	94 82	79 59	14 11	53 84
14	91 93	21 80	20 91	19 52	71 42	20 08	35 28	35 91
15	14 88	89 71	48 84	32 22	13 63	78 02	72 97	61 13
16	90 97	01 01	01 59	01 39	96 96	74 40	66 88	44 01
17	85 29	86 31	19 55	37 86	04 56	98 35	78 82	29 06
18	22 06	90 91	84 64	81 65	63 51	54 35	94 73	79 48
19	53 69	65 88	46 28	13 02	00 17	27 22	69 94	45 92
20	66 42	58 45	92 71	40 83	59 73	64 04	42 92	38 84
21	91 09	52 38	46 27	17 09	95 13	43 45	80 94	98 07
22	88 72	65 18	16 79	80 82	75 47	29 46	95 09	91 01
23	86 63	79 92	70 95	28 32	95 25	95 25	81 18	30 16
24	46 70	03 50	46 40	34 78	59 35	93 12	24 69	37 37
25	25 50	24 15	01 82	38 36	21 15	71 91	31 52	52 51

TABLE A TABLE OF RANDOM NUMBERS *(Continued)*

III.	1-4	5-8	9-12	13-16	17-20	21-24	25-28	29-32
1	81 18	43 25	80 75	71 96	48 88	91 61	85 37	12 99
2	30 49	80 30	38 91	62 72	55 48	26 82	31 66	09 47
3	47 11	18 80	57 00	16 38	30 49	09 74	15 77	52 68
4	00 31	08 24	87 17	23 04	98 29	30 98	14 95	03 98
5	24 25	33 96	83 92	12 02	80 63	80 41	49 40	62 49
6	93 06	35 71	31 25	38 38	01 14	75 19	72 11	92 52
7	45 33	21 67	02 64	92 85	74 74	82 52	68 19	23 78
8	44 45	81 92	53 10	78 28	17 76	84 74	60 37	34 08
9	95 75	30 09	08 93	87 71	68 63	71 84	96 91	86 02
10	13 99	84 18	43 72	24 15	33 59	11 97	95 69	24 43
11	46 15	99 32	14 87	23 20	44 45	84 19	35 65	20 97
12	74 00	80 33	52 32	96 66	57 94	25 86	63 40	57 70
13	59 47	85 05	51 89	52 12	75 48	61 99	99 11	48 07
14	84 22	38 10	92 68	34 81	15 10	03 01	30 59	79 03
15	58 57	59 12	18 06	15 63	10 13	58 18	49 60	64 86
16	13 95	54 28	92 78	95 64	71 57	33 08	46 27	18 31
17	77 32	46 87	82 56	39 99	81 79	87 57	12 82	24 62
18	52 58	01 90	82 77	99 53	71 20	22 50	41 28	99 16
19	73 26	35 54	15 57	67 00	28 84	47 57	10 46	36 12
20	86 83	39 22	37 11	93 21	01 40	14 71	20 98	40 01
21	12 70	37 38	82 24	18 01	45 58	72 29	70 45	18 50
22	41 31	79 58	62 06	58 38	72 38	13 63	51 70	38 27
23	24 40	09 84	72 10	84 64	50 20	85 02	37 12	46 34
24	99 76	09 25	14 33	44 93	20 35	18 69	50 99	24 73
25	29 74	34 46	74 62	58 08	68 78	60 16	83 98	86 80

IV.								
1	81 66	77 08	85 13	41 90	28 77	60 13	59 55	15 05
2	32 42	50 81	08 01	90 10	43 28	89 50	05 42	05 95
3	11 50	29 08	16 39	76 94	70 66	53 71	23 03	41 38
4	93 65	51 22	55 70	86 61	59 58	64 86	29 00	65 26
5	33 43	08 85	65 24	55 58	73 81	50 89	58 50	57 14
6	59 87	34 45	41 02	32 95	98 53	54 21	95 66	01 13
7	15 52	12 97	14 84	23 98	01 13	04 26	20 79	37 97
8	75 02	13 24	27 60	50 77	97 95	06 07	52 64	86 71
9	14 52	99 22	31 39	70 43	58 93	43 52	61 47	51 75
10	59 22	81 73	42 89	96 56	46 75	39 39	37 10	70 70
11	12 07	71 71	64 45	62 43	10 27	48 73	88 72	28 81
12	51 12	39 78	31 47	77 80	94 00	57 79	54 64	08 44
13	25 52	91 93	90 49	05 47	33 28	40 98	13 85	92 30
14	30 63	79 24	45 48	15 83	74 00	62 15	54 44	71 02
15	84 21	76 60	10 79	48 36	76 51	74 52	04 72	96 63
16	02 22	20 00	82 36	33 25	60 45	80 70	93 29	33 61
17	77 24	43 56	80 48	10 29	45 33	44 14	19 76	26 11
18	62 79	14 00	69 80	81 05	33 52	93 34	10 31	55 89
19	76 87	06 38	98 93	50 50	06 33	95 12	52 93	92 72
20	18 90	97 14	95 07	88 30	69 06	75 51	75 45	32 65
21	17 27	41 02	29 40	58 56	16 66	35 38	58 29	04 01
22	94 73	05 06	86 43	55 58	55 70	43 35	53 80	43 49
23	68 94	03 23	79 95	89 33	64 51	68 19	87 45	29 13
24	96 34	44 78	24 04	72 98	17 07	13 28	49 56	68 69
25	53 55	99 99	13 59	25 54	41 52	27 17	05 74	28 03

TABLE A TABLE OF RANDOM NUMBERS *(Continued)*

V.	1-4	5-8	9-12	13-16	17-20	21-24	25-28	29-32
1	61 81	17 50	68 00	35 10	30 90	59 71	09 95	01 14
2	78 95	64 65	24 82	14 05	27 63	33 96	10 41	88 70
3	84 28	44 68	07 47	21 46	56 81	32 87	28 40	40 50
4	92 33	63 98	99 22	09 21	97 18	10 03	79 46	17 13
5	15 79	75 50	29 36	12 37	63 39	02 47	57 02	97 17
6	80 16	09 75	22 28	35 25	53 57	72 64	09 98	63 50
7	68 20	33 03	43 73	80 96	21 13	97 61	90 37	35 77
8	55 26	85 04	30 60	68 10	73 53	89 35	58 45	83 23
9	60 00	37 51	42 89	52 32	46 00	57 02	71 97	44 16
10	59 69	31 20	16 37	66 34	99 76	07 23	40 85	64 91
11	84 42	33 66	58 54	17 16	45 73	67 20	09 27	90 96
12	57 46	65 19	78 34	57 12	77 45	54 65	17 17	30 90
13	78 17	51 47	69 22	41 48	01 99	66 46	00 28	21 74
14	27 66	33 21	49 11	24 15	33 70	06 95	04 67	98 56
15	82 54	98 27	81 86	77 35	87 56	32 72	60 90	26 75
16	33 06	79 71	73 57	96 74	85 94	36 97	87 79	82 00
17	77 94	61 11	69 61	78 78	36 51	45 21	82 94	39 22
18	87 15	49 66	56 55	34 99	05 26	45 35	59 83	55 47
19	24 98	52 45	79 85	15 67	32 21	29 94	98 90	02 27
20	05 66	15 23	83 66	24 98	06 75	60 69	64 26	58 24
21	84 90	70 29	01 36	90 78	56 40	61 00	58 40	75 37
22	49 50	30 71	87 38	70 10	80 71	12 54	60 76	62 13
23	27 53	95 47	04 78	61 85	56 15	71 76	25 31	96 39
24	56 17	07 83	96 29	88 39	67 86	98 23	95 03	82 62
25	41 67	05 42	29 18	54 76	71 82	04 81	82 63	00 23

VI.								
1	59 63	29 38	35 59	02 05	68 55	03 47	95 17	41 48
2	11 94	21 16	54 90	92 80	80 91	71 76	54 03	81 00
3	97 62	76 15	96 67	40 92	96 85	18 84	70 89	87 05
4	07 02	78 73	21 58	12 95	04 96	66 91	42 11	11 86
5	13 40	90 19	67 10	34 15	62 59	21 10	30 33	62 02
6	45 48	25 79	20 34	96 33	32 28	66 64	75 37	10 31
7	41 01	61 47	38 60	24 86	08 42	37 27	16 65	20 86
8	49 89	46 70	52 25	93 67	00 76	10 78	72 86	63 18
9	83 90	36 74	50 40	54 90	71 78	80 65	48 29	01 80
10	94 53	17 86	40 96	47 83	02 53	42 68	21 55	96 13
11	73 09	63 85	30 66	78 74	91 82	64 54	81 95	50 35
12	42 47	76 36	12 49	28 49	39 61	49 16	69 42	98 75
13	36 77	54 32	75 18	69 35	54 26	99 87	86 21	94 49
14	65 15	36 03	51 29	64 85	08 56	58 27	32 20	94 21
15	25 25	78 24	08 31	20 26	82 28	54 65	30 00	70 33
16	61 33	06 70	01 51	70 43	38 01	21 94	08 89	77 54
17	59 55	83 53	92 99	57 59	52 24	07 21	82 57	97 42
18	54 63	59 93	45 39	87 90	49 18	71 36	19 88	79 46
19	42 26	36 54	15 06	45 23	61 47	59 31	30 37	08 64
20	84 41	06 96	39 93	14 09	66 32	58 24	55 92	10 76
21	77 00	01 88	33 81	04 10	26 96	38 07	08 44	34 62
22	92 12	15 04	18 73	45 98	64 20	60 93	35 60	29 32
23	59 19	35 73	78 60	18 44	27 97	71 69	44 70	30 76
24	83 17	41 05	62 05	10 84	51 42	68 31	85 75	87 63
25	74 41	13 36	38 79	52 23	86 67	34 06	15 75	62 86

TABLE A TABLE OF RANDOM NUMBERS *(Continued)*

VII.	1-4	5-8	9-12	13-16	17-20	21-24	25-28	29-32
1	38 72	32 95	74 46	76 67	68 09	89 22	60 57	82 64
2	11 82	49 29	40 38	95 27	81 69	93 47	66 37	58 35
3	95 03	45 03	70 47	62 85	98 97	30 77	42 07	56 81
4	27 08	80 32	56 82	94 55	08 77	65 06	96 89	25 01
5	85 28	95 84	92 94	24 58	44 70	29 77	51 65	30 77
6	24 20	59 89	51 58	35 01	11 42	28 86	74 92	98 59
7	43 99	91 57	41 65	31 50	92 26	73 23	69 01	73 61
8	60 40	15 37	73 20	69 70	55 57	15 29	04 80	55 84
9	46 21	83 42	72 91	77 61	43 03	99 51	61 55	95 52
10	65 16	18 75	61 81	63 65	90 58	08 61	91 93	12 82
11	94 09	28 63	60 84	61 72	33 05	21 22	62 33	08 64
12	73 29	16 49	23 94	04 08	56 16	50 82	71 54	41 85
13	40 89	53 52	52 24	71 12	67 11	81 57	14 17	41 61
14	17 26	53 28	16 02	11 05	00 16	90 00	92 59	31 17
15	97 21	81 57	15 12	72 51	77 72	39 16	37 90	14 82
16	95 97	68 82	11 37	52 15	14 17	56 10	38 80	19 31
17	13 74	46 34	72 76	97 12	89 94	79 35	53 65	57 47
18	96 84	74 54	51 90	61 57	86 10	63 29	91 78	63 30
19	07 23	87 56	36 47	49 13	08 05	60 03	41 28	97 32
20	07 30	14 39	08 34	67 36	77 48	42 73	38 26	76 46
21	79 92	18 02	31 87	99 28	29 12	86 94	42 93	92 15
22	67 55	96 52	20 18	23 86	06 70	04 88	18 43	14 06
23	79 27	82 62	05 89	98 67	55 36	11 38	00 68	15 01
24	09 79	13 97	33 12	37 47	52 10	37 21	15 39	17 63
25	54 08	82 23	41 24	83 94	36 43	69 56	55 29	06 16

VIII.								
1	55 75	40 42	28 16	15 56	85 90	22 52	87 75	06 06
2	53 93	89 69	57 62	00 50	26 55	37 24	65 06	64 78
3	84 57	05 82	38 34	07 71	97 01	91 60	45 25	33 38
4	03 05	40 70	93 45	19 14	58 44	31 03	79 85	07 37
5	65 05	52 06	67 02	47 09	67 69	07 49	67 57	91 35
6	50 47	01 71	99 25	67 79	31 22	22 06	54 11	75 95
7	60 33	68 59	90 87	16 12	28 04	74 19	58 23	70 65
8	72 59	23 82	08 63	50 57	22 08	93 19	04 14	97 89
9	90 76	09 09	51 31	94 39	66 42	80 81	52 57	75 01
10	55 08	70 65	08 89	97 03	10 80	17 66	02 34	05 62
11	75 34	83 06	54 74	25 75	21 66	70 25	56 88	94 85
12	35 12	26 18	99 18	94 62	15 26	67 31	20 64	11 94
13	31 34	34 41	42 27	82 69	18 17	43 38	34 60	27 82
14	11 06	90 62	46 36	29 49	61 61	59 17	06 05	30 10
15	45 39	91 03	75 07	44 92	10 88	83 66	36 68	60 00
16	25 39	41 17	27 38	47 37	53 47	24 61	14 26	63 71
17	09 89	05 38	91 35	66 23	35 07	24 25	70 97	04 84
18	05 08	88 91	65 32	98 80	67 94	01 96	67 42	37 23
19	25 77	61 05	66 95	67 51	95 05	31 65	11 81	80 88
20	80 27	99 51	62 04	97 57	64 38	74 62	66 96	88 31
21	77 63	69 24	66 53	10 06	53 68	30 75	58 69	66 09
22	60 36	08 06	17 43	89 37	19 21	29 61	41 54	17 80
23	72 25	43 32	87 36	99 15	39 77	47 45	28 97	90 13
24	82 42	03 82	40 22	17 69	64 42	39 68	44 19	83 56
25	41 38	49 33	89 76	67 76	87 32	94 70	01 81	90 14

TABLE B TABLE OF AREAS UNDER A NORMAL CURVE*

THE USE OF TABLE B

The use of Table B requires that the raw score be transformed into a z-score and that the variable be normally distributed.

The values in Table B represent the proportion of area in the standard normal curve which has a mean of 0, a standard deviation of 1.00, and a total area also equal to 1.00.

Since the normal curve is symmetrical, it is sufficient to indicate only the areas corresponding to positive z-values. Negative z-values will have precisely the same proportions of area as their positive counterparts.

Column B represents the proportion of area between the mean and a given z.

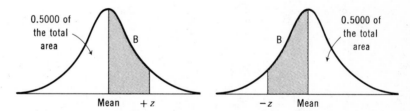

Column C represents the proportion of area beyond a given z.

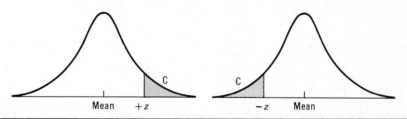

*Source: Table A, Richard P. Runyon and Audrey Haber, *Fundamentals of Behavioral Statistics*, Second Edition, Addison-Wesley Publishing Company, Reading, Massachusetts, 1971. Reprinted by permission.

TABLE B Table of Areas under a Normal Curve *(Continued)*

(A) z	(B) area between mean and z	(C) area beyond z	(A) z	(B) area between mean and z	(C) area beyond z	(A) z	(B) area between mean and z	(C) area beyond z
0.00	.0000	.5000	0.55	.2088	.2912	1.10	.3643	.1357
0.01	.0040	.4960	0.56	.2123	.2877	1.11	.3665	.1335
0.02	.0080	.4920	0.57	.2157	.2843	1.12	.3686	.1314
0.03	.0120	.4880	0.58	.2190	.2810	1.13	.3708	.1292
0.04	.0160	.4840	0.59	.2224	.2776	1.14	.3729	.1271
0.05	.0199	.4801	0.60	.2257	.2743	1.15	.3749	.1251
0.06	.0239	.4761	0.61	.2291	.2709	1.16	.3770	.1230
0.07	.0279	.4721	0.62	.2324	.2676	1.17	.3790	.1210
0.08	.0319	.4681	0.63	.2357	.2643	1.18	.3810	.1190
0.09	.0359	.4641	0.64	.2389	.2611	1.19	.3830	.1170
0.10	.0398	.4602	0.65	.2422	.2578	1.20	.3849	.1151
0.11	.0438	.4562	0.66	.2454	.2546	1.21	.3869	.1131
0.12	.0478	.4522	0.67	.2486	.2514	1.22	.3888	.1112
0.13	.0517	.4483	0.68	.2517	.2483	1.23	.3907	.1093
0.14	.0557	.4443	0.69	.2549	.2451	1.24	.3925	.1075
0.15	.0596	.4404	0.70	.2580	.2420	1.25	.3944	.1056
0.16	.0636	.4364	0.71	.2611	.2389	1.26	.3962	.1038
0.17	.0675	.4325	0.72	.2642	.2358	1.27	.3980	.1020
0.18	.0714	.4286	0.73	.2673	.2327	1.28	.3997	.1003
0.19	.0753	.4247	0.74	.2704	.2296	1.29	.4015	.0985
0.20	.0793	.4207	0.75	.2734	.2266	1.30	.4032	.0968
0.21	.0832	.4168	0.76	.2764	.2236	1.31	.4049	.0951
0.22	.0871	.4129	0.77	.2794	.2206	1.32	.4066	.0934
0.23	.0910	.4090	0.78	.2823	.2177	1.33	.4082	.0918
0.24	.0948	.4052	0.79	.2852	.2148	1.34	.4099	.0901
0.25	.0987	.4013	0.80	.2881	.2119	1.35	.4115	.0885
0.26	.1026	.3974	0.81	.2910	.2090	1.36	.4131	.0869
0.27	.1064	.3936	0.82	.2939	.2061	1.37	.4147	.0853
0.28	.1103	.3897	0.83	.2967	.2033	1.38	.4162	.0838
0.29	.1141	.3859	0.84	.2995	.2005	1.39	.4177	.0823
0.30	.1179	.3821	0.85	.3023	.1977	1.40	.4192	.0808
0.31	.1217	.3783	0.86	.3051	.1949	1.41	.4207	.0793
0.32	.1255	.3745	0.87	.3078	.1922	1.42	.4222	.0778
0.33	.1293	.3707	0.88	.3106	.1894	1.43	.4236	.0764
0.34	.1331	.3669	0.89	.3133	.1867	1.44	.4251	.0749
0.35	.1368	.3632	0.90	.3159	.1841	1.45	.4265	.0735
0.36	.1406	.3594	0.91	.3186	.1814	1.46	.4279	.0721
0.37	.1443	.3557	0.92	.3212	.1788	1.47	.4292	.0708
0.38	.1480	.3520	0.93	.3238	.1762	1.48	.4306	.0694
0.39	.1517	.3483	0.94	.3264	.1736	1.49	.4319	.0681
0.40	.1554	.3446	0.95	.3289	.1711	1.50	.4332	.0668
0.41	.1591	.3409	0.96	.3315	.1685	1.51	.4345	.0655
0.42	.1628	.3372	0.97	.3340	.1660	1.52	.4357	.0643
0.43	.1664	.3336	0.98	.3365	.1635	1.53	.4370	.0630
0.44	.1700	.3300	0.99	.3389	.1611	1.54	.4382	.0618
0.45	.1736	.3264	1.00	.3413	.1587	1.55	.4394	.0606
0.46	.1772	.3228	1.01	.3438	.1562	1.56	.4406	.0594
0.47	.1808	.3192	1.02	.3461	.1539	1.57	.4418	.0582
0.48	.1844	.3156	1.03	.3485	.1515	1.58	.4429	.0571
0.49	.1879	.3121	1.04	.3508	.1492	1.59	.4441	.0559
0.50	.1915	.3085	1.05	.3531	.1469	1.60	.4452	.0548
0.51	.1950	.3050	1.06	.3554	.1446	1.61	.4463	.0537
0.52	.1985	.3015	1.07	.3577	.1423	1.62	.4474	.0526
0.53	.2019	.2981	1.08	.3599	.1401	1.63	.4484	.0516
0.54	.2054	.2946	1.09	.3621	.1379	1.64	.4495	.0505

TABLE B TABLE OF AREAS UNDER A NORMAL CURVE *(Continued)*

(A)	(B) area between mean and	(C) area beyond	(A)	(B) area between mean and	(C) area beyond	(A)	(B) area between mean and	(C) area beyond
z	z	z	z	z	z	z	z	z
1.65	.4505	.0495	2.22	.4868	.0132	2.79	.4974	.0026
1.66	.4515	.0485	2.23	.4871	.0129	2.80	.4974	.0026
1.67	.4525	.0475	2.24	.4875	.0125	2.81	.4975	.0025
1.68	.4535	.0465	2.25	.4878	.0122	2.82	.4976	.0024
1.69	.4545	.0455	2.26	.4881	.0119	2.83	.4977	.0023
1.70	.4554	.0446	2.27	.4884	.0116	2.84	.4977	.0023
1.71	.4564	.0436	2.28	.4887	.0113	2.85	.4978	.0022
1.72	.4573	.0427	2.29	.4890	.0110	2.86	.4979	.0021
1.73	.4582	.0418	2.30	.4893	.0107	2.87	.4979	.0021
1.74	.4591	.0409	2.31	.4896	.0104	2.88	.4980	.0020
1.75	.4599	.0401	2.32	.4898	.0102	2.89	.4981	.0019
1.76	.4608	.0392	2.33	.4901	.0099	2.90	.4981	.0019
1.77	.4616	.0384	2.34	.4904	.0096	2.91	.4982	.0018
1.78	.4625	.0375	2.35	.4906	.0094	2.92	.4982	.0018
1.79	.4633	.0367	2.36	.4909	.0091	2.93	.4983	.0017
1.80	.4641	.0359	2.37	.4911	.0089	2.94	.4984	.0016
1.81	.4649	.0351	2.38	.4913	.0087	2.95	.4984	.0016
1.82	.4656	.0344	2.39	.4916	.0084	2.96	.4985	.0015
1.83	.4664	.0336	2.40	.4918	.0082	2.97	.4985	.0015
1.84	.4671	.0329	2.41	.4920	.0080	2.98	.4986	.0014
1.85	.4678	.0322	2.42	.4922	.0078	2.99	.4986	.0014
1.86	.4686	.0314	2.43	.4925	.0075	3.00	.4987	.0013
1.87	.4693	.0307	2.44	.4927	.0073	3.01	.4987	.0013
1.88	.4699	.0301	2.45	.4929	.0071	3.02	.4987	.0013
1.89	.4706	.0294	2.46	.4931	.0069	3.03	.4988	.0012
1.90	.4713	.0287	2.47	.4932	.0068	3.04	.4988	.0012
1.91	.4719	.0281	2.48	.4934	.0066	3.05	.4989	.0011
1.92	.4726	.0274	2.49	.4936	.0064	3.06	.4989	.0011
1.93	.4732	.0268	2.50	.4938	.0062	3.07	.4989	.0011
1.94	.4738	.0262	2.51	.4940	.0060	3.08	.4990	.0010
1.95	.4744	.0256	2.52	.4941	.0059	3.09	.4990	.0010
1.96	.4750	.0250	2.53	.4943	.0057	3.10	.4990	.0010
1.97	.4756	.0244	2.54	.4945	.0055	3.11	.4991	.0009
1.98	.4761	.0239	2.55	.4946	.0054	3.12	.4991	.0009
1.99	.4767	.0233	2.56	.4948	.0052	3.13	.4991	.0009
2.00	.4772	.0228	2.57	.4949	.0051	3.14	.4992	.0008
2.01	.4778	.0222	2.58	.4951	.0049	3.15	.4992	.0008
2.02	.4783	.0217	2.59	.4952	.0048	3.16	.4992	.0008
2.03	.4788	.0212	2.60	.4953	.0047	3.17	.4992	.0008
2.04	.4793	.0207	2.61	.4955	.0045	3.18	.4993	.0007
2.05	.4798	.0202	2.62	.4956	.0044	3.19	.4993	.0007
2.06	.4803	.0197	2.63	.4957	.0043	3.20	.4993	.0007
2.07	.4808	.0192	2.64	.4959	.0041	3.21	.4993	.0007
2.08	.4812	.0188	2.65	.4960	.0040	3.22	.4994	.0006
2.09	.4817	.0183	2.66	.4961	.0039	3.23	.4994	.0006
2.10	.4821	.0179	2.67	.4962	.0038	3.24	.4994	.0006
2.11	.4826	.0174	2.68	.4963	.0037	3.25	.4994	.0006
2.12	.4830	.0170	2.69	.4964	.0036	3.30	.4995	.0005
2.13	.4834	.0166	2.70	.4965	.0035	3.35	.4996	.0004
2.14	.4838	.0162	2.71	.4966	.0034	3.40	.4997	.0003
2.15	.4842	.0158	2.72	.4967	.0033	3.45	.4997	.0003
2.16	.4846	.0154	2.73	.4968	.0032	3.50	.4998	.0002
2.17	.4850	.0150	2.74	.4969	.0031	3.60	.4998	.0002
2.18	.4854	.0146	2.75	.4970	.0030	3.70	.4999	.0001
2.19	.4857	.0143	2.76	.4971	.0029	3.80	.4999	.0001
2.20	.4861	.0139	2.77	.4972	.0028	3.90	.49995	.00005
2.21	.4864	.0136	2.78	.4973	.0027	4.00	.49997	.00003
						4.50	.4999966	.0000034
						5.00	.4999997	.0000003
						5.50	.4999999	.0000001

TABLE C STUDENT'S t-DISTRIBUTION*

	Level of Significance for one-tailed test					
	.10	.05	.025	.01	.005	.0005
	Level of Significance for two-tailed test					
df	.20	.10	.05	.02	.01	.001
1	3.078	6.314	12.706	31.821	63.657	636.619
2	1.886	2.920	4.303	6.965	9.925	31.598
3	1.638	2.353	3.182	4.541	5.841	12.941
4	1.533	2.132	2.776	3.747	4.604	8.610
5	1.476	2.015	2.571	3.365	4.032	6.859
6	1.440	1.943	2.447	3.143	3.707	5.959
7	1.415	1.895	2.365	2.998	3.499	5.405
8	1.397	1.860	2.306	2.896	3.355	5.041
9	1.383	1.833	2.262	2.821	3.250	4.781
10	1.372	1.812	2.228	2.764	3.169	4.587
11	1.363	1.796	2.201	2.718	3.106	4.437
12	1.356	1.782	2.179	2.681	3.055	4.318
13	1.350	1.771	2.160	2.650	3.012	4.221
14	1.345	1.761	2.145	2.624	2.977	4.140
15	1.341	1.753	2.131	2.602	2.947	4.073
16	1.337	1.746	2.120	2.583	2.921	4.015
17	1.333	1.740	2.110	2.567	2.898	3.965
18	1.330	1.734	2.101	2.552	2.878	3.922
19	1.328	1.729	2.093	2.539	2.861	3.883
20	1.325	1.725	2.086	2.528	2.845	3.850
21	1.323	1.721	2.080	2.518	2.831	3.819
22	1.321	1.717	2.074	2.508	2.819	3.792
23	1.319	1.714	2.069	2.500	2.807	3.767
24	1.318	1.711	2.064	2.492	2.797	3.745
25	1.316	1.708	2.060	2.485	2.787	3.725
26	1.315	1.706	2.056	2.479	2.779	3.707
27	1.314	1.703	2.052	2.473	2.771	3.690
28	1.313	1.701	2.048	2.467	2.763	3.674
29	1.311	1.699	2.045	2.462	2.756	3.659
30	1.310	1.697	2.042	2.457	2.750	3.646
40	1.303	1.684	2.021	2.423	2.704	3.551
60	1.296	1.671	2.000	2.390	2.660	3.460
120	1.289	1.658	1.980	2.358	2.617	3.373
∞	1.282	1.645	1.960	2.326	2.576	3.291

*Adapted from Table III of R. A. Fisher and F. Yates, *Statistical Tables for Biological, Agricultural and Medical Research*, 1948 Edition (Edinburgh and London: Oliver & Boyd Limited) by permission of the authors and publishers.

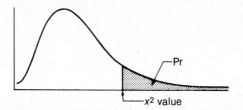

x² value

TABLE D Values of Chi-Square*

			Probabilities				
df†	.99	.975	.95	.90	.80	.70	.50
1	.000157	.000982	.00393	.0158	.0642	.148	.455
2	.0201	.0506	.103	.211	.446	.713	1.386
3	.115	.216	.352	.584	1.005	1.424	2.366
4	.297	.484	.711	1.064	1.649	2.195	3.357
5	.554	.831	1.145	1.610	2.343	3.000	4.351
6	.872	1.237	1.635	2.204	3.070	3.828	5.348
7	1.239	1.690	2.167	2.833	3.822	4.671	6.346
8	1.646	2.180	2.733	3.490	4.594	5.527	7.344
9	2.088	2.700	3.325	4.168	5.380	6.393	8.343
10	2.558	3.247	3.940	4.865	6.179	7.267	9.342
11	3.053	3.816	4.575	5.578	6.989	8.148	10.341
12	3.571	4.404	5.226	6.304	7.807	9.034	11.340
13	4.107	5.009	5.892	7.042	8.634	9.926	12.340
14	4.660	5.629	6.571	7.790	9.467	10.821	13.339
15	5.229	6.262	7.261	8.547	10.307	11.721	14.339
16	5.812	6.908	7.962	9.312	11.152	12.624	15.338
17	6.408	7.564	8.672	10.085	12.002	13.531	16.338
18	7.015	8.231	9.390	10.865	12.857	14.440	17.338
19	7.633	8.907	10.117	11.651	13.716	15.352	18.338
20	8.260	9.591	10.851	12.443	14.578	16.266	19.337
21	8.897	10.283	11.591	13.240	15.445	17.182	20.337
22	9.542	10.982	12.338	14.041	16.314	18.101	21.337
23	10.196	11.689	13.091	14.848	17.187	19.021	22.337
24	10.865	12.401	13.848	15.659	18.062	19.943	23.337
25	11.524	13.120	14.611	16.473	18.940	20.867	24.337
26	12.198	13.844	15.379	17.292	19.820	21.792	25.336
27	12.879	14.573	16.151	18.114	20.703	22.719	26.336
28	13.565	15.308	16.928	18.939	21.588	23.647	27.336
29	14.256	16.047	17.708	19.768	22.475	24.577	28.336
30	14.953	16.791	18.493	20.599	23.364	25.508	29.336

*Adapted from Table IV of R. A. Fisher and F. Yates, *Statistical Tables for Biological, Agricultural and Medical Research,* 1948 edition (Edinburgh and London: Oliver & Boyd Limited), by permission of the authors and publishers.
†For larger values of df, the expression $\sqrt{2x^2} - \sqrt{df^2 - 1}$ may be used as a normal deviate with unit variance.

TABLE D VALUES OF CHI-SQUARE *(Continued)*

df	.30	.20	.10	.05	.025	.01	.001
			Probabilities				
1	1.074	1.642	2.706	3.841	5.024	6.635	10.827
2	2.408	3.219	4.605	5.991	7.378	9.210	13.815
3	3.665	4.624	6.251	7.815	9.348	11.345	16.268
4	4.878	5.989	7.779	9.488	11.143	13.277	18.465
5	6.064	7.289	9.236	11.070	12.832	15.086	20.517
6	7.231	8.558	10.645	12.592	14.449	16.812	22.457
7	8.383	9.803	12.017	14.067	16.013	18.475	24.322
8	9.524	11.030	13.362	15.507	17.535	20.090	26.125
9	10.656	12.242	14.684	16.919	19.023	21.666	27.877
10	11.781	13.442	15.987	18.307	20.483	23.209	29.588
11	12.899	14.631	17.275	19.675	21.920	24.725	31.264
12	14.011	15.812	18.549	21.026	23.337	26.217	32.909
13	15.119	16.985	19.812	22.362	24.736	27.688	34.528
14	16.222	18.151	21.064	23.685	26.119	29.141	36.123
15	17.322	19.311	22.307	24.996	27.488	30.578	37.697
16	18.418	20.465	23.542	26.296	28.845	32.000	39.252
17	19.511	21.615	24.769	27.587	30.191	33.409	40.790
18	20.601	22.760	25.989	28.869	31.526	34.805	42.312
19	21.689	23.900	27.204	30.144	32.852	36.191	43.820
20	22.775	25.038	28.412	31.410	34.170	37.566	45.315
21	23.858	26.171	29.615	32.671	35.479	38.932	46.797
22	24.939	27.301	30.813	33.924	36.781	40.289	48.268
23	26.018	28.429	32.007	35.172	38.076	41.638	49.728
24	27.096	29.553	33.196	36.415	39.364	42.980	51.179
25	28.172	30.675	34.382	37.652	40.646	44.314	52.620
26	29.246	31.795	35.563	38.885	41.923	45.642	54.052
27	30.319	32.912	36.741	40.113	43.194	46.963	55.476
28	31.391	34.027	37.916	41.337	44.461	48.278	56.893
29	32.461	35.139	39.087	42.557	45.722	49.588	58.302
30	33.530	36.250	40.256	43.773	46.979	50.892	59.703

TABLE E DISTRIBUTION OF F^*; $Pr = 0.05$

df_2 \ df_1	1	2	3	4	5	6	8	10
1	161.4	199.5	215.7	224.6	230.2	234.0	238.9	241.9
2	18.51	19.00	19.16	19.25	19.30	19.33	19.37	19.40
3	10.13	9.55	9.28	9.12	9.01	8.94	8.85	8.79
4	7.71	6.94	6.59	6.39	6.26	6.16	6.04	5.96
5	6.61	5.79	5.41	5.19	5.05	4.95	4.82	4.74
6	5.99	5.14	4.76	4.53	4.39	4.28	4.15	4.06
7	5.59	4.74	4.35	4.12	3.97	3.87	3.73	3.64
8	5.32	4.46	4.07	3.84	3.69	3.58	3.44	3.35
9	5.12	4.26	3.86	3.63	3.48	3.37	3.23	3.14
10	4.96	4.10	3.71	3.48	3.33	3.22	3.07	2.98
11	4.84	3.98	3.59	3.36	3.20	3.09	2.95	2.85
12	4.75	3.89	3.49	3.26	3.11	3.00	2.85	2.75
13	4.67	3.81	3.41	3.18	3.03	2.92	2.77	2.67
14	4.60	3.74	3.34	3.11	2.96	2.85	2.70	2.60
15	4.54	3.68	3.29	3.06	2.90	2.79	2.64	2.54
16	4.49	3.63	3.24	3.01	2.85	2.74	2.59	2.49
17	4.45	3.59	3.20	2.96	2.81	2.70	2.55	2.45
18	4.41	3.55	3.16	2.93	2.77	2.66	2.51	2.41
19	4.38	3.52	3.13	2.90	2.74	2.63	2.48	2.38
20	4.35	3.49	3.10	2.87	2.71	2.60	2.45	2.35
21	4.32	3.47	3.07	2.84	2.68	2.57	2.42	2.32
22	4.30	3.44	3.05	2.82	2.66	2.55	2.40	2.30
23	4.28	3.42	3.03	2.80	2.64	2.53	2.37	2.27
24	4.26	3.40	3.01	2.78	2.62	2.51	2.36	2.25
25	4.24	3.39	2.99	2.76	2.60	2.49	2.34	2.24
26	4.23	3.37	2.98	2.74	2.59	2.47	2.32	2.22
27	4.21	3.35	2.96	2.73	2.57	2.46	2.31	2.20
28	4.20	3.34	2.95	2.71	2.56	2.45	2.29	2.19
29	4.18	3.33	2.93	2.70	2.55	2.43	2.28	2.18
30	4.17	3.32	2.92	2.69	2.53	2.42	2.27	2.16
40	4.08	3.23	2.84	2.61	2.45	2.34	2.18	2.08
60	4.00	3.15	2.76	2.53	2.37	2.25	2.10	1.99
80	3.96	3.11	2.72	2.48	2.33	2.21	2.05	1.95
120	3.92	3.07	2.68	2.45	2.29	2.17	2.02	1.91
∞	3.84	3.00	2.60	2.37	2.21	2.10	1.94	1.83

*Adapted from Table V of R. A. Fisher and F. Yates, *Statistical Tables for Biological, Agricultural and Medical Research*, 1948 edition (Edinburgh and London: Oliver & Boyd, Limited) by permission of the authors and publishers.

TABLE E   DISTRIBUTION OF F; Pr = 0.05 *(Continued)*

df_1 / df_2	12	15	20	30	40	60	120	∞
1	243.9	245.9	248.0	250.1	251.1	252.2	253.3	254.3
2	19.41	19.43	19.45	19.46	19.47	19.48	19.49	19.50
3	8.74	8.70	8.66	8.62	8.59	8.57	8.55	8.53
4	5.91	5.86	5.80	5.75	5.72	5.69	5.66	5.63
5	4.68	4.62	4.56	4.50	4.46	4.43	4.40	4.36
6	4.00	3.94	3.87	3.81	3.77	3.74	3.70	3.67
7	3.57	3.51	3.44	3.38	3.34	3.30	3.27	3.23
8	3.28	3.22	3.15	3.08	3.04	3.01	2.97	2.93
9	3.07	3.01	2.94	2.86	2.83	2.79	2.75	2.71
10	2.91	2.85	2.77	2.70	2.66	2.62	2.58	2.54
11	2.79	2.72	2.65	2.57	2.53	2.49	2.45	2.40
12	2.69	2.62	2.54	2.47	2.43	2.38	2.34	2.30
13	2.60	2.53	2.46	2.38	2.34	2.30	2.25	2.21
14	2.53	2.46	2.39	2.31	2.27	2.22	2.18	2.13
15	2.48	2.40	2.33	2.25	2.20	2.16	2.11	2.07
16	2.42	2.35	2.28	2.19	2.15	2.11	2.06	2.01
17	2.38	2.31	2.23	2.15	2.10	2.06	2.01	1.96
18	2.34	2.27	2.19	2.11	2.06	2.02	1.97	1.92
19	2.31	2.23	2.16	2.07	2.03	1.98	1.93	1.88
20	2.28	2.20	2.12	2.04	1.99	1.95	1.90	1.84
21	2.25	2.18	2.10	2.01	1.96	1.92	1.87	1.81
22	2.23	2.15	2.07	1.98	1.94	1.89	1.84	1.78
23	2.20	2.13	2.05	1.96	1.91	1.86	1.81	1.76
24	2.18	2.11	2.03	1.94	1.89	1.84	1.79	1.73
25	2.16	2.09	2.01	1.92	1.87	1.82	1.77	1.71
26	2.15	2.07	1.99	1.90	1.85	1.80	1.75	1.69
27	2.13	2.06	1.97	1.88	1.84	1.79	1.73	1.67
28	2.12	2.04	1.96	1.87	1.82	1.77	1.71	1.65
29	2.10	2.03	1.94	1.85	1.81	1.75	1.70	1.64
30	2.09	2.01	1.93	1.84	1.79	1.74	1.68	1.62
40	2.00	1.92	1.84	1.74	1.69	1.64	1.58	1.51
60	1.92	1.84	1.75	1.65	1.59	1.53	1.47	1.39
80	1.88	1.80	1.70	1.60	1.54	1.49	1.41	1.32
120	1.83	1.75	1.66	1.55	1.50	1.43	1.35	1.25
∞	1.75	1.67	1.57	1.46	1.39	1.32	1.22	1.00

TABLE E DISTRIBUTION OF F; $Pr = 0.01$ *(Continued)*

df_1 df_2	1	2	3	4	5	6	8	10
1	4052	4999.5	5403	5625	5764	5859	5982	6056
2	98.50	99.00	99.17	99.25	99.30	99.33	99.37	99.40
3	34.12	30.82	29.46	28.71	28.24	27.91	27.49	27.23
4	21.20	18.00	16.69	15.98	15.52	15.21	14.80	14.55
5	16.26	13.27	12.06	11.39	10.97	10.67	10.29	10.05
6	13.75	10.92	9.78	9.15	8.75	8.47	8.10	7.87
7	12.25	9.55	8.45	7.85	7.46	7.19	6.84	6.62
8	11.26	8.65	7.59	7.01	6.63	6.37	6.03	5.81
9	10.56	8.02	6.99	6.42	6.06	5.80	5.47	5.26
10	10.04	7.56	6.55	5.99	5.64	5.39	5.06	4.85
11	9.65	7.21	6.22	5.67	5.32	5.07	4.74	4.54
12	9.33	6.93	5.95	5.41	5.06	4.82	4.50	4.30
13	9.07	6.70	5.74	5.21	4.86	4.62	4.30	4.10
14	8.86	6.51	5.56	5.04	4.69	4.46	4.14	3.94
15	8.68	6.36	5.42	4.89	4.56	4.32	4.00	3.80
16	8.53	6.23	5.29	4.77	4.44	4.20	3.89	3.69
17	8.40	6.11	5.18	4.67	4.34	4.10	3.79	3.59
18	8.29	6.01	5.09	4.58	4.25	4.01	3.71	3.51
19	8.18	5.93	5.01	4.50	4.17	3.94	3.63	3.43
20	8.10	5.85	4.94	4.43	4.10	3.87	3.56	3.37
21	8.02	5.78	4.87	4.37	4.04	3.81	3.51	3.31
22	7.95	5.72	4.82	4.31	3.99	3.76	3.45	3.26
23	7.88	5.66	4.76	4.26	3.94	3.71	3.41	3.21
24	7.82	5.61	4.72	4.22	3.90	3.67	3.36	3.17
25	7.77	5.57	4.68	4.18	3.85	3.63	3.32	3.13
26	7.72	5.53	4.64	4.14	3.82	3.59	3.29	3.09
27	7.68	5.49	4.60	4.11	3.78	3.56	3.26	3.06
28	7.64	5.45	4.57	4.07	3.75	3.53	3.23	3.03
29	7.60	5.42	4.54	4.04	3.73	3.50	3.20	3.00
30	7.56	5.39	4.51	4.02	3.70	3.47	3.17	2.98
40	7.31	5.18	4.31	3.83	3.51	3.29	2.99	2.80
60	7.08	4.98	4.13	3.65	3.34	3.12	2.82	2.63
80	6.96	4.88	4.04	3.56	3.25	3.04	2.74	2.55
120	6.85	4.79	3.95	3.48	3.17	2.96	2.66	2.47
∞	6.63	4.61	3.78	3.32	3.02	2.80	2.51	2.32

TABLE E DISTRIBUTION OF F; Pr $= 0.01$ *(Continued)*

df_1 / df_2	12	15	20	30	40	60	120	∞
1	6106	6157	6209	6261	6287	6313	6339	6366
2	99.42	99.43	99.45	99.47	99.47	99.48	99.49	99.50
3	27.05	26.87	26.69	26.50	26.41	26.32	26.22	26.13
4	14.37	14.20	14.02	13.84	13.75	13.65	13.56	13.46
5	9.89	9.72	9.55	9.38	9.29	9.20	9.11	9.02
6	7.72	7.56	7.40	7.23	7.14	7.06	6.97	6.88
7	6.47	6.31	6.16	5.99	5.91	5.82	5.74	5.65
8	5.67	5.52	5.36	5.20	5.12	5.03	4.95	4.86
9	5.11	4.96	4.81	4.65	4.57	4.48	4.40	4.31
10	4.71	4.56	4.41	4.25	4.17	4.08	4.00	3.91
11	4.40	4.25	4.10	3.94	3.86	3.78	3.69	3.60
12	4.16	4.01	3.86	3.70	3.62	3.54	3.45	3.36
13	3.96	3.82	3.66	3.51	3.43	3.34	3.25	3.17
14	3.80	3.66	3.51	3.35	3.27	3.18	3.09	3.00
15	3.67	3.52	3.37	3.21	3.13	3.05	2.96	2.87
16	3.55	3.41	3.26	3.10	3.02	2.93	2.84	2.75
17	3.46	3.31	3.16	3.00	2.92	2.83	2.75	2.65
18	3.37	3.23	3.08	2.92	2.84	2.75	2.66	2.57
19	3.30	3.15	3.00	2.84	2.76	2.67	2.58	2.49
20	3.23	3.09	2.94	2.78	2.69	2.61	2.52	2.42
21	3.17	3.03	2.88	2.72	2.64	2.55	2.46	2.36
22	3.12	2.98	2.83	2.67	2.58	2.50	2.40	2.31
23	3.07	2.93	2.78	2.62	2.54	2.45	2.35	2.26
24	3.03	2.89	2.74	2.58	2.49	2.40	2.31	2.21
25	2.99	2.85	2.70	2.54	2.45	2.36	2.27	2.17
26	2.96	2.81	2.66	2.50	2.42	2.33	2.23	2.13
27	2.93	2.78	2.63	2.47	2.38	2.29	2.20	2.10
28	2.90	2.75	2.60	2.44	2.35	2.26	2.17	2.06
29	2.87	2.73	2.57	2.41	2.33	2.23	2.14	2.03
30	2.84	2.70	2.55	2.39	2.30	2.21	2.11	2.01
40	2.66	2.52	2.37	2.20	2.11	2.02	1.92	1.80
60	2.50	2.35	2.20	2.03	1.94	1.84	1.73	1.60
80	2.41	2.28	2.11	1.94	1.84	1.75	1.63	1.49
120	2.34	2.19	2.03	1.86	1.76	1.66	1.53	1.38
∞	2.18	2.04	1.88	1.70	1.59	1.47	1.32	1.00

TABLE F TRANSFORMATION OF r TO Z'*

r	.000	.001	.002	.003	.004	.005	.006	.007	.008	.009
.000	.0000	.0010	.0020	.0030	.0040	.0050	.0060	.0070	.0080	.0090
.010	.0100	.0110	.0120	.0130	.0140	.0150	.0160	.0170	.0180	.0190
.020	.0200	.0210	.0220	.0230	.0240	.0250	.0260	.0270	.0280	.0290
.030	.0300	.0310	.0320	.0330	.0340	.0350	.0360	.0370	.0380	.0390
.040	.0400	.0410	.0420	.0430	.0440	.0450	.0460	.0470	.0480	.0490
.050	.0501	.0511	.0521	.0531	.0541	.0551	.0561	.0571	.0581	.0591
.060	.0601	.0611	.0621	.0631	.0641	.0651	.0661	.0671	.0681	.0691
.070	.0701	.0711	.0721	.0731	.0741	.0751	.0761	.0771	.0782	.0792
.080	.0802	.0812	.0822	.0832	.0842	.0852	.0862	.0872	.0882	.0892
.090	.0902	.0912	.0922	.0933	.0943	.0953	.0963	.0973	.0983	.0993
.100	.1003	.1013	.1024	.1034	.1044	.1054	.1064	.1074	.1084	.1094
.110	.1105	.1115	.1125	.1135	.1145	.1155	.1165	.1175	.1185	.1195
.120	.1206	.1216	.1226	.1236	.1246	.1257	.1267	.1277	.1287	.1297
.130	.1308	.1318	.1328	.1338	.1348	.1358	.1368	.1379	.1389	.1399
.140	.1409	.1419	.1430	.1440	.1450	.1460	.1470	.1481	.1491	.1501
.150	.1511	.1522	.1532	.1542	.1552	.1563	.1573	.1583	.1593	.1604
.160	.1614	.1624	.1634	.1644	.1655	.1665	.1676	.1686	.1696	.1706
.170	.1717	.1727	.1737	.1748	.1758	.1768	.1779	.1789	.1799	.1810
.180	.1820	.1830	.1841	.1851	.1861	.1872	.1882	.1892	.1903	.1913
.190	.1923	.1934	.1944	.1954	.1965	.1975	.1986	.1996	.2007	.2017
.200	.2027	.2038	.2048	.2059	.2069	.2079	.2090	.2100	.2111	.2121
.210	.2132	.2142	.2153	.2163	.2174	.2184	.2194	.2205	.2215	.2226
.220	.2237	.2247	.2258	.2268	.2279	.2289	.2300	.2310	.2321	.2331
.230	.2342	.2353	.2363	.2374	.2384	.2395	.2405	.2416	.2427	.2437
.240	.2448	.2458	.2469	.2480	.2490	.2501	.2511	.2522	.2533	.2543
.250	.2554	.2565	.2575	.2586	.2597	.2608	.2618	.2629	.2640	.2650
.260	.2661	.2672	.2682	.2693	.2704	.2715	.2726	.2736	.2747	.2758
.270	.2769	.2779	.2790	.2801	.2812	.2823	.2833	.2844	.2855	.2866
.280	.2877	.2888	.2898	.2909	.2920	.2931	.2942	.2953	.2964	.2975
.290	.2986	.2997	.3008	.3019	.3029	.3040	.3051	.3062	.3073	.3084
.300	.3095	.3106	.3117	.3128	.3139	.3150	.3136	.3172	.3183	.3195
.310	.3206	.3217	.3228	.3239	.3250	.3261	.3272	.3283	.3294	.3305
.320	.3317	.3328	.3339	.3350	.3361	.3372	.3384	.3395	.3406	.3417
.330	.3428	.3439	.3451	.3462	.3473	.3484	.3496	.3507	.3518	.3530
.340	.3541	.3552	.3564	.3575	.3586	.3597	.3609	.3620	.3632	.3643
.350	.3654	.3666	.3677	.3689	.3700	.3712	.3723	.3734	.3746	.3757
.360	.3769	.3780	.3792	.3803	.3815	.3826	.3838	.3850	.3861	.3873
.370	.3884	.3896	.3907	.3919	.3931	.3942	.3954	.3966	.3977	.3989
.380	.4001	.4012	.4024	.4036	.4047	.4059	.4071	.4083	.4094	.4106
.390	.4118	.4130	.4142	.4153	.4165	.4177	.4189	.4201	.4213	.4225
.400	.4236	.4248	.4260	.4272	.4284	.4296	.4308	.4320	.4332	.4344
.410	.4356	.4368	.4380	.4392	.4404	.4416	.4429	.4441	.4453	.4465
.420	.4477	.4489	.4501	.4513	.4526	.4538	.4550	.4562	.4574	.4587
.430	.4599	.4611	.4623	.4636	.4648	.4660	.4673	.4685	.4697	.4710
.440	.4722	.4735	.4747	.4760	.4772	.4784	.4797	.4809	.4822	.4835
.450	.4847	.4860	.4872	.4885	.4897	.4910	.4923	.4935	.4948	.4961
.460	.4973	.4986	.4999	.5011	.5024	.5037	.5049	.5062	.5075	.5088
.470	.5101	.5114	.5126	.5139	.5152	.5165	.5178	.5191	.5204	.5217
.480	.5230	.5243	.5256	.5279	.5282	.5295	.5308	.5321	.5334	.5347
.490	.5361	.5374	.5387	.5400	.5413	.5427	.5440	.5453	.5466	.5480

*From Albert E. Waugh, *Statistical Tables and Problems* (New York: McGraw-Hill Book Company, 1952, table A11, pp. 40–41, by permission of the author and publisher.

r	.000	.001	.002	.003	.004	.005	.006	.007	.008	.009
.500	.5493	.5506	.5520	.5533	.5547	.5560	.5573	.5587	.5600	.5614
.510	.5627	.5641	.5654	.5668	.5681	.5695	.5709	.5722	.5736	.5750
.520	.5763	.5777	.5791	.5805	.5818	.5832	.5846	.5860	.5874	.5888
.530	.5901	.5915	.5929	.5943	.5957	.5971	.5985	.5999	.6013	.6027
.540	.6042	.6056	.6070	.6084	.6098	.6112	.6127	.6141	.6155	.6170
.550	.6184	.6198	.6213	.6227	.6241	.6256	.6270	.6285	.6299	.6314
.560	.6328	.6343	.6358	.6372	.6387	.6401	.6416	.6431	.6446	.6460
.570	.6475	.6490	.6505	.6520	.6535	.6550	.6565	.6579	.6594	.6610
.580	.6625	.6640	.6655	.6670	.6685	.6700	.6715	.6731	.6746	.6761
.590	.6777	.6792	.6807	.6823	.6838	.6854	.6869	.6885	.6900	.6916
.600	.6931	.6947	.6963	.6978	.6994	.7010	.7026	.7042	.7057	.7073
.610	.7089	.7105	.7121	.7137	.7153	.7169	.7185	.7201	.7218	.7234
.620	.7250	.7266	.7283	.7299	.7315	.7332	.7348	.7364	.7381	.7398
.630	.7414	.7431	.7447	.7464	.7481	.7497	.7514	.7531	.7548	.7565
.640	.7582	.7599	.7616	.7633	.7650	.7667	.7684	.7701	.7718	.7736
.650	.7753	.7770	.7788	.7805	.7823	.7840	.7858	.7875	.7893	.7910
.660	.7928	.7946	.7964	.7981	.7999	.8017	.8035	.8053	.8071	.8089
.670	.8107	.8126	.8144	.8162	.8180	.8199	.8217	.8236	.8254	.8273
.680	.8291	.8310	.8328	.8347	.8366	.8385	.8404	.8423	.8442	.8461
.690	.8480	.8499	.8518	.8537	.8556	.8576	.8595	.8614	.8634	.8653
.700	.8673	.8693	.8712	.8732	.8752	.8772	.8792	.8812	.8832	.8852
.710	.8872	.8892	.8912	.8933	.8953	.8973	.8994	.9014	.9035	.9056
.720	.9076	.9097	.9118	.9139	.9160	.9181	.9202	.9223	.9245	.9266
.730	.9287	.9309	.9330	.9352	.9373	.9395	.9417	.9439	.9461	.9483
.740	.9505	.9527	.9549	.9571	.9594	.9616	.9639	.9661	.9684	.9707
.750	.9730	.9752	.9775	.9799	.9822	.9845	.9868	.9892	.9915	.9939
.760	.9962	.9986	1.0010	1.0034	1.0058	1.0082	1.0106	1.0130	1.0154	1.0179
.770	1.0203	1.0228	1.0253	1.0277	1.0302	1.0327	1.0352	1.0378	1.0403	1.0428
.780	1.0454	1.0479	1.0505	1.0531	1.0557	1.0583	1.0609	1.0635	1.0661	1.0688
.790	1.0714	1.0741	1.0768	1.0795	1.0822	1.0849	1.0876	1.0903	1.0931	1.0958
.800	1.0986	1.1014	1.1041	1.1070	1.1098	1.1127	1.1155	1.1184	1.1212	1.1241
.810	1.1270	1.1299	1.1329	1.1358	1.1388	1.1417	1.1447	1.1477	1.1507	1.1538
.820	1.1568	1.1599	1.1630	1.1660	1.1692	1.1723	1.1754	1.1786	1.1817	1.1849
.830	1.1870	1.1913	1.1946	1.1979	1.2011	1.2044	1.2077	1.2111	1.2144	1.2178
840	1.2212	1.2246	1.2280	1.2315	1.2349	1.2384	1.2419	1.2454	1.2490	1.2526
.850	1.2561	1.2598	1.2634	1.2670	1.2708	1.2744	1.2782	1.2819	1.2857	1.2895
.860	1.2934	1.2972	1.3011	1.3050	1.3089	1.3129	1.3168	1.3209	1.3249	1.3290
.870	1.3331	1.3272	1.3414	1.3456	1.3498	1.3540	1.3583	1.3626	1.3670	1.3714
.880	1.3758	1.3802	1.3847	1.3892	1.3938	1.3984	1.4030	1.4077	1.4124	1.4171
.890	1.4219	1.4268	1.4316	1.4366	1.4415	1.4465	1.4516	1.4566	1.4618	1.4670
.900	1.4722	1.4775	1.4828	1.4883	1.4937	1.4992	1.5047	1.5103	1.5160	1.5217
.910	1.5275	1.5334	1.5393	1.5453	1.5513	1.5574	1.5636	1.5698	1.5762	1.5825
.920	1.5890	1.5956	1.6022	1.6089	1.6157	1.6226	1.6296	1.6366	1.6438	1.6510
.930	1.6584	1.6659	1.6734	1.6811	1.6888	1.6967	1.7047	1.7129	1.7211	1.7295
.940	1.7380	1.7467	1.7555	1.7645	1.7736	1.7828	1.7923	1.8019	1.8117	1.8216
.950	1.8318	1.8421	1.8527	1.8635	1.8745	1.8857	1.8972	1.9090	1.9210	1.9333
.960	1.9459	1.9588	1.9721	1.9857	1.9996	2.0140	2.0287	2.0439	2.0595	2.0756
.970	2.0923	2.1095	2.1273	2.1457	2.1649	2.1847	2.2054	2.2269	2.2494	2.2729
.980	2.2976	2.3223	2.3507	2.3796	2.4101	2.4426	2.4774	2.5147	2.5550	2.5988
.990	2.6467	2.6996	2.7587	2.8257	2.9031	2.9945	3.1063	3.2504	3.4534	3.8002

TABLE G   TABLE OF SQUARES AND SQUARE ROOTS OF NUMBERS

k	$\sqrt{k}$	$\sqrt{10k}$	k^2	k	$\sqrt{k}$	$\sqrt{10k}$	k^2
1	1	3.16228	1	51	7.14143	22.5832	2601
2	1.41421	4.47214	4	52	7.2111	22.8035	2704
3	1.73205	5.47723	9	53	7.28011	23.0217	2809
4	2	6.32456	16	54	7.34847	23.2379	2916
5	2.23607	7.07107	25	55	7.4162	23.4521	3025
6	2.44949	7.74597	36	56	7.48331	23.6643	3136
7	2.64575	8.3666	49	57	7.54983	23.8747	3249
8	2.82843	8.94427	64	58	7.61577	24.0832	3364
9	3	9.48683	81	59	7.68115	24.2899	3481
10	3.16228	10	100	60	7.74597	24.4949	3600
11	3.31662	10.4881	121	61	7.81025	24.6982	3721
12	3.4641	10.9545	144	62	7.87401	24.8998	3844
13	3.60555	11.4018	169	63	7.93725	25.0998	3969
14	3.74166	11.8322	196	64	8	25.2982	4096
15	3.87298	12.2474	225	65	8.06226	25.4951	4225
16	4	12.6491	256	66	8.12404	25.6905	4356
17	4.12311	13.0384	289	67	8.18535	25.8844	4489
18	4.24264	13.4164	324	68	8.24621	26.0768	4624
19	4.3589	13.784	361	69	8.30662	26.2679	4761
20	4.47214	14.1421	400	70	8.3666	26.4575	4900
21	4.58258	14.4914	441	71	8.42615	26.6458	5041
22	4.69042	14.8324	484	72	8.48528	26.8328	5184
23	4.79583	15.1658	529	73	8.544	27.0185	5329
24	4.89898	15.4919	576	74	8.60233	27.2029	5476
25	5	15.8114	625	75	8.66025	27.3861	5625
26	5.09902	16.1245	676	76	8.7178	27.5681	5776
27	5.19615	16.4317	729	77	8.77496	27.7489	5929
28	5.2915	16.7332	784	78	8.83176	27.9285	6084
29	5.38516	17.0294	841	79	8.88819	28.1069	6241
30	5.47723	17.3205	900	80	8.94427	28.2843	6400
31	5.56776	17.6068	961	81	9	28.4605	6561
32	5.65685	17.8885	1024	82	9.05539	28.6356	6724
33	5.74456	18.1659	1089	83	9.11043	28.8097	6889
34	5.83095	18.4391	1156	84	9.16515	28.9828	7056
35	5.91608	18.7083	1225	85	9.21954	29.1548	7225
36	6	18.9737	1296	86	9.27362	29.3258	7396
37	6.08276	19.2354	1369	87	9.32738	29.4958	7569
38	6.16441	19.4936	1444	88	9.38083	29.6648	7744
39	6.245	19.7484	1521	89	9.43398	29.8329	7921
40	6.32456	20	1600	90	9.48683	30	8100
41	6.40312	20.2485	1681	91	9.53939	30.1662	8281
42	6.48074	20.4939	1764	92	9.59166	30.3315	8464
43	6.55744	20.7364	1849	93	9.64365	30.4959	8649
44	6.63325	20.9762	1936	94	9.69536	30.6594	8836
45	6.7082	21.2132	2025	95	9.74679	30.8221	9025
46	6.78233	21.4476	2116	96	9.79796	30.9839	9216
47	6.85565	21.6795	2209	97	9.84886	31.1448	9409
48	6.9282	21.9089	2304	98	9.89949	31.305	9604
49	7	22.1359	2401	99	9.94987	31.4643	9801
50	7.07107	22.3607	2500	100	10	31.6228	10000

TABLE G *(continued)*

k	$\sqrt{k}$	$\sqrt{10k}$	k^2	k	$\sqrt{k}$	$\sqrt{10k}$	k^2
101	10.0499	31.7805	10201	151	12.2882	38.8587	22801
102	10.0995	31.9374	10404	152	12.3288	38.9872	23104
103	10.1489	32.0936	10609	153	12.3693	39.1152	23409
104	10.198	32.249	10816	154	12.4097	39.2428	23716
105	10.247	32.4037	11025	155	12.4499	39.37	24025
106	10.2956	32.5576	11236	156	12.49	39.4968	24336
107	10.3441	32.7109	11449	157	12.53	39.6232	24649
108	10.3923	32.8634	11664	158	12.5698	39.7492	24964
109	10.4403	33.0151	11881	159	12.6095	39.8748	25281
110	10.4881	33.1662	12100	160	12.6491	40	25600
111	10.5357	33.3167	12321	161	12.6886	40.1248	25921
112	10.583	33.4664	12544	162	12.7279	40.2492	26244
113	10.6301	33.6155	12769	163	12.7671	40.3733	26569
114	10.6771	33.7639	12996	164	12.8062	40.4969	26896
115	10.7238	33.9116	13225	165	12.8452	40.6202	27225
116	10.7703	34.0588	13456	166	12.8841	40.7431	27556
117	10.8167	34.2053	13689	167	12.9228	40.8656	27889
118	10.8628	34.3511	13924	168	12.9615	40.9878	28224
119	10.9087	34.4964	14161	169	13	41.1096	28561
120	10.9545	34.641	14400	170	13.0384	41.2311	28900
121	11	34.7851	14641	171	13.0767	41.3521	29241
122	11.0454	34.9285	14884	172	13.1149	41.4729	29584
123	11.0905	35.0714	15129	173	13.1529	41.5933	29929
124	11.1355	35.2136	15376	174	13.1909	41.7133	30276
125	11.1803	35.3553	15625	175	13.2288	41.833	30625
126	11.225	35.4965	15876	176	13.2665	41.9524	30976
127	11.2694	35.6371	16129	177	13.3041	42.0714	31329
128	11.3137	35.7771	16384	178	13.3417	42.19	31684
129	11.3578	35.9166	16641	179	13.3791	42.3084	32041
130	11.4018	36.0555	16900	180	13.4164	42.4264	32400
131	11.4455	36.1939	17161	181	13.4536	42.5441	32761
132	11.4891	36.3318	17424	182	13.4907	42.6615	33124
133	11.5326	36.4692	17689	183	13.5277	42.7785	33489
134	11.5758	36.606	17956	184	13.5647	42.8952	33856
135	11.619	36.7423	18225	185	13.6015	43.0116	34225
136	11.6619	36.8782	18496	186	13.6382	43.1277	34596
137	11.7047	37.0135	18769	187	13.6748	43.2435	34969
138	11.7473	37.1484	19044	188	13.7113	43.359	35344
139	11.7898	37.2827	19321	189	13.7477	43.4741	35721
140	11.8322	37.4166	19600	190	13.784	43.589	36100
141	11.8743	37.55	19881	191	13.8203	43.7035	36481
142	11.9164	37.6829	20164	192	13.8564	43.8178	36864
143	11.9583	37.8153	20449	193	13.8924	43.9318	37249
144	12	37.9473	20736	194	13.9284	44.0454	37636
145	12.0416	38.0789	21025	195	13.9642	44.1588	38025
146	12.083	38.2099	21316	196	14	44.2719	38416
147	12.1244	38.3406	21609	197	14.0357	44.3847	38809
148	12.1655	38.4708	21904	198	14.0712	44.4972	39204
149	12.2066	38.6005	22201	199	14.1067	44.6094	39601
150	12.2474	38.7298	22500	200	14.1421	44.7214	40000

TABLE G *(continued)*

k	$\sqrt{k}$	$\sqrt{10k}$	k^2	k	$\sqrt{k}$	$\sqrt{10k}$	k^2
201	14.1774	44.833	40401	251	15.843	50.0999	63001
202	14.2127	44.9444	40804	252	15.8745	50.1996	63504
203	14.2478	45.0555	41209	253	15.906	50.2991	64009
204	14.2829	45.1664	41616	254	15.9374	50.3984	64516
205	14.3178	45.2769	42025	255	15.9687	50.4975	65025
206	14.3527	45.3872	42436	256	16	50.5964	65536
207	14.3875	45.4973	42849	257	16.0312	50.6952	66049
208	14.4222	45.607	43264	258	16.0624	50.7937	66564
209	14.4568	45.7165	43681	259	16.0935	50.892	67081
210	14.4914	45.8258	44100	260	16.1245	50.9902	67600
211	14.5258	45.9347	44521	261	16.1555	51.0882	68121
212	14.5602	46.0435	44944	262	16.1864	51.1859	68644
213	14.5945	46.1519	45369	263	16.2173	51.2835	69169
214	14.6287	46.2601	45796	264	16.2481	51.3809	69696
215	14.6629	46.3681	46225	265	16.2788	51.4782	70225
216	14.6969	46.4758	46656	266	16.3095	51.5752	70756
217	14.7309	46.5833	47089	267	16.3401	51.672	71289
218	14.7648	46.6905	47524	268	16.3707	51.7687	71824
219	14.7986	46.7974	47961	269	16.4012	51.8652	72361
220	14.8324	46.9042	48400	270	16.4317	51.9615	72900
221	14.8661	47.0106	48841	271	16.4621	52.0577	73441
222	14.8997	47.1169	49284	272	16.4924	52.1536	73984
223	14.9332	47.2229	49729	273	16.5227	52.2494	74529
224	14.9666	47.3286	50176	274	16.5529	52.345	75076
225	15	47.4342	50625	275	16.5831	52.4404	75625
226	15.0333	47.5395	51076	276	16.6132	52.5357	76176
227	15.0665	47.6445	51529	277	16.6433	52.6308	76729
228	15.0997	47.7493	51984	278	16.6733	52.7257	77284
229	15.1327	47.8539	52441	279	16.7033	52.8205	77841
230	15.1658	47.9583	52900	280	16.7332	52.915	78400
231	15.1987	48.0625	53361	281	16.7631	53.0094	78961
232	15.2315	48.1664	53824	282	16.7929	53.1037	79524
233	15.2643	48.2701	54289	283	16.8226	53.1977	80089
234	15.2971	48.3735	54756	284	16.8523	53.2917	80656
235	15.3297	48.4768	55225	285	16.8819	53.3854	81225
236	15.3623	48.5798	55696	286	16.9115	53.479	81796
237	15.3948	48.6826	56169	287	16.9411	53.5724	82369
238	15.4272	48.7852	56644	288	16.9706	53.6656	82944
239	15.4596	48.8876	57121	289	17	53.7587	83521
240	15.4919	48.9898	57600	290	17.0294	53.8516	84100
241	15.5242	49.0918	58081	291	17.0587	53.9444	84681
242	15.5563	49.1935	58564	292	17.088	54.037	85264
243	15.5885	49.295	59049	293	17.1172	54.1295	85849
244	15.6205	49.3964	59536	294	17.1464	54.2218	86436
245	15.6525	49.4975	60025	295	17.1756	54.3139	87025
246	15.6844	49.5984	60516	296	17.2047	54.4059	87616
247	15.7162	49.6991	61009	297	17.2337	54.4977	88209
248	15.748	49.7996	61504	298	17.2627	54.5894	88804
249	15.7797	49.8999	62001	299	17.2916	54.6809	89401
250	15.8114	50	62500	300	17.3205	54.7723	90000

Index